Advances in the Biosciences **4**

Advances in the Biosciences

The Schering Symposia and Workshop Conferences
are conducted and sponsored by
Schering AG, 1 Berlin 65, Müllerstraße 170

Advances in the Biosciences 4

Schering Symposium
on Mechanisms Involved in Conception
Berlin, March 13 to 15, 1969

Editor: Gerhard Raspé

Pergamon Press · Vieweg

Oxford · London · Edinburgh · New York
Toronto · Sidney · Paris · Braunschweig

Pergamon Press Ltd., Headington Hill Hall, Oxford
4 & 5 Fitzroy Square, London W. 1
Pergamon Press (Scotland) Ltd., 2 & 3 Teviot Place, Edinburgh 1
Pergamon Press Inc., Maxwell House, Fairview Park, Elmsford, New York 10523
Pergamon of Canada Ltd., 207 Queen's Quay West, Toronto 1
Pergamon Press (Aust.) Pty. Ltd., 19a Boundary Street, Rushcutters Bay,
N. S. W. 2011, Australia
Pergamon Press S. A. R. L., 24 rue des Écoles, Paris 5e
Vieweg + Sohn GmbH., Burgplatz 1, Braunschweig

Editorial Assistance: *Werner Schröder*

1970

Library of Congress Catalog Card No. 74–76593
Set by Friedr. Vieweg + Sohn GmbH, Braunschweig
Printed by E. Hunold, Braunschweig
Bookbinder: W. Langelüddecke, Braunschweig
Cover design: Herbert W. Kapitzki, Frankfurt
Printed in Germany

Pergamon Press
08 017546 5 H hard cover
08 017547 3 S soft cover

Vieweg
7686 gebunden
7683 kartoniert

Table of Contents

D. R. S. Kirby

David Kirby died on 11th November 1969 after a considerable display of fortitude following the consequences of a dreadfully disabling accident.

His sincerity, warmth of character and other assets are seldom found combined in one person.

The stimulating characteristic style of his lectures evoked responses in many audiences. The challenging quality of his research adeptness, of his experimental work and incisiveness of his scientific interpretations provided a constant stimulus to many people engaged in research. His work on the implantation of embryos, the immunology of reproduction and hormonal control of early pregnancy was proving of ever-increasing interest and significance to an understanding of relations between mother and fetus, and his early death is a major loss to these fields of research.

David Kirby was born on 18th January, 1932, in Buckinghamshire, England, was married, and had three children. He first studied zoology at King's College, Durham, and carried out his doctoral research in the Department of Zoology at Oxford University. He remained in Oxford for his postdoctoral work first as a Senior Research Officer with support from the Medical Research Council, and since 1967 as a lecturer in the University. He was a Fellow of St. Catharine's College.

Opening of the Symposium

G. Raspé

Schering AG, 1 Berlin 65, Germany

Ladies and Gentlemen,

I have the privilege of welcoming you as participants in the third Schering Symposium on 'Advances in Biosciences' which is supposed to deal with 'Mechanisms Involved in Conception'.
Before inviting the Chairman of the first section to take the chair I have to perform three tasks.

My first and very pleasant task is to introduce the winners of the „Schoeller-Junkmann-Preis 1969" who were announced during the Annual Meeting of the German Endocrine Society last week. The First Prize was awarded to Dr. *J. L. H. O'Riordan* from the Middlesex Hospital, London, for his work on:
'Human Parathyroid Hormone.

1) Isolation and Characterization,
2) Immunological Properties,
3) Hyperparathyroidism in Chronic Renal Failure.'

The Second Prize was awarded to Dr. *G. Jütting,* Department of Gynecology and Obstetrics, of the Medical Faculty of the Rheinisch-Westfälische Technische Hochschule in Aachen for his work on:

'Hormonale Enzyminduktion im Myometrium (Beispiel einer Östrogenwirkung am Erfolgsorgan)'.

Unfortunately, Dr. *O'Riordan* was prevented from attending this Symposium, but I am pleased to show you at least his photograph now. You will also see the photo of Dr. *Jütting* who is among the audience.

The second task is to inform those of you who are participating in the Schering Symposium for the first time about the objectives we had in mind when we started these gatherings two years ago. Our thoughts were based on three essentials:

1) There is obviously no need for more large-size meetings dealing with topics in the field of experimental biology or endocrinology.
2) There seems to be some lack in communication between people working in experimental sciences and those concerned with clinical work.
3) Research workers, even if concerned with very related problems, do not meet very often as a group.

In some cases, we have learned that even in these days people who have been exchanging letters and reprints for years had never met.

After the first meeting which consisted of a number of topics, we decided to concentrate on one topic which may be divided into no more than four sub-topics. This proved to be a reasonable concept judged from the comments we received after the second meeting concerned with 'Biodynamics and Mechanism of Action of Steroid Hormones'.

Besides arranging these annual Symposia, we have started to invite smaller numbers of research workers for Workshop Conferences dealing with particular subjects in experimental biology. The first meeting of this kind took place in December, 1968, and was dealing with the topic *'In vitro* versus *in vivo';* the second meeting is scheduled for late spring and will deal with 'Pharmacokinetics'.
After some discussions with the participants of the first meeting, we have decided to publish Proceedings the first of which are just out of print after some – but as we might say – unavoidable delay. Without any doubt, we shall be able to speed up the printing process with the aid of the lecturers and the participants in the discussions.

The third task consists in explaining to you why we did choose the topic 'Mechanisms involved in Conception' for this meeting. This laboratory has devoted a great deal of its activities to the study of steroid and polypeptide hormones, and their action, since the thirties. As far as the therapeutic application of hormones is concerned, the use of steroid hormones for hormonal contraception has become a very important aspect in the recent years. It has to be stated, however, that although clinical and experimental studies are performed by numerous groups and their results reported continously in the literature, the therapeutic approaches till now are almost entirely based on empirical conclusions. I would like to mention here the experiments of *Casida* et al., who as early as 1947 described the effect of progesterone on the fertilisation process. Let me also cite à remark of *Short,* published in the Annual Review of Physiology where he explained, that from the currant concepts in endocrinology only a few are new: 'Plus ça change, plus c'est la même chose'.

Let me now say a few words about the subtopics, which were selected by Dr. *Neumann* and his group who also invited the speakers.

After the discovery by *Chang* and *Austin* in 1951 of the capacitation process, importance of this mechanism for conception was soon realized.

A large number of papers has been published since 1951 (a bibliography published by the Reproduction Research Information Service Ltd. cites 79 papers up to 1967) out of which the contributions of *Bedford, Soupart* and *Williams* should be particularly mentioned.

Sperm capacitation will be the topic of this morning. The importance of the status of the cervical mucus with respect to sperm penetration has been realized as early as 1949. It may, however, be that the use of hormones for conception in man has tempted many people to believe that cervical mucus is the decisive factor which now seems to be an over-statement. The significance of the cervical mucus and sperm transport will be discussed this afternoon.

The first attempt to fertilize mammalian ova in vitro was performed in 1912 with rat ova by *Long,* and by *Champy* and *Pincus* et al. in 1927 and 1930 using rabbit ova. The complexity of the uterine milieu and its influence on egg development and the existing inter-dependence with egg transport, including the timing and the role of steroid hormones, was realized only in the recent years.

Egg transport, egg-development and the nidation process will be the topics of tomorrow.

The experiments of *Loeb* performed in 1923 showed that after removal of the uterus of guinea-pigs the corpus luteum can persist for a long time, which led to the postulation of a luteolytic factor. *Short* in one of his papers stated:

'Scientifically speaking, we have only just begun to start thinking about the corpus luteum and its relationship to the uterus. We have much to learn and the lesson is going to be an exciting one with promise of many discoveries of practical significance.'

These aspects will be discussed on the third day (Saturday) of this Symposium.

May I now ask You, Dr. *Austin,* to take the chair. Thank You.

Sperm Capacitation - Biological Significance in various Species

C. R. Austin

Physiological Laboratory, University of Cambridge, Cambridge, England

Summary: Species in which the need for sperm capacitation can be inferred from observations on the time relations of fertilization *in vivo,* or the requirements for fertilization *in vitro,* now include frog, rabbit, rat, mouse, golden hamster, Chinese hamster, ferret, sheep, cow, pig, monkey and man. Basically the function of capacitation appears to be that it unblocks the acrosome reaction, which in turn permits the spermatozoon to pass through egg investments with the aid of lytic enzymes. The investments in question are the zona pellucida and, except in the sheep and cow in which the egg appears to have a naked zona at the time of sperm entry, the matrix of the cumulus oophorus. The nature of capacitation remains unknown. While evidence in some species tends to favour involvement of a specific agent in the female tract, in others it supports the conception of an endogenous change in the spermatozoon promoted by a precise set of environmental conditions.

Under this title, it seems right for me to say briefly what I think capacitation is, and then to see what evidence we have that capacitation plays a role in the reproductive pattern of different animals. Basically, capacitation – as Dr. *Chang* said in his first paper on this topic – is a physiological change that fits the spermatozoon for fertilization. Later we modified this a little, pointing out that the specific task which the spermatozoon needs to be made fit for is the penetration of the egg investments, the matrix of the cumulus oophorus and the zona pellucida. But the nature of capacitation has proved a hard nut to crack. Even now, 18 years later, there is not much we can add with confidence to the broad definition. The idea that capacitation involves the removal of a surface coating from the spermatozoon received strong support with the discovery of the decapacitation factor by Dr. *Chang,* and from the subsequent intensive investigations on this factor by him and Dr. *Bedford,* as well as by Dr. *Williams* with Dr. *Dukelow,* Dr. *Brackett* and other colleagues. More recently, intriguing new light has been shed on this problem by Dr. *Ericsson,* with his studies on tetracycline fluorescence and the effect of incubating spermatozoa with eosinophils. Removal of a surface coat may well result in increased permeability of the cell and thus in increased metabolic turnover – further points investigated by Dr. *Williams* and his team – and these changes evidently lead to the structural alteration characterized as the acrosome reaction, which has been so fruitfully studied

Manuscript received: 28 March 1969

by Dr. *Bedford.* It is reasonable to suppose that the enzymes released from the acrosome by this reaction are the means that permit the spermatozoon to penetrate the cumulus and the zona pellucida. I think most people would agree with this interpretation, though parts are tentative and many details still lacking.

Now, how widespread in the animal kingdom is this need for spermatozoa to undergo further change, after release from the male, before acquiring fertilizing capacity? The possibility of its existence in invertebrate organisms cannot be ruled out, especially in those with internal fertilization. It is equally true that we have not yet direct positive evidence for any species of invertebrate, but there are some suggestive observations. It has been noted that in certain anaspidean mollusks, nudibranchs and crabs the eggs can be fertilized *in vitro* with spermatozoa from the female seminal receptacle but not with those from the male sperm duct [7, 9, 15a]. These systems deserve further research since they could present a simpler version of the problem as it exists in mammals.

Species in which the need for sperm capacitation can be inferred from direct evidence – that is, from time relations of fertilization *in vivo,* or from the requirements for fertilization *in vitro* – are all vertebrates, namely the frog, rabbit, rat, mouse, golden hamster, Chinese hamster, ferret, sheep, pig, cow, monkey and man.

In the frog, spermatozoa evidently need to pass through the jelly coat, or come into contact with soluble components of the coat, before they can proceed with fertilization. Extracts of oviduct but of no other tissue tested have the same effect upon the spermatozoa, and the time required for capacitation is only about 10 minutes. The active agent appears to be one of the soluble antigenic components of the jelly coat. If eggs are treated with the appropriate antibodies, they become unfertilizable to freshly released spermatozoa but not to those that have come in contact with jelly coat or oviduct extract. Attempts to identify the specific antigen are now in progress. These observations are all attributable to *C. A. Shivers* (quoted by *Shaver* [21]; and personal communication, 1969).

Data on capacitation in the rabbit are derived both from *in vivo* and *in vitro* studies, and indicate that the process takes about 5 hours in the Fallopian tube and 12 hours in the uterus [6, 11]. As yet there does not appear to be any repeatable method for inducing full capacitation *in vitro.* The report by *Smith* [23] in which she claimed to have observed the penetration of spermatozoa and the formation of pronuclei, after incubation of the spermatozoa *in vitro* with fragments of tubal mucosa, still lacks either confirmation or refutation, but certainly deserves reinvestigation. *Smith's* preparation was such that spermatozoa must have made frequent contact with mucosal cells, and the importance of cell contact in capacitation has been stressed by *Soupart & Orgebin-Crist* [24] and more recently by *Ericsson* [14] and *Hamner & Sojka* [15].

Comparatively little work has been done with the rat in recent years. Earlier observations with delayed mating showed clearly enough, I believe, that capacitation was needed in this species and that it took some 2 to 3 hours [1], but as yet there is no report of successful fertilization *in vitro.* The difficulty is emphasised by *Toyoda & Chang* [28] in a recent paper; however they found that removal of the zona pellucida permitted fertilization with uncapacitated spermatozoa. This observation is of special interest in relation to current ideas on the mechanism of gamete fusion. Descriptions given by *Barros & Franklin* [4] nicely support the interpretation made by *Piko & Tyler* [20] to the effect that fusion with the egg is initiated in the region of the postnuclear cap of the spermatozoon. Consequently, it can be inferred that an acrosome reaction, though directly involved in the penetration of egg investments, is not a prerequisite for actual gamete fusion in mammals, as it is in the invertebrates that have been studied. Thus, where there is no need for an acrosome reaction (as with zona-less eggs) there is no need for capacitation.

Early attempts to determine capacitation time in the mouse were in vain, owing it seemed either to there being no need for capacitation in the mouse or to the very short time involved [8]. Recently, *Whittingham* [30] recovered uterine spermatozoa 1 to 2 hours after coitus and placed them with eggs held *in vitro* in a special medium [29]. *Whittingham* achieved a good rate of fertilization, but he does not say whether he tried epididymal spermatozoa. We can guess that he probably did and obtained no success with them, but as a result can only infer that capacitation is probably necessary in the mouse.

In the golden hamster, earlier work by *Strauss* [25] indicated a 4-hour capacitation period *in vivo.* Since then it has been shown that the process takes about 4 hours *in vitro* also, and is induced more effectively by contact with follicular fluid than tubal secretions [2, 3, 31–33]. Recent findings by *Bavister* [5] point up the importance for capacitation of the general properties of the culture medium, and imply that these features are as critical as the possible presence of a specific agent. Hydrogen-ion concentration and the nature of the gaseous phase both have distinct effects; thus the incidence of sperm penetration showed a peak of 80 % at pH 7.6 with air gassing and at pH 7.8 with CO_2 gassing. Protein appears to be an essential component; indeed, acrosome reactions could be seen in a large proportion of spermatozoa incubated in heated serum alone, though no fertilization occurred. At all pH levels, higher rates of fertilization were obtained when the recovered tubal contents were diluted with at least two volumes of culture medium, and results were still good with dilution at 200-fold or after repeated washing of the cumulus masses. These results do not favour the idea of a capacitation agent but rather suggest involvement of an inhibitory *fertilizin-like* factor in the hamster, such as described by *Thibault & Dauzier* [27] in the rabbit. As yet, dialysis experiments have yielded equivocal results.

The studies made with the Chinese hamster are those reported by *Pickworth & Chang* [19] and are unusual in that, in order to spare scarce animals, the Chinese hamster spermatozoa were incubated with only golden hamster fluids and then placed with the Chinese hamster eggs. Otherwise conditions and time relations were similar to those described for the golden hamster.

For data on the ferret the source is the work of *Chang & Yanagimachi* [10]. They induced ovulation with sterile matings and recovered eggs at different times after introducing epididymal spermatozoa into the ovarian capsule. Only after intervals of 3.5 hours or more were penetrated eggs observed, though spermatozoa adhering to the zona surface were seen earlier.

Observations on the sheep and the pig stem originally from *Dauzier* and *Thibault* [12] and *Thibault* [26], who succeeded in obtaining the fertilization *in vitro* of three sheep eggs and one pig egg (out of totals of 51 and 13, respectively) with uterine spermatozoa (recovered 7 hours or longer after coitus), having previously failed to get fertilization with freshly ejaculated spermatozoa. *In vivo* data on the sheep were obtained by *Mattner* [18] who injected spermatozoa into the Fallopian tubes of ewes that had just ovulated. Animals killed at 0.5 1 or 1.25 hours later yielded only unpenetrated eggs, but eggs recovered at 1.5 hours and later were nearly all undergoing fertilization. For the pig, *Hunter & Dziuk* [16] recently reported recovering many penetrated eggs at 2 hours after artificial insemination (performed after ovulation).

The observations that have been made on the cow are those of *R. G. Edwards* and as yet unpublished. They are simply that when ejaculated spermatozoa were added to some cow eggs *in vitro* they failed to penetrate, but when the cells were incubated for 6 or 7 hours in follicular fluid, penetration and pronucleus formation were seen.

Evidence for the rhesus monkey is as yet a trifle slim – *Marston & Kelly* [17] injected spermatozoa into the uterine lumen of mid-cyclic animals that had ovulated, and they flushed out a total of three penetrated eggs 6 hours later. One of the eggs was pronucleate.

Recent work on human eggs that I would like to mention is that of *Edwards, Bavister & Steptoe* [13], who seminated oocytes matured *in vitro* with washed ejaculated spermatozoa and incubated the preparation for periods ranging from 6 to 31 hours. Eggs studied at 6 hours were unpenetrated, but at 7 hours onwards stages of penetration and pronucleus formation were seen. These data are consistent with the need for a capacitation interval but are as yet too few to constitute good evidence. If supported, they would denote that capacitation *in vitro* of human spermatozoa requires essentially the same conditions as that of hamster spermatozoa.

By way of summary, I think we can say that we have convincing evidence for the occurrence of capacitation in the frog. rabbit, rat, both hamsters, ferret, sheep and pig, and suggestive data for mouse, cow, monkey and man. Full capacitation *in vitro* has been achieved only with the hamsters, and at least in the golden hamster the process may not actually require a specific agent, but does depend on rather precise environmental conditions. This inference may appear to be at variance with ideas on the existence of a specific agent in the rabbit, but the difference may well be one of degree rather than of kind – the rabbit spermatozoon does take longer to capacitate than any of the others.

References

[1] *Austin, C. R. & Braden, A. W. H.* (1954). Time relations and their significance in the ovulation and penetration of eggs in rats and rabbits. Australian J. Biol. Sci. **7**, 179.

[2] *Barros, C.* (1968). *In vitro* capacitation of golden hamster spermatozoa with fallopian tube fluid of the mouse and rat. J. Reprod. Fertil., **17**, 203.

[3] *Barros, C. & Austin, C. R.* (1967). *In vitro* fertilization and the sperm acrosome reaction in the hamster. J. Exper. Zool., **166**, 317.

[4] *Barros, C. & Franklin, L. E.* (1968). Behavious of the gamete membranes during sperm entry into the mammalian egg. J. Cell Biol., **37**, C13.

[5] *Bavister, B. D.* (1969). Environmental factors important for *in vitro* fertilization in the hamster. J. Reprod. Fertil. , **18**, 544.

[6] *Bedford, J. M.* (1967). Experimental requirement for capacitation and observations on ultrastructural changes in rabbit spermatozoa during fertilization. J. Reprod. Fertil., Suppl. 2, 35.

[7] *Beeman, R. D.* (1966) The biology of reproduction in *Phyllaplysa taylori* Dall, 1900 (Gastropoda: Opisthobranchia: Anaspidea). Ph. D. Diss., Stanford Univ.

[8] *Braden, A. W. H. & Austin, C. R.* (1954) Fertilization of the mouse egg and the effect of delayed coitus and of hot-shock treatment. Australian J. Biol. Sci., **7**, 552.

[9] *Brown, G. G.* (1966) Ultrastructural studies of sperm morphology and sperm-egg interaction in the decapod *Callinectes sapidus.* J. Ultrastr. Res., **14**, 425.

[10] *Chang, M. C. & Yanagimachi, R.* (1963). Fertilization of ferret ova by deposition of epididymal sperm into the ovarian capsule with special reference to the fertilizable life of ova and the capacitation of sperm. J. Exp. Zool. **154**, 175.

[11] *Dauzier, L. & Thibault, C.* (1956) Recherche expérimentale sur la maturation des gamètes males chez les mammifères, par l'étude de la fécondation *in vitro* de l'oeuf de lapine. III. International Congress on Animal Reproduction. Section 1, 58.

[12] *Dauzier, L. & Thibault, C.* (1959). Données nouvelles sur la fécondation *in vitro* de l'oeuf de la Lapine et de la Brebis. C. R. Acad. Sci., **248,** 2655.

[13] *Edwards, R. G., Bavister, B. D. & Steptoe, P. C.* (1969). Early stages of fertilization *in vitro* of human oocytes matured *in vitro.* Nature, **221,** 632.

[14] *Ericsson, R. J.* (1967). Technology, physiology and morphology of spermatozoa capacitation. J. Reprod. Fert., Suppl. 2, 65.

[15] *Hamner, C. E. & Sojka, N. J.* (1968). Requirements for capacitation of rabbit sperm. Nature, **220,** 1042.

[15a] *Holman, A.* (1966) Personal communication.

[16] *Hunter, R. H. F. & Dziuk, P. J.* (1968). Sperm penetration of pig eggs in relation to the timing of ovulation and insemination. J. Reprod. Fertil., **15,** 199.

[17] *Marston, J. H. & Kelly, W. A.* (1968). Time relationships of spermatozoon penetration into the egg of the rhesus monkey. Nature, **217,** 1073.

[18] *Mattner, P. E.* (1963). Capacitation of ram spermatozoa and penetration of the ovine egg. Nature, **199,** 772.

[19] *Pickworth, S. & Chang, M. C.* (1969). Fertilization of Chinese hamster eggs *in vitro.* J. Reprod. Fertil. (in press).

[20] *Piko, L. & Tyler, A.* (1964). Fine structural studies of sperm penetration in the rat. 5th Internat. Congr. Anim. Reprod. Artif. Insem., Trento, **2,** 372.

[21] *Shaver, J. R.* (1966). Immunobiological studies of the jelly-coats of anuran eggs. Amer. Zoologist, **6,** 75.

[22] *Smith, A. U.* (1951). Fertilization *in vitro* of the mammalian egg. Biochem. Soc. Symp. No. 7, 3.

[23] *Soupart, P. & Orgebin-Crist, M. C.* (1966). Capacitation of rabbit spermatozoa delayed *in vivo* by double ligation of uterine horn. J. Exp. Zool., **163,** 311.

[24] *Strauss, F.* (1956). The time and place of fertilization of the golden hamster egg. J. Embryol. Exp. Morph., **4,** 42.

[25] *Thibault, C.* (1959). Analyse de la fécondation de l'oeuf de la truie après accouplement ou insemination. artificielle. Ann. Zootech. Ser. D, **8** (suppl.)., 165.

[26] *Thibault, C. & Dauzier, L.* (1960) "Fertilisins" et fécondation *in vitro* de l'oeuf de lapine. C. R. Acad. Sci., Paris, **250,** 1358.

[27] *Toyoda, Y. & Chang, M. C.* (1968). Sperm penetration of rat eggs *in vitro* after dissolution of zona pellucida by chymotrypsin. Nature, **220,** 589.

[28] *Whitten, W. K. & Biggers, J. D.* (1968). Complete development *in vitro* of the pre-implantation stages of the mouse in a simple chemically defined medium. J. Reprod. Fertil. **17,** 399.

[29] *Whittingham, D. G.* (1968). Fertilization of mouse eggs *in vitro.* Nature, **220,** 592.

[30] *Yanagimachi, R.* (1966). Time and process of sperm penetration into hamster ova *in vivo* and *in vitro.* J. Reprod. Fertil., **11,** 359.

[31] *Yanagimachi, R.* (1969). *In vitro* capacitation of hamster spermatozoa by follicular fluid. J. Reprod. Fertil. **18,** 275.

[32] *Yanagimachi, R. & Chang, M. C.* (1964). *In vitro* fertilization of golden hamster ova. J. Exper. Zool. **156,** 361.

Discussion

J. P. Bennett: Could you comment on the recent work of *Stambaugh,* at the University of Pennsylvania School of Medicine, who showed that trypsin could in fact remove the zona pellucida. Do you think therefore that the sperm acrosome, or the outer sperm coat may possess a trypsin-like substance which acts on the zona pellucida to allow penetration?

C. R. Austin: Yes, *Buckley & Stambaugh* have demonstrated a macromolecular complex of trypsin and hyaluronidase in extracts of acrosome. That is consistent with the idea that the acrosome contains the necessary enzymic equipment to get the sperm through the cumulus and zona. What, specifically, is your point?

J. P. Bennet: Well, previously, hyaluronidase was suggested as the active member in denudation of the egg and approach of the sperm to the egg, now trypsin is implicated in the process as a new member of the acrosomal complex. In what relationship would you put the two together?

C. R. Austin: I cannot see any further significance, except of course that we now have an enzyme which will make a hole in the zona, which hyaluronidase could not do. So the trypsin may be involved there. It is a bit difficult to explain because the acrosome – not to anticipate Dr. *Bedford* too much – the acrosome has gone by the time the sperm reaches the zona, so presumably its content of trypsin has too. Unless some of this tryptic enzyme is associated more closely with the cell surface and is not extracted.

M. C. Chang: Could you, according to your experience, pragmatically speak on the application of these observations? Do you think from the practical or medical point of view that study of capacitation can be used to devise some contraceptive methods?

C. R. Austin: Yes, I should think it is highly probable, but I have no suggestions to make. I think the closer we can identify specific steps in the process of fertilization, the better chance we have of devising a block to each specific step. I will leave you with that thought.

Hormonal Regulation of Sperm Capacitation

M. C. Chang

The Worcester Foundation for Experimental Biology Shrewsbury, Massachusetts, USA

Summary: Published and unpublished work concerning the relationship between capacitation of rabbit sperm in the female tract and various hormones has been reviewed. It has been reported that capacitation of sperm in the uterus cannot occur in hypophysectomized rabbits unless 200–300 μg of LH is administered, and that treatment of estrous rabbits with 25–75 but not 100–400 i.u. of HCG improves uterine capacitation of sperm. These observations, however, have not been confirmed by other workers nor demonstrated by other reasoning. Although the results obtained by investigators varied greatly mainly due to variations in the time of tubal insemination, it appears that capacitation may still occur in the uterus of ovariectomized rabbits but that estrogen treatment can improve the efficiency of capacitation. Despite the inconsistency of experimental results obtained by different investigators there is much evidence to support the view that capacitation of sperm in the uterus is inhibited in pseudopregnant and progesterone-treated rabbits. The degree and the mode of inhibition is still, however, unclear. Intravaginal insemination at the time of ovulation with sperm incubated in the uterus of progesterone-treated rabbits has shown that capacitation was significantly inhibited. By contrast a similar experiment using sperm incubated in estrogen-treated rabbits showed no facilitation of capacitation.

It has been reported that capacitation of rabbit sperm in the Fallopian tubes can be achieved in pseudopregnant rabbits and in ovariectomized rabbits either treated with estrogen or with progesterone, but cannot be achieved in ovariectomized rabbits given no hormones, confirmation of these reports is still lacking. Further study of hormonal regulation of sperm capacitation in other species is urged.

Introduction

Deposition of ejaculated or epididymal sperm into the Fallopian tubes of rabbits at various times before or after ovulation and deposition of sperm recovered from the uterus at various times after mating have shown that fertilization of eggs can occur only when sperm have been in either the tube or the uterus for about 6 hours [1, 8, 9]. This phenomenon has also been observed in the rat [1, 2, 24] and termed capacitation of sperm by *Austin* [2].

The fertilizing capacity of rabbit sperm developed in the uterus, however, can be lost when sperm recovered from the uterus are treated with seminal plasma of various species, but can be regained after a further period in the female tract [10].

Manuscript received: 14 March 1969

These two processes are termed decapacitation and recapacitation [5, 26, 29]. The complexities and difficulties of the experimental procedures used for the assay of capacitation have been discussed [12].

It is perhaps appropriate to quote here the following sentences of *Mann* [20]. "In the course of 15 years the problem of sperm capacitation has lost little of its initial interest, complexity and perplexity. The whole phenomenon remains a puzzle, and an object of theories, speculations and even controversies, while the original discoverers observe a highly benevolent and tolerant attitude towards those who have followed in their footsteps. How unlike that is from *Edison,* for instance, who once said "I don't sue imitators, I ruin them". On the contrary, the arguments on sperm capacitation continue to be conducted in the spirit of General de Gaulle's famous motto that "to be great is to sustain a great quarrel".

It is not the purpose of this article to ruin the imitators or to sustain a great quarrel. It is intended to review critically published and unpublished work concerning the hormonal regulation of sperm capacitation in the hope of advancing our knowledge in animal reproduction. This paper also reports a simple experiment dealing with the effect of estrogen and progesterone on capacitation of rabbit sperm.

Effects of gonadotropic hormones on the capacitation of rabbit sperm

Incubation of rabbit sperm in the uterus of hypophysectomized rabbits was performed by *Soupart* [28] who found that none of 11 eggs were fertilized following tubal insemination of the incubated sperm. Capacitation of sperm in the uterus of hypophysectomized rabbits was achieved (50 – 70 % fertilization) however, when 200–300 μg of LH (NIH S11 ovine) were given to the hypophysectomized rabbits. Since the proportion of eggs fertilized varies a great deal following tubal insemination and since only 11 eggs were examined in this study, achievement of capacitation in the uterus of hypophysectomized animals may still be possible.

Although capacitation of sperm can be achieved in the uterus of immature rabbits, a significantly higher number of eggs (69 %) were fertilized by sperm incubated in the uterus of immature rabbits injected with 28–40 iu. of sheep pituitary extract than without injection (34 %, *Chang* [11]). Since none of 47 eggs were fertilized by tubal insemination of sperm incubated in the uterus of pseudopregnant rabbits treated with gonadotropin in this study [11], it seems that inhibition of capacitation in pseudopregnant uterus was not affected by administration of gonadotropin.

Soupart [28] has reported, however, that capacitation of rabbit sperm in the uterus of estrous rabbits was much improved if HCG was given to the rabbits. This was shown by an increase of eggs fertilized from 27 % to 74.5 % when 25 i.u. to 75 i.u.

of HCG was given to the rabbits before incubation of sperm. A high dose of HCG, from 100 to 400 i.u., however, caused a reduction of the proportion of eggs fertilized (1–14 %) by the incubated sperm. Such an inhibition of capacitation in the uteri of rabbits injected with a high dose of HCG, was not confirmed by *Bedford* [4] who observed that 76 % of 53 eggs were fertilized when sperm recovered from the uterus of rabbits injected with 500–1000 i.u. of HCG were deposited into the tubes of other rabbits that had received an ovulation injection 13 1/2 hours earlier.

Moreover, in the experiment of *Soupart* [28] a very high proportion of eggs (82.3–100 %) were fertilized in rabbits which received HCG in his "capacitation system", thus the sperm incubated in the uterus of rabbits treated with various doses of HCG were undoubtedly all capacitated. When the sperm recovered from the uterus were deposited into the Fallopian tubes of other rabbits that had ovulated recently in his "capacitation test system", the proportion of eggs fertilized varied greatly. Obviously the variation of fertilization rates was not due to whether the sperm were capacitated or not, but due to factors involved at tubal insemination, such as the physiological condition of the recipient rabbits, the physiological condition of sperm and eggs, and the chance of their meeting.

Capacitation of sperm in the uterus of estrous and ovariectomized rabbits

Fig. 1 presents the results concerning the capacitation of sperm in the uterus of estrous rabbits as reported by several investigators. It can be seen that the highest fertilization rate (97 %) was obtained by *Noyes, Walton & Adams* [25] and the lowest rate (15.4 %) by *Soupart* [28]. In these two experiments the important difference is that tubal insemination was done at the time of ovulation (9.5–10.5 hours after ovulation injection) by *Noyes* et al. but about 3 hrs. after ovulation (13 hrs. after ovulation injection) by *Soupart.* This indicates that the time of tubal insemination in relation to the time of ovulation plays a very important role in determining the proportion of eggs fertilized. Although a high proportion of eggs (64 %) were fertilized in *Bedford's* experiment when tubal insemination was done 5 hours after ovulation, this can be interpreted as due to a better quality and quantity of sperm obtained following mating. Moreover, it can be seen that in experiments conducted in the same laboratory the fertilization rate may vary from 33 % to 74 % [16, 18] even though the time of tubal insemination did not vary much.

The results of capacitation in the uterus of ovariectomized rabbits with estrogen or progesterone treatment are presented in Fig. 2. Capacitation of sperm was possible in the uterus of ovariectomized rabbits as reported by *Chang* (37 %) [11] and by *Bedford* (36 %) [4], but a high proportion of fertilized eggs (79 %) was obtained by *Noyes* et al. [25] when tubal insemination was done earlier than other investigators. By contrast, in the experiments of both *Soupart* [28] and *Hamner* et al. [16]

Fig. 1. Capacitation of sperm in the uterus of estrous rabbits.

Sperm used	Lenght of incubation (in hrs.)	Time of tubal insem. (hrs. after ovul. inj.)	Total No. of eggs examined	No. of eggs fert.	% of fert. eggs	Investigators
Ejaculated	12–13	12–13	95	60	63	*Chang* [11]
Ejaculated	12–13	12–13	31	18	58	*Chang* [11]
Ejaculated & Centrifuged	6	9.5–10.5	61	59	97	*Noyes* et al. [25]
Epididymal	10	13	97	15	15.4	*Soupart* [28]
Ejaculated & Centrifuged	10	11.5–13	114	55	49	*Hamner* et al. [16]
Ejaculated & Centrifuged	7–8	12–13	46	15	33	*Hamner* et al. [16]
Ejaculated & Centrifuged	6–7	12–13	12	5	42	*Hamner* et al. [16]
Ejaculated & Centrifuged	10	12.5–13	39	29	74	*Hamner & Sojka* [18]
Mating	12–13	15	22	14	64	*Bedford* [4]

Fig. 2. Capacitation of sperm in the uterus of ovariectomized rabbits under different treatments.

Treatment of rabbits	Sperm used	Length of incubation (in hrs.)	Time of tubal insem. (hrs. after ovul. inj.)	Total eggs examin.	No. of eggs. fert.	% of eggs fert.	Investigators
No treatment	Ejaculated	12–13	12–13	41	15	37	*Chang* [11]
Gonadotropin	Ejaculated	12–13	12–13	42	11	26	*Chang* [11]
Estrogen	Ejaculated	12–13	12–13	48	16	33	*Chang* [11]
Progesterone	Ejaculated	12–13	12–13	36	3	8.4	*Chang* [11]
No treatment	Ejaculated & Centrifuged	5.5–7	9.5–10.5	223	176	79	*Noyes* et al. [25]
No treatment	Epididymal	10	13	97	4	4.2	*Soupart* [28]
Estrogen	Epididymal	10	13	138	75	54.4	*Soupart* [28]
Cottonseed Oil	Ejaculated & Centrifuged	6–11	11.5–13	34	3	9	*Hamner* et al. [16]
Estrogen	Ejaculated & Centrifuged	6–11	11.5–13	91	6	7	*Hamner* et al. [16]
Progesterone	Ejaculated & Centrifuged	6–11	11.5–13	29	0	0	*Hamner* et al. [16]
Estrogen	Ejaculated & Centrifuged	10–11	12.5–13	99	85	86	*Hamner & Sojka* [18]
No treatment	Ejaculated	12	12	39	14	36	*Bedford* [4]
No treatment	Ejaculated	12	15	21	0	0	*Bedford* [4]

capacitation was not achieved in the uterus of ovariectomized rabbits. Capacitation of sperm in the uterus of ovariectomized rabbits treated with various doses of estrogens was neither improved in the experiment of *Chang* [11] nor in the experiment of *Hammer* et al. [16], but definite improvement was observed in the experiment of *Soupart* [28] and in the experiment of *Hamner & Sojka* [18].

In the experiment of *Soupart* [28] the proportion of fertilized eggs was significantly higher (38–54 %) following tubal insemination of sperm incubated in the uterus of ovariectomized rabbits treated with 0.5–2.5 μg/kg of estradiol benzoate than without treatment (4.2 % of 96 eggs). He also presented data to show a significant decrease of fertilized eggs (12 %) when sperm were incubated in the uterus of ovariectomized rabbits treated with a higher dose of estrogen (3–3.5 μg/kg).

It has been reported [11] that the proportion of fertilized eggs was not significantly increased following tubal insemination of sperm incubated in the uterus of immature rabbits treated with high doses of estrogen. Similarly the proportion of fertilized eggs was not increased when sperm were incubated in the uterus of ovariectomized rabbits treated with estrogen. Following treatment of ovariectomized rabbits with progesterone, capacitation of sperm in their uterus was definitely inhibited in both the experiments of *Chang* [11] and the experiments of *Hamner* et al. [16].

From this account it appears that results obtained by different investigators vary greatly. The variation was due not only to the work being conducted by different investigators, since even the same investigator was unable to confirm his earlier results [16, 18]. At any rate, it can be said here that capacitation is probably possible in the uterus of ovariectomized rabbits and estrogen treatment may improve the ability of the uterus to capacitate sperm.

Effects of estrogen on the capacitation of rabbit sperm in the Fallopian tubes was studied by *Bedford* [4] who reported that following the deposition of sperm in the tube for 12 hrs., none of 53 eggs were fertilized when placed in the tubes of ovariectomized rabbits, but most of the eggs were fertilized in the tubes of unilaterally ovariectomized (90 % of 30 eggs) or estrogen treated ovariectomized rabbits (100 % of 20 eggs).

Capacitation of sperm in the uterus of pseudopregnant rabbits

Fig. 3 summarizes the results concerning the capacitation of sperm in the uterus of pseudopregnant rabbits as reported by several investigators. It has been reported that capacitation of rabbit sperm cannot be achieved in the uterus of 6–14 days pseudopregnant rabbits [11]. Evidence to support the inability to capacitate sperm in the 7–9 day, but not in the 3 day pseudopregnant uterus was also shown by the presence or absence of fluorescence in the sperm labelled with tetracycline, a technique developed by *Ericsson* [15]. The inability to capacitate sperm in the uterus of pseudopregnant rabbits was also confirmed by *Soupart* [28] and *Bedford* [3].

Fig. 3. Capacitation of sperm in the uterus of pseudopregnant and progesterone-treated rabbits.

Rabbits for incubation of sperm	Sperm used	Length of incubation (in hrs.)	Time of tubal insem. (hrs. after ovul. inj.)	Total eggs exam.	No. of eggs fert.	% of eggs fert.	Investigators
6–14 day Pseudopregnant	Ejaculated	12–13	12–13	84	2	2.4	*Chang* [11]
9-day Pseudopregnant	Epididymal	10	13	36	2	5.6	*Soupart* [28]
13-day Pseudopregnant	Epididymal	10	13	36	0	0	*Soupart* [28]
7–10 day Pseudopregnant	Epididymal	14–20	12.25–12.75	90	6	7.7	*Bedford* [3]
7–10 day Pseudopregnant	Epididymal	14–20	9–10	52	33	63.5	*Bedford* [3]
Estrous Progesterone	Ejaculated	12–13	12–13	47	1	2.1	*Chang* [11]
Immature Proegesterone	Ejaculated	12–13	12–13	32	1	3.1	*Chang* [11]
Ovariectomized Progesterone	Ejaculated	12–13	12–13	36	3	8.4	*Chang* [11]
Ovariectomized Progesterone	Ejaculated & Centrifuged	10	11.5–13	29	0	0	*Hamner* et al. [16]
Estrous Progesterone	Ejaculated	12–13	12–13	30	7	23	*Chang* [13]
Estrous Medroxy-progesterone acetate	Ejaculated	12–13	12–13	25	10	40	*Chang* [13]
Estrous Chlormadinone acetate	Ejaculated	12–13	12–13	44	20	45	*Chang* [13]

Changes in the metabolism of rabbit sperm after incubation in the uterus of estrous rabbits [19, 22] were not observed after incubation in the pseudopregnant uterus [7] but were observed by *Murdoch & White* [23] when sperm were incubated in the uterus of rabbits treated with progesterone.

In the experiment of *Bedford* [3], incubation of sperm in the uterus was much longer (14–20 hrs) and a high proportion of eggs (63.5 %) were fertilized when tubal insemination was performed 1 to 0 hrs. before the estimated time of ovulation. This implies that capacitation of sperm in the uterus of pseudopregnant rabbits was possible if the time of incubation was longer and if tubal insemination was at the time of ovulation but not after ovulation.

It is known that capacitation of rabbit sperm can be achieved if they are incubated for 17 hrs. in the uterus of rats [6]. It has been reported that capacitation of rabbit sperm was possible in the uterus of estrous or ovariectomized rats, or in the uterus of estrous or anestrous dogs even when the incubation time was only 6–8 hours; capacitation of rabbit sperm, however, was not achieved in the uterus of pseudopregnant rats or dogs [16].

The possibility of partial capacitation of sperm in the uterus and then completed in the Fallopian tubes was suggested by *Chang* [9]. The capacitation of rabbit sperm in the isolated bladder, colon, anterior chamber of eye or glandula vesicularis was reported by *Noyes* et al. [25]. *Bedford* [3] reported, however, that only partial capacitation can be achieved in the anterior chamber of the eye, colon or seminal vesicles because fertilization was possible only when sperm incubated in these sites were deposited in the tubes of rabbits long before ovulation but not at the time of ovulation. Moreover, *Hamner & Sojka* [17] reported that a higher proportion of eggs were fertilized by sperm incubated in the uterus or oviduct than by the sperm incubated in the bladder or colon and that a higher proportion of eggs were fertilized when the incubated sperm were deposited into the tubes at the time of ovulation (10 hr. after ovulation injection) than 2 hrs. after ovulation (12 hrs. after ovulation injection). All these observations point to the difficulty and ambiguity in determining the achievement of capacitation if the time of tubal insemination was not well controlled and if the preparations to induce ovulation were not the same.

Capacitation of rabbit sperm in the uterus of rabbits treated with progestational compounds

Fig. 3 also presents the results concerning the capacitation of sperm in the uterus of rabbits treated with progestational compounds. It has been reported that the inhibition of capacitation in the uterus of pseudopregnant rabbits could not be corrected by treatment with gonadotropin or high doses of estrogen. Capacitation of sperm in the uterus of immature, ovariectomized or estrous rabbits, however, was almost inhibited by subcutaneous injection of a very high dose of progesterone (25 mg. macrocrystalline progesterone) 3–4 days before the incubation of sperm in their uterus [11]. The inhibition of capacitation in the uterus of ovariectomized rabbits treated with progesterone was also reported by *Hamner* et al. [16].

In a later study rabbit sperm were injected into the uterus or Fallopian tube of estrous rabbits treated daily with 2 mg. progesterone, 4 mg. medroxyprogesterone acetate, or 1 mg. of chlormadinone acetate for 3 days [13]. When the incubated sperm were deposited into the Fallopian tubes of other rabbits about 2–3 hours after ovulation, a relatively low proportion of eggs were fertilized (23–45 %). To a certain degree, these results are in disagreement with the results obtained in a previous experiment [11], which showed that capacitation was inhibited in the uterus of estrous, immature, or ovariectomized rabbits under a very high dose of progesterone treatment (only 2–8 % eggs fertilized). One of the obvious reasons for such discrepancy is that a single high dose of progesterone (25 mg) was administered in the early experiment while 3 modest doses of progesterone or synthetic progestins were administered in the later experiment. Moreover, in the early study sheep pituitary was used to induce ovulation, many of the uterine horns were ligated in both ends, and .9 % NaCl was used to flush sperm from the excised uterus. In the later experiment HCG was used to induce ovulation, uterine horns were not ligated and Ringer's solution was used to flush sperm from the uterus of anesthetized animals. In view of the fact that a relatively low proportion of eggs (23–45 %) were fertilized following tubal insemination of sperm recovered from the uterus of estrous rabbits treated with various progestins as compared with samples recovered from uteri of estrous rabbits (58 %–63 % [11]; 97 % [25]) it appears that capacitation of sperm in the uterus is inhibited to a certain degree by treatment of rabbit with progestins.

Although capacitation of sperm cannot be accomplished in the uterus of pseudopregnant rabbits, it can be achieved in the Fallopian tubes of pseudopregnant rabbits [11]. A later study [13] revealed, however, that there was no significant difference in the proportion of eggs fertilized by the sperm incubated in the uterus or in the Fallopian tube of animals treated with modest doses of progesterone or synthetic progestins. *Hamner* et al. [16] reported that none of 29 eggs were fertilized when capacitated sperm were placed in the tubes of pseudopregnant rabbits that had been induced to ovulate recently, but 50 % of 24 eggs were fertilized when placed in the tubes of estrous rabbits with newly ovulated eggs. In this case the failure of fertilization in the tube of pseudopregnant rabbits is not necessarily due to the fact that the capacitated state of sperm was reversed in the pseudopregnant rabbits as the authors suggested. It is more probably due to the irregularity of egg transport in the pseudopregnant or progesterone-treated rabbits [13, 14].

By incubation of sperm for 12 hours in the tube of ovariectomized rabbits treated with progesterone (25–50 mg) before the transfer of eggs into their tubes, *Bedford* [4] obtained a high proportion of eggs fertilized (68 % of 22 eggs). It is of interest to mention here that in *Bedford's* experiment no fertilization occurred in the tube of ovariectomized rabbits when 10 million of sperm were deposited but 82 % of 17 eggs were fertilized when 0.1–0.4 million of sperm were deposited. In this respect, the great variations obtained by investigators may also be due to the difficulty to control the number of sperm deposited into the tube.

Despite the inconsistency of the experimental results obtained by different investigators, it may be said that there is strong evidence to support the view that capacitation is inhibited in the uterus of pseudopregnant or progesterone-treated animals but to what degree and by what mechanism is still not clear.

Vaginal insemination at the time of ovulation with sperm incubated in the uterus of estrous, estrogen- or progesterone-treated rabbits

Most of the experiments discussed above were performed by deposition of sperm into the tubes, at the time of ovulation or soon after. It was thought that the variations of results may have been due to the surgical procedures of tubal insemination. Finally a simple experiment was performed as follows: Ejaculated sperm were incubated for 12 hours in the uterus of estrous rabbits treated with estrogen (Feeding 0.5 mg Ethinyl estradiol for 3–4 days) or with progesterone (Feeding 4 mg medroxyprogesterone acetate for 3–4 days). The incubated samples were recovered and inseminated without centrifugation intravaginally to another group of rabbits at the time of ovulation (10 hours after ovulation injection). The results presented in Fig. 4 shows clearly that the proportion of eggs fertilized was significantly higher ($P < .01$) by sperm incubated in the uterus of estrous rabbits or estrogen-treated rabbits than by sperm incubated in the progesterone-treated rabbits. The proportion of eggs fertilized was very much the same whether ejaculated sperm or sperm incubated in the uterus of progesterone-treated rabbits were used. It is obvious that sperm incubated in the uterus of progesterone-treated rabbits behaved like the uncapacitated ejaculated sperm. Thus the inhibition of capacitation in the uterus of progesterone-treated rabbits was again demonstrated.

It is possible that low doses of various synthetic progestins used recently as effective oral contraceptives (Chlormadinone acetate [21]; Noregestrel [27]) may inhibit sperm capacitation in humans. It is known, however, that fertilization can be inhibited by disturbance of sperm and egg transport beside the inhibition of sperm

Fig. 4. Capacitation of sperm in the uterus of progesterone or estrogen treated rabbits (Vaginal insemination 10 hrs. after ovulation injection).

Vaginal insem. of sperm	No. of rabbits used	No. of sperm ins. in M. Avg. (Range)	Total No. of eggs recovered	No. of fertilized eggs (%)
Incubated in the uterus of progesterone-treated rabbits	7	0.9 (.2–3.8)	58	29 (50 %)
Incubated in the uterus of estrogen-treated rabbits	6	0.6 (.056–.80)	51	43 (84 %)
Incubated in the uterus of estrous rabbits	4	0.52 (.29–.80)	40	38 (95 %)
Ejaculated	9	80 (20–188)	94	48 (51 %)

capacitation, and that the degeneration of eggs occurs during pseudopregnancy or under progesterone-treatment in the rabbit [13, 14]. Direct experimental studies to determine which one of these disturbances occurs by administration of progestational compounds in humans and other species is therefore urged.

References

[1] *Austin, C. R.* (1951). Observations on the penetration of sperm into the mammalian egg. Aust. J. Sci. Res., B., **4,** 581–596.

[2] *Austin, C. R.* (1952). The "capacitation" of the mammalian egg. Nature, Lond., **170,** 326.

[3] *Bedford, J. M.* (1967). Experimental requirement for capacitation and observations on ultra-structural changes in rabbit spermatozoa during fertilization. J. Reprod. Fert. Suppl. **2** (1961), 35–48.

[4] *Bedford, J. M.* (1969). Hormonal aspects of capacitation. Submitted to J. Endocrin.

[5] *Bedford, J. M. & M. C. Chang* (1962). Removal of decapacitation factor from seminal plasma by high-speed centrifugation. Amer. J. Physiol. **202,** 178–181.

[6] *Bedford, J. M. & R. Shalkovsky* (1967). Species specificity of sperm capacitation in the rabbit. J. Reprod. Fert. **13,** 361–364.

[7] *Brackett, B. G.* (1968). Respiration of spermatozoa after in utero incubation in estrous and pseudopregnant rabbits. 6th Int. Congress Anim. Reprod. & A. I., Paris, page 8.

[8] *Chang, M. C.* (1951). Fertilizing capacity of spermatozoa deposited into the Fallopian tubes. Nature, **168,** 697.

[9] *Chang, M. C.* (1955). Development of fertilizing capacity of rabbit spermatozoa in the uterus. Nature, Lond. **175,** 1036.

[10] *Chang, M. C.* (1957). A detrimental effect of seminal plasma on the fertilizing capacity of sperm. Nature, Lond. **179,** 258–259.

[11] *Chang, M. C.* (1958). Capacitation of rabbit spermatozoa in the uterus with special reference to the reproductive phases of the female. Endocrinology, **63,** 619–628.

[12] *Chang, M. C.* (1967). Summation of the Workshop: Capacitation of spermatozoa and endocrine control of spermatogenesis. J. Reprod. Fert. Suppl. 2 (1967) 145–147.

[13] *Chang, M. C.* (1967). Effects of progestrone and related compounds on fertilization, transportation and development of rabbit eggs. Endocrinology **81,** 1251–1260.

[14] *Chang, M. C.* (1969). Fertilization, transportation and degeneration of eggs in the pseudopregnant or progesterone-treated rabbits. Endocrinology **84,** 356–361.

[15] *Ericsson, R. J.* (1967). Technology, physiology and morphology of sperm capacitation. J. Reprod. Fert. Suppl. **2** (1967) 65–74.

[16] *Hamner, C. E., J. P. Jones & N. J. Sojka* (1968). Influence of the hormonal state of the female on fertilizing capacity of rabbit spermatozoa. Fert. and Ster. **19,** 137–143.

[17] *Hamner, C. E. & N. J. Sojka* (1967). Capacitation of rabbit spermatozoa: Species, specificity and organ specificity. Proc. Soc. Exp. Biol. Med. **124,** 689–691.

[18] *Hamner, C. E. & N. J. Sojka* (1968). Requirement for capacitation of rabbit sperm. Nature, Lond. **220,** 1042–1043.

[19] *Hamner, C. E. & W. L. Williams* (1963). Effect of the female reproductive tract on sperm metabolism in the rabbit and fowl. J. Reprod. Fert. **5,** *143.*

[20] *Mann, T.* (1967). Capacitation for reproductive research. J. Reprod. Fert. Suppl. **2** (1967) 149–155.

[21] *Martinez-Manautou, J., J. Jiner-Velazques, R. Aznar-Ramos, M. Lozano-Balderas & W. H. Rudel* (1967). Continuous administration of 500 μg of chlormadinone acetate as a method of regulating fertility without inhibiting ovulation. Proc. of 8th Int. Con. Int. Plan. Parent. Fed., SAntiago, Chile, 241–252.

[22] *Mounib, M. S. & M. C. Chang* (1964). Effect of *in utero* incubation on the metabolism of rabbit spermatozoa. Nature, Lond. **201,** 943.

[23] *Murdoch, R. N. & I. G. White* (1967). The metabolism of labelled glucose by rabbit spermatozoa after incubation *in utero.* J. Reprod. Fert. **14,** 213–223.

[24] *Noyes, R. W.* (1953). The fertilizing capacity of spermatozoa. West. J. Surg. Obst. Gynec., **61,** 342–349.

[25] *Noyes, R. W., A. Walton & C. E. Adams* (1958). Capacitation of rabbit spermatozoa. J. Endocrin. **17,** 374–380.

[26] *Pinsker, M. C. & W. L. Williams* (1968). Spermatozoa decapacitation factor (DF) in human seminal plasma. Proc. Soc. Exp. Biol. Med., **129,** 446–448.

[27] *Roland, M.,* (1968). Endocervical inhibition of sperm capacitation by Norgestrel contraception. Int. J. Fert., **13,** 390–398.

[28] *Soupart, P.* (1967). Studies on the hormonal control of rabbit sperm capacitation. J. Reprod. Fert. Suppl. 2 (1967) 49–63.

[29] *Weinman, D. E. & W. L. Williams* (1964). Mechanism of capacitation of rabbit spermatozoa. Nature, Lond. **203,** 423–424.

Discussion

P. Soupart: After Dr. *Chang's* remarks, I feel practically obliged to defend my own position. With your permission, Mr. Chairman, I would like to show a few figures.

It seems quite obvious that if we want to approach the problem of hormonal control of sperm capacitation on a quantitative basis, the first thing we need is an experimental system. Fig. 5

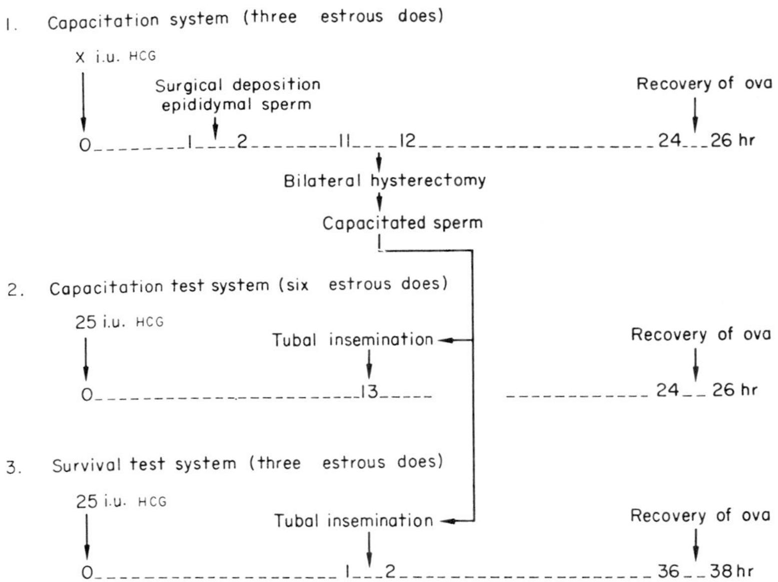

Fig. 5. Experimental design for the study of rabbit sperm capacitation *in vivo.*

describes the system which we have been using for the past few years. You take an animal which serves as an incubator for sperm and you incubate sperm in the uterus. This animal can be treated any way your choose in your particular experiment. After 10 hours incubation, which is normally the period in the rabbit which separates the time of mating from the beginning of ovulation, you recover the sperm and you test them. You test them by tubal insemination in an animal which has been injected with HCG in order to ovulate. If you recall the work of *Harper,* ovulation in the rabbit is not an instantaneous process, it takes a few hours to reach completion (from 10 to 13 hrs or more after the ovulating injection). For practical reasons, you want to recover a significant number of eggs, so you do your tubal insemination at 13 hours after the ovulation inducing injection.

At that time, presumably, the mucin coat will deposit very soon and the viability of the egg will begin to decline. Since it takes several hours for uncapacitated sperm to become capacitated, you are playing there on a time factor which may be favorable for testing the degree of capacitation achieved by previous treatment of sperm.

We were also concerned with a method of evaluating the possible damage suffered by sperm during the experimental treatment. To achieve this end, we devised another system, which we call the survial test system. It is exactly the same as the capacitation test system, except that the tubal insemination is done shortly after the ovulation injection, allowing plenty of time for sperm to become capacitated if it were not as a result of initial incubation. Using this procedure, if you get a negative result in terms of the capacitation test system, a positive result in the survial test system will tell you that your experimental treatment has not damaged the fertilizing ability of the spermatozoa.

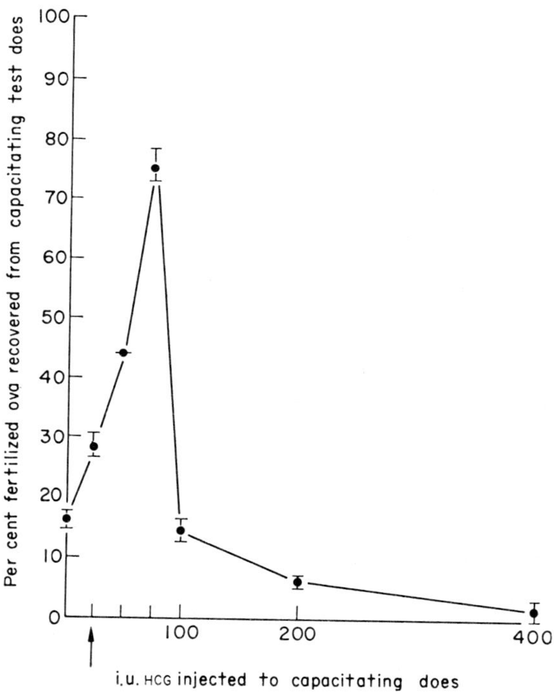

Fig. 6
Effects of a single dose of HCG on sperm capacitation in the uterus of intact oestrous does over a 10 hr incubation period. The arrow indicates the standard dose of 25 I.U. HCG that results in ovulation without failure.

Fig. 6 illustrates a study of the relationship between intensity of gonadotropin (HCG) stimulus and the efficiency of capacitation in the uterus. We found that sperm, after 10 hours incubation in the uterus of a rabbit not injected at all with HCG, are able to fertilize a few ova in the capacitation test system. The percentage of success is somewhere between 15 and 20 %. If we inject 25 units of HCG, the minimum ovulating dose, we get an increase in the response, a stimulation of the capacitation process by HCG. But we also find that we need 3 times as much – 75 units of HCG, to get the optimal response. For higher doses – 100 I.U. HCG and over – we observe a reversal of the situation, apparently an inhibitory effect. These effects are obtained using exogenous gonadotropin, not of pituitary origin, and we were concerned to see if we could get the same effects using pituitary gonadotropin. Fig. 7 shows the effect of purified LH from NIH, in rabbits that were hypophysectomized. Also shown here are the results of the survival

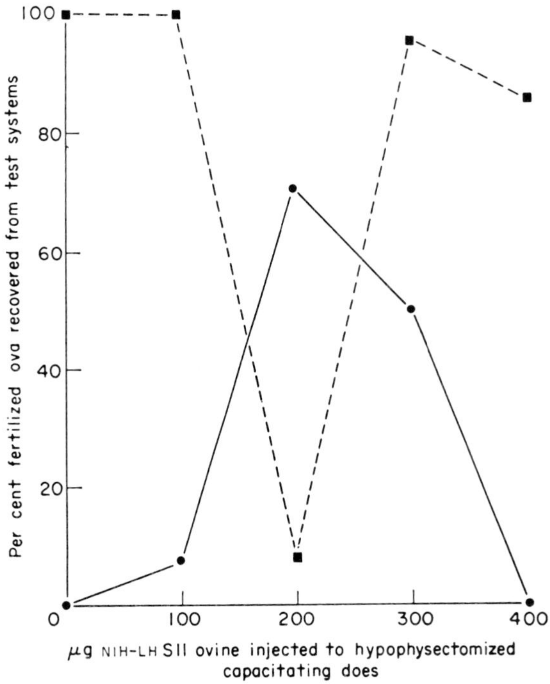

Fig. 7
Effects of a single dose of a purified LH preparation (NIH – LH S 11 ovine) on sperm capacitation in the uterus of hypophysectomized does over a 10 hr. incubation period. ■ survival test; ● capacitation test.

tests corresponding to the capacitation tests. You see that with no stimulation with LH, you get no fertilization success in your capacitation test system, but sperm survival is excellent. When you start injecting LH, you reach an optimal situation for about 200 μg of the preparation. Corresponding to that optimal response, you get the lowest survival. It has been a constant finding in those experiments that good, effective capacitation is always associated with a shortening of the fertilizing ability and the life-span of the spermatozoa. And the contrary is true also – if you do not achieve capacitation by the experimental treatment the spermatozoa survive longer.

Our next question was (Fig. 8): do we need mediation through the ovary or are there effects (stimulation followed by inhibition) obtained by direct effect of the gonadotropins on the uterine environment? Fig. 8 shows the results of similar experiments done one castrated rabbits, receiving increasing doses of HCG, up to 600 I. U. As you can see, the efficiency of capacitation is very poor, and the survial, as a rule, is excellent. Since in the pre-ovulatory period the hormone that dominates the scene is estrogen, we were wondering if we could reproduce the situation, usingestrogen injected into castrated animals.

Fig. 9 will show you what happens if you use castrated rabbits and inject them in order to create increasing levels of estrogen impregnation. We found that we needed about 1 μg of estradiol benzoate per kg of body weight to obtain the same efficiency of capacitation after 10 hours incubation in the uterus that we would get with optimal HCG of LH stimulation. Once again, when capacitation efficiency is good, the survival is minimal; if capacitation efficiency is poor the survival is extended.

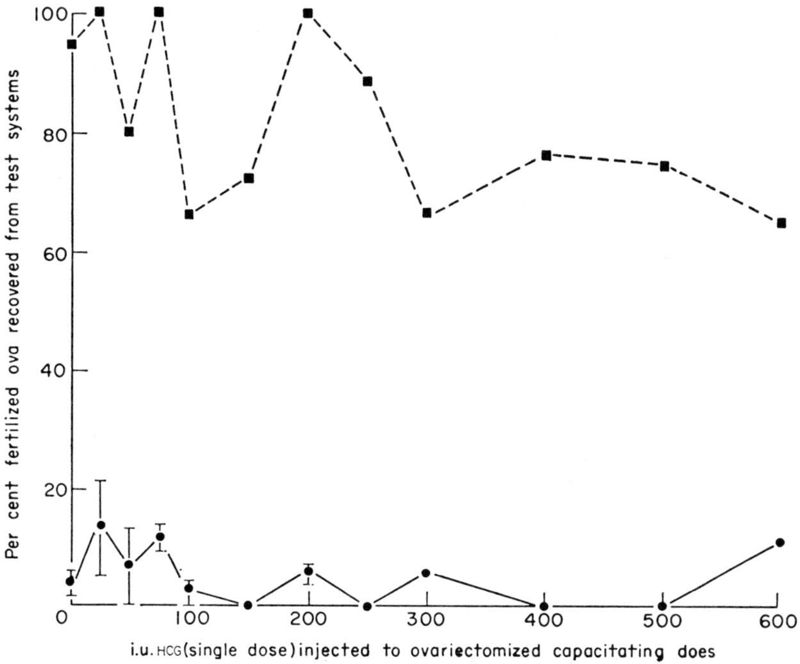

Fig. 8. Effects of a single dose of HCG on sperm capacitation in the uterus of ovariectomized does over a 10 hr. incubation period. Vertical bars indicate the range of variation observed in replicate experiments.
■ survival test; ● capacitation test.

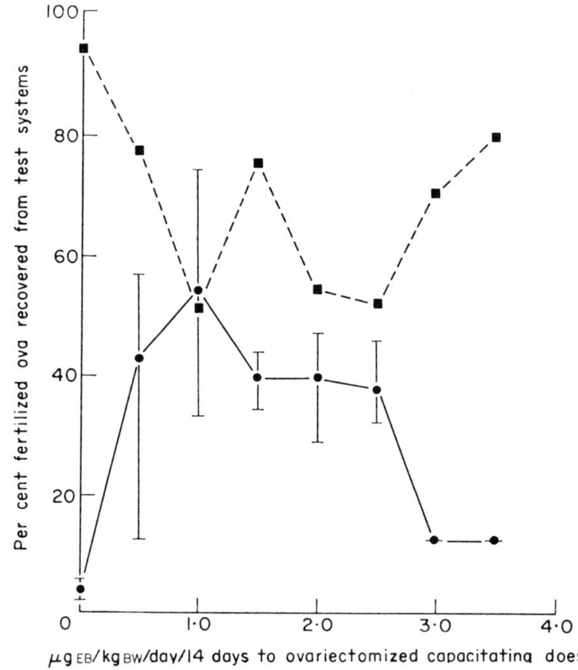

Fig. 9. Effects of increasing levels of estradiol benzoate on sperm capacitation in the uterus of ovariectomized does over a 10 hr. incubation period. Vertical bars indicate the range of variation observed in replicate experiments.
■ average survival test; ● capacitation test.

In all situations described so far you have conditions which can be divided into two categories: conditions under which the stimulation of the capacitation process is not very good, and conditions under which it is better. If you study the distribution of development stages reached by the eggs at time of recovery from the capacitation test system, from the view point of good or poor conditions for capacitation, you find that distribution of development stages is significantly different in the two situations (Fig. 10). When the conditions for capacitation have

Fig. 10. Distribution of development stages of ova in capacitation tests.

	Control (*Orgebin-Crist*, unpublished)	Capacitation tests (This study)	
Autopsy	25 hr after HGG	25 to 26 hr after HGG	
Condition of spermatozoa from distal tail of epididymis	In-utero insemination 1 hr after 50 i.u. HGG, fully capacitated	Tubal insemination 13 hr after 25 i.u. HGG using spermatozoa incubated in utero for 10 hr under conditions expected	
		To induce capacitation	Not to induce capacitation
No. of experiments	6	24	23
No. of ova recovered	81	1033	1053
% fertilized	87.6	46.5 (29.2-78.0)*	8.9 (3.9-18.9)*
% uncleaved	19.7	20.4	63.8
2 cell	76.0	78.8	35.1
3 cell	2.8	0.6	0
4 cell	1.4	0.2	1.1

* Average and range

been effectively stimulating the process, you find that the percentage of eggs still at the pronuclear stage, when recovered 25 to 26 hours after the ovulating stimulus, does not exceed 20 %. If the conditions were poor, as shown in the last column of Fig. 10, you find up to 63 % of ova still at the pronuclear stage. This means delayed fertilization, which cannot be explained otherwise than by capacitation having occurred in your capacitation test animal. This is what I call "super-imposed capacitation". It results in some kind of a blank in the *in vivo* test method, that must be taken into account. Therefore, there is a basic level of success in a capacitation test that I consider on the basis of those data to be about 20 % fertilized ova, which is meaningless to me, because it can very well be due to superimposed capacitation occurring in the *in vivo* test system. Of course, this is one of the defects of the *in vivo* experimental approach. There is another possible method that nobody to my knowledge has used so far, which is testing for capacitation in an *in vitro* fertilization system. We will come to it later. During the pre-ovulatory period in the rabbit, estrogen is not the only hormone dominating the scene. There is another one: 20 α-hydroxy-progesterone. So far, according to *Hilliard* and co-workers, the only role which has been assigned to 20 α hydroxy progesterone is to serve as a positive feedback on the pituitary to cintinue the release of LH.

The first appearance of LH in the system triggers the synthesis and release of the 20 α hydroxy progesterone by the interstitial cells of the ovary. We were wondering if 20 α hydroxy progesterone is not perhaps also playing a role in the hormonal control of sperm capacitation. In other words, the situation could be a little bit more complicated than we had thought so far.

Fig. 11 illustrates the effects of increasing ratios of 20 *a* OH progesterone to Estradiol benzoate on the effeciency of capacitation in the uterus of costrated does. The castrates, all primed with 1 μg of estradiol benzoate per kilo of body weight, are used for *in utero* incubation of epididymal spermatozoa over a 10-hr period. At the time of sperm deposition in the uterus a single injection of 20 *a* OH progesterone is given in order to mimic the output of this hormone by the ovary, output which is normally triggered by LH release. The doses tested are: 0, 500, 1000, 1500, 2000 and 3000 μg 20 *a* OH progesterone per kilo of body weight, added to a constant level of 1 μg estradiol benzoate per kilo of body weight. As shown by Fig. 11, ratios of 20 *a* OH progesterone to estradiol benzoate ranging from 0/1 to 1500/1 μg per KBW result in good stimulation of capacitation. For higher ratios, a reversal of the action is observed, which I regard as a strong inhibition.

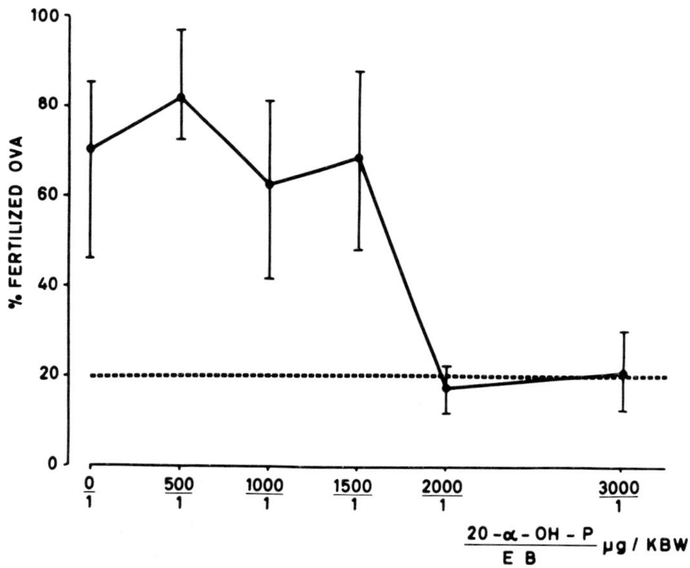

Fig. 11. Results of capacitation tests after 10 hr. incubation of sperm in uteri of ovariectomized does, stimulated by increasing ratios of 20 α OH progesterone to estradiol benzoate. Dotted line: upper limit of error inherent to *in vivo* capacitation test method.

The dotted line indicates what I consider as the upper limit of error inherent to the *in vivo* testing method. This strong inhibition suggests to me an explanation for the inhibitory phase of the curve that has been observed with LH and HCG. If you overstimulate an ovary with gonadotropins, you may very well increase the output of 20 *a* hydroxy progesterone to such an extent that the anti-estrogenic effect of this progestational compound will interfere with the stimulation of capacitation in the uterus.

There is a question which is still not completely clear in my mind. Is there a significant difference between the effeciency of capacitation with estrogen stimulation alone and the efficiency of stimulation by estrogen combined with 20 *a* hydroxy progesterone? I have no statistical data to present yet, but Fig. 12 may show what I mean. You will remember that Fig. 11 showed a

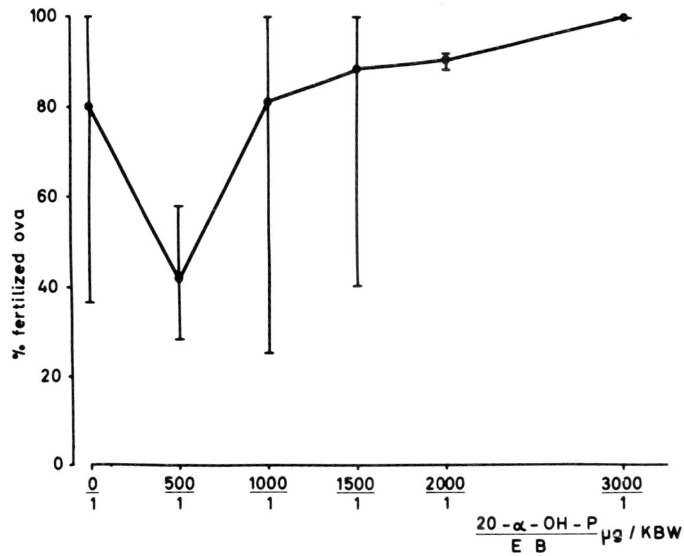

Fig. 12. Results of survival tests after 10 hr. incubation of sperm *in uteri* of ovariectomized does, stimulated by increasing ratios of 20 α OH progesterone to estradiol benzoate.

ratio of 20 *a* OH progesterone to estradiol benzoate of 500/1 to have the best stimulation effect. On this figure you have the survival test curve corresponding to the capacitation test curve you have just seen. There is a maximal depression of survival for the same ratio. This means to me that under such conditions of hormonal stimulation, when you get the best capacitation efficiency associated with the most depressed survival test, you have probably achieved the best conditions for stimulating your system. Fig. 13 summarizes the numerical data on capacitation ability of the rabbit uterus stimulated by estrogen alone and by combination of estrogen and 20 *a* OH progesterone. The *in vivo* approach may not permit to discriminate effectively between the effect of estrogen alone and estrogen plus 20 *a* hydroxy progesterone on the uterus, because the capacitation test system itself is an environment fully activated in terms of capacitation efficiency and may very well be able to complete a process only partially achieved in the uterus. What I suggest as the next step in this study is to incubate sperm in uteri stimulated by estrogen alone and test them for their capacity to fertilize *in vitro*. Next, do a similar experiment, having incubated the sperm in uteri stimulated by what we may consider as the optimal ratio of 20 *a* hydroxy progesterone to estrogen. This approach will probably provide an answer to my question. As you very well know estrogen as well as progesterone control the synthesis of specific proteins, which we may tentatively call capacitation enzymes. If estrogen alone is needed to achieve full capacitation efficiency, *in vitro* fertilization test will be positive, because the process will have been fully completed. On the other hand, if a certain ratio of 20 *a* hydroxy progesterone to estrogen is needed to achieve completely the capacitation process, the *in vitro* fertilization test would be negative when using sperm incubated in an environment stimulated by estrogen alone, but positive if stimulation is achieved using an optional ratio of the two hormones. I hope I will be able to present the results of such

Fig. 13. Effects of 20 α OH progesterone on capacitation of rabbit sperm *in utero*. Castrated does primed with 1 μg estradiol benzoate per kilo of body weight.

20 α OH / EB μg/KBW	No. of expts	Capacitation tests			Survival tests	
		ova recovered	% fertilized ova mean and range	% pronuclear ova mean and range +	ova recovered	% fertilized ova mean and range
0/1	6	204	70 (45.5–85.7)	6.0 (5.9–27.3)	86	80 (36.4–100)
500/1	4	189	82* (71.7–96.9)	3.9 (0–29.4)	57	42** (27.8–57.9)
1,000/1	7	247	62* (41.5–81.4)	7.8 (0–27.3)	94	81 (25.0–100)
1,500/1	4	218	69* (47.6–88.0)	8.7 (0–20.0)	59	88 (40.0–100)
2,000/1	2	97	17*** (11.4–22.6)	11.8 (8.3–20.0)	21	91 (89.0–93.0)
3,000/1	2	104	21*** (12.7–30.6)	26.0 (25.0–28.5)	19	100 (100–100)

+ Upper normal value at standard time for ovum recovery : 20.4 %
* NS : $.4 < P < .5$
** NS : $P > 0.05$
*** Highly significant : $P < 0.01$

investigations very soon. So far my investigations have been concerned with the hormonal control of capacitation in the rabbit uterus. I plan to investigate the response of the oviduct in a similar manner. From previous observations (*P. Soupart, J. Reprod.* Fertil. Supp. 2, 49–63, 1967) it can already be anticipated that the response of the oviduct to various hormonal stimuli may be different than that of the uterus.

J. M. Bedford: I would like very briefly and frankly to disagree with both Dr. *Chang* and Dr. *Soupart.* Two small points: Firstly, I have tried to repeat these experiments with HCG (this nice convincing curve) and I must say I can get no effect with HCG on capacitation in doses of up to 1000 units. The only difference in our experimental methods is that I use vaginal insemination or actual mating, whereas I know you put the sperm into the uterus. I am not sure why you put the sperm into the uterus directly.

P. Soupart: I did not discuss the details of the experimental procedure but the reason for depositing sperm directly in the uterus of incubating animals is as follows. We use 3 rabbits to incubate sperm in 6 uterine horns. After 10 hrs. incubation we recover incubated sperm from those 6 uterine horns and we use them for tubal insemination in 6 capacitation test animals and 3 survival test animals. This represents a total of 18 oviducts to inseminante and requires that a significant number of sperm be recovered from uteri of animals used for the initial incubation. To insure such a recovery we use sperm from the distal tail of the epididymis from one or several males, which permitts surgical deposition in the incubating uteri of a much larger number of sperm than if sperm from ejaculates were used.

J. P. Bedford: Nevertheless, in our hands, following normal mating or vaginal insemination, one cannot get any effect with HCG – that is just one brief point. The second point vis à vis Dr. *Chang's* comments on hormonal control. After a long series of experiments on the rabbit – and this I think would interest our hosts – I tend to doubt that one can obtain complete inhibition of capacitation in the rabbit with steroids. In the ovariectomized rabbit capacitation will ultimately occur in the uterus, but not for about 20 hrs., which is a very long time. There is no doubt from Dr. *Soupart's* work that estrogen does stimulate the capacitating activity in the uterus; and Dr. *Chang* has shown that progesterone does depress capacitating activity in the uterus. But remember that we have a Fallopian tube and this is where fertilization occurs. In the Fallopian tube of the ovariectomized rabbit, the capacitating activity is depressed in the absence of estrogen. However, the estrogen deficient Fallopian tube will still capacitate significant numbers of sperm, considerably more than the number that normally get up into the Fallopian tube after mating. Thus the capacitating potential of the Fallopian tube is only reduced but not destroyed in the absence of estrogen. The ovariectomized or the normal Fallopian tube under the influence of progesterone will still capacitate sperm; even in the ovariectomized animal given large doses of progesterone, 15 or 20 times the amount of progesterone required to maintain pregnancy, one can still obtain capacitation in the Fallopian tube. So at least in the rabbit it seems doubtful that one can obtain complete inhibition of capacitation with steroids. There may be species differences of course; in the monkey Fallopian tube, for instance, the cilia require exogenous estrogen, or the ovarian secretions. In the rabbit they donot, in the absence of the ovaries the cilia will go on for many months quite happily. So there may be important species differences here.

M. C. Chang: Dr. *Soupart,* your introduction of survival test and the effect of 20 α hydroxy progesterone is of interest but you did not defend yourself by answering my criticism. There is only one point I would like to make to Dr. *Bedford's* regarding the Fallopian tube. Dr. *Bedford* did some studies with the progesterone-treated animals and found that sperm can be capacitated in the tube of rabbits treated with progesterone as I have reported in the pseudopregnant rabbits in 1958. Even in the uterus, as you have done, if you incubate for a longer time, not 12 hrs. but 20 to 24 hrs. you get good capacitation.

Morphological Aspects of Sperm Capacitation in Mammals

J. M. Bedford

Department of Anatomy and International Institute for Human Reproduction, Columbia University College of Physicians and Surgeons, New York, N. Y., USA

Summary: Observations in the phase contrast and electron microscope fail to reveal any structural changes in rabbit sperm which can be interpreted as a morphological concomitant of capacitation. In the vicinity of the egg, however, capacitated rabbit and hamster sperm undergo an acrosome reaction characterized by fusion and vesicle formation between the plasma and outer acrosomal membranes. By contrast, non-capacitated (ejaculate or epididymal) rabbit sperm show no such response when adhering to the ovum *in vivo.* The changes involved in capacitation may, therefore, confer upon the sperm the ability to undergo the acrosome reaction in response to a stimulus from the egg/granulosa cell mass. The "true" vesiculation acrosome reaction differs generally from the acrosomal breakdown associated with non-specific degeneration or that caused by harmful agents in which the plasma membrane and acrosome rupture and are lost randomly as separate entities.

Preliminary experiments show that rabbit follicular fluid brings about the discharge of the acrosome in a majority of uterine sperm, but not in epididymal sperm. Yet, in most uterine sperm first exposed to seminal plasma, the head plasma membrane and acrosome remain intact in the presence of follicular fluid. This suggests that the seminal decapacitation factor (DF) may act to mask receptor sites at the capacitated sperm surface or in some other way to stabilize the cell membrane which overlies the sperm head.

Introduction

After their release from the male, the spermatozoa of some and perhaps all mammals undergo a period of further maturation in the female. This finally confers upon them the capacity to pass through the granulosa cell barrier around the ovum, and to penetrate the zona pellucida. Our present knowledge of the requirement for capacitation in different species of mammals is reviewed by *Austin* elsewhere in this symposium. As yet, we do not have any clear understanding of the changes which occur in sperm as a concomitant of capacitation; the more obvious possibilities involve either metabolic or structural changes in the sperm. There is evidence of change in the respiratory patterns of rabbit sperm after residence for some hours in the female tract [15, 21, 22, 28]. However, the functional significance of such metabolic changes in the sperm cell and their relation to capacitation still remain obscure. In examination of uterine and tubal sperm, or sperm in the act of penetrating the zona pellucida, no change has been detected in the fine structure of the

Manuscript received: 29 April 1969

mitochondria or other elements of the tail region. By contrast, certain organelles of the sperm head, in particular the plasma membrane and the acrosome, undergo profound structural changes before true penetration of the zona pellucida begins. The present discussion of the morphological aspects of capacitation is therefore centred around the sperm head and its organelles, as seen at different phases of approach to and interaction with the egg, in both normal and experimental situations. Major attention is paid to the rabbit and to the hamster, since this aspect of capacitation has been studied most intensively in these two species.

The morphology of capacitated sperm

The basic ultrastructure of the mammalian sperm has now been rather extensively documented [7, 8, 14, 18, 19, 23, 24, 26].

In a previous study [7] comparison of the ultrastructure of uterine and epididymal/ejaculated rabbit sperm failed to bring to light any change which might readily be interpreted as a concomitant of the functional change involved in capacitation, though in osmiumfixed sperm there was a suggestion of a relatively greater swelling of the plasma membrane over the acrosome. In that study, samples were collected from the uterus at 10 hours after mating, on the assumption that capacitation is completed in this site in about 6 hours. More recent data show, however, that rabbit sperm do not acquire the capacity to penetrate eggs in the uterus until about 10–11 hours after insemination [12]. The possibility that morphological changes do occur in rabbit sperm during capacitation has been reexamined, therefore, using samples recovered from the uterus or Fallopian tube between 13–15 hours after insemination. These sperm were fixed in 3 % glutaraldehyde in a phosphate buffer for about 30 minutes, followed by post-osmication and embedding in Epon 812. Thin sections were stained with uranyl acetate and lead citrate.

These more recent studies using current methods of specimen preparation allow essentially the same conclusions as those reported previously. There is no ultrastructural change in the heads of the majority of rabbit sperm recovered from the uterus of oviduct 13–15 hours after mating, and therefore no direct ultra-structural concomitant of capacitation, *per se.* The great majority of uterine and oviducal sperm remain quite intact and show no change in the physical proportions of the organelles (Fig. 1). A few sperm manifest obvious signs of degeneration, as reported in a previous light microscope study [6], and both intact and degenerating sperm have been observed within uterine leucocytes (Fig. 2). In cases in which degeneration of the acrosome was evident, the plasma membrane and the acrosome usually appeared to be disintegrating in a haphazard way, as separate structures (Fig. 3). This contrasts with the coordinated breakdown and fusion between the outer

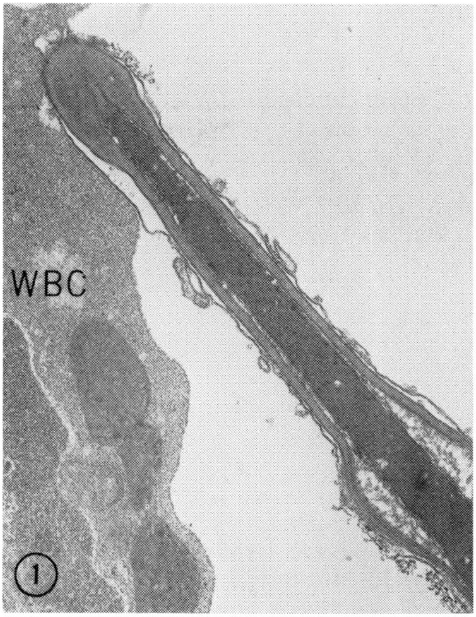

Fig. 1

Rabbit sperm recovered from the uterus 13 1/2 hours after coitus. The sperm is sticking to a white blood cell and shows no change in the acrosome or plasma membrane. Note the particulate debris which adheres to this part of the surface, but not to the membrane which covers the tail (not shown).
Magnification X 20,500

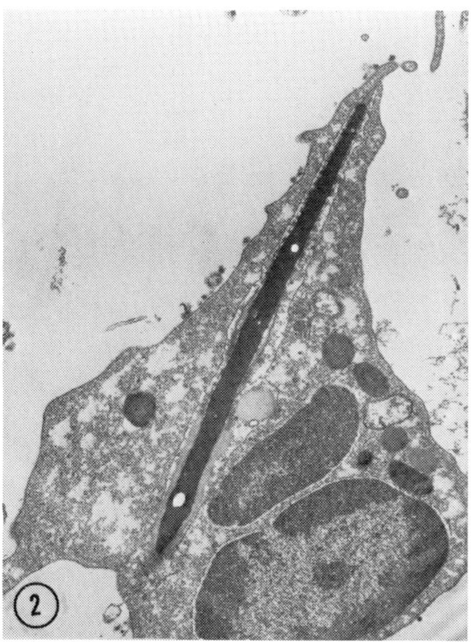

Fig. 2

Rabbit sperm nucleus inside a white blood cell – recovered from the uterus 13 1/2 hours after coitus. In this case the outer acrosomel elements have been lost from the sperm head, but intact sperm are also ingested by cells in this uterine environment.
Magnification X 12,400

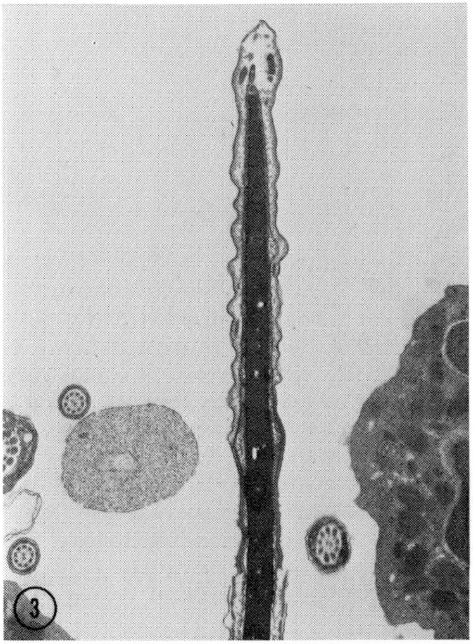

Fig. 3

Rabbit sperm recovered from the uterus 14 hours after coitus. The plasma membrane has been shed over the region of the acrosome, which itself is beginning to swell prior to degenerative rupture.
Magnification X 15,800

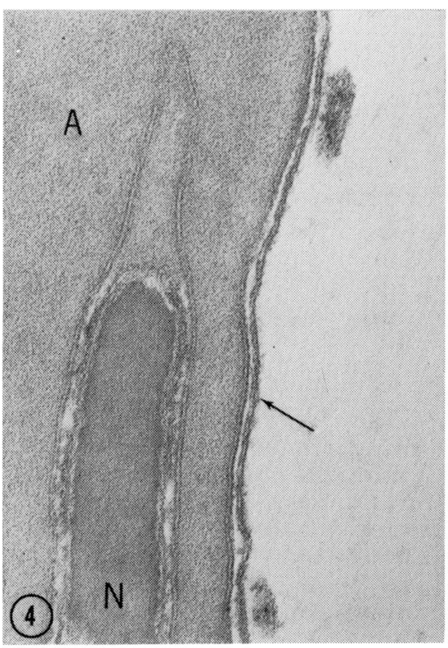

Fig. 4

Rabbit sperm recovered from the uterus 14 hours after coitus. The plasma membrane (arrowed) displays the tri-laminar 'unit' membrane arrangement.

A acrosome; N nucleus.
Magnification X 81,000

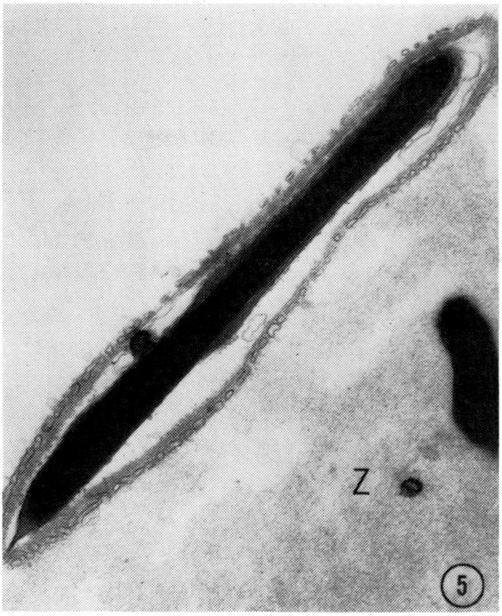

Fig. 5

Hamster sperm capacitated *in vitro,* has begun to undergo the acrosome reaction at the surface of the zona pellucida (Z). Note the ring of small vesicles which envelop the dense nucleus. Some of the acrosome content remains within this shroud and appears as a homogeneous inner coat to the vesiculated boundary.
Magnification X 19,000

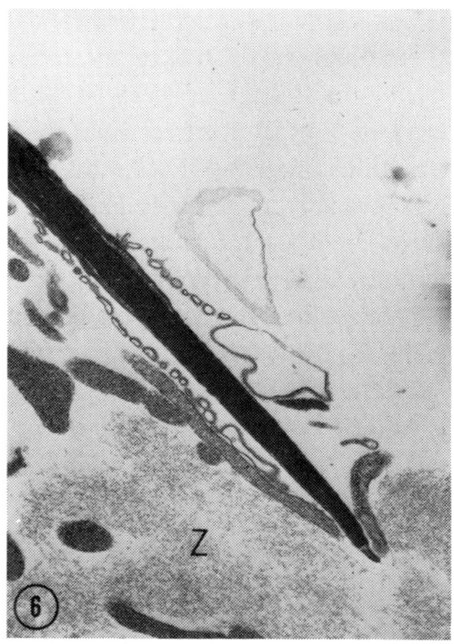

Fig. 6 & 7

Rabbit sperm undergoing the acrosome vesiculation reaction, at the zona surface. Fig. 7 depicts a sperm which was inseminated into the oviduct about 13 hours before ovulation. The sperm in Fig. 8 was observed at the surface of an egg collected 12 1/2 hours after normal coitus.
Magnification X 14,000; X 12,000

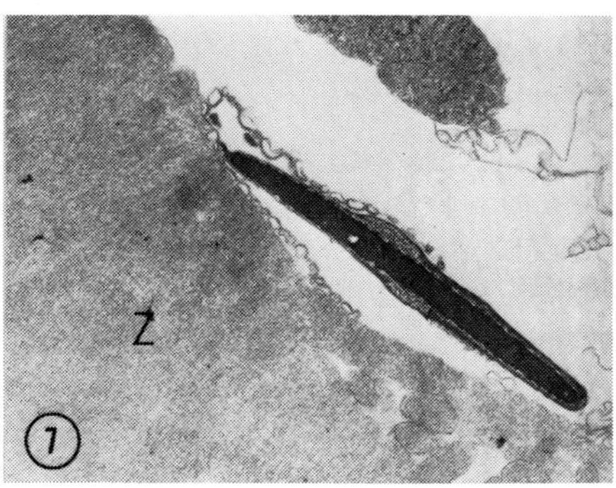

acrosomal and the plasma membranes seen during the true acrosome reaction (Figs. 5–7). These differences are discussed below in a consideration of the "true" and "false" acrosome reaction.

The use of uranyl acetate and lead, with good fixation, allows clear visualization of the tri-laminar "unit" membrane arrangement of the acrosomal and plasma membranes (Fig. 4). This laminated impression appears equally clearly in these membranes in ejaculate sperm and in sperm recovered from the uterus more than 13 hours after coitus. It is not yet clear whether any change occurs in the fine structure of the plasma membrane during capacitation; the question of possible alteration in electron density of the outer lamina is currently being studied using different techniques of membrane visualization in the electron microscope.

The acrosome reaction

In studies on the guinea pig and several murine rodents, *Austin* and *Bishop* [2] observed that sperm which were penetrating the egg appeared to have lost the acrosome. These authors also commented on the fact that motile sperm which lacked the acrosome were seen in tubal fluids in the hamster and guinea pig. While it was recognized that acrosomal detachment may occur as a post-mortem change, these observations evoked the conclusion that the acrosome becomes modified during the sojourn of sperm in the female, and that this modification constitutes capacitation. By contrast, light microscope observations on rabbit sperm in various situations suggested that in this species the loss of the acrosome from uterine sperm was probably a manifestation of a non-physiological rupture or degeneration of the acrosome not directly associated with capacitation [6]. More recent observations with the light microscope [29], and with the electron microscope [5, 9, 11] have confirmed that the acrosome becomes detached before penetration of the zona pellucida begins. It is now clear, however, that the acrosome can be discharged from the sperm head either as a "true" or a "false" acrosome reaction. The "true" acrosome reaction appears to occur in the rabbit in most cases only in the vicinity of the egg [11] but can take place quite commonly in tubal or follicular fluids in the hamster (*Barros* and *Franklin* – personal communication).

The events which occur during the "true" acrosome reaction have been described for the hamster and rabbit [4, 11] and involve a progressive breakdown and fusion between the plasma membrane and outer acrosome membrane, to form a series of vesicles. Spaces or ports appear between these vesicles, and it is through these ports that the acrosome content is presumed to escape in the early stages of the reaction. The vesicular remnants, the product of acrosomal and plasma membrane fusion, are lost before the sperm begins to penetrate the zona to any depth (Fig. 8) and, although they may be discharged before the sperm reaches the zona during its

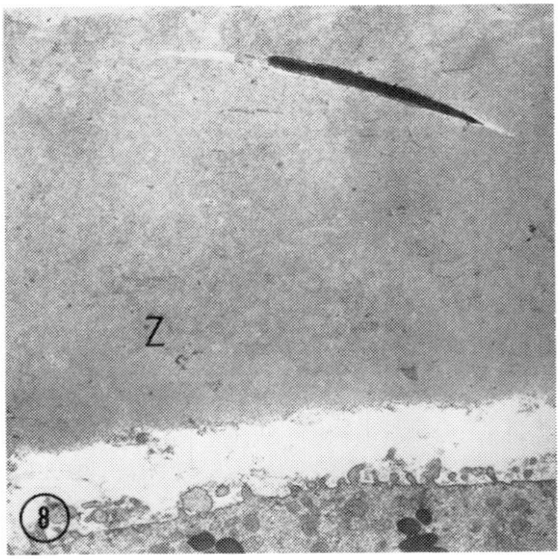

Fig. 8

Sperm passing through the zona of a rabbit egg. Note that the acrosomal vesicles have been discharged, leaving the bared inner membrane of the acrosome in apposition to the substance of the zona. The sperm is traversing the zona typically at an oblique angle.
Magnification X 3,600

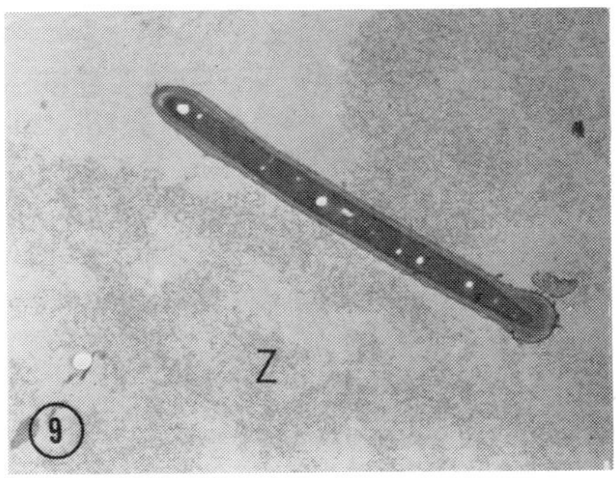

Fig. 9

Rabbit sperm aged *in vitro* for 10 hours before its insemination into an oviduct about 1/2 hour after ovulation. Recovered 1 1/4 hours later adhering to the zona surface. Notice that, in contrast to the interaction of capacitated sperm with the egg (Fig. 6 & 7), this sperm shows no sign of impending acrosomal breakdown and still retains the plasma memebrane tightly apposed to the underlying acrosome.
Magnification X 9,800

Fig. 10

Vesiculated remnants of the rabbit acrosome adhering to the zona surface after discharge from the sperm, which presumably has penetrated the zona.
Magnification X 16,800

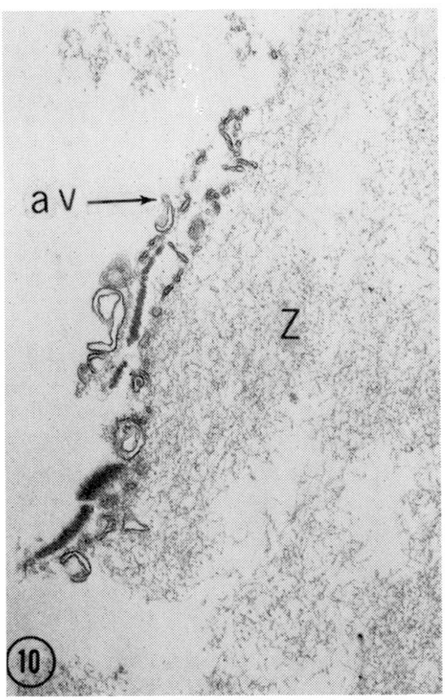

passage through the cumulus, sperm heads surrounded by the vesiculating acrosome are seen commonly on the zona surface in the rabbit and hamster (Figs. 5–7). The vesiculated acrosomal remnants sometimes adhere to the zona surface after entry of the sperm (Fig. 10).

The "false" reaction describes the changes in the acrosome which occur in testicular. epididymal, ejaculate, and capacitated sperm, as a result of non-specific degeneration. It is true to say that the acrosome cap in the guinea-pig and hamster sperm is much less stable than that of rabbit sperm, and tends to undergo the breakdown or degeneration seen in the "false" reaction with relative ease. This reaction has been distinguished in the electron microscope from the "true" reaction, in the hamster [20] and can easily be detected in the rabbit. In this type of change, both the acrosomal and the overlying plasma membranes rupture and are lost as separate entities in a random manner. In some degenerating rabbit uterine sperm, the plasma membrane is shed, though the acrosome remains more or less intact (Fig. 3).

An interesting example of the false acrosome reaction is seen in samples of uterine or ejaculated rabbit sperm treated with fresh homologous serum, which is known to be spermicidal [16]. Sperm fixed after suspension in heated serum remain quite intact, but 98 % of those suspended in fresh serum undergo rupture and loss of the outer acrosomal elements and the plasma membrane over the head region. In almost all of these cases, this rupture occurs as a false reaction in which the plasma membrane alone takes part in vesicle formation, or forms a membranous whorl (Fig. 11);

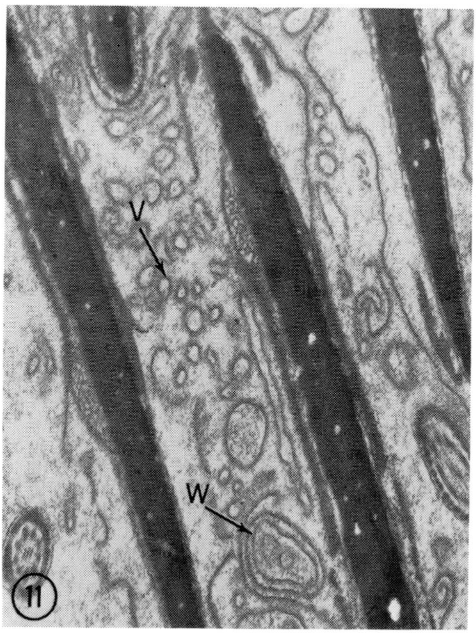

Fig. 11

Ejaculated rabbit sperm exposed to fresh homologous serum. The acrosome has ruptured in most sperm (a), and the plasma membrane has formed small vesicles (v), or membranous whorls (w). Magnification X 19,800

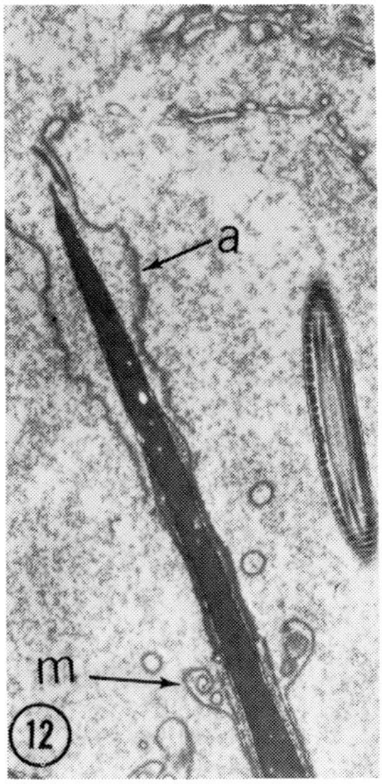

Fig. 12
Rabbit sperm treated with fresh serum; the plasma membrane overlying the acrosome has been discharged leaving its posterior remnant (m) over the post-acrosomal region of the head. The outer membrane of the acrosome (a), has been disrupted at the apex. Magnification × 14,000

in fresh serum the acrosome itself is ruptured in a reaction which often develops first at the apex of the acrosome (Fig. 12). Although the false reaction usually takes place in the random way described here, it is important to note that acrosome degeneration similar to the true reaction has been observed in testicular, epididymal, ejaculated and uterine sperm, and in ejaculate sperm exposed to fresh serum (Figs. 13, 14). However, this mode of acrosomal degeneration which mimics the true acrosome reaction, is seen only very rarely (in less than 1 % of degenerating sperm), in contrast to capacitated sperm in the vicinity of the egg in which rupture involving fusion between the plasma and acrosomal membranes appears to be the rule.

Significance of capacitation in development of the acrosome reaction

No gross structural change occurs in rabbit sperm as a concomitant of capacitation in the uterus or Fallopian tube. Since many of these sperm in the vicinity of the egg then undergo the true acrosome reaction described above, it has seemed possible that this reaction might be stimulated by substances emanating from the egg mass in a way similar to that by which the acrosome reaction is triggered in sea urchin sperm by egg fertilisins. As an extension of this notion, it seemed a reasonable speculation that the process of capacitation might involve changes in the sperm surface, which allow the sperm to respond to hypothetical "fertilisin"-like substances by undergoing the acrosome reaction. This possibility has been tested in the following

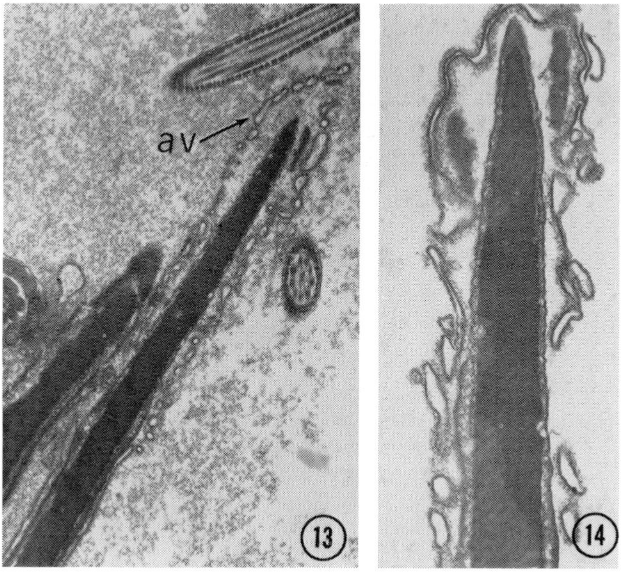

Fig. 13 & 14. Two examples in which vesiculation has occurred between the plasma membrane and outer acrosomal membrane (av), as seen during the true acrosome reaction at the surface of the ovum. In these rare cases, however, this vesiculation was produced as a consequence of exposure of ejaculate sperm to fresh serum (Fig. 13), and during non-specific degeneration of an ejaculated sperm (Fig. 14).
Magnification X 15,000; X 22,000

manner. Sperm samples collected from the uterus 13–14 hours after mating were concentrated by centrifugation at about 1000 x G for 3–5 minutes. Aliquots of 0.025 ml of these concentrated uterine sperm samples were then inseminated by way of a flank incision into the ampulla of the Fallopian tube in rabbits which had received an ovulating injection of HCG about 11 hours previously (i.e., 0–1 hours after ovulation). The ova were recovered by flushing the tubes 1 $^1/_4$ hours later and were fixed with glutaraldehyde and osmium, and prepared for electron microscopy, as described above. This procedure was repeated using epididymal or ejaculated sperm which before insemination had been incubated for 10–12 hours *in vitro* in Tyrode + 5 % heated rabbit serum. In a third group, epididymal or ejaculate sperm were instilled into the Fallopian tube about 3 hours before administration of an ovulating injection; the ova were then collected from these tubes about 15 $^1/_2$ hours after insemination and fixed for electron microscopy.

The results of these experiments indicate that rabbit sperm do not undergo the acrosome reaction in the vicinity of the egg without prior exposure for some hours to the environment of the female tract. In those cases in which eggs were exposed to sperm taken from the uterus or to those resident in the tube for 12 hours before ovulation, many sperm sticking to the egg had lost the outer acrosome membrane and its content, and were in the process of passing through the zona (after 1 $^1/_4$ hours allowed for egg/sperm interaction). Many others on the surface of the zona displayed the typical vesiculation characteristic of the true acrosome reaction (see Figs. 5–7), and in a few cases the adherent sperm remained intact without any sign of impending acrosomal breakdown. By contrast, ejaculate or epididymal sperm aged *in vitro* before tubal insemination were never observed to undergo the acrosome reaction described above, and in most cases were found to be quite intact, adherent to the zona surface by the plasma membrane which itself had remained closely apposed to the underlying acrosome (Fig. 9). In some cases the intact sperm head had burrowed into the loosely dispersed material of the zona surface, but in none had these sperm made significant inroads into the dense material of the zona proper. The most striking feature in this last group, however, was the difficulty experienced in finding sperm associated with the eggs, even after insemination of several million sperm into the tube only 1 $^1/_4$ hours previously. With very occasional exceptions it was possible to find these sperm adhering only over areas of the zona from which the corona cells had already been lost. These findings are summarized in Fig. 15.

Fig. 15

Uterine / Tubal { 13–14 hours capacitated sperm	- - - - +	
13 hour tubal sperm near ovum	+ + + + -	(True reaction)
13 hour uterine sperm near ovum	+ + + + -	(True reaction)
Ejac./Epididymal sperm near ovum	- - - - -	

+ Acrosome rupture; – Acrosome and plasma membrane unchanged.

Effects of follicular fluid on capacitated sperm and their modification by seminal plasma

Recent reports have shown that hamster sperm may undergo capacitation *in vitro* when incubated in follicular fluid [3]. Since there is also an implication from work in the hamster and the rabbit (see above) that substances associated with the egg mass will stimulate the acrosome reaction, it seemed of interest to find out whether follicular fluid will induce any morphological reaction in the rabbit sperm head. To this purpose, in each of two experiments, five female rabbits were injected intramuscularly with 150 units of PMS and after 60 hours were injected intravenously with 50 iu of HCG. Two hours before the estimated time of ovulation the females were killed, follicular fluid was withdrawn from all ovarian follicles of significant size and was diluted by 50 % with Tyrode's solution. Uterine sperm obtained about 13 hours after mating were concentrated by brief centrifugation and mixed 1:1 by volume with the follicular fluid preparation. This mixture was incubated at 37 °C for 30 minutes, before centrifugation and fixation in the usual way for electron microscopy. The same procedure was repeated using epididymal sperm, and also uterine sperm pellets previously mixed 2:1 with rabbit seminal plasma before exposure to the follicular fluid, at the same concentration.

In both experiments performed, follicular fluid had a marked effect on the acrosomes of uterine sperm. When fixed after 30 minutes incubation in follicular fluid, 70–95 % of uterine sperm had lost the acrosome, whereas sperm in the untreated control samples were intact in most cases (see Fig. 16). By contrast, epididymal

Fig. 16

Ejac. sperm and fresh serum	95 % acrosome rupture (False reaction)
Uterine sperm and fresh serum	95 % acrosome rupture (False reaction)
Ejac./Epididymal sperm and follicular fluid	– – – – –
13 hour uterine sperm and follicular fluid	+ + + + –
DF-treated uterine sperm and follicular fluid	+ – – – –
Untreated uterine sperm (control)	+ – – – –

\+ Acrosome rupture; – Acrosome and plasma membrane unchanged.

sperm were not noticeably affected by incubation with follicular fluids. Furthermore, the plasma membrane and acrosome of most sperm in uterine samples treated with seminal plasma remained intact after a 30 minute exposure to follicular fluid It is not clear, as yet, whether the acrosome reaction in samples incubated with follicular fluid matches the "true" acrosome reaction at the surface of the ovum, since after 30 minutes' incubation in follicular fluid most uterine sperm had completely lost he acrosomal remnants. It is interesting to note that whereas fresh normal serum causes the rupture of the acrosome as a "false" reaction in almost 100 % of ejaculated sperm, these sperm are largely unaffected by follicular fluid. This suggests that the active elements in fresh serum are not acting in the same manner as those present in rabbit follicular fluid. These findings are summarized in Fig. 16.

Discussion and conclusions

In the absence of critical observations in other species, this discussion has necessarily been centred around the sperm of rabbits and murine rodents. It seems clear that there is no obvious structural change in the organelles of the rabbit sperm head during the process of capacitation *per se.* However, the question of change in the ultrastructure of the sperm plasma membrane warrants further study with sophisticated techniques after subjection of sperm to different environments whose capacitation potential is known. No evidence has been forthcoming to support the conclusions of *Bernstein* [13] that a progressive erosion of the acrosomal content and generalized loosening of the plasma membrane over head and tail occur concomitantly with capacitation. This latter view is rendered unlikely also by the fact that the medium to which the sperm were exposed in his experiments (B amylase or cervical mucus) is not known to bring about any significant degree of capacitation. The hint of increased loosening of the plasma membrane in uterine sperm [7] has not been confirmed using newer methods of fixation, following which the majority of 13–14 hour rabbit uterine sperm display the cell membrane lying in close apposition to the underlying acrosome.

The suggestion has been made that the changes in capacitation allow the sperm to respond in an appropriate manner (i.e., by undergoing the acrosome reaction) to stimulating substances emanating from the ovum [11, 25]. The experimental results discussed here support this idea. Ejaculate or epididymal sperm seen at the egg surface show no signs of ultra-structural change; on the other hand, uterine sperm and sperm resident in the oviduct for 12 hours before ovulation often displayed the typical acrosomal vesiculation reaction discussed by *Barros* et al. [4], or had lost the acrosome and were beginning to penetrate the zona pellucida. One of the most striking features of these experiments was the difficulty experienced in locating rabbit sperm in association with the granulosa cells or zona surface some 1 $^1/_4$ hours after tubal insemination of several million non-capacitated ejaculated or epididymal sperm. This observation is in agreement with previous experiments [1,10] which indicate that as well as allowing their passage through the zona, capacitation also confers upon sperm the ability to penetrate the barrier posed by the surrounding cumulus and corona cells, thus facilitating contact with the zona surface early in the fertile life of the ovum.

It is apparent that the coordinated breakdown of the acrosomal complex in the vicinity of the egg prior to its penetration (the true reaction) usually differs from the acrosomal disintegration which occurs during degeneration (the false reaction), or following exposure to fresh serum or to treatments used to induce release of acrosomal enzymes [27]. Notwithstanding this, the true reaction is evidently not an absolute prerogative of the sperm stimulated by substances emanating from the egg, and the classical "true" reaction involving localized fusion and vesiculation between plasma and outer acrosomal membranes has been noted on rare occasions in testicular, epididymal, ejaculate and uterine spermatozoa.

Although the decapacitation effect of seminal plasma is well documented, its mode of action has remained uncertain. The experiments outlined here show that treatment of rabbit uterine sperm with seminal plasma tends to protect the plasma membrane and acrosome from the disruptive effects of fresh follicular fluids; a similar effect of seminal plasma on hamster spermatozoa in the presence of follicular fluid has been observed in the phase microscope, by *Gwatkin* (personal communication). This suggests that the decapacitation factor may possibly act to "stabilize" the plasma membrane, or to mask active receptor sites on the membrane surface. Decapacitation induced by seminal plasma is thought to be reversible since DF-treated sperm samples replaced in the female tract may then ultimately fertilize after a period of recapacitation [17]. While this may be so, it must be remembered that we have no knowledge at present of the proportion of sperm which are completely capacitated in a sample recovered after 12 hours in the uterus. The possibility cannot be ruled out, therefore, that during supposed "recapacitation" some sperm finally become completely capacitated for the first time, and it is these sperm which fertilize when the DF-treated sperm sample is re-inseminated into second (recapacitating?) female.

Acknowledgement: The electron micrograph in Fig. 5 was taken by Dr. *C. Barros* and Dr. *L. E. Franklin,* to whom the author is indebted for its use. Thanks are due to Miss *E. Pena* and Mrs. *M. J. Bent* for technical assistance. Supported by NIH grant HD-03623.

References

[1] *Austin, C. R.* (1960). Capacitation and release of hyaluronidase. J. Reprod. Fert. **1.** 310.

[2] *Austin, C. R. & Bishop, M. W. H.* (1958). Role of the rodent acrosome and perforatorium in fertilization. Proc. Roy. Soc. **B 148,** 241.

[3] *Barros, C. & Austin, C. R.* (1967). *In vitro* fertilization and the sperm acrosome reaction in the hamster. J. Exp. Zool. **166,** 317.

[4] *Barros, C., Bedford, J. M., Franklin, L. E. & Austin, C. R.* (1967). Membrane vesiculation as a feature of the mammalian acrosome reaction. J. Cell Biol. **34,** C1.

[5] *Barros, C. & Franklin, L. E.* (1968). Behaviour of the gamete membranes during sperm entry into the egg. J. Cell Biol. **37,** C13.

[6] *Bedford, J. M.* (1963). Morphological reaction of spermatozoa in the female reproductive tract of the rabbit. J. Reprod. Fert. **6,** 245.

[7] *Bedford, J. M.* (1964). Fine structure of the sperm head in ejaculate and uterine spermatozoa of the rabbit. J. Reprod. Fert. **7,** 221.

[8] *Bedford, J. M.* (1967a). Observations on the fine structure of spermatozoa of the Bush Baby (Galago senegalensis), the African Green Monkey (Cercopithecus aethiops) and Man. Am. J. Anat. **121,** 443.

[9] *Bedford, J. M.* (1967b). Experimental requirement for capacitation and observations on ultrastructural changes in rabbit spermatozoa during fertilization. J. Reprod. Fert. Suppl. **2,** 35.

[10] *Bedford, J. M.* (1967c). Importance of capacitation for establishing contact between sperm and eggs in the rabbit. J. Reprod. Fert. **13,** 365.

[11] *Bedford, J. M.* (1968). Ultrastructural changes in the sperm head during fertilization in the rabbit. Am. J. Anat. **123,** 329.

[12] *Bedford, J. M.* (1969). Limitations of the uterus in development of the fertilizing ability (capacitation) of spermatozoa. J. Reprod. Fert. Supplement Nr. 8.

[13] *Bernstein, M. H.* (1966). Modifications of sperm structure in capacitation. J. Cell Biol. **31,** 12 A (Abstract).

[14] *Blom, E. & Birch-Andersen, A.* (1965). The ultrastructure of the bull sperm. II. The sperm head. Nord. Vet. **17,** 193.

[15] *Brackett, B.* (1968). Respiration of spermatozoa after *in utero* incubation in estrous and pseudopregnant rabbits. VI Congr. Reprod. and Artif. Insem. Paris, Abstract, p. 8.

[16] *Chang, M. C.* (1947). Effects of serum on spermatozoa. J. Gen. Physiol. **30,** 321.

[17] *Chang, M. C.* (1957). A detrimental effect of seminal plasma on the fertilizing capacity of sperm. Nature, **179,** 258.

[18] *Fawcett, D. W.* (1965). The anatomy of the mamalian spermatozoon with particular reference to the guinea pig. Z. Zellforsch. **67,** 279.

[19] *Fawcett, D. W. & Ito, S.* (1965). The fine structure of bat spermatozoa. Am. J. Anat. **116,** 567.

[20] *Franklin, L. E., Barros, C. & Fussell, E. N.* (1968). Fine structure of the acrosomal region of golden hamster spermatozoa and the acrosome reaction. J. Cell Biol. **39,** 46 A (Abstract).

[21] *Hamner, C. E. & Williams, W. L.* (1964). Effect of the female reproductive tract on sperm metabolism in the rabbit and fowl. J. Reprod. Fert. **5,** 143.

[22] *Mounib, S. & Chang, M. C.* (1964). Effect of *in utero* incubation on the metabolism of rabbit spermatozoa. Nature, **201,** 943.

[23] *Nicander, L. & Bane, A.* (1962). Fine structure of boar spermatozoa. Z. Zellforsch. **57,** 390.

[24] *Nicander, L. & Bane, A.* (1966). Fine structure of the sperm head in mammals with particular reference to the acrosome and subacrosomal substance. Z. Zellforsch. **72,** 496.

[25] *Piko, L. & Tyler, A.* (1964). Ultrastructure of the acrosome and the early events of sperm penetration in the rat. Am. Zool. **4,** 287 (Abstract).

[26] *Saacke, R. G. & Almquist, J. O.* (1964). Ultrastructure of bovine spermatozoa. I. The head of normal ejaculated sperm. Am. J. Anat. **115,** 143.

[27] *Srivastava, P. N.* (1965). Biochemistry of the sperm acrosome with special reference to sialic acid. Ph. D. Thesis, University of Cambridge.

[28] *White, I. G. & Murdoch, R. N.* (1968). The effect of the uterine environment on the metabolism of rabbit spermatozoa. (Title only) J. Reprod. Fert. **17,** 409.

[29] *Yanagimachi, R.* (1966). Time and process of sperm penetration into the hamster ova, *in vivo* and *in vitro.* J. Reprod. Fert. **11,** 359.

Discussion

R. G. Edwards: I would like to know whether you think the analogy between the acrosome and the lysosome has any validity to these studies. The analogy was made by *Brachet* and *Allison* a few years ago, and it seems to me there are similarities; do you accept this or is this not so?

J. M. Bedford: I really do not think I am in a position to accept it or not accept it, because I do not know enough about the lysosome reaction. I do know that people in Cambridge have recently been doing studies on the content of the lysosome and the acrosome, comparing them, and I am really not very familiar with the results.

R. G. Edwards: When you say you find that serum gives you a false reaction, we know that there is an antibody in the components in the follicular fluid. Why should the reaction between spermatozoa and serum and that around the egg be different; you see we have serum components there. You have drawn the differences, but would these be meaningful similarities?

J. M. Bedford: Yes, you know, I did the work with fresh serum, after finding the follicular fluid effect, because I was aware of the serum components in follicular fluid. Since follicular fluid seems to have most of the serum components, I thought this might be a similar a reaction but there *is* a difference here, you see, because the fresh serum affects almost 100 % of normal ejaculate sperm, whereas in the rabbit, the ejaculated sperm do not seem to react to follicular fluid. There is also a difference in the response of the uterine sperm which also do not generally undergo the true acrosome reaction after exposure to fresh serum.

Methods for Measurement of Sperm Capacitation in Different Species

R. J. Ericsson

Fertility Research, The Upjohn Company Kalamazoo, Michigan, USA

Summary: A fluorometric method for measuring capacitation – removal of tetracycline HCl (T-HCl) from sperm – serves as an adjunct to fertilization as a method to detect physiologic changes in mammalian sperm. T-HCl is a supravital stain and it does not alter sperm motility or potential fertility. Fluorescence of live and dead sperm is not of equal intensity. The latter have brighter fluorescence due to greater uptake of T-HCl. Stain specificity makes the equatorial segment of live sperm discernible; and head membrane changes associated with aging are easily observed. A correlation exists between removal of T-HCl from rabbit sperm and their ability to fertilize. Work with ^{3}H-T-HCl shows the absence of fluorescence is due to loss of T-HCl and not quenching. Loss of T-HCl from sperm *in vivo* before functional capacitation is given as evidence against this technique. However, it is agreed that T–HCl loss occurs where sperm can be capacitated and retained under those conditions not normal for capacitation. Loss of T-HCl from bull sperm in cows of different hormonal conditions fits the pattern of the rabbit. The majority of natural and synchronized estrous cows remove T-HCl whereas essentially none of the luteal or progestogen (MGA®) [1] cows do. Two-thirds of the cows which conceive retain fluorescence on sperm after 5 hours in utero. This may mean that capacitation too soon (loss of fluorescence) before ovulation will result in lower fertility. Incubation of some enzymes with sperm *in vitro* fails to remove T-HCl. Success in removal occurs with admixture of sperm and heterologous leucocytes. Mule eosinophils take up T-HCl from rabbit, bull, and monkey sperm after several hours at 35 °C. Eosinophils also have the ability to initiate capacitation of rabbit sperm *in vitro*. These diverse data support the T-HCl technique as a practical tool to study capacitation.

Introduction

Almost two decades have passed since it was shown that some mammalian sperm require a physiologic change before penetrating ova. During this era numerous studies have helped to explain the needs of sperm for capacitation. As courageous and enlightening as these experiments are, we still rely almost exclusively on fertilized ova as the means to denote sperm changes. This biolocigal necessity restricts research to laboratory and domestic species with known ovulation times. It is imperative that capacitation

---

Manuscript received: 21 April 1969

[1]) MGA ® is a registered trademark of The Upjohn Company for melengestrol acetate.

measurement be divorced from the recently ovulated ovum so research in this field can expand to meet its expectations. Technology and laboratory sophistication are such that other reliable methods to measure sperm function are available. The need is for a simple, practical yet dependable method; one that can bring capacitation to the humanist and fundamental biologist. Sperm maturation, female reproductive cyclicity, and fertilization *in vivo* and *in vitro* encompass capacitation. New knowledge in capacitation should help to speed research in these peripheral areas.

The goal is to separate functional sperm from those which have not undergone a change – without relying on the ovum. To achieve this end we might consider the sperm head plasma membrane and substances which coat it. Evidence associactes capacitation with enzymatic alteration or removal of substances from the plasma membrane [6, 9, 20]. Speculation by the author and others and supported in part with evidence (*see* [22]) is that capacitation starts with removal of coating substances and ends with ovum penetration. To achieve the latter sperm must make available their acrosomal enzymes. Ultrastructural changes in sperm which precede penetration first involve vesiculation of the plasma and outer acrosome membranes [2] followed by loss of these vesicular elements and content of the acrosome cap [5]. The plasma membrane must be modified early in capacitation to allow the above changes to occur. The task is to discover how and when the plasma membrane is modified. A technique which incorporates membrane differences (this includes coating substances) eliminates many of the prior restrictions to capacitation research.

The antibiotic tetracycline HCl (T-HCl) is a fluorophor which acts as a supravital stain when mixed with mammalian sperm [16]. T-HCl uptake is greater in dead sperm, which is thought due to a postmortem change in membrane permeability. Stain specificity makes the equatorial segment of live sperm discernible and provides a useful procedure to study head membrane changes associated with aging [17]. It was learned later that sperm lose the bound T-HCl primarily under conditions known to capacitate them. The remainder of the paper is devoted to the status, flexibility, and reliability of the T-HCl technique for measuring sperm capacitation.

The fluorometric method

Procedures for labeling sperm with T-HCl have been published [13, 16]. In brief, T-HCl is mixed with semen in micrograms per million sperm quantities and minutes later the sperm are separated by centrifugation. This antibiotic does not alter sperm motility or potential fertility although it readily binds to but is not easily lost from sperm. These sperm, when resuspended in a physiological solution, are ready for all types of experimentation. The method relies on removal of T-HCl from sperm to denote a physiologic change. A fluorescence microscope of high quality is necessary to accurately use the technique.

Sperm from several species were first tested in the uterus of rabbits and seen to keep fluorescence only in does with functional corpora lutea (Fig. 1). Capacitation was checked by exposing these sperm to ova of recipient does. Only the nonfluorescent sperm had capacitated [13]. Fertility of labeled sperm left in the original does is good

Fig. 1. Removal of T-HCL from sperm in the rabbit uterus.

Species of sperm	No. of does	Hours in utero	Uterine condition	Sperm fluorescence*
Rabbit	2	2	Estrus	+
	2	3	Estrus	–
	38	4 to 8	Estrus	–
	2	6	Pseudopregnant (Day 3)	–
	37	6 to 12	Pseudopregnant (Days 7-9)	+
Bull	1	6	Estrus	–
	2	6	Pseudopregnant (day 8)	+
Human	5	6	Estrus	–
	3	6	Pseudopregnant (Day 8)	+

* + = fluorescence
– = no fluorescence

– 100 % of 18 ova recovered after 48 hours cleaved and term does kindled normal live young [14]. The use of ^{3}H-T-HCl showed that loss of fluorescence is indeed removal of T-HCl from sperm [14]. Rabbit sperm were labeled with ^{3}H-T-HCl and exposed to various media *in vitro*. The 0.5 ml of 150 x 10^6 sperm and substance were left at room temperature for one hour and then washed four times (*see* [18] for details). All substances, in the pH range of sperm survival, failed to remove T-HCl when compared to the Tyrode's solution control (Fig. 2). Sperm mixed with ^{3}H-T-HCl and T-HCl in different order have the same per cent of radioactivity – which points to a less than equilibrium concentrations of T-HCl molecules (Fig. 2).

The original work with T-HCl-labeled bull sperm in cows gave results comparable to the rabbit [14]. An expanded study was designed to learn if capacitation was associated with the lower fertility at synchronized estrus. MGA (a potent progestogen)

Fig. 2. Removal of ^{3} H-T-HCl from rabbit sperm *in vitro.*

0.1 μg/10^6 Sperm		150 x 10^6 Sperm Samples		
Condition	% DPM Washed sperm	1.90 μg ^{3}H-T-HCL	T-HCL (μg)	% DPM Washed sperm
Tyrode's Solution	38.2	lst*	0	6.2
Peroxidase (2 mg/ml)	39.3	lst	0.6	6.3
Rabbit Plasma	28.1	lst	60.0	6.8
Citrate - PO_4 Buffer (pH 5.68)	16.0	2nd	0	5.9
5-Hydroxytryptamine (2 mg/ml)	39.8	2nd	0.6	5.9
Histamine (2 mg/ml)	23.1	2nd	60.0	7.6
Nuclease (2 mg/ml)	38.0	same	0	6.2
Cholinesterase (2 mg/ml)	38.7	same	0.6	7.3
Uterine Homogenate (125 mg/ml)	31.6	same	60.0	6.3
Bradykinin (1 mg/ml)	37.7			
Citrate - PO_4 Buffer (pH 4.01)	17.4			

*Refers to when ^{3}H-T-HCl was mixed with sperm in reference to T-HCl.

Fig. 3. Removal of T-HCL from bull sperm placed in cows.

Condition	No. of cows	Time in utero (hr.)	Number of cows removing T-HCl	% of cows removing T-HCl
Estrus	45	5.0	23	51
Synchronized estrus (MGA)	21	5.0	13	62
Luteal	38	5.0	1	3
MGA for 18 days	21	5.0	0	0

served as the means to synchronize estrus and as a control to the normal luteal cows [21]. A marked difference in ability to remove T-HCl from sperm exists between cows in estrus and those in a natural or artificial luteal phase (Fig. 3). Sperm from an additional nine estrous cows lost fluorescence when stored at 5 °C for 48-72 hours; none of the sperm samples taken from cows in the luteal groups lost fluorescence. Twenty-four of the estrous cows conceived and 16 (67 %) of them still had fluorescent sperm after five hours in utero.

Many attempts *in vitro* to remove T- HCl from sperm failed [14, 16]. The best non-physiologic method found was to lower pH and wash the sperm [Fig. 2]. Initial success came when sperm were placed with mule leucocytes. Extensive experimentation with types of leucocytes, various species of leucocytes and sperm, and various ratios and species in combination showed that homo- and heterologous leucocytes can remove T-HCl from mammalian sperm (Fig. 4). T-HCl lost from sperm was transferred to the leucocytes – as measured by fluorescence microscopy. Phagocytosis of sperm occurred with greater frequency when incubated with total leucocytes than with eosinophils alone. A relationship between loss of T-HCl *in vitro* and capacitation was established for the rabbit [15]. Sperm incubated with eosinophils (sperm in time lost fluorescence) and then placed in recently ovulated recipient does were able to fertilize ova. Times of insemination after sterile mating in which these sperm fertilized ova showed that partial capacitation occurred (Fig. 5).

Fig. 4. Removal *in vitro* of T-HCL from sperm with leucocytes.

Leucocytes	Sperm	Ratio L : S	Time removed (Hr)	Sperm motility
Rat (E)	Rabbit	4 : 1	2	60 % 3°
Rabbit	Rabbit	4 : 1	5	5 % 3°
Human	Human	5 : 1	2	30 % 2°
Human	Rabbit	10 : 1	2	20 % 2°
Rabbit	Bull	4 : 1	6	None
Mule (E)	Bull	2 : 1	2	30 % 3°
Mule (E)	Monkey	2 : 1	3	50 % 3°
Mule (E)	Rabbit	2 : 1	2	70 % 3°
		1 : 1	2	70 % 3°

All controls fluoresced
(E) Eosinophils
Degree of motility – 0 to 4

Fig. 5. *In vitro* capacitation of rabbit sperm with total leucocytes and eosinophils.

Treatment	Hrs. after mating sperm deposited	No. Ova recovered	No. Ova cleaved	% Ova cleaved
Rabbit Leucocytes	11.5 to 12	88	0	0.0
Rat Eosinophils	12 to 13	43	1	2.3
Rat Eosinophils	5 to 5.5	13	5	38.5
Mule Eosinophils	9 to 10	64	12	18.7
Mule Eosinophils	10 to 11	68	15	22.0
Mule Eosinophils	11 to 12	73	0	0.0
Nontreated Controls	5 to 13*	302	0	0.0

* 91.8 % of the females received sperm 10–12 hours after mating.

Discussion

The fluorometric method of measuring capacitation has been tried in a small number of laboratories. Results obtained agree with the original work. There is, however, disagreement on the value of this technique. *Vaidya, Bedford, Glass* and *Morris* [26] point out that T-HCl removal in the rabbit occurs much sooner than functional capacitation. They also state that in situations (immature, ovariectomized, pseudopregnant or progesterone-treated females) where partial capacitation may take place the fluorescence remains. Even though it is known that in these situations capacitation takes longer or is never fully achieved [4, 10, 11, 19]. Other work showed that primate sperm lost fluorescence when incubated for 12 hours in uteri of hamsters, rats, mice and rabbits [12]. An *Bedford* and *Shalkovsky* [7] were able to bring about partial capacitation of rabbit sperm in the estrous rat uterus. The rabbit oviduct is capable of capacitation under conditions that the uterus is not [10]. In no condition was T-HCl retained on sperm incubated in the oviduct [13, 26].

The cattle experiments re-emphasized two points and established a third. Cows act like does with regard to reproductive phase in loss of sperm fluorescence. And treating sperm with T-HCl does not prevent fertility. The new point was loss of T-HCl *in vitro* from sperm taken, along with uterine fluids, out of the uterus of some estrous cows. This would not be significant except none of the sperm taken from luteal cows lost fluorescence *in vitro*. Cows which took T-HCl off sperm within five hours in utero were expected to have a higher conception rate than those in which fluorescence remained. The opposite happened. Lower fertility may result if sperm capacitate (loss of fluorescence) too soon before ovulation thus lowering the numbers of sperm and their fertile status. After capacitation sperm are more acceptable to uterine leucocytes [3]. Perhaps some cows, particularly those with synchronized estrus, change sperm to their functional ability much earlier than do other cows. This would require a longer "fertile sperm to ovulation time" and give leucocytes added opportunities for phagocytosis. *Brown* and *Nellor* [8] state that estrogen secretion is directly correlated with the appearance of eosinophilic granulated cells in the uterus. And *Zimbelman* [27] suggests cows fed progestogens are in a "hyperestrogenic" condition.

Ericsson [15] suggests a role for eosinophils in capacitation and gives a sequence of events necessary to bring about this phenomenon. Suffice it to say, eosinophils have some important characteristics for their role. They are well supplied with granules which contain a number of hydrolytic enzymes [1] and they respond to antigen and antigen-antibody complex reactions [23, 25]. Capacitation *in vitro* with eosinophils and the possible involvement of capacitation in lower fertility of synchronized cows was learned through the fluorometric method of measuring capacitation. T-HCl can help in placing sperm into functional and nonfunctional groups. Removal of T-HCl from sperm would seem to signify that sperm have undergone a physiologic change but not necessarily complete capacitation.

References

[1] *Archer, G. T. & Hirsch, J. G.* (1963) Isolation of granules from eosinophil leucocytes and study of their enzyme content. J. Exp. Med. **118,** 227.

[2] *Barros, C., Bedford, J. M., Franklin, L. E. & Austin, C. R.* (1967) Membrane vesiculation as a feature of the mammalian acrosome reaction. J. Cell Biol. **34,**Cl.

[3] *Bedford, J. M.* (1965) Effect of environment on phagocytosis of rabbit spermatozoa. J. Reprod. Fert. **9,** 249.

[4] *Bedford, J. M.* (1967) Experimental requirement for capacitation and observations on ultra-structural changes in rabbit spermatozoa during fertilization. J. Reprod. Fert. Suppl. **2,** 35.

[5] *Bedford, J. M.* (1968) Ultrastructural changes in the sperm head during fertilization in the rabbit. Am. J. Anat. **123,**329.

[6] *Bedford, J. M. & Chang, M. C.* (1962) Removal of decapacitation factor from seminal plasma by high-speed centrifugation. Am. J. Physiol. **202, 179.**

[7] *Bedford, J. M. & Shalkovsky, R.* (1967) Species-specificity of sperm capacitation in the rabbit. J. Reprod. Fert. **13,** 361.

[8] *Brown, J. E. & Nellor, J. E.* (1968) A relationship between ovarian and genital tract eosinophilic granulated cells. Anat. Rec. **162,** 15.

[9] *Chang, M. C.* (1957) A detrimental effect of seminal plasma on the fertilizing capacity of sperm. Nature **179,** 258.

[10] *Chang, M.C.* (1958) Capacitation of rabbit spermatozoa in the uterus with special reference to the reproductive phases of the female. Endocrinology **63,** 619.

[11] *Chang, M. C.* (1969) Fertilization, transportation and degeneration of eggs in pseudopregnant or progesterone-treated rabbits. Endrocrinology **84,** 356.

[12] *Dukelow, W. R.* (1968) Capacitation regulation as a means of population control. Primate News **6,** 3.

[13] *Ericsson, R. J.* (1967 a) A fluorometric method for measurement of sperm capacitation. Proc. Soc. Exp. Biol. Med. **125,** 1115.

[14] *Ericsson, R. J.* (1967b) Technology, physiology, and morphology of spermatozoa capacitation. J. Reprod. Fert. Suppl. **2,** 65.

[15] *Ericsson, R. J.* (1969) Capacitation in vitro of rabbit sperm with mule eosinophils. Nature **221,** 568.

[16] *Ericsson, R. J. & Baker, V. F.* (1967) Binding of tetracycline to mammalian spermatozoa. Nature **214,** 403.

[17] *Ericsson, R. J. & Buthala, D. A.* (1970) Head membranes of live and dead sperm from the same and different species. Biol. Reprod. (in prep.).

[18] *Ericsson, R. J., Cornette, J. C. & Buthala, D. A.* (1967) Binding of sex steriods to rabbit sperm. Acta endocrin. **56,** 424.

[19] *Hamner, C. E. & Sojka, N. J.* (1968) Requirements for capacitation of rabbit sperm. Nature **220,** 1042.

[20] *Kirton, K. T. &Hafs, H. D.* (1965) Sperm capacitation by uterine fluid or beta-amylase in vitro. Science **150,** 618.

[21] *Lauderdale, J. W. & Ericsson, R. J.* (1970) Hormonal conditions affecting ability of uterus of cattle to remove tetracycline hydrochloride from bull sperm. Biol. Reprod. (in press).

[22] *Piko, L.* (1967) Immunological phenomena in the reproductive process. Int. J. Fert. **12,** 377.

[23] *Roberts, A. N.* (1966) Rapid uptake of tritiated antigen by mouse eosinophils. Nature **210,** 266.

[24] *Ross, R. & Klebanoff, S. J.* (1966) The eosinophilic leukocyte: fine structure studies of changes in the uterus during the estrous cycle. J. Exp. Med. **124,** 653.

[25] *Sabesin, S. M.* (1962) A funciton of the eosinophil: phagocytosis of antigen-antibody complexes. Proc. Soc. Exp. Biol. Med. **112,** 667.

[26] *Vaidya, R. A., Bedford, J. M., Glass, R. H. & Morriss, J. McL.* (1969) Evaluation of the removal of tetracycline fluorescence from spermatozoa as a test for capacitation in the rabbit. J. Reprod. Fert. (in press).

[27] *Zimbelman, R. G.* (1966) Effect of progestagens on ovarian and pituitary activity in the bovine. J. Reprod. Fert. Suppl. **1,** 9.

Discussion

R. G. Edwards: May I suggest, *Dr. Ericsson,* that here again we are deadling with an antibody, which reacts with spermatozoa. Now if this is true, all that you do with leukocytes, is to add an antibody, which is an opsonin and which reacts with the acrosome. So I would suggest that what you really do here is create an immunological destruction of the acrosome – of the non-specific type II that *Bedford* described. Do you think this is possible?

R. J. Ericsson: I think it is possible; however, when we tried uterine leukocytes, either intact or sonocated, we never were able to remove tetracycline HCl. And eosinophils, according to some scientists who spend their lives working with them, function either to remove antigens by themselves or antigen-antibody complexes. Perhaps the eosinophils are releasing enzymes which remove a substance or the sperm-coating antigen which stabilizes the plasma membrane, instead of what you suggested might occur. It may involve the acrosome reaction. Also, we got our best response with eosinophils as compared with total leukocytes, because of the problem of phagocytosis which occurs quite readily. Another point on follicular fluid – there is a recent paper by *Brown & Nellor* (Anat. Rec. **162** 15, 1968), where they found that estrogens are directly correlated with the appearance of eosinophilic granules in the follicles and throughout the female tract.

Decapacitation Factor and Capacitation

W. L. Williams, R. T. Robertson and W. R. Dukelow

Department of Biochemistry, University of Georgia, Athens, Georgia, USA

Summary: Decapacitation factor (DF) is a substance occurring in seminal plasma that reversibly blocks the union of sperm and ovum. Using inhibition of fertilization as an assay, an isolation scheme for obtaining reasonably pure DF from rabbit seminal plasma has been developed. The procedure utilizes ultracentrifugation, digestion with Pronase, n-butanol extraction, and column chromatography on Sephadex G–10 eluted with water. The single active elution area from the G–10 column with an absorption maximum of 233 mμ was applied to a Silica Gel H thin layer plate and developed in a 7:3, ethanol:water solvent system. After developing the plate the DF activity is concentrated in a short U.V. absorbing, ninhydrin positive area with an absorption maximum of 233 mμ and an R_f of 0.35. The minimum effective dose level of this area indicates a purification of about 200-fold over whole rabbit seminal plasma.

DF activity is stable to iodination but unstable at pH 10. Strong hydrogenation using palladium black as the catalyst destroys DF activity and U.V. absorption. The ionic character of DF is supported by its water solubility and the zwitter-ion absorption of purified preparations in the infrared. The fact that acid treatment of a Sephadex G–10 column by a procedure which exposes carboxyl groups excluded the DF activity completely tends to substantiate this hypothesis. DF was not extractable into chloroform:methanol reagent and was not adsorbed on deactivated charcoal.

DF is the only substance known to specifically block union of egg and sperm and may exert this effect by inhibition of the sperm enzyme that digests a passage through the corona cell layer of the egg. Part of the capacitation process appears to involve removal or destuction of DF. DF may function to prevent sperm from penetrating cells or tissues in the male.

Introduction

It has been over 18 years since the capacitation phenomena was discovered independently by Drs. *Austin* and *Chang* [1, 2]. Previous papers at this symposium have documented the progress made during this period of time relating to the physiological aspects of capacitation. I would like to devote this section of the program to the substance found in the seminal plasma of mammals which has been termed "Decapacitation Factor" (DF).

In 1957 *Chang* [3] found that capacitated sperm could be "decapacitated" (i.e. the capacitation phenomenon could be functionally reversed) by incubation of the sperm in the seminal plasma of humans, rabbits and bulls. Since that discovery, the

Manuscript received: 5 May 1969

possibility has been considered that part of the process of capacitation of sperm may involve destruction or removal from sperm of DF. That this, at present, is more than just a possibility is supported by the following observations:

(1) Capacitated sperm are more permeable to sugars and dyes and metabolize faster than ejaculated sperm as if something had been removed from surface membranes.

(2) Capacitated sperm take up DF from solution during the decapacitation process [4].

(3) DF inhibits and apparently reacts with a sperm acrosomal enzyme involved in removal of the corona radiata of the ovum [5, 6].

(4) Recapacitation of decapacitated sperm requires as much time in the oviduct as the original capacitation [7]. Because of these observations and because DF is the only naturally-occurring and non-toxic substance that prevents union of the gametes, we have attempted to identify and purify this substance. The initial work on purification of DF by ultracentrifugation and properties of DF was reported by *Bedford* and *Chang* in 1962 [3]. Since then a major effort of our research group at the University of Georgia has been directed towards the origin, purification, isolation and identification of decapacitation factor.

Biological assay of decapacitation factor

We have previously described the biological assay of this substance [8, 9]. This assay consists of recovering capacitated sperm from the uterus of a doe (capacitator) 11 hr. after mating. These sperm are divided into treatment and control tubes and the treatment sperm are incubated with the sample to be tested. The amount of protein or dry matter which is incubated with each 10^5 sperm is predetermined and the minimum level that will decapacitate the sperm is determined on crucial samples. For initial screening of seminal plasma, a level of 5 mg of seminal protein per 10^5 sperm is normally used. The incubations are for 20 minutes at 37 °C. The sperm are then washed and 2.5 x 10^4 sperm are surgically inseminated into the infundibular end of one oviduct 12 hr after an ovulatory dose of 2.5 mg of luteinizing hormone was administered to the ovum donor animals intravenously. The contralateral oviduct always receives an equal number of capacitated sperm which have not been incubated with the DF sample in question. In cases of suspected sperm death, and to insure that the decapacitated sperm are capable of a second capacitation *in utero* (recapacitation), in some experiments sperm were injected into the uterine horn of a second doe 6 hr after LH injection (i.e. approximately 4 hr pre-ovulation). In the contralateral uterine horn of the recapacitator a similar number of twice-washed ejaculated sperm are injected to be sure of the doe's ability to capacitate sperm. All does are killed 26 hr or less after ovulation when fertile ova would be expected to be in the four-cell stage of cleavage [10]. Fertilized ova should be found in the control oviducts of the ovum donors and in both sides of the recapacitator. If DF is present in the test sample, unfertilized ova will be recovered from the test oviducts.

Origin of decapacitation factor in the male

Austin and *Chang* [1, 2] demonstrated that epididymal sperm must be capacitated. *Bedford* and *Chang* [3] showed that the seminal plasma of vasectomized males contained DF. This evidence suggested that sperm do not contact the DF until the time of ejaculation, however work from our laboratory shown in Fig. 1 demonstrated that DF is clearly present in epididymal fluid [11]. We also have evidence to

Fig. 1a. Decapacitation with epididymal fluid [11].

Additions to capacitated spermatozoa	Ova recovered	Fertile
None	50	40 (80 %)
Epididymal fluid	38	0

Fig. 1b. Recapacitation of decapacitated sperm.

Sperm treatment	Ova recovered	Fertile
Freshly ejaculated, washed	13	8 (61 %)
Decapacitation with epididymal fluid	16	15 (94 %)

indicate its presence in testicular secretions. This indicates that DF can be found distributed throughout the male reproductive tract and supports the hypothesis that capacitation is the inactivation or removal of DF from the sperm.

Presence of DF in various species

By the assay previously described we have demonstrated the presence of DF in the seminal plasma of the bull, rabbit, stallion and boar [9]. DF is absent in rooster seminal plasma, however its presence in dog seminal plasma is questionable (Fig. 2). In primates, DF can be found in rhesus monkey and human seminal plasma (Fig. 3), although higher assay levels are required for demonstration in the human than in any other species tested [12]. In a separate study [13] the sediment obtained from

Fig. 2. DF activity in the seminal plasma of various species [9].

Species	Demonstrated need for capacitation in literature	No. rabbits	% and No. of ova fertilized	
			Capac. sperm. control	Capac. sperm. + DF
Bull*	+	57	54.8 (172)	0 (134)
Rabbit	+	12	87.4 (48)	0 (38)
Monkey†	+	4	70.0 (10)	0 (12)
Stallion		7	85.2 (27)	0 (26)
Boar	+	7	35.8 (28)	0 (19)
Human		16	53.3 (62)	24.1 (54)
Dog		7	63.2 (19)	32.3 (31)
Rooster	–	6	66.7 (24)	42.2 (19)

* Assayed as a DF pellet, 1 mg protein/10^5 spermatozoa.

† Assayed at 2 mg protein/10^5 spermatozoa.

Fig. 3. Decapacitation factor activity of primate seminal plasma [12].

Sample	Amount per 10^5 sperm	No. of oviducts	Total ova recovered	% fertile
Control	None	8	38	65.8
Rhesus seminal plasma	5 mg protein	3	18	0
Human seminal plasma	7 mg protein	2	9	0
Human ultracentrifuged sediment	1 mg solids	1	6	0

human seminal plasma by ultracentrifugation was examined for DF activity at several different levels. The amount necessary to decapacitate 100,000 sperm appeared to be above 2 mg of sediment solids (Fig. 4). In 1967 we postulated that the presence of DF in the seminal plasma of a species was an indirect indication of the need for capacitation in that species based on information then available on capacitation requirements in various animals [9].

Enzymatic destruction of DF

In our first experiments designed to identify the biochemical nature of DF, we attempted to destroy its activity by incubation of either seminal plasma or an ultracentrifuged sediment with various enzymes [9]. The results are shown in Fig. 5.

Fig. 4. Decapacitation of rabbit sperm with a pellet obtained by ultracentrifugation of human seminal plasma [13].

Treatment of capacitated sperm	No. of oviducts	Total ova recovered	Ova fertilized (%)
None	2	8	88
Decapacitation attempted with 2 mg sediment/10^5 sperm	2	11	27
None	4	21	67
Decapacitation with 5 mg sediment/10^5 sperm	4	26	0
None	4	9	78
Decapacitation with 7 mg sediment/10^5 sperm	4	22	0

Fig. 5. Enzymatic destruction of DF activity [9].

Enzyme	% and No. of ova recovered		
	Capac. sperm. controi	Capac. sperm. + DF + enzyme	Capac. sperm. + sem. plasma + enzyme
α-Amylase	59.5 (37)	5.9 (17)	–
β-Amylase	50.8 (116)	26.2 (42)*	53.4 (15)†
Glucose oxidase	63.7 (11)	0 (13)	–
Hyaluronidase	37.6 (8)	0 (4)	–
Lysozyme	62.2 (37)	0 (10)	0 (15)
V. *cholera* filtrate	57.1 (49)	–	0 (9)
Pronase	46.2 (13)	0 (6)	0 (4)

All enzymes tested at 100 μg/5 mg protein except V. C. filtrate tested at 34 μg due to toxicity.

* 6 to 12 hr incubation at 37 °C.

† 10 min incubation at 37 °C.

Hyaluronidase, glucose oxidase, Pronase and lysozyme failed to destroy DF activity. Resistance to Pronase digestion indicates that DF is not a protein or a polypeptide of significant size. Since a few peptide bonds such as those involving glycine are resistant to Pronase, digestion with Pronase does not eliminate the possibility that such bonds are present in the DF molecule. β-amylase destroyed DF activity in rabbit seminal plasma after 10 minutes incubation. Six to 13 hr incubation were required to destroy DF activity using the ultracentrifuged sediment. α-amylase partially destroyed DF activity but not as effectively as β-amylase. This suggested that DF may contain an α 1-4 glucoside bond. Our report [14] of this information in 1965, as well as reports on increased amylase activity in the female reproductive tract at the time of capacitation [15] that same year, resulted in several experiments in other laboratories to utilize β-amylase in studies of *in vitro* capacitation [16] and for improving the fertility of semen used in bovine artifical insemination. While such work is of importance, there is yet to be developed a successful, consistent method of *in vitro* capacitation using β-amylase. In addition, studies in our laboratory have failed to demonstrate maltose or free glucose in samples of DF treated with β-amylase. When we obtained β-amylase of the highest purity from Dr. *A. K. Balls,* it failed to destroy DF activity. These factors, plus the known heterogeneity of many commercial β-amylase preparations, have led us to conclude that the initial destruction of DF in rabbit seminal plasma was perhaps due to unknown contaminants in the commercial β-amylase used. We have had occasional destruction of DF activity by bacterial growth and it is possible that the aerobic spore-forming bacteria of the Bacillus type possess enzymes that will destroy DF.

Attempts to use sialic acid and sialyl lactose to decapacitate capacitated sperm or to prevent decapacitation using DF were unsuccessful [9].

Stability of DF

To determine the heat stability and effect of low ionic strength buffers on DF an ultracentrifuged bull DF sediment was dried at 37 °C *in vacuo* for 4 hr or at 100 °C for 9 hr (Fig. 6). Such treatment did not alter DF activity [17]. These studies demonstrate that DF is unusually stable for a biologically active substance in that it can be dissolved in water and dried at temperatures up to 100° without loss of biological activity.

Fig. 6. Effect of dehydration on decapacitation factor activity in rabbits [17].

Treatment	No. of ova	No. of ova fertilized	% Fertile
Capacitated sperm	128 (26)	71	55.5
Capacitated sperm, DF in liquid	33 (9)	0	0
Capacitated sperm, dehydrated DF*	37 (11)	0	0
Capacitated sperm, dehydrated DF†	10 (2)	0	0
Capacitated sperm, dehydrated DF‡	10 (4)	0	0
Capacitated sperm, dehydrated DF, recapacitated *in utero*	14 (3)	2	14.3
Capacitated sperm (control in recapacitation experiment)	14 (3)	13	92.8

Numbers in ( ) represent number of oviducts.

* Dehydrated *in vacuo* for 4 hr at 37 °C.

† Dehydrated for 9 hr at 100 °C.

‡ Dehydrated *in vacuo* for 4 hr at 37 °C, stored for 28 days at 37 °C.

Purification of DF

Before entering into a discussion of isolation studies I might provide an overall summary of the present system for the purification of DF so that it might clarify our discussion. At the present time, we produce our purest DF by:

a) Several ultracentrifugations of seminal plasma at 105,000 x g for 4 hrs with resuspension in water each time.
b) Treatment of the sediment with Pronase, followed by ultracentifugation to remove the now inactive sediment.
c) Extraction of inactive material from the supernatant with butanol.
d) Molecular sieving on a Sephadex G–10 column.
e) Migration of the active DF on a thin layer chromatographic plate with ethanol: water (7:3).

The discovery [14] that DF was resistant to prolonged proteolysis by the *Streptomyces* protease, Pronase, provided a valuable purification step in that after ultracentrifugation rabbit DF was removed from the high molecular weight sediment having been rendered soluble by proteolysis. The best DF fractions contain small amounts of ninhydrin-reacting amino acids but not enough that these could be a part of a molecule with a molecular weight as low as that of DF. It is rather surprising that with very clear evidence that DF is resistant to digestion with Pronase and that most ninhydrin-reacting material can be extracted with butanol that we still find occasional reports that DF is a protein [18].

Molecular sieving on a Sephadex G–10 column after Pronase digestion and butanol extraction resulted in further purification. Referring to Fig. 7, the DF activity is

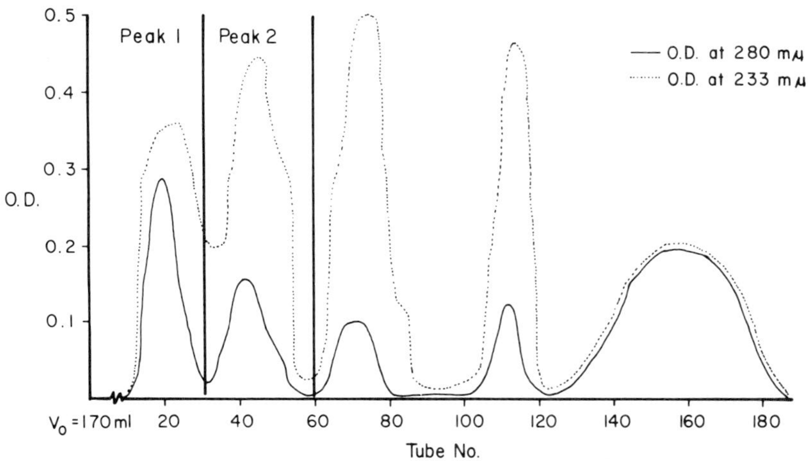

Fig. 7. Typical elution pattern obtained from pronase digested sediment from rabbit seminal plasma on Sephadex G-10 column (2.5 x 100 cm).

located entirely in the second peak in a normal column. If the Sephadex column is first washed with water at pH 2.0, the activity is found in the first peak. Activity is not found in any other peak. Molecular sieving on columns of various pore sizes has allowed us to estimate the molecular weight of DF at 300 to 500 [19].

We are faced with our own conflicting results as to the extractability of DF from aqueous solutions into chloroform-methanol mixtures. In the first experiments [20] after saturation of the organic phase with a large excess of water very little methanol was retained in the chloroform phase. Extracting with this organic phase caused denaturation of proteins present in the DF pellet in the aqueous phase. The organic phase had activity but it may have resulted from suspended denatured protein with DF attached. More recent experiments in which special care was taken to clarify the organic phase and test the sediment separately indicated that no DF activity was extracted [21]. The extractability with chloroform:methanol is thus still in question and conflicts will have to be resolved by further experiments.

The material collected in the active peak from the Sephadex G–10 column was placed on a TLC silica gel H plate and developed with ethanol and water (7:3 by volume). Three spots were visualized by U. V. light or iodine vapor treatment. The DF activity was located in the second spot with an R_f of 0.35. The other spots were at the origin and solvent front so there was a clear separation of DF activity. When the second spot was tested for the presence of steroids or phosphate using phosphomolybdic acid the results were negative. After extraction from the plate iodination of material from the second spot by the mild techniques used for the iodination of proteins did not effect DF activity indicating that no SH groups or double bonds were involved in DF activity. UV and IR spectra of this spot indicates the presence of a conjugated carbonyl group [21].

As purification proceeds the UV absorption of active fractions at 280 mμ is decreased and absorption at 233 mμ becomes more distinct. The TLC spot containing DF activity has an absorption maximum at 233 mμ. At present there are some indications that adsorption at 233 mμ in purified samples is an indication that DF activity is present. DF activity is destroyed by both mild alkali (pH 10 for 20 minutes) and mild acid (pH 4, 30 minutes) treatment. This indicates an essential, readily-hydrolyzed bond in DF but this property rules out the presence of an essential glycoside bond [21].

At this point it would perhaps be well to summarize progress in the purification of DF and to close by speculating on the possible biological significance of DF.

Our best preparations are active at 50 to 100 μg per 10^5 sperm (Fig. 8). When we consider that whole seminal plasma must be present at a level of 10,000 μg per 10^5 sperm it is apparent that about a 200-fold purification has been achieved [21]. This

Fig. 8. Minimum effective dose of related DF preparations [21]

Preparation	Mg of solids to decapacitate 10^5 sperm	References
Whole rabbit seminal plasma	10 mg	*Weinman* [22]
Sediment from single ultracentrifugation	1 mg	*Weinman* [22] *Pinsker* [23]
Ultracentrifuged sediment after 5 resuspensions in water	.32 mg	*Pinsker & Williams* [13]
Active fractions from G–10 column	.1 mg	*Robertson* [21]
Active TLC area	.05 mg	*Robertson* [21]

means that 99.5 % of the impurities have been removed. We have many indications that our best DF fractions are approaching complete purity; thus, the most difficult task is nearly complete. This leaves the way open for structural determination by use of the mass spectrometer and other physical and chemical means.

It is tempting to speculate on the biological function of DF. One that has been suggested is to prolong the fertilizable life of the sperm in the female tract since it is known that rabbit sperm, once capacitated, have a definite limitation of their lifespan. The existance of DF and the phenomenon of capacitation may confer survival value on a species especially a spontaneously ovulating species because DF prevents immediate fertilization upon initial contact of sperm and egg. The fertilizing sperm must at least have the vigor to survive in the female tract for several hours.

Dr. *Pierre Soupart* has offered an attractive hypothesis. After capacitation sperm become penetrating cells, not only penetrating ova but also uterine cells, tissues and phagocytes. The presence of DF on sperm and the excess of DF in the fluids of the male tract might function to prevent sperm from penetrating cells and tissues in the male with pathological results.

DF will block union of the gametes *in vitro* and inhibit the corona removing enzyme obtained from sperm acrosome preparations [5, 6]. The effect on this enzyme may be the first clue as to the molecular mode of action of DF.

Acknowledgement: This research was supported by National Institutes of Health Research Grant HD GM 02799 and Training Grant 5 T01 HD-00140 from National Institute of Child Health and Human Development and Career Development Award 2 K3 GM 4831 from National Institute of General Medical Sciences.

References

[1] *Austin, C. R.* (1951): Observations on the penetration of the sperm into the mammalian egg. Aust. J. Sci. Res. B. **4,** 581.

[2] *Chang, M. C.* (1951): Fertilizing capacity of spermatozoa deposited into the fallopian tubes. Nature **168,** 697.

[3] *Chang, M. C.* (1957): A detrimental effect of seminal plasma on the fertilizing capacity of sperm. Nature **179,** 258.

Bedford, J. M. & Chang, M. C. (1962): Removal of decapacitation factor from seminal plasma by high-speed centrifugation. Amer. J. Physiol. **202, 179.**

[4] *Pinsker, M. C., Caster, W. O., Robertson, R. T. & Williams, W. L.* (1968): Sperm decapacitation factor. Fed. Proc. **27,** 567.

[5] *Zaneveld, L. J. D., McRorie, R. A. & Williams, W. L.* (1968): A sperm enzyme that removes the corona radiata from ova and its inhibition by decapacitation factor. Fed. Proc. **27,** 567.

[6] *Zaneveld, L. J. D., McRorie, R. A. & Williams, W. L.* (1969): Enzymatic removal of the corona radiata of the ovum. Fed. Proc. **28,** 705.

[7] *Dukelow, W. R., Chernoff, H. N. & Williams, W. L.* (1967): Fertilizable life of the rabbit ovum relative to sperm capacitation. Am. J. Physiol. **213, 1397**

[8] *Williams, W. L., Hamner, C. E., Weinman, D. E. & Brackett, B. G.* (1964): Experiments on *in vitro* fertilization. Proc. 5th International Congress on Animal Reproduction and Artificial Insemination (Trento, Italy) 7, 288.

[9] *Dukelow, W. R., Chernoff, H. N. & Williams, W. L.* (1967): Properties of decapacitation factor and presence in various species. J. Reprod. Fert. **14,** 393.

[10] *Austin, C. R.* (1963): Acrosome loss from the rabbit spermatozoa in relation to entry into the egg. J. Reprod. Fertil. **6,** 313.

[11] *Weinman, D. E. & Williams, W. L.* (1964): Mechanism of capacitation of rabbit spermatozoa. Nature **203,** 423.

[12] *Dukelow, W. Richard & Chernoff, H. N.* (1969): Primate sperm survival and capacitation in a foreign uterine environment. Amer. J. Physiol. **216, 682.**

[13] *Pinsker, M. Craig & Williams, Williams L.* (1968): Spermatozoan decapacitation factor (DF) in human seminal plasma. Proc. Soc. Exp. Biol. Med. **129,** 446.

[14] *Dukelow, W. Richard, Williams, William L. & Chernoff, Harvey* (1965): Biochemical nature of a naturally-occurring sperm antifertility factor. Fed. Proc. **24,** 700.

[15] *Dukelow, W. Richard, Williams, W. L. & Chernoff, H. N.* (1965): Studies on the biochemical nature of decapacitation factor in seminal plasma. J. Dairy Sci. **48,** 807.

[16] *Kirton, K. T. & Hafs, H. D.* (1965): Sperm capacitation by uterine fluid or beta-amylase *in vitro.* Science **150,** 618.

[17] *Dukelow, W. Richard, Chernoff, H. N. & Williams, William L.* (1966): Stability of spermatozoan decapacitation factor. Amer. J. Physiol. 211, 826.

[18] *Hunter, A. & Nornes, H.* (1968): Relationship between sperm coating antigens and the decapacitation factor (DF) of rabbit seminal plasma. Proc. 6th International Congress on Animal Reproduction and Artificial Insemination (Paris, France) 8, 12.

[19] *Pinsker, M. Craig & Williams, William L.* (1967): Properties of a spermatozoan antifertility factor. Archives of Biochemistry and Biophysics **122,** 111.

[20] *Chernoff, H. N. & Dukelow, W. R.* (1969): Decapacitation factor purification with lipid solvents. J. Reprod. Fertil. **18**, 191.

[21] *Robertson, R. T.* (1969): Fractionation and characterization of decapacitation factor. M. S. Thesis, University of Georgia, Athens, Ga.

[22] *Weinman, D. E.* (1966): Physiological processes which certain animal spermatozoa undergo preliminary to formation of the zygote. Ph. D. Thesis, University of Georgia, Athens, Ga.

[23] *Pinsker, M. C.* (1968): Properties, purification and occurrance of decapacitation factor. Ph. D. Thesis, University of Georgia, Athens, Ga.

Discussion

J. P. Bennett: In view of your last comment concerning the action of DF in possible capacitated sperms in the male. Have you tried the effects of capacitated sperms, say in the vagina? Do they in fact penetrate the cervical tissue, for example? Dr. *Sandro* in Australia has done some work suggesting penetration of sperm in the cervix, which was indirectly indicated in some of the cervical sections.

W. R. Dukelow: We have not investigated this, Dr. *Bennett.* I have read reports of sperm penetration in the uterus but not in the cervix.

M. C. Chang: I enjoyed your paper very much. Do you think the decapacitated sperm and the uncapacitated sperm are the same or different? If they are different, what are the differences? If they are the same, why should they be the same?

W. R. Dukelow: I do not know if they are the same. The only evidence in this regard would be the required time of 10 hours to recapacitate decapacitated sperm in the oviduct and the identical time of 10 hours to capacitate freshly ejaculated sperm.

It would be very nice if we would coat the capacitated sperm with DF and show that it has the same structure as the uncapacitated sperm. But this has not been possible. The work by Dr. *Nicander,* reported by Dr. *Bedford* this morning, may allow us to do this.

D. W. Schomberg: How did you localize the DF on your thin layer chromatogram?

W. R. Dukelow: Mr. *Robertson* does this work and has used three means of identifying the spots

1) ultraviolet light,
2) 1 % ninhydrin, or
3) iodine vapor.

G. A. Overbeek: I wonder if any of the speakers of this morning would care to comment on the possible role that inhibition of capacitation could play in the effect of the progestational component in oral contraceptives. In looking at the various data that have been presented, I was not too much impressed by the inhibiting effects of the progestationals that have been used. Two questions:

firstly, is anything known about the capacitation effect during the secretory part of the mentrual cycle; secondly, to what extent may the duration of treatment with the progestational component affect the results – it was not always clear whether the period of treatment with these components was not too short. I imagine that long treatment would make a considerable difference.

C. R. Austin: I think perhaps this question is more appropriate for Dr. *Chang.*

M. C. Chang: Thank you very much for this question. It is rather complicated. Firstly, we know for sure that shorter treatment, as shown in my earlier work published in 1958 that progesterone treatment definitely inhibits capacitation of sperm in the uterus of rabbits. I demonstrated this with a very high dose of progesterone, 25 mg, ten times the normal dose. In my 1967 paper on the progestational compounds, I used a normal dose, 2 mg, where the effect was not so clear-cut. But as I mentioned this morning in connection with Dr. *Bedford's* work, that long time of incubation of sperm in the pseudopregnant rabbits will capacitate the sperm. Thus it depends how long the sperms stay there. If they stay longer they might get capacitated. It not only depends on how long we treated the rabbits with progestational compounds it also depends upon what dosage we use and what compounds we use. It also depends on how long sperm were incubated in the uterus. Moreover, in earlier study of capacitation, it is a qualitative measure. Now the capacitation of sperm have to be measured in the quantitative terms. It is possible that a large portion of sperm in an ejaculate can not be capacitated under certain conditions, some sperm may be capacitated more slowly in another situation, and an individual sperm may be only partially capacitated under another condition. So, as I have mentioned in my paper, we need more study on different species for the determination of inhibition of capacitation by administration of progestational compounds.

H. G. Goslar: Dr. *Dukelow,* you have described that iodination did not destroy the activity. Have you seen, perhaps, a decoloration of the iodine solutions? Have you investigated the behavior of blocking agents, for instance, sulfhydryl or free amino blocking agents?

W. R. Dukelow: In answer to your first question. No, we have not done this with the solutions. In reference to the second question, sulfhydryl groups have little effect on capacitation. Dr. *Ken Gould,* of our group, has found that 0.01 M mercaptoethanol will remove the zona pellucida in non fertile ova but not in fertile ova. We have done some work on blocking agents of sulfhydryl groups in relationship to the sperm enzymes that are involved in removing the corona radiata, and they may play a role here, but in the enzyme action and not in the decpacitation factor itself.

In vitro Fertilization of Mammalian Ova

B. G. Brackett

Division of Reproductive Biology, Department of Obstetrics and Gynecology, University of Pennsylvania School of Medicine, Philadelphia, Pennsylvania, USA

Summary: Although claims made before the last 2 decades are not acceptable by the rigorous criteria which have evolved for proof of *in vitro* fertilization, significant advances have been made as a result of these efforts. Major contributions have been made, primarily through use of the rabbit, since the discovery of the requirement for sperm capacitation in 1951. *Thibault, Dauzier & Wintenberger* confirmed the need for sperm capacitation as a prerequisite of fertilization *in vitro*. *Chang* was the first to establish indisputably that mammalian ova could be fertilized *in vitro* by subsequent development of *in vitro* fertilized ova into live young following surgical transfer to host animals. Advances have been made through the development of simple chemically defined media and the defining of physical conditions which enable one to consistantly achieve fertilization of a large proportion of rabbit ova *in vitro*. Meaningful controlled *in vitro* fertilization experiments can now be executed. Effects of various treatments such as washing of uterine spermatozoa and tubal ova have been assessed by their fertilizing performance *in vitro*. Fertilization of rabbit ova *in vitro* can be inhibited by trypsin inhibitors which inhibit the sperm acrosomal enzyme responsible for penetration of the zona pellucida. In addition, *in vitro* fertilization provides a means by which rabbit sperm capacitation can be assayed in the absence of the complications of the presently used *in vivo* assay. A large proportion of recently ovulated rabbit ova removed from the ovarian surface can be fertilized *in vitro*. Rabbit follicular oocytes can now be matured and fertilized *in vitro*.

In the hamster, penetration through the zona pellucida and into the vitellus by a spermatozoon has been directly observed *in vitro*. It is possible in the hamster to capacitate spermatozoa and fertilize ova in a single *in vitro* system.

In vitro fertilization is now possible in the mouse. The authenticity of the resulting fertilization in this species has also been proven by the birth of live young.

Preliminary efforts have been made to fertilize ova of other mammalian species *in vitro* but convincing evidence of success is lacking. Hopefully, many of the techniques of handling the gametes as well as the physical and biochemical conditions which permit fertilization of the ova of the above mentioned laboratory species will prove applicable in further experimentation on the *in vitro* fertilization of ova in domestic animals and primates.

In vitro fertilization of mammalian ova

Reports of *in vitro* fertilization of mammalian ova appear in the scientific literature as early as 1878 [64]. Fertilization of mammalian ova *in vitro* has been the subject of several excellent reviews [2, 5, 6, 26, 28, 29, 68, 80]. The most commonly studied mammalian species has been the rabbit, although success of *in vitro* fertilization has

---

Manuscript received: 15 March 1969

also been claimed in the guinea pig, rat, hamster, mouse, and man. Contributions of major significance have been forthcoming since the discovery of sperm capacitation in 1951 [2, 25]. Capacitation is some form of physiological change that the spermatozoon must undergo before it is capable of penetrating the ovum [3]. Sufficient evidence has accumulated to support the authenticity of fertilization *in vitro* for only three mammalian species, the hamster, the mouse, and the rabbit.

Fertilization is the union of the gametes which begins with penetration of the ovum by a spermatozoon and includes formation, growth, and syngamy of the pronuclei with the union of their respective chromosome groups on the first cleavage spindle, and ends with the metaphase of the first cleavage mitosis or cleavage to the 2 cell stage of the fertilized ovum. The qualification *in vitro* (literally, in glass) implies that the fertilization process takes place outside the animal body and under controllable conditions. The purpose of this paper is to review the literature on the *in vitro* fertilization of ova of various species with special emphasis on recent investigations.

Investigations of fertilization in vitro before the discovery of sperm capacitation

Schenk [64] treated rabbit and guinea pig ovarian ova with spermatozoa *in vitro* and noted formation of a polar body and following culture, the division of the ovum. *Onanoff* [53] stated that rabbit and guinea pig ova, taken from the uterus, could be fertilized *in vitro,* and subsequently develop either when cultured or when implanted into the abdominal cavity of females or males either of the same or of another species. It is presently known that rabbit and guinea pig ova are already unfertilizable when they reach the uterus. *Long* [46] described the breakup of the cumulus and the formation of the second polar body in rat ova incubated with spermatozoa *in vitro. Frommolt* [41] observed shrinkage of the vitellus of rabbit ova after their exposure to spermatozoa *in vitro. Krassovskaja* [44, 45] claimed *in vitro* fertilization of rabbit ova with rat and rabbit spermatozoa. The ova were collected from the oviducts 17 hours after copulation with a vasectomized buck. Such ova were probably near the end of their fertile life. According to *Smith* [68] the gametes were handled at room temperature and cleavages which began 16 or 17 hours after insemination were probably due to parthenogenetic activation caused by cooling. *Yamane* [91] published pictures supporting a claim of sperm penetration into the vitellus of a rabbit ovum. However, since the spermatozoa pictured showed no change toward pronucleus formation, they were probably artifacts.

Extensive experiments on *in vitro* fertilization were carried out by *Pincus* [56, 57] and *Pincus & Enzmann* [59, 60]. After the addition of ejaculated or epididymal spermatozoa to ovarian or tubal ova they noted shrinkage of the vitellus, extrusion of the second polar body, presence of spermatozoa in the perivitelline space and vitellus as seen in histological sections, formation of pronuclei, segmentation of ova, and birth of young. However, *Pincus & Enzmann* [61],Pincus [57], and *Pincus & Shapiro* [62] have also recorded extrusion of the second polar body, formation of pronuclei, segmentation of ova and birth of young when ova, in the absence of spermatozoa, were subjected to supra- or subnormal tempatatures, hyper- or hypotonic solutions, butyric acid, or culturing in a moist chamber. *Thibault* [75–77] has studied artificial activation by cold and noted the formation of a single, diploid nucleus or two nuclei closely resembling pronuclei. Following cold treatment the first polar body may divide so that the ovum appears to have two normal polar bodies. Regarding the claim of live young resulting from the transfer of their *in vitro* fertilized ova, *Chang & Pincus* [29] point out that the development of transferred ova may have been due to adherent spermatozoa effecting fertilization within the oviducts.

Moricard & Bossu [52] and *Moricard* [49, 50] found that if ova still contained in pieces of oviduct were treated with spermatozoa under a layer of oil and incubated for seven hours sperm heads could be identified in the zona pellucida, perivitelline space, and vitellus. No penetration was observed under aerobic conditions in which methylene blue remained oxidized or in the absence of oviduct. They concluded that a reducing potential existed in the normal oviduct and that this is an important requirement for sperm penetration. Later *Moricard* [51] reported a microcinematographic study in which he observed sperm penetration of the zona pellucida and formation of pronuclei. About 30 % of the ova incubated with uterine spermatozoa under paraffin oil at 36.5 °C to 37 °C were penetrated by sperm cells.

Smith [68, 69] after incubating rabbit ova and ejaculated spermatozoa at 37 °C with pieces of oviduct observed spermatozoa inside the zona pellucida and vitellus, pronuclei formation, and cleavage. She remarked that the process seemed slower than *in vivo*. Without the pieces of oviduct in the preparation, sperm penetration through the zona pellucida did not occur.

Venge [83] treated follicular and tubal rabbit ova with semen and after about 3 $\frac{1}{2}$ hours washed the ova and transferred them to the oviducts of does that had been mated previously to vasectomized bucks. Young of both sexes resulted. However, in the same report, *Venge* stated that he could not repeat his initial results in later attempts.

Criteria for in vitro fertilization

Since the discovery of capacitation investigators have become more conscious of the difficulty of settling on suitable criteria for *in vitro* fertilization. Cleavage of ova has been widely used as a criterion for the achievement of fertilization *in vitro* (Figs. 1 and 2). However, unfertilized mammalian ova, both *in vivo* and *in vitro*,

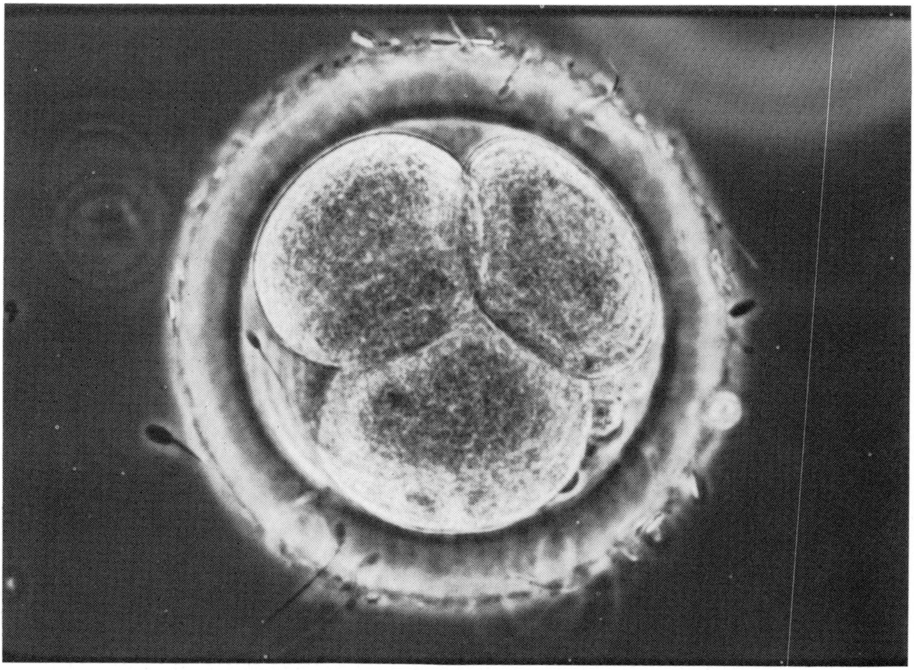

Fig. 1. 4-cell stage ovum, fertilized and cultured *in vitro.* (From *Brackett* [21]).

may undergo degenerative fragmentation, superficially resembling the cleavage of fertilized ova [1, 8, 24, 43, 77]. Cleaved ova that are fertilized have chromatin in each blastomere which can be stained, whereas fragments of cytoplasm that may appear to be blastomeres are usually devoid of chromatin. Ova may also be activated to parthenogenetic development, especially by abnormal environmental conditions [58]. One or the other of these processes might have been responsible for early claims, i. e. before the discovery of capacitation, in which the ova were incubated with ejaculated or epididymal spermatozoa.

An excellent way to prove that *in vitro* fertilization has occurred is to obtain genetically marked young of both sexes after transfer of the fertilized ova into a host. The possibility of parthenogenetic development (of the males at least) is thereby excluded. The ova should be in at least the 2-cell stage before transfer to eliminate the possibility of fertilization by spermatozoa adhering to the ova in the oviduct of the host animal. Also the host animal must be induced to ovulate by hormonal injection or by a vasectomized male known to be sterile. It is most likely that the early workers mentioned above did not give adequate attention to one of these details. *Chang* [27] was the first to provide proof of *in vitro* fertilization by development of *in vitro* fertilized rabbit ova into young following transfer to a host doe. Other investigators have subsequently proven the achievement of authentic *in vitro* fertilization and cleavage of rabbit ova [15, 21, 65, 80], and mouse ova [85] by the development of fertilized ova to young following transfer to host animals (Fig. 3).

Although it is desirable to prove that fertilization has occurred *in vitro* by development of ova to live young, it is not always practical, especially in the case of human experiments. Under such circumstances an ideal proof that fertilization has occurred *in vitro* would be direct observation of the process; such an observation could be documented by time lapse microcinematography.

The presence of a spermatozoon inside the perivitelline space of an ovum is evidence that the ovum is undergoing fertilization, provided the experimental treatment has not caused breaks in the zona pellucida. Histological sectioning of ova can result in sperm heads being translocated by the microtome, which may cause the ovum to appear to have been penetrated by a sperm cell. According to *Austin* [5] several investigators have probably been misled by such pitfalls.

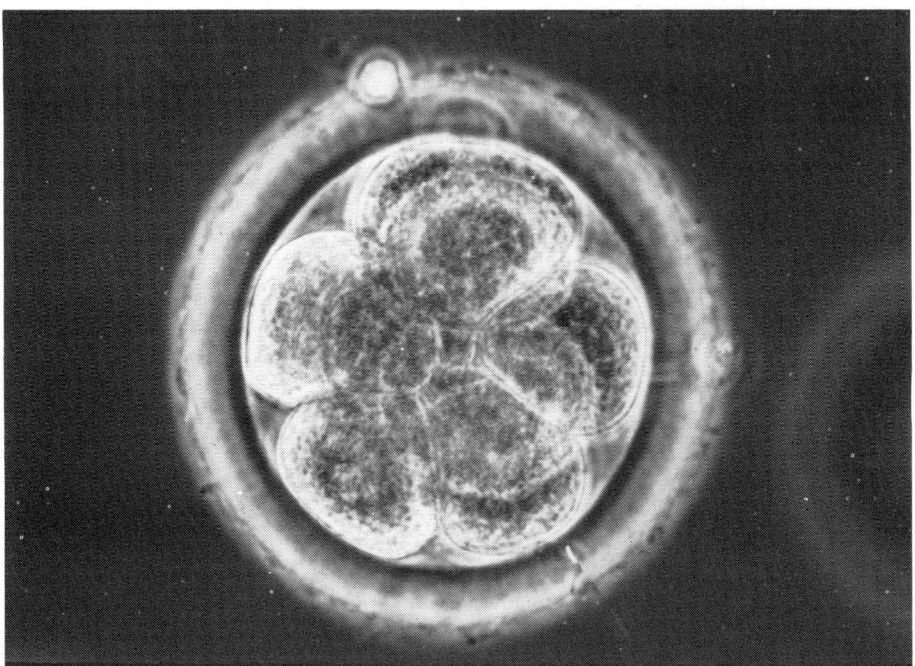

Fig. 2. 8-cell stage ovum, fertilized and cultured *in vitro.* (From *Brackett* [21]).

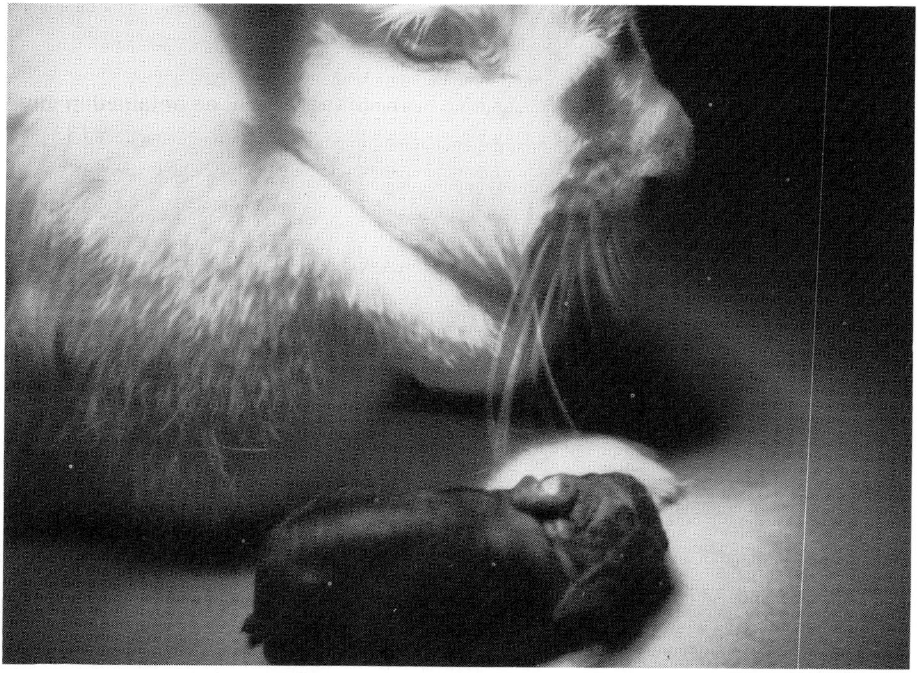

Fig. 3. Normal young developed from an ovum fertilized *in vitro* by a spermatozoon from a black buck (From *Brackett* [21]).

A simple control in which ova are subjected to exactly the same experimental treatment as those to be fertilized but with no exposure to sperm cells can eliminate the possibility of confusing fertilization with parthenogenetic activation if conditions are such that cleavage does not occur in the no sperm control. This type of control is especially important when an experimental procedure is being developed.

Investigations of fertilization in vitro after the discovery of sperm capacitation

Rabbit ova

Thibault, Dauzier & Wintenberger [81] reported that 3.8 % of the ova incubated with fresh ejaculated spermatozoa were activated compared with 86.9 % when uterine spermatozoa were used. The same workers [31] operating at 38–39 °C in order to completely avoid parthenogenetic activation placed ova and spermatozoa in Locke's solution and cultured some ova in serum to observe cleavage. They confirmed the need for sperm capacitation as a prerequisite of fertilization *in vitro. Dauzier & Thibault* [32] reported that 50 of 197 ova (25.4 %) were fertilized by uterine spermatozoa recovered 12 hours after mating in contrast to the fertilization of only 3 of 89 ova (3.3 %) incubated with uterine spermatozoa recovered 8 hours after mating; and no fertilization resulted from the use of uterine spermatozoa recovered 6 hours after mating. They also reported that capacitation could be obtained in any part of the female reproductive tract including the vagina. The same workers [33] reported that 28 of 91 ova (30 %) incubated with capacitated spermatozoa were fertilized *in vitro* when they added the spermatozoa to the ova within 5 to 15 minutes after recovery. They obtained some fertilized ova in 72 % of their experiments. Also, they presented *in vitro* fertilization evidence which suggested the requirement for sperm capacitation in sheep. *Thibault & Dauzier* [79], *Dauzier & Thibault* [34], later postulated a sperm repulsive substance on ova analogous to the "fertilizin" of certain invertebrate ova but different from the "fertilizin" of mammalian ova described by *Bishop & Tyler* [16]. By washing the ova in Locke's solution for 25 to 157 minutes before adding capacitated spermatozoa, 353 (66 %) of 532 ova were fertilized and some fertilized ova were obtained in 95 % of such experiments [80]. Without the washing period, there was a 2–3 hour latent period before the spermatozoon could penetrate the ovum. "Antifertilizin" was proposed in the female reproductive tract which neutralizes the "fertilizin" of the ova enabling sperm penetration to occur immediately after ovulation *in vivo.* The French workers based their claims of fertilization largely on histological studies from 3–7 hours after putting the gametes together. In 1961, *Thibault & Dauzier* procured young after transfer of *in vitro* fertilized ova that had cleaved in culture. *Bomsel-Helmreich & Thibault* [18] have induced tripoloidy in rabbit ova with colchicine. Colchicine suppresses the expulsion of the second polar body during fertilization, provided it is introduced before sperm penetration and removed one hour later. Following surgical transfer to host does the resulting heteroploid embryos die between day 12 and day 16 of gestation [17].

Thibault [78] has compared the fertilization process in the cow, sheep, and rabbit and has divided the process into 5 stages. The first stage includes penetration of the zona pellucida, sticking of the sperm head to the vitelline membrane, rotation of the second maturation spindle and rupture of the cortical granules previously described by *Flechon & Thibault* [39]. In the second stage the sperm head begins changing, telophase of the second maturation division is achieved, the second polar body is extruded and the perivitelline space is formed. In stage 3, male and female pronuclei form. The detached midpiece of the spermatozoon remains for a while in the vicinity of the male pronucleus and a sperm aster radiating around the proximal centriole is conspicuous in rabbit ova. The female pronucleus forms after the male pronucleus and can be identified by an unequal distribution of DNA. In stage four, the smaller female pronucleus and the male pronucleus come closer together. Duplication of DNA occurs largely in this phase [54]. In stage five the pronuclei undergo syngamy and the chromosomes appear.

Investigations of *in vivo* and *in vitro* fertilization revealed that 13.5–50 % of the fertilizations were abnormal when rabbit ova were fertilized 9 hours after ovulation [78]. Most of the abnormal fertilizations were digynic. Furthermore, digyny was observed when freshly ovulated ova were fertilized by aged spermatozoa and it was concluded that the aging of male gametes appears to have the same consequences of fertilization as the aging of the ovum itself.

In 1959, *Chang* [27] reported the fertilization of 21 % of the ova incubated with capacitated spermatozoa in Krebs-Ringer bicarbonate solution with glucose added. By using acidic saline containing heated serum and glucose for the *in vitro* fertilization medium, 42 of 74 ova (57 %) were fertilized *in vitro.* Later *Bedford & Chang* [15] reported attempts to raise the proporiton of ova fertilized in their system. No benefit was gained by washing the ova. Best results were obtained with recently ovulated ova used immediately after recovery.
Suzuki & Mastroianni [73] reported that 63.7 % of 79 rabbit ova incubated in tubal fluid with capacitated spermatozoa were fertilized *in vitro* when the gametes were incubated under paraffin oil equilibrated with 5 % CO_2-in-air. *Suzuki* [72] has reported development of 21.5 % of 65 *in vitro* fertilized ova to at least the 16-cell stage in a medium consisting of 90 % Medium 199 and 10 % rabbit serum with 1.0 mg lactic acid added per ml. *Suzuki & Mastroianni* [74] have recently reported successful *in vitro* fertilization in tubal fluid of 28.9 % of 74 ova recovered from ovarian follicles 12 hours after injection of the rabbit with HCG. With our present knowledge it appears that a follicular rabbit oocyte could be taken from the ovary, then fertilized and cultured to the blastocyst stage *in vitro* and surgically transplanted into the uterus of a hormonally prepared recipient doe where development could be allowed to continue to birth.

In the development of a repeatable method for *in vitro* fertilization attention has been focused on those physical and biochemical conditions which allow the gametes to unite. The author achieved consistent fertilization of rabbit ova by the method of *Chang* [27], and *Bedford & Chang* [15] when the gametes were handled initially and incubated at 37 or 39 °C under an atmosphere of 96–97 % relative humidity [86]. Better results were obtained when the gametes were also covered with paraffin oil, and when 0.6–2.0 ml of fresh uterine fluid was recovered with the capacitated spermatozoa and incubated in the medium. Fifty-four of 73 ova (73 %) were fertilized *in vitro* [22]. The fresh uterine fluid accounted for 20–40 % of the fertilization medium in these experiments.

Further experiments were directed toward eliminating the requirement of high relative humidity and the presence of biologic fluids of unknown composition in the fertilization medium [23]. Crystalline bovine albumin was substitited for serum in the medium. Results of previous experiments indicated that a pH of 7.8 was optimal for *in vitro* fertilization. Therefore, sodium bicarbonate was added to maintain this pH under an atmosphere of 5 % CO_2-in-air at 37 °C. Equilibration of the medium and paraffin oil in 5 % CO_2-in-N_2 seemed to enhance fertilization and early development *in vitro*. These innovations eliminated the need for high relative humidity as well as the need for serum and fresh uterine fluid in the medium. Results of experiments in which fresh uterine fluid was the only component of the fertilization medium of unknown composition were quite similar to those experiments in which no measurable volume of uterine fluid was recovered with the capacitated spermatozoa (Fig. 4).

Fig. 4. Comparison of *in vitro* fertilization with and without fresh uterine fluid in the medium [23].

Experiment	Fertilization %	Total No. of ova
I. Medium with 18–36 % of total volume contributed by Fresh Uterine Fluid recovered with cap. sp.	67	45
II. Defined medium with no measurable volume of Fresh Uterine Fluid recovered with cap. sp.	65	54

Fig. 5. Composition of the medium for *in vitro* fertilization of rabbit ova [21].

Component	g/l	mM
NaC1	6.550	112.00
KC1	0.300	4.02
$CaC1_2 \cdot 2H_2O$	0.330	2.25
$NaH_2PO_4 \cdot H_2O$	0.113	0.83
$MgC1_2 \cdot 6H_2O$	0.106	0.52
$NaHCO_3$	3.104	37.00
Glucose	2.500	13.90
Crystalline bovine albumin	3.000	
Penicillin, sodium salt	0.031	
Distilled water to 1000 ml		

The medium for *in vitro* fertilization of rabbit ova which is in current use in the author's laboratory (Fig. 5) consists of a simple salt solution, which is a modification of the salt solution originally used by *Hammond* [42] and referred to as an acidic saline by *Chang* [27], plus 50 units of penicillin G per ml and with crystalline bovine albumin, glucose and sodium bicarbonate added just before use [21]. The osmolarity of this synthetic medium is 280 mOs and the pH at 37 °C under a CO_2-in-air atmosphere is 7.8. Ova are incubated with capacitated spermatozoa for 5 hours in this medium. Then the ova are removed from the sperm suspension and cultured under an air atmosphere through early cleavage stages in acidic saline plus 50 units penicillin per ml, with heated rabbit serum which comprises 10 % by volume of the final solution and 0.25 % glucose. The sodium chloride content of acidic saline is 8.8 gm/liter which brings the osmolarity up to 287 mOs. The pH of this culture medium is 7.2.

Knowledge of obtaining and handling the gametes, the physical requirements for maintaining gamete viability, and a simple defined medium in which sperm penetration of the ovum can take place has made possible the execution of controlled *in vitro* fertilization experiments (Fig. 6). Ova from one oviduct of each ovum donor

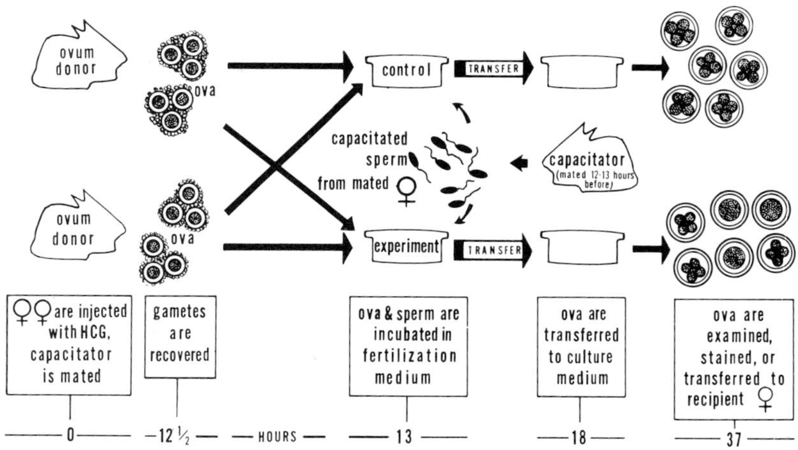

Fig. 6. Schematic diagram of a controlled *in vitro* fertilization experiment (From *Brackett* [21]).

can be placed in one tissue culture dish and ova from the other oviduct can be placed in the other dish. The sperm suspension recovered in 4.0 ml of medium is divided, and approximately one half of the sperm cells can be incubated with each group of ova. This distribution of gametes from the same animals allows comparison of results from two different treatments in each experiment. In this manner experiments have been carried out to assess the effect of washing the secretions of the female reproductive tract from rabbit ova and spermatozoa prior to their incubation together [21] (Fig. 7) to test the ability of trypsin inhibitors to inhibit fertilization [70] and to compare the fertilizing ability of uterine spermatozoa of different ages (*Brackett, Rocha & Seitz,* unpublished data).

In experiments carried out to assess the effects of washing the gametes, ova in cumulus were washed by transferring them through three different 1 ml aliquots of the medium under paraffin oil; and capacitated spermatozoa were washed by centrifugation at 395 g for 8 minutes, removing the supernatant medium, and resuspending the cells to 4.0 ml volume in fresh medium. Washing both ova and spermatozoa caused a statistically significant decrease in the proportion of ova that could be fertilized *in vitro* when neither type gamete was washed (Fig. 7, Experiment I). This

Fig. 7. Effect of washing the gametes on fertilization of rabbit ova *in vitro* [21].

Experiment		Fertilization %	Total No. of ova
I.	Effect of washing both gametes		
	unwashed ova + unwashed sperm	28	67
	washed ova + *washed* sperm	9	43
II.	Effect of washing the ova		
	unwashed ova + unwashed sperm	27	59
	washed ova + unwashed sperm	38	66
III.	Effect of washing the sperm		
	unwashed ova + unwashed sperm	54	37
	unwashed ova + *washed* sperm	27	51
IV.	Effect of centrifuging the sperm		
	unwashed ova + unwashed sperm	62	55
	unwashed ova + *washed* sperm + *washings*	42	66
V.	Effect of washing both gametes compared with the effect of washing the sperm only		
	washed ova + washed sperm	48	42
	unwashed ova + washed sperm	44	45
VI.	Effect of adding sperm washings		
	washed ova + washed sperm	47	32
	washed ova + washed sperm + sperm washings from a previous experiment	40	30
VII.	Effect of adding ampullary oviduct fluid to washed sperm		
	washed ova + washed sperm	61.1	18
	washed ova + washed sperm + oviduct fluid	54.6	11

decrease was accounted for by washing only the spermatozoa (Experiments III and V). An increase in the fertilization rate after washing the ova alone was not statistically significantly different from that of the controls (Experiments II). Likewise, a decrease in the fertilization rate following centrifuging and resuspending the sperm cells in the original solution was not statistically significant (Experiment IV). Neither washings of uterine spermatozoa from previous experiments (Experiment VI) nor ampullary oviduct fluid (Experiment VII), when added to washed spermatozoa, were able to restore the fertilizing ability to the level expected for unwashed spermatozoa. An acceptable explanation for these observations remains to be elucidated.

Possible explanations might be that an extremely labile soluble substance in the uterine environment may be removed from the sperm cells by washing, or the lower ability of the spermatozoa to fertilize ova might be accounted for by the changes in the O_2 and/or CO_2 tensions in the vicinity of the sperm cells during the experimental procedures. Nevertheless, these results indicate that fertilization is possible in a defined medium after removal, by washing, of soluble substances carried from the female reproductive tract by the ova and spermatozoa.

Stambaugh & Buckley [71] reported isolation of zona pellucida dissolution enzymes from the rabbit sperm head and found hyaluronidase and trypsin-like activity present. Experiments were conducted to test the ability of crystalline ovomucoid and soybean trypsin inhibitors to inhibit *in vitro* fertilization [70]. A direct relationship was found between the percent inhibition of the trypsin-like enzyme of the rabbit sperm acrosome by the inhibitor and the percent inhibition of fertilization by the inhibitor. Only 4 of 105 ova were fertilized when soybean trypsin inhibitor was added to the medium (1.0 mg of inhibitor per ml of medium) while 39 of 96 ova (41 %) were fertilized in the control dish with no added inhibitor. These results support the contention that the trypsin-like component of the acrosome is essential for sperm penetration of the zona pellucida.

Recent experiments have been done to compare the fertilizing ability of spermatozoa recovered from the uterus of mated does at varying intervals after coitus (*Brackett, Rocha & Seitz,* unpublished data). Preliminary results indicate that 18 hour uterine spermatozoa can fertilize a greater proportion of ova than 13 hour uterine spermatozoa are capable of fertilizing. The ova were recovered from the oviduct of does induced to ovulate by HCG injection 13 hours previously. In addition, *in vitro* fertilization is being used as an assay for the detection of complete functional sperm capacitation [20]. The authenticity of the accomplishment of *in vitro* capacitation, in the opinion of the author, should be documented by successful fertilization *in vitro,* in the absence of complications of presently used *in vivo* assays.

Also in progress are experiments to observe the discrete events of the fertilization process, i. e. sperm penetration, formation of pronuclei, syngamy, and cleavage, and the time sequence in which these events usually occur under defined *in vitro* conditions (*Brackett,* unpublished data). The observation of sperm penetration through the zona pellucida is not possible in rabbit ova due to the surrounding corona and cumulus cells. Preliminary observations have revealed actively motile spermatozoa inside the zona pellucida when ova were examined 3 hours post insemination. Cleavage to the 2-cell stage has been observed when ova were examined 15 hours, 50 minutes after insemination. Some of these observations are recorded on 16 mm film by time lapse microcinematography.

The highest fertilization rate obtained to date has been achieved by using recently ovulated rabbit ova recovered from the ovarian surface and capacitated spermatozoa recovered from the uterus of a doe mated 12 hours before [65]. The medium used in these experiments was prepared by adding heated serum to the defined medium (Fig. 5) to the extent of 20 % by volume of the final solution. Also the paraffin oil was equilibrated in 5 %CO_2-in-air instead of 5 %CO_2-in-N_2 just before the experiment. The inclusion of serum in the fertilization medium eliminated the necessity of transferring the recently penetrated ova for development through the early cleavage stages. By this procedure 117 of 143 ova (81 %) were fertilized *in vitro*. These results demonstrate that any conditioning of rabbit ova prior to fertilization by direct physical contact with the oviduct is unnecessary, or has been adequately duplicated by the *in vitro* conditions. Further experimentation using this system is underway to ellucidate the contributions of the oviduct to sperm capacitation.

Hamster ova

Yanagimachi & *Chang* [89, 90] were the first to fertilize rodent ova *in vitro*. Enlarged sperm heads or male pronuclei in the vitellus along with the tails of the fertilizing spermatozoa were taken as evidence of fertilization. Best results were obtained when Tyrode's solution was used as the fertilization medium under paraffin oil. Since fertilization was achieved with epididymal spermatozoa and hamster spermatozoa reportedly require capacitation as a prerequisite of fertilization [30], these investigators may be credited with the achievement of sperm capacitation and fertilization in a single *in vitro* system. They reported the *in vitro* fertilization of 134 of 207 hamster ova (64.7 %) when spermatozoa recovered from the uterus of females mated 4–5 hours previously were used. When they used spermatozoa recovered from females mated 0.5 hours previously, 52 of 100 ova were fertilized. When epididymal spermatozoa were used, 80 of 180 ova (44.4 %) were fertilized. Judging from the development of male pronuclei it was concluded that epididymal spermatozoa were the slowest to penetrate ova, while 0.5 hour uterine spermatozoa were slower in penetrating ova than were 4–5 hour uterine spermatozoa. The incidence of polyspermy and of supplementary spermatozoa was much higher than was found to occur normally *in vivo* [4, 7].

Yanagimachi [87] observed the process of sperm penetration through the zona pellucida on one occasion. He observed that the acrosome or at least the outer acrosome membrane of the fertilizing spermatozoon was absent before the spermatozoon started to penetrate the zona pellucida. The spermatozoon penetrated the zona at an angle to the surface and the time required was 3–4 minutes. Only 1–2 seconds was required to traverse the perivitelline space and the head of the spermatozoon lay flat on the vitelline surface and sank into the vitellus without allowing movement of the sperm tail.

Barros & Austin [10, 11] extended these observations on the *in vitro* fertilization of hamster ova. After washing ova in cumulus to remove tubal fluids before insemination, only 5 of 109 ova (4.6 %) were undergoing fertilization, while 54 of 110 unwashed ova (49 %) from the same donors were found to be undergoing fertilization. Epididymal spermatozoa were used in these experiments. When ova were recovered from large ovarian follicles that were nearly ready to ovulate, 94 of 141 ova (66.5 %) were undergoing fertilization when examined after incubation at 37 °C for 6–10 hours. The incidence of polyspermy in these follicular ova was higher than found in tubal ova.

The acrosomal reaction was supported *in vitro* by tubal fluid with or without follicular contributions and follicular fluid without tubal fluid; 95 % of randomly chosen vigorously motile spermatozoa were observed to have undergone the acrosomal reaction in follicular fluid alone [11]. Follicular contents recovered 9–10 hours after HCG injection induced the acrosome reaction in 97 % of epididymal spermatozoa, while follicular contents recovered 4 hours after HCG induced the acrosome reaction in only 43 % of epididymal spermatozoa [11]. This suggests that the responsible agent or agents are produced or accumulate in the follicle mainly in the late preovulatory phase. *Barros & Austin* [12] concluded that secretions in the female tract could be dispensed with for fertilization. When *Barros* [9] incubated half the spermatozoa with tubal fluid containing ova in saline for 4 hours, sperm penetration was observed to occur 2.5–3 hours earlier than when spermatozoa were incubated in saline or control media without tubal fluid. *Yanagimachi* [88] has recently found that hamster epididymal spermatozoa could be fully capacitated by incubation in Tyrode's solution with follicular fluid for 3 hours. After this treatment, sperm penetration occurred most often between 30 and 50 minutes after insemination. Hamster follicular fluid was the best medium for effecting capacitation, and mouse follicular fluid was better than rat follicular fluid, while rabbit follicular fluid was totally ineffective. By using *in vitro* fertilization, excellent observations of the acrosome reaction and behavior of the gamete membranes during sperm entry into the ovum at the ultrastructural level have been possible [13, 14, 40].

Mouse ova

Brinster & Biggers [19] obtained fertilization and development of mouse ova to the blastocyst in the explanted oviduct cultured in *Brinster's* medium for ovum culture with 3.16×10^{-4} M pyruvate and 1 mg/ml glucose [19]. The ampulla containing ova was incubated at 32 to 37 °C with a suspension of epididymal spermatozoa for 15 to 30 minutes, or 2 to 5 μl of sperm suspension was introduced into the fimbrial end of the oviduct with a fine pipet before the ampulla was cut free of the lower two-thirds of the oviduct. The ampulla was cultured for 4 days before examination of the ova. Of 1,748 ova, 175 (10 %) developed into morula and blastocysts.

Pavlok [55] using a modification of this method, found that a decrease of glucose in his medium from 500 mg % to 66 mg %, produced a negative effect on fertilization and initial development of the ova. A lowering of pH from near 8.0 to near 7.0 caused a decrease in the fertilization of ova 16 to 17 hours after HCG injection but ova fertilized 13 to 15 hours after HCG injection showed no such effect. Fertilizing capacity of spermatozoa recovered from the uterus 6 1/4 to 7 hours post coitus was lower than that of spermatozoa recovered 1 to 1 1/2 hours post coitus. The highest fertilization rate 95 (66.4 %) of 143 ova, was obtained by spermatozoa of the latter type. Ova were examined after 18 to 22 hours or after 48 hours of culture. Ova with 2 or more pronuclei, ova in the process of syngamy, and ova with 2 or more blastomeres were classified as fertilized.

Whittingham [85] has recently succeeded in fertilizing mouse ova *in vitro* in a simple defined medium. Spermatozoa were recovered from the uterus approximately 1 to 2 hours post coitus. Proof of fertilization was obtained by subsequent development of some *in vitro* fertilized ova to 17 day old fetuses following transfer of the ova to host females. The defined medium used was a slight modification of the medium used successfully by *Whitten & Biggers* [84] to culture mouse zygotes to the blastocyst stage. The highest yield of 2 cell ova observed at the end of 24 hours of culture was obtained when oocytes from F_1 hybrids (C57B1 × Balb/c) were inseminated with spermatozoa from Swiss males; 65 (40.9 %) of 159 ova were fertilized *in vitro*. *Whittingham* pointed out that with the fertilization of mouse ova *in vitro* and the development of mouse zygotes to the blastocyst stage [84] it should now be possible to obtain all stages of preimplantational development in simple chemically defined conditions *in vitro* for the first time in a mammalian species.

Rat ova

Toyota & Chang [82] point out that authentic reports of successful *in vitro* fertilization of rat ova are still lacking. None of 844 ova were penetrated in attempts to fertilize rat ova *in vitro* by the procedures of *Yanagamachi & Chang* [89] using various media with added heated rat serum, follicular, bursal or uterine fluid. However, it was found that if the zona pellucida was dissolved by treatment with proteolytic enzymes, epididymal spermatozoa as well as spermatozoa recovered from the uterus could penetrate the vitelline surface, activate the ova and lead to the formation of pronuclei. Following removal of the zona pellucida by treatment with 0.02 % chymotrypsin 34 (14.5 %) of 235 zona – free ova of immature superovulated rats were penetrated by epididymal spermatozoa while 30 of 100 zona-free ova from superovulated immature rats were penetrated by uterine spermatozoa; 17 (53 %) of 32 zona-free ova which were spontaneously ovulated were penetrated by epididymal spermatozoa and 18 (47 %) of 38 zona-free ova which were spontaneously ovulated were penetrated by uterine spermatozoa. Such zona-free ova were unable to undergo development beyond the early pronuclear stages. *Toyota & Chang* [82] stated that either capacitation of spermatozoa could be achieved in a short time in the presence of chymotrypsin or that the capacitation of spermatozoa is necessary only for the penetration of the zona pellucida; and this experiment demonstrated that the passing of spermatozoa through the zona pellucida is not a prerequisite for the fusion of gametes and the formation of pronuclei.

Human ova

There are only a few reports in the scientific literature in which fertilization of human ova *in vitro* is claimed. A need for sperm capacitation in the human remains to be demonstrated. *Rock & Menkin* [63] and *Menkin & Rock* [48] cultured human oocytes *in vitro* for 24 hours to induce maturation and then cultured them for an additional 45 hours in the presence of human spermatozoa. Four of 138 ova cleaved into 2 to 3 cells. It is doubtful that these ova were really fertilized since the oocytes were allowed to remain at room temperature for several hours before the beginning of the maturation incubation and spermatozoa were added at room temperature for an hour before the second incubation period. In addition, 2 of the 3 cleaved ova which were described in detail were recovered from cultures that were grossly contaminated by bacteria [48]. *Shettles* [66, 67] incubated human follicular and tubal ova in follicular fluid or serum with semen and pieces of tubal mucosa. He reported sperm penetration through the zona pellucida and development of one ovum, matured and fertilized *in vitro,* to the morula stage. However, the published pictures are not convincing evidence of such an accomplishment.

Edwards [36] has successfully matured human oocytes *in vitro* from the post-dictyate stages to the extrusion of the first polar body by culturing them 36 to 43 hours. When human oocytes were matured and then exposed to experimentally treated spermatozoa, *Edwards, Donahue, Baramki, & Jones* [38] reported that 4 of 56 oocytes were possibly fertilized by washed spermatozoa, 1 of 14 by spermatozoa after incubation in the rabbit uterus, and 2 of 20 by spermatozoa in the presence of endosalpinx. Their evidence for fertilization was the presence of pronuclei, sperm tail, or the second polar body. These criteria are insufficient for proof that the early part of the fertilization process was underway. Furthermore, their published pictures of pronuclei without identification of a sperm tail in the vitellus as well as polar bodies is not convincing evidence of fertilization.

Edwards, Bavister, & Steptoe [37] have fashioned attempts to fertilize human ova *in vitro* after successful hamster *in vitro* fertilization experiments [10–12, 90]. They used a medium based on Tyrode's solution, but with 3.0 mg bicarbonate, 9.0 μg sodium pyruvate, 2.5 mg bovine serum albumen, 20 μg phenol red and 100 units of penicillin per ml of the final medium. Follicular oocytes were cultured in follicular fluid or follicular fluid diluted with various culture media. Spermatozoa were washed and sometimes were treated with follicular fluid. Of 56 ova inseminated 20 were found to still be in dictyotene having failed to mature *in vitro*. Of the 34 matured ova spermatozoa were present inside the perivitelline space of 5 ova and in 4 of these ova the spermatozoa were motile when examined. Seven ova had pronuclei but only 2 of these ova could be considered to be undergoing normal fertilization, i. e. with 2 pronuclei and 2 polar bodies present in each. The other 5 ova had at least 3 pronuclei. Structures of a size similar to mid-pieces were seen in 3 ova after fixation. This evidence suggests that human follicular oocytes can be matured to metaphase II of meiosis and that human spermatozoa can be capacitated *in vitro* by exposure to follicular fluid. From this investigation and that of *Shettles* [66] it seems that the mechanism of human sperm capacitation may be similar to that of the hamster. Although present evidence is suggestive, [35, 47] a requirement for capacitation of primate spermatozoa as a prerequisite for fertilization remains to be established. The evidence recently put forth by *Edwards* et al. [37] supporting the achievement of early stages of the fertilization process, i. e. sperm penetration and formation of pronuclei, in at least a few human ova is the most convincing to date. However, the achievement *in vitro* of complete fertilization, including syngamy, and subsequent cleavage of human ova remains to be documented in a scientifically acceptable manner.

References

[1] *Austin, C. R.* Fertilization of the rabbit egg. Nature **166,** 407, 1950.

[2] *Austin, C. R.* Observations of the penetration of the sperm into the mammalian egg. Austr. J. Sci. Res. **B4,** 581, 1951.

[3] *Austin, C. R.* The capacitation of the mammalian sperm. Nature **170,** 326, 1952.

[4] *Austin, C. R.* Ovulation, fertilization and early cleavage in the hamster (Mesocricetus auratus). J. R. micr. Soc. **75,** 1401, 1956.

[5] *Austin, C. R.* Fertilization of mammalian eggs *in vitro.* Inter. Rev. Cytology **12,** 337, 1961.

[6] *Austin, C. R. & Bishop, M. W. H.* Fertilization in mammals. Biol. Rev. **32,** 296, 1957.

[7] *Austin, C. R. & Bishop, M. W. H.* Capacitation of mammalian sperm. Nature **181,** 851, 1958.

[8] *Bacsich, P. & Wyburn, G. M.* Parthenogenesis of matretic ova in the rodent ovary. J. Anat. **79,** 177, 1945.

[9] *Barros, C. In vitro* capacitation in golden hamster spermatozoa. Anat. Rec. **160,** 310, 1968.

[10] *Barros, C. & Austin, C. R. In vitro* fertilization of golden hamster ova. Anat. Rev. **157,** 209, 1967.

[11] *Barros, C. & Austin, C. R. In vitro* acrosomal reaction of golden hamster spermatozoa. Anat. Rev. **157,** 348, 1967.

[12] *Barros, C. & Austin, C. R. In vitro* fertilization and the sperm acrosome reaction in the hamster. J. Exp. Zool. **166,** 317, 1967c.

[13] *Barros, C., Bedford, J. M., Franklin, L. E. & Austin, C. R.* Membrane vesculation as a feature of the mammalian acrosome reaction. J. Cell. Biol. **34,** C1, 1967.

[14] *Barros, C. & Franklin, L. E.* Behavior of the gamete membrane during sperm entry into the mammalian egg. J. Cell. Biol. **37,** C13, 1968.

[15] *Bedford, J. M. & Chang, M. C.* Fertilization of rabbit ova *in vitro.* Nature **193,** 898, 1962.

[16] *Bishop, D. W. & Tyler, A.* Fertilizins of mammalian eggs. J. Exp. Zool. **132,** 575, 1956.

[17] *Bomsel-Helmreich, O.* Heteroploidy and embryonic death. In. Ciba Symposium on Preimplantation Stages of Pregnancy, *G. E. W. Wolstemholme & M. O'Connor,* ed., pg. 246, London, J. and A. Churchill, Ltd., 1965.

[18] *Bomsel-Helmreich, O. & Thibault, C.* Fecondation *in vitro* en presence de colchicine et polyploidie experimentale chez le lapine. Ann. Biol. Anim. Bioch. Biophys. **2,** 13, 1962.

[19] *Brinster, R. L. & Biggers, J. D. In vitro* fertilization of mouse ova within the explanted fallopian tube. J. Reprod. Fertil. **10,** 277, 1965.

[20] *Brackett, B. G.* Respiration of spermatozoa after in utero incubation in estrus and pseudopregnant rabbits. VIth Internat. Congr. Anim. Reprod. and A. I., 1 B, Paris, 1968.

[21] *Brackett, B. G.* Effects of washing the gametes on fertilization *in vitro.* Fertil. and Steril. **20,** 127, 1969.

[22] *Brackett, B. G. & Williams, W. L. In vitro* fertilization of rabbit ova. J. Exp. Zool. **160,** 271, 1965.

[23] *Brackett, B. G. & Williams, W. L.* Fertilization of rabbit ova in a defined medium. Fertil. and Steril. **19,** 144, 1968.

[24] *Chang, M. C.* Cleavage of unfertilized ova in mature ferrets. Anat. Rec. **108,** 31, 1950.

[25] *Chang, M. C.* Fertilizing capacity of spermatozoa deposited in the fallopian tubes. Nature **168,** 697, 1951.

[26] *Chang, M. C.* some aspects of mammalian fertilization in *A. Tyler, C. B. Metz & R. C. Borstel,* edit. The Beginning of Embryonic Development. Page 109. American Association for the Advancement of Science. 1957.

[27] *Chang, M. C.* Fertilization of rabbit ova *in vitro.* Nature **184,** 466, 1959.

[28] *Chang, M. C.* *In vitro* fertilization of mammalian eggs. J. Anim. Sci. 27, Suppl. **1**, 15, 1968.

[29] *Chang, M. C. & Pincus, G.* Physiology of fertilization in mammals. Physiol. Rev. **31**, 1, 1951.

[30] *Chang, M. C. & Sheaffer, D.* Number of spermatozoa ejaculated at copulation, transported into the female tract and present in the male tract of the golden hamster. J. Heredity **48**, 107, 1957.

[31] *Dauzier, L., Thibault, C. & Wintenberger, S.* La fecondation *in vitro* de l'oeuf de la lapine. C. R. Acad. Sci., Paris, **238**, 844, 1954.

[32] *Dauzier, L. & Thibault, C.* Recherches experimentale sur la maturation des gametes male chez les mammiferes, par l'etude de la fecondation *in vitro* de l'oeuf de lapine. Proc. IIIrd Int. Congr. Anim. Reprod., Cambridge, Section 1, p. 58, 1956.

[33] *Dauzier, L. & Thibault, C.* Donnees nouvelles sur la fecondation *in vitro* de l'oeuf de la lapine et de la brebis. Compt. Rend. Acad. Sci., **248**, 2655, 1959.

[34] *Dauzier, L. & Thibault, C.* La fecondation *in vitro* de l'oeuf de lapine. Proc. IVth Internat. Congr. Anim. Reprod., The Hague, p. 727, 1961.

[35] *Dukelow, W. R. & Chernoff, H. N.* Primate sperm capacitation. Fed. Proc. **27**, 567, 1968.

[36] *Edwards, R. G.* Maturation *in vitro* of human ovarian oocytes. Lancet **2**, 926, 1965.

[37] *Edwards, R. G., Bavister, B. G. & Steptoe, T. C.* Early stages of fertilization *in vitro* of human oocytes matured *in vitro*. Nature **221**, 632, 1969.

[38] *Edwards, R. G., Donahue, R. P., Baramki, T. A. & Jones, H.* Preliminary attempts to fertilize human oocytes Matured *in vitro*. Amer. J. Obstet. Gynec. **96**, 192, 1966.

[39] *Flechon, J. E. & Thibault, C.* Modifications ultrastructural de l'ovocyte de la lapine au sur de l'activition. J. Microsc. **3**, 34, 1964.

[40] *Franklin, L. E., Barros, C. & Fussell, E. N.* Defined structure of the acrosomal region of golden hamster spermatozoa and the acrosome reaction. Presented at the 8th Annual Meeting for the Society for Cell Biology, 1968.

[41] *Frommolt, G.* Die Befruchtung und Furchung des Kanincheneies im Filf. Zbl. Gynak. **58**, 7, 1934.

[42] *Hammond, J.* Recovery and culture of tubal mouse ova. Nature **163**, 28, 1949.

[43] *Kingery, H. M.* So called parthenogenesis in the white mouse. Biol. Bull., Woods Hole **27**, 240, 1914.

[44] *Krassovskaja, O. V.* Fertilization of the rabbit egg outside the organism. Arch. Russ. Anat. Histol. Embryol. **13**, 415, 1934, as cited by *Smith* (1951).

[45] *Krassovskaja, O. V.* Cytological studies of the heterogenous fertilization of the egg of rabbit outside the organism. Acta. Zool. **16**, 449, 1935, as cited by *Smith* (1951).

[46] *Long, J. A.* Studies on early stages of development in rats and mice. Univ. Calif. Publ. Zool. **9**, 105, 1912.

[47] *Marston, J. H. & Kelly, W. A.* Time relationship of spermatozoon penetration into the egg of the rhesus monkey. Nature, **217**, 1073, 1968.

[48] *Menkin, M. S. & Rock, J.* *In vitro* fertilization and cleavage of human ovarian eggs. Amer. J. Obstet. Gynec. **55**, 440, 1948.

[49] *Moricard, R.* Penetration *in vitro* du spermaotozide dans le ovule des mammiferes et niveau du potentiel d'oxydoreduction tubaire. C. R. Soc. Franc. Gynec. **19**, 226, 1949.

[50] *Moricard, R.* Penetration of the spermatozoon into the mammalian ovum oxydo-potential level. Nature, **165**, 763, 1950.

[51] *Moricard, R.* Observation of *in vitro* fertilization in the rabbit. Nature **173**, 1140, 1954.

[52] *Moricard, R. & Bossu, J.* Premieres etudes du passage du spermatozoid autravers de la membrane pellucida d'ovocytes de lapine fecondes *in vitro.* Bull. Acad. Nat. Med. **133, 659, 1949.**

[53] *Onanoff, J.* Recherches sur la fecondation et la gestation des mammiferes. C. R. Soc. Biol. Paris, **45**, 719, 1893.

[54] *Oprescu, S., & Thibault, C.* Duplication de l'ADN dans les oeufs de lapine avant la segmentation. Ann. Biol. Anım. Bioch. Biophys. **5**, 151, 1965.

[55] *Pavlok, A.* Fertilization of mouse ova *in vitro.* I. Effect of some factors on fertilization. J. Reprod. Fert. **16**, 401, 1968.

[56] *Pincus, G.* Observations on the living eggs of the rabbit. Proc. Roy. Soc. B, **107**, 132, 1930.

[57] *Pincus, G.* The comparative behavior of mammalian eggs *in vivo* and *in vitro.* IV. The development of fertilized and artificially activated rabbit eggs. J. Exp. Zool. **82**, 85, 1939.

[58] *Pincus, G.* The Eggs of Mammals. MacMillan Co, New York, 1936.

[59] *Pincus, G. & Enzmann, E. V.* Can mammalian eggs undergo normal development *in vitro.* Proc. Nat. Acad. Sci. **20**, 121, 1934.

[60] *Pincus, G. & Enzmann, E. V.* The comparative behavior of mammalian eggs *in vivo* and *in vitro.* I. The activation of ovarian eggs. J. Exp. Med. **62**, 665, 1935.

[61] *Pincus, G. & Enzmann, E. V.* The comparative behavior of mammalian eggs *in vivo* and *in vitro.* II. The activation of the tubal eggs of the rabbit. J. Exp. Zool. **73**, 195, 1936.

[62] *Pincus, G. & Shapiro, H.* The comparative behavior of mammalian eggs *in vivo* and *in vitro.* VII. Further studies on the activation of rabbit eggs. Proc. Amer. Phil. Soc. **83**, 631, 1940.

[63] *Rock, J. & Menkin, M. F.* *In vitro* fertilization and cleavage of human ovarian eggs. Science **100**, 105, 1944.

[64] *Schenk, S. L.* Das Säugethierei künstlich befruchtet ausserhalb des Mutterthieres. Mitt. Embr. Inst. K. K. Univ. Wien. **1**, 107, 1878.

[65] *Seitz, H. M., Jr., Brackett, B. G. & Mastroianni, L.* *In vitro* fertilization of ovulated rabbit ova recovered from the ovary. Submitted for publication, 1969.

[66] *Shettles, L. B.* Observations on human follicular and tubal ova. Amer. J. Obstet. Gynec. **66**, 235, 1953.

[67] *Shettles, L. B.* A morula stage of human ova developed *in vitro.* Fertil. and Steril. **6**, 287, 1955.

[68] *Smith, A. U.* Fertilization *in vitro* of the mammalian egg. Biochem. Soc. Symp. **7**, 3, 1951.

[69] *Smith, A. U.* In discussion after paper by *Venge* (1953).

[70] *Stambaugh, R., Brackett, B. G. & Mastroianni, L.* Inhibition of the *in vitro* fertilization of rabbit ova by trypsin inhibitors. Biology of Reproduction in press, 1969.

[71] *Stambaugh, R. & Buckley, J.* Zona pellucida dissolution enzymes of the rabbit sperm head. Science **161**, 585, 1968.

[72] *Suzuki, S. In vitro* cultivation of rabbit ova following *in vitro* fertilization in tubal fluid. Cytologia **31,** 416, 1966.

[73] *Suzuki, S. & Mastroianni, L. In vitro* fertilization of rabbit ova in tubal fluid. Amer. J. Obstet. Gynec. **93,** 465, 1965.

[74] *Suzuki, S. & Mastroianni, L. In vitro* fertilization of rabbit follicular oocytes in tubal fluid. Fertil. Steril. **19,** 716, 1968.

[75] *Thibault, C.* La parthenogenese experimentale chez le lapin. C. R. Acad. Sci., Paris, **224,** 297, 1947.

[76] *Thibault, C.* L'activation et la regulation de l'ovocyte parthenogenetique de lapine. C. R. Soc. Biol., Paris, **142,** 495, 1948.

[77] *Thibault, C.* L.oeuf des mammiferes. Son development parthenogenetique. Ann. Sci. nat. Zool. Ser. **11,** 136, 1949.

[78] *Thibault, C.* Analyse comparee de la fecondation et de ses anomalies chez la brebis, la vache et la lapine. Ann. Biol. Anim. Biochem. Biophys. **7,** 5, 1967.

[79] *Thibault, C. & Dauzier, L.* "Fertilisines" et fecondation *in vitro* de l'oeuf de lapine. C. R. Acad. Sci., Paris, **250,** 1358, 1960.

[80] *Thibault, C. & Dauzier, L.* Analyse des conditions de la fecondation *in vitro* de l'oeuf de la lapine. Ann. Biol. Anim. Bioch. Biophys. **1,** 277, 1961.

[81] *Thibault, C., Dauzier, L. & Wintenberger, S.* Etude cytologique de la fecondation *in vitro* de l'oeuf de la lapine. C. R. Soc. Biol., Paris, **148,** 789, 1954.

[82] *Toyota, Y. & Chang, M. C.* Sperm penetration of rat eggs *in vitro* after dissolution of zona pellucida with cymotrypsin. Nature **220,** 589, 1968.

[83] *Venge, O.* Experiments on fertilization of rabbit ova *in vitro* with subsequent transfer to alien does. In Mammalian Germ Cells. *G. E. W. Wolstenholme, M. P. Cameron & J. S. Freeman,* edit. Page 243, London, Churchill Ltd., 1953.

[84] *Whitten, W. K. & Biggers, J. E.* Complete development *in vitro* of the preimplantation stages of the mouse ova in a simple chemically defined medium. J. Reprod. Fertil. **17,** 399, 1968.

[85] *Whittingham, D. G.* Fertilization of mouse eggs *in vitro.* Nature **220,** 592, 1968.

[86] *Williams, W. L., Hamner, C. E., Weinman, D. E. & Brackett, B. C.* Capacitation of rabbit spermatozoa and initial experiments on *in vitro* fertilization. Proc. Vth. Internat. Congr. Reprod., Trento, VII, 288, 1964.

[87] *Yanagimachi, R.* Time and process of sperm penetration into hamster ova *in vivo* and *in vitro.* J. Reprod. Fert. **11,** 359, 1966.

[88] *Yanagimachi, R. In vitro* capacitation of hamster spermatozoa by follicular fluid. J. Reprod. Fertil. **18,** 275, 1969.

[89] *Yanagimachi, R. & Chang, M. C.* Fertilization of hamster eggs *in vitro.* Nature **200,** 281, 1963.

[90] *Yanagimachi, R. & Chang, M. C. In vitro* fertilization of hamster ova. J. Exp. Zool. **156,** 361, 1964.

[91] *Yamane, J.* Kausal-analytische Studien über die Befruchtung des Kanincheneies. I. Die Dispersion der Follikelcellen und die Ablösung der Zellen der Corona radiata des Eies durch Spermotozon. Cytologia, Tokyo, **6,** 233, 1935.

Discussion

M. C. Chang: May I ask a question? It is a nasty one. In one set of your data you have showed that washed eggs and washed sperm get 9 % of fertilization; in another set you have showed that washed sperm and washed eggs get 48 % fertilization. Moreover in another set of your data you have showed that unwashed eggs and washed sperm get 27 % of fertilization; in another set unwashed eggs and washed sperm get 44 % of fertilization – can you explain this discrepancy?

B. G. Brackett: It was as much of a surprise to me as it was to you when you recognized that here. These experiments are controlled experiments; the control ova and sperm undergo the same conditions of handling and incubation as do the experimentally – treated gametes. In some experiments we got more eggs fertilized; in some we got fewer fertilized; but since we did controlled experiments, the comparative data is meaningful.

We have many things yet to learn, especially concerning the atmosphere. I mentioned oxygen and CO_2 tensions. In addition, all of you who have worked with rabbits realize that some times of year you have much better rabbits than at other times; the effect of light and temperature on the animals and on their reproductive performance are possible reasons for the variation in results. Therefore, I feel it is most important to do controlled experiments such as those we have done in my laboratory.

J. M. Bedford: One question on the trypsin inhibitors, which looked quite convincing at the higher levels. When you were putting one mg of this inhibitor per ml of medium you mention controls, but in your controls, have you also put in a mg of something else? Because I wonder when you add that much of substance whether it also changes the osmolarity of the medium enough to affect the fertilization.

B. G. Brackett: No, it would not change the osmolarity, since the inhibitor is a protein with a high molecular weight. We have found that the range of osmolarity in which fertilization will take place is very wide.

J. M. Bedford: Can you inactivate the inhibitor, to make this a tight control?

B. G. Brackett: This could be done, but it is certainly unnecessary.

Biochemical and Biophysical Properties of Cervical Mucus in Different Hormonal States

G. F. B. Schumacher

Department of Obstetrics and Gynecology, The University of Chicago, The Chicago Lying-in Hospital, Chicago, Illinois, USA

Summary: There is a wide variation in anatomy and physiology of the uterine cervix among the mammalian species. Cervix and cervical secretions play a role in reproductive processes in some species, they may not in others. Analytical studies on the structure and composition of cervical secretions are somewhat limited by the amounts of available material or by the methods which have been employed.

The content of the cervical canal is a mixture of semi-solid gel substances, soluble components and also cells. These secretions derive predominantly from the cervical glands but also from other parts of the internal female genital tract, which are target tissues for sex hormones. Cervical secretions undergo dramatic changes under the influence of endogenous or exogenous hormones. In humans, as well as in certain other species, the cervical mucus is being exposed to and, more or less successfully, negotiated by sperm cells.

The peculiar rheological properties of cervical mucus, which are well known to the clinicians can be attributed to the network structure of cross linked epithelial glycoprotein filaments representing a hydrogel with mesh sizes that vary according to the hormonal situation. Current experimental data suggests that the overall structure of the polydispers mucin consists of a polypeptide backbone with polysaccharide side chains. The aqueous cavities between the micelles allow free movement of soluble substances and also of sperms, which migrate preferably along the filaments.

The water retained in the open network of the gel contains a great number of soluble and diffusible components, macromolecular constitutents as well as low molecular weight organic substances and electrolytes. Some of the macromolecules are serum proteins, others are locally produced components. Microdiffusion techniques in gel are particularly suitable for the quantitative assessment of proteins and certain enzymes in this type of secretion. It seems interesting that locally produced enzymes such as muramidase (lysozyme) show similar quantitative changes under the influence of hormones when compared with some of the serum proteins present in cervical mucus, such as α1-Antitrypsin, Albumin and others. Low values are observed under the influence of estrogen, high values under the influence of progesterone or progestational agents (ovulatory cycles and cycles under the sequential regimen). This might be partially a result of an increase or decrease in water content. However, the differences are considerably high and the question as to why the mucus-gel is taking up water or why it is losing water under the influence of these steroids still remains to be answered.

Manuscript received: 15 March 1969

There is a variety of anatomical and functional differences in the uterine cervix among the mammalian species. The cervical canal in many species is lined by a columnar epithelium; in others, such as the mouse, by a stratified squamous epithelium continuous with that of the vagina [22]. Moisture is absent during estrus according to *Snell* [59]. Guinea pigs, for example, show squamous epithelium covered by mucus producing columnar cells in the lower portion of the cervix. They produce tenacious mucus in proestrus which changes to a thinner consistency in late estrus [24].

In rats, mice, horses and swine the semen is ejaculated directly into the uterus, whereas in rabbits, ruminants and primates the semen is deposited in the vagina. In the latter group the cervix, with its secretions, acts as barrier and reservoir for sperms, whereas in the first group of species, the uteral tubal junction is believed to have a similar function [1, 17, 18].

The production of uterine secretions is under endocrine control. The biophysical and biochemical properties of cervical secretions change dramatically under the influence of estrogens and progestagens, adding, therefore, more variables to the multiplicity of the biological phenomena in mammalian cervix mucus. In this brief review it seems, therefore, opportune to highlight some of the biophysical and biochemical properties of human cervical secretions supplemented by certain findings in animal material.

Physiology

Cervical mucus is a mixture of secretions and transsudates derived from the cervical glands and mucosa. Mucus is continuously being released in varying quantities dependent upon the hormonal situation. Contributions from endometrium, oviducts, follicles and peritoneum are added, to an unknown extent. The mucus production increases during the proliferative phase or under estrogen administration. The secretion is profuse, watery, clear, alkaline and favorable for sperm penetration. During the postovulatory phase or under administration of progestational agents the mucus is scanty, thick, opaque, less alkaline or acid, with an increased number of leucocytes, and unfavorable for sperm penetration.

Rheological properties

Physical properties such as viscosity or consistency, flow elasticity, spinnbarkeit, plasticity, thixotrophy, tack, stickiness and adhesiveness have been described [3, 7–10, 13]. *Platt* [39] summarizes these phenomenologic entities as "bio viscosity" of cervical mucus. The assessment of spinnbarkeit, which is the capacity of mucus to be drawn into threads [10, 13], is a widely used office procedure together with the fern-test and microscopic examination of the cervical mucus for the estimation of estrogenic activity and ovulation timing [11, 12]. The assessment of the length of mucus threads drawn between a slide and a cover slip is certainly not a very accurate method. It gives, however, together with other methods, considerable information on the hormonal state. Determination of the index of consistency are more accurate. A consistometer [8] allows one to measure the time, in seconds, necessary to move a mucus column in a capillary of constant size under definite pressure [3, 7, 8, 55]. Consistency and spinnbarkeit show opposite courses during the cycle: spinnbarkeit increases and consistency decreases with increasing estrogenic activity before ovulation. After ovulation consistency increases and spinnbarkeit decreases. *Beller* and *Vogler* [3] observed the minimum values of the consistency index on the day preceding ovulation.

Overall composition

The quantitative overall composition of cervical mucus is listed in Fig. 1. Epithelial glycoprotein and probably a cross-linking protein are the elements of a semisolid gel which is responsible for the rheologic properties of the mucus. There are cellu-

MACROMOLECULAR COMPONENTS		LOW MOLECULAR WEIGHT COMPONENTS	CELLULAR COMPONENTS
SEMISOLID GEL COMPONENTS	SOLUBLE MACROMOLECULES		
EPITHELIAL GLYCOPROTEIN, CROSS-LINKING GLYCOPROTEIN ?	LOCALLY PRODUCED PROTEIN, ENZYMES, ACTIVATORS, INHIBITORS, POLYSACCHARIDES, ETC CERTAIN SERUM PROTEINS	ELECTROLYTES, CARBOHYDRATES, AMINO ACIDS, LIPIDS, UREA, ETC	EXFOLIATED CELLS, LEUCOCYTES, DEBRIS SPERMS, ERYTROCYTES BACTERIA

Fig. 1. Cervical mucus constituents.

lar elements between the meshes of the gel as well as low molecular weight constituents, electrolytes and macromolecules in solution. Considerable differences exist between the species in regard to the absence or presence of constituents. Fig. 2, which is taken from a publication by *Odeblad* [32] illustrates the basic composition of human cervical mucus during the cycle and under oral contraception.

It demonstrates that the sodium chloride concentration is constant during the cycle, as has also been published by *Herzberg, Jöel* and *Katchalsky* [21]. Mucoide, as gel-forming structures, and soluble proteins (as well as other substances), show

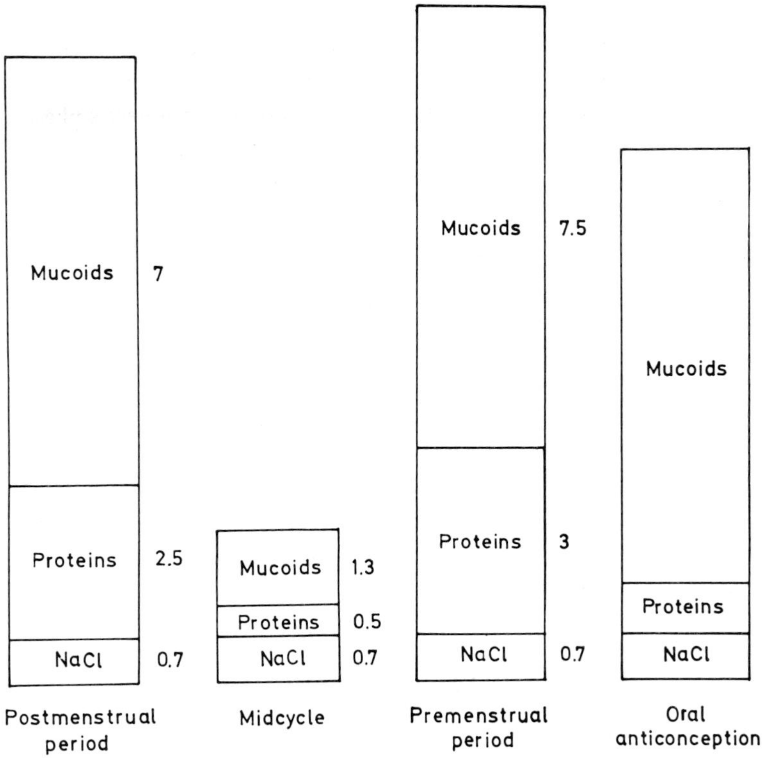

Fig. 2. Average basic composition of human cervical mucus during some different functional states. Figures to the right indicate percentage (weight/weight), balance water. (From *Odeblad* [32], p. 73).

a minimum at midcycle and high values during the postmenstrual and premenstrual period. The range of total dry substance is from 2 % at midcycle to a maximum of 12 % during the luteal phase [32]. The corresponding figures of the water content are between 90 % and 98 % [14, 40].

Physicochemical aspects of the gel structure

The rate of sperm migration in ovulatory mucus is much higher than one would expect according to the gross-viscosity of this material. *Odeblad* has studied this problem to a great extent with nuclear magnetic resonance techniques [32–36] which are non-destructive spectographic techniques, based on radio frequency absorption by atomic nuclei in a magnetic field. Information on the degree and type of hydration has been obtained as well as information on hydrogen bonding (predominant in midcycle mucus), and on electric dipolar binding of water (luteal phase mucus). Ninety-eight percent of the water is free in midcycle mucus. The internal mobility of water molecules indicates a low viscosity of the aqueous phase. This is characteristic for the network structure of this hydrogel. Fig. 3 is taken

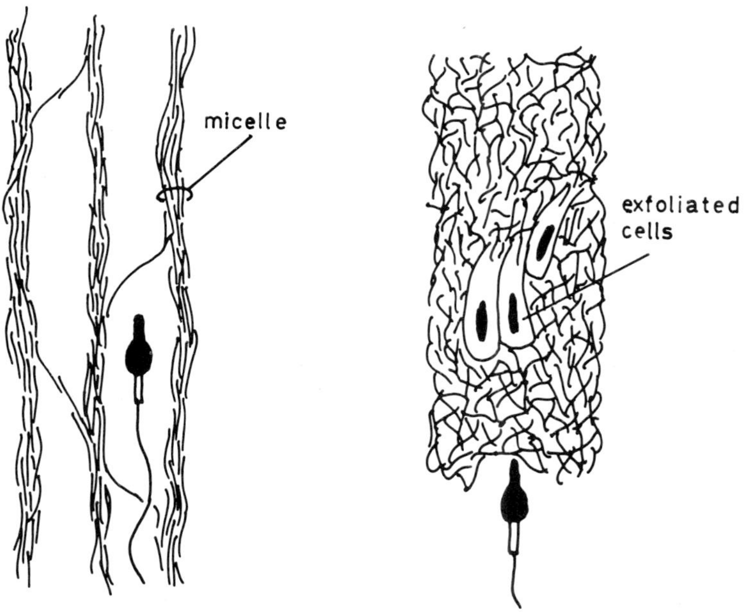

Fig. 3. Model of cervical mucus showing the ovulatory (estrogenic) type mucus to the left and gestagenic type mucus to the right. The left model indicates the micells and the inter-micellar space with a spermatozoid. The right model indicates the dense network without micellar structure which does not permit entrance of spermatozoids. (From *Odeblad* [32] p. 61).

from a recent publication by *Odeblad* [32] illustrating his thoughts about the molecular biology of the cervical mucus gel. In midcycle mucus the interconnected glycoprotein micelles are arranged in parallel. Their distances vary between one and ten microns (average 3 microns). The aqueous cavities between the micelles allow the movement of sperm along the lines of strain [14] and also allow for the diffusion of soluble substances. There is some experimental evidence that the sperm migration is supported by thermal undulations [35] of the chains of micelles. The micelles are bundled in long rods, each probably originating from separate glands [36].

The mucus gel has a different structure during the luteal phase or under the influence of progestagens. The network is dense and the mesh size is much smaller, about 0.3 microns according to *Odeblad's* model.

Similar pictures have been obtained by staining very thin cervical mucus threads on slides which have been pulled out by glass capillaries. Fig. 4 shows the mucus threads which were obtained from the same individual under the sequential regimen, after the intake of estrogen for 15 days (Fig. 4a) and after the administration of a combination of estrogen and progestagen (C-Quens) for two days (Fig. 4b).

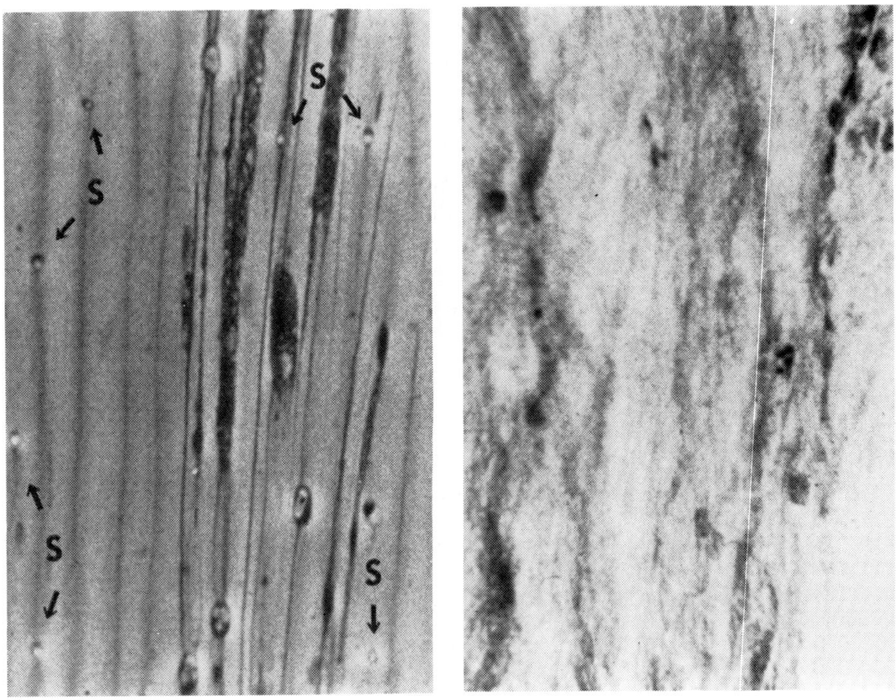

a) b)

Fig. 4. Threads of cervical mucus, which have been pulled out by a very thin glass capillary and then positioned on a microscopic slide. The air dried threads were fixed in 95 % ethanol for 10 min., stained with GIEMSA solution for 10–15 min. and rinsed with water.

Fig. 4 a) Shows a thread of mucus taken from a patient after 15 days of estrogen intake, (0.08 mg. Mestranol).

Fig. 4 b) Shows a thread of mucus taken from the same patient after two days of subsequent administration of estrogen-gestagen combination (0.08 mg Mestranol plus 1.0 mg Chlormadinone acetate, "C-Quens", *E. Lilly).*

There is no doubt that drying and staining causes certain artifacts; however, the basic structure of this mucus gel under strain is visible. The distance between the filaments appears much larger as it can be estimated by comparison with the size of leucocytes and sperms. The difference between the estrogenic and the gestagenic mucus is striking. The filaments seem to be short and swollen, the network is denser and more leucocytes, but no sperms, are present.

The "tricot-like macromolecular gel arrangement" [32] explains some of the rheological properties and biological functions of the cervical mucus and is in good agreement with current concepts of the chemical structure of the glycoprotein gel.

Biochemical aspects of the gel structure

The structure elements of the mucus gel consist of a polydisperse system [15] of long glycoprotein molecules with a high degree of branching. The polysaccharide moiety is with 70–90 % very large. Galactose, N-acetyl-glucosamine, N-acetyl-galactosamine, N-acetyl-neuraminic acid and fucose are the main sugar constitutents [14]. There are species differences; human material, for example, contains more fucose but less neuraminic acid if compared with bovine material. Differences between the thin estrous mucus and the thick pregnancy mucus are not striking. However, the neuraminic acid content is higher in the pregnancy mucus [14, 16].

Roberts and *Gibbons* [28] and *Gibbons* and *Mattner* [14] propose, on the basis of their intensive studies, that the overall structure of the glycoprotein consists of a polypeptide backbone with numerous branched sugar chains. These are attached to the long polypeptide chain via O-seryl and O-threonyl glycosidic linkages. Neuraminic acid and fucose are at the ends of the sugar chains. The glycoprotein is resistant to proteolytic enzymes.

There is certain evidence that the glycoprotein is associated with a cross-linking protein [14]. Digestion of human cervical mucus with trypsin or chymotrypsin results in loss of spinnbarkeit and loss of the gel structure [31]. Cross linkages between the glycoprotein molecules may involve the neuraminic acid residues as well as hydrogen bonds or other types of intermolecular forces [14, 43]. The degree of cross-linking is probably a very important factor in regard to the physico-chemical properties of the mucus gel. Fig. 5, which is taken from a publication by *Gibbons* and *Mattner* [14], illustrates the possible macromolecular structure and its implications: coherence and elasticity as the most striking properties of this biological material. *Odeblad's* concept of the micellar network structure and *Gibbon's* concept of the molecular structure support each other considerably.

This idea of a mucus structure, which is best characerized by *Odeblad's* formulation of a "tricot-like molecular gel arrangement", is of consequence for the conception

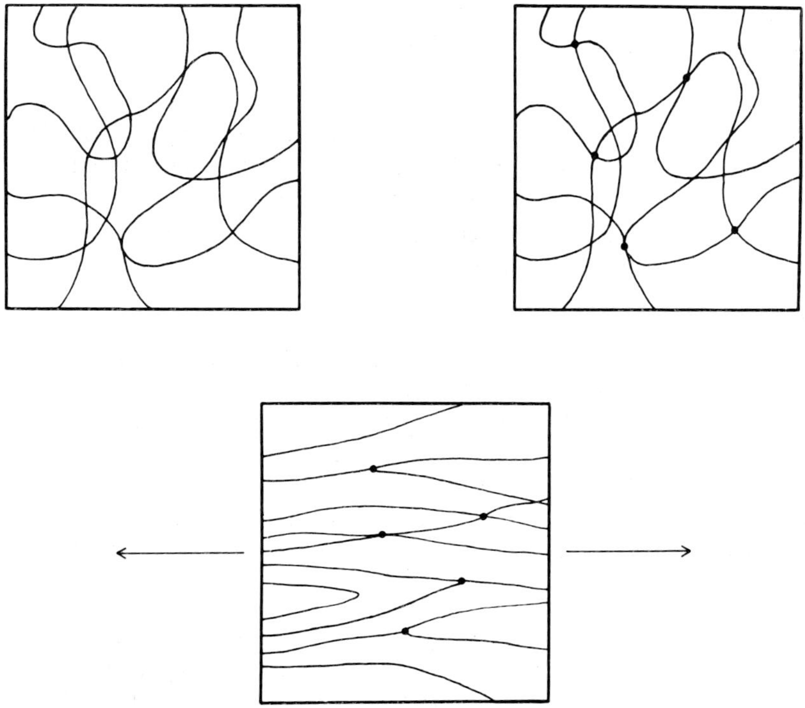

Fig. 5. Molecular anisotrophy arising as a result of stress in a random cross-linked network of flexible threadlike molecules. (From *Gibbons* and *Mattner* [14]).

of mechanisms of sperm migration and the role of the cervix as a sperm reservoir [1, 6, 14, 17, 18, 28, 63]. The sperms must have mechanisms available which enable them to enter this system and to leave it on their way to the oviduct. Proteolytic enzymes [28, 61] and their inhibitors may play a role in these physiological processes [2, 45–48].

Soluble constituents of lower molecular weight

Many soluble substances of lower molecular weight are present in cervical secretions. Sodium chloride represents the major part of it and, together with potassium-ions, is responsible for the crystallization phenomenon, which is known as "fern" reaction [41, 26]. The sodium chloride does not show significant changes of its approximately isotonic concentration during the cycle [32, 21]. The proportion between sodium chloride and organic material in dried mucus, however, shows a sharp maximum at midcycle with a high salt content.

Electrolytes, such as bicarbonate, magnesium, copper and other ions are also present. They can influence osmolarity and pH as well as the physical properties of the macromolecular components and their interactions [39].

Organic compounds, such as degradation products of proteins and polysaccharides, aminoacids, sugars, lipids and their metabolites, are present in lower concentration and may be of biological significance as well.

Soluble macromolecular constituents

Some of the macromolecular constituents of the cervical mucus are locally produced either in the cervical canal or in other tissues of the internal female genital tract. Other components are derived from the serum. There is some experimental evidence that small amounts of acid mucopolysaccharides [64] are present in this material, probably keratosulphates [66], since uronic acids are absent [56]. It is not clear whether or not this group of substances participates in the gel formation. Small amounts of glycogen [40] probably derive from endometrial secretions [14, 32].

Several enzymes are being produced locally; such as, alkaline phosphatase, acid phosphatase, nonspecific esterase, amylase, muramidase (lysozyme) and others [14, 42, 49, 50, 51, 57, 65]. In contrast to findings in the endometrium, cyclic variations of phosphatases and esterases in the human endocervical epithelium could not be identified by histochemical methods [42, 65]. However, a recent report by *Smith, Hunter, Spadoni* and *Herrmann* [58] disclosed quantitative changes of alkaline phosphatase in cervical mucus during the cycle: low preovulatory values, followed by a postovulatory rise. Administration of chlormadinone acetate also increases the values. Alkaline phosphatase appears in the vaginal mucus of the cow prior to estrus. The activity in the abundant estrous secretions is low; but increases between the 8th and 12th day post-estrum [62]. A fall in mucus amylase levels in human specimens occurs at the time of the thermal shift in basal body temperature, according to *Skerlavay, Epstein* and *Sobrero* [57].

Investigations by *Beller* and *Weiss* [4] revealed quantitative changes of the heat stable activator of the fibrinolytic enzyme system. The plasminogen activator shows high values before and after menstruation and very low or zero values at midcycle. Administration of stilbestrol was followed by a significant decrease of this factor. Cervix mucus also contains proactivator, but only very small amounts of plasminogen. Plasminogen activator appears to be secreted in the glandular epithelium [5] of the cervix. Non-specific proteolytic activity at a pH of 7.5 or 8.2 has been observed only occasionally [4, 51].

Another locally produced enzyme in cervical secretions is lysozyme, which is also called muramidase [46, 47, 49, 50]. The latter name should be preferred in order to avoid confusion with lysosomes. Muramidase is one of a class of enzymes capable of hydrolysing the structurally important beta [1–4] linkage between N-acetyl-muramic acid and N-acetyl-glucosamine in the cell walls of certain bacteria [23, 52]. It is conceivable, however, that this protein has functions of biological significance, based on its character as a basic protein which has a tendency to build complexes with acid substances [46, 49, 25].

Muramidase seems to be secreted within the area of the mucus producing tissue elements of the endocervix. Diffusible lysozyme, per mg wet weight of tissue, is highest in the lower portion of the cervix, if compared with the upper portion. The endometrium and the Fallopian tubes show relatively little muramidase activity [49, 50].

Micro-radial diffusion in agar gel containing Micrococcus Lysodeikticus has been found especially suitable for quantitative determinations of this enzyme in cervical mucus [46, 47, 37]. Muramidase secretion in the human uterine cervix appears to be under endocrine control. Fig. 6 shows the summary of observations during 12 cycles which were considered as being ovulatory cycles according to the biphasic BBT curves and typical changes in mucorrhea, spinnbarkeit, fern test and cellularity. Ovulation was induced by gonadotropins in 4 cycles. Conception occured in 3 of these cycles, 2 of them after treatment with gonadotropins. Muramidase, spinnbarkeit and BBT curves are arranged in such a manner that the last day of low BBT values is determined as day zero. Muramidase values during the last few days before the rise of the BBT are extremely low, being in most instances at a level which corresponds to the average serum values. The graph shows increased muramidase levels as early as 24 hours after zero time. These postovulatory values are extremely pronounced in some cases, rising between 5 and 50 times compared with preovulatory values.

Seven cycles under the sequential regimen of oral contraception have been investigated thus far. Fig. 7 demonstrates that estrogen administration results in lowering the postmenstrual values of muramidase and that the subsequent intake of additional gestagens is followed by a significant increase. However, the level of muramidase values during the time of estrogen intake is not as low when compared with preovulatory values during the ovulatory cycles. The increase seems to be therefore, not so pronounced. The observations also reveal that it takes more than 12–15 hours, but less than 36–40 hours after the first gestagen intake (reinforced by the second tablet 24 hours later) until changes in the muramidase level become apparent. It is not known whether muramidase interferes directly or indirectly with reproductive processes.

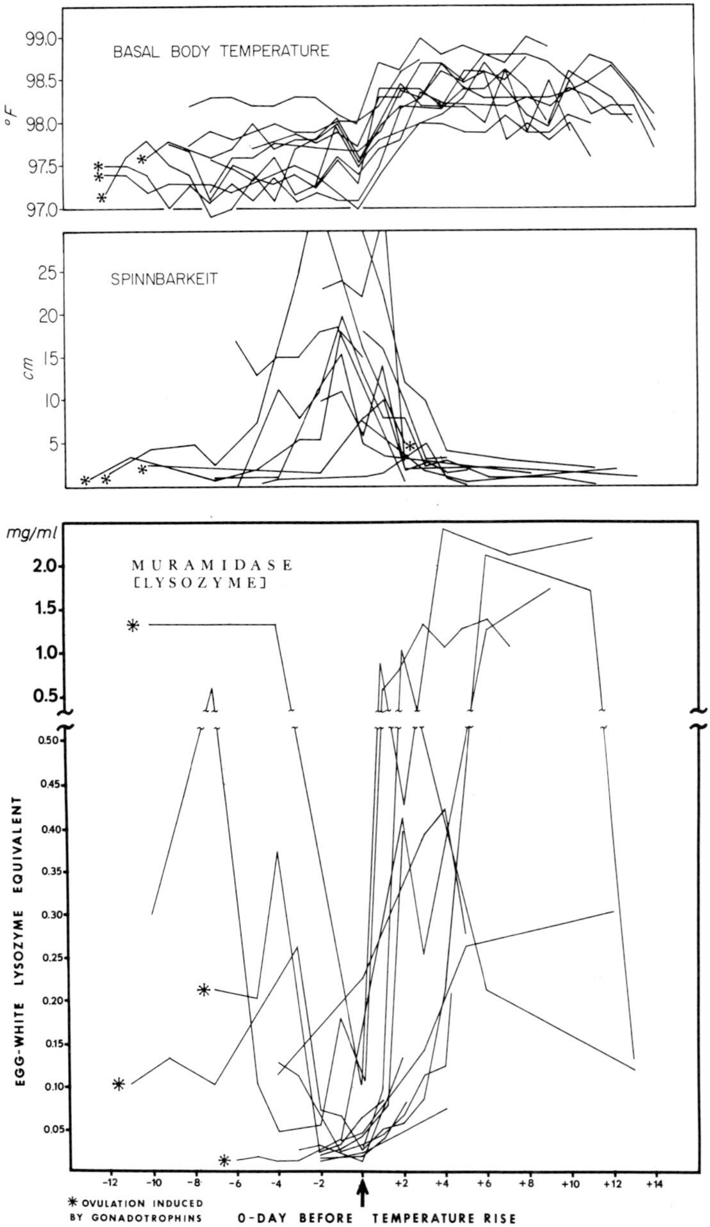

Fig. 6. Muramidase (Lysozyme) in cervical mucus, spinnbarkeit and BBT during 12 cycles which were considered as being ovulatory cycles according to the biphasic BBT curves and typical changes in mucorrhea, spinnbarkeit, fern test and cellularity. The last day of low BBT before the rise was determined as 0-day and the curves were arranged accordingly.

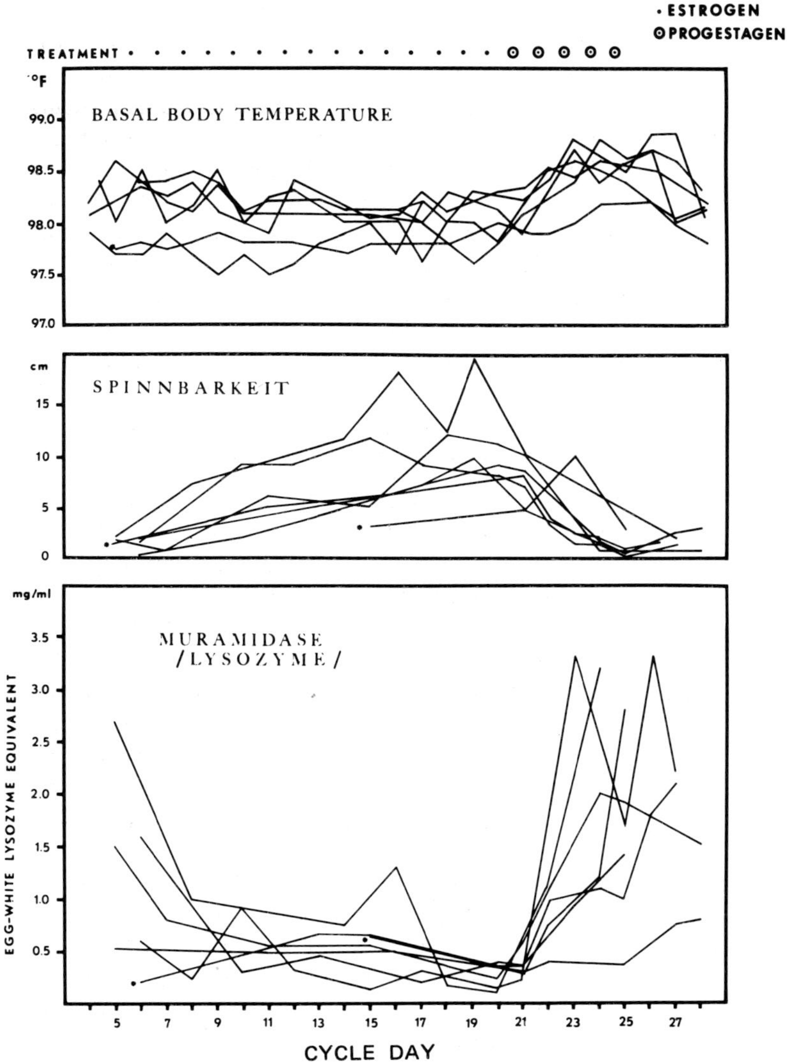

Fig. 7. Muramidase (Lysozyme) in cervical mucus, spinnbarkeit and BBT during 7 cycles under the sequential regimen of oral contraception ("C-Quens", *E. Lilly* and "Oracon", *Mead-Johnson*). The days of estrogen intake and subsequent administration of estrogen and progestagen are indicated on the top of the graph.

Protein and glycoproteins immunochemically related to, or identical with, serum proteins are present in the aqueous phase of the cervical mucus gel [20, 29, 30, 45, 48]. Microimmunoelectrophoresis and microimmunodiffusion techniques are suitable methods for the qualitative and quantitative analysis of soluble proteins in cervical secretions. The application of antisera against total human serum proteins allows a gross-orientation of the composition; whereas the use of specific antisera against individual serum proteins enables the investigator to identify the components. Radial immunodiffusion in gels allows the quantitative assessment of these components in body fluids using standard solutions of purified antigens or standardized and stabilized reference sera (*Mancini, Carbonara* and *Heremans* [27]). The Behringwerke AG, Marburg, Germany, (*Schultze, Schwick, Störiko* and *Coworkers* [43, 54, 53, 60]) did pioneer work in this field and in making a variety of these immunochemical tools commercially available. Radial diffusion methods in gel are particularly suitable for quantitative microanalysis of cervical secretions because of their rheological properties [46, 47].

Several individual proteins have been identified in cervical secretions such as prealbumin, albumin, α_1-lipoprotein, orosomucoid, α_1-antitrypsin, haptoglobin, ceruloplasmin, transferrin, β_{1C}/β_{1A}-Globulin (C_3-Complement), immunoglobulin A (IgA), immunoglobulin g (IgG) and occasionally traces of macroglobulins [20, 29, 45–48]. The number of protein species appearing in immunoelectrophoresis is normally not more than five or six. However, this depends to a certain extent on the quality of the antiserum. More precipitin lines appear under pathological conditions. Results on quantitative changes of serum proteins are obtained from a small number of cases and they are incomplete in regard to the number of proteins normally present. They give, however, useful information, especially when compared with the results on locally produced constituents such as muramidase.

Data have been obtained on albumin (M. W. 60.000) and on α_1-Antitrypsin (M. W. 45.000), a powerful trypsin-chymotrypsinelastase inhibitor [54, 44, 19, 2], during a number of ovulatory cycles and cycles under the sequential regimen of oral contraception as described in the results of the muramidase studies. Fig. 8 demonstrates the courses of albumin together with the courses of spinnbarkeit and BBT in 10 ovulatory cycles, partially identical with the cycles in Fig. 6. The midcycle minimum of albumin shows values of less than 0.6 mg/ml, whereas 10 to 50 times higher levels have been observed before and after day zero. In some instances BBT increase and increase in albumin coincide. In a few cases, however, the increase of albumin in the cervical mucus begins before the BBT shift. Whether this is significant remains to be investigated in more cycles. At least no decreasing values have been observed after the BBT shift. *Moghissi* [48] reported similar results based on electrophoretic studies. Fig. 9 shows a more clearly cut picture of the cycles during the sequential regimen. Albumin decreases from high postmenstrual levels within a few days and stays at lower levels during the time of estrogen intake. Additional adminstration of gestagens results, in most instances, in a significant albumin increase.

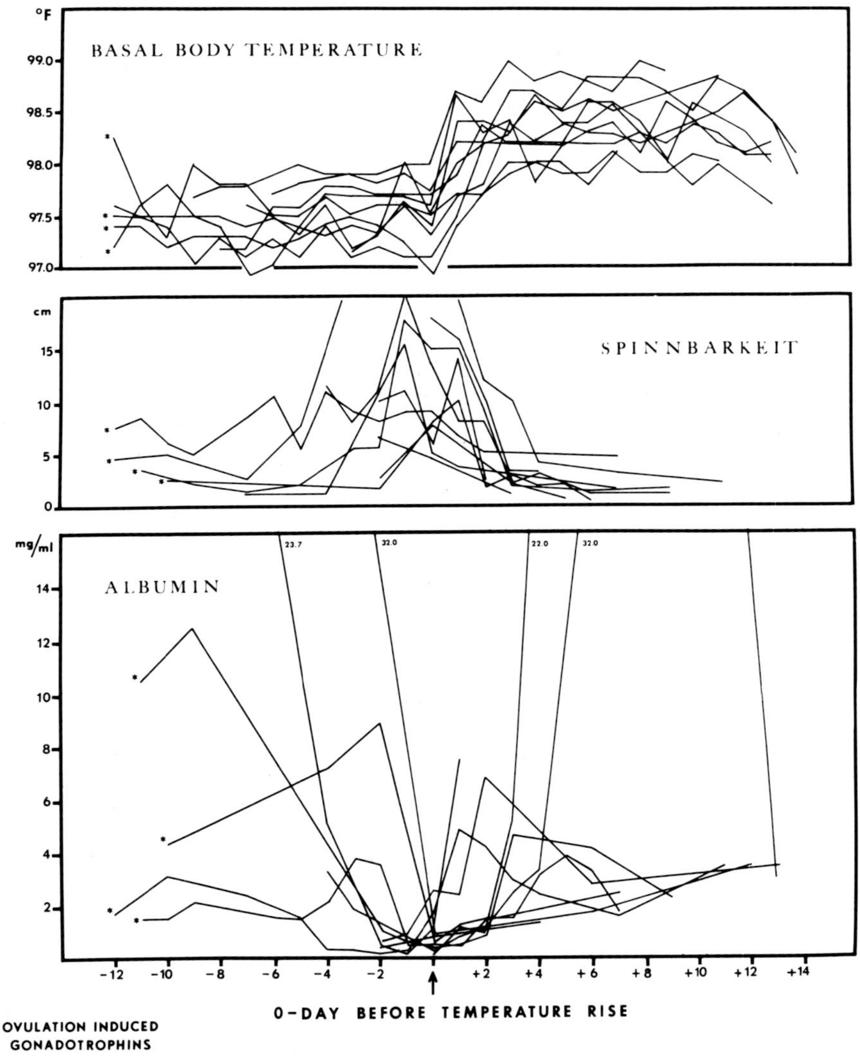

Fig. 8. Albumin in cervical mucus spinnbarkeit and BBT during ten cycles which were considered as being ovulatory. This material is partially identical with the one in Fig. 6 and Fig. 10. The arrangement of the curves is the same.

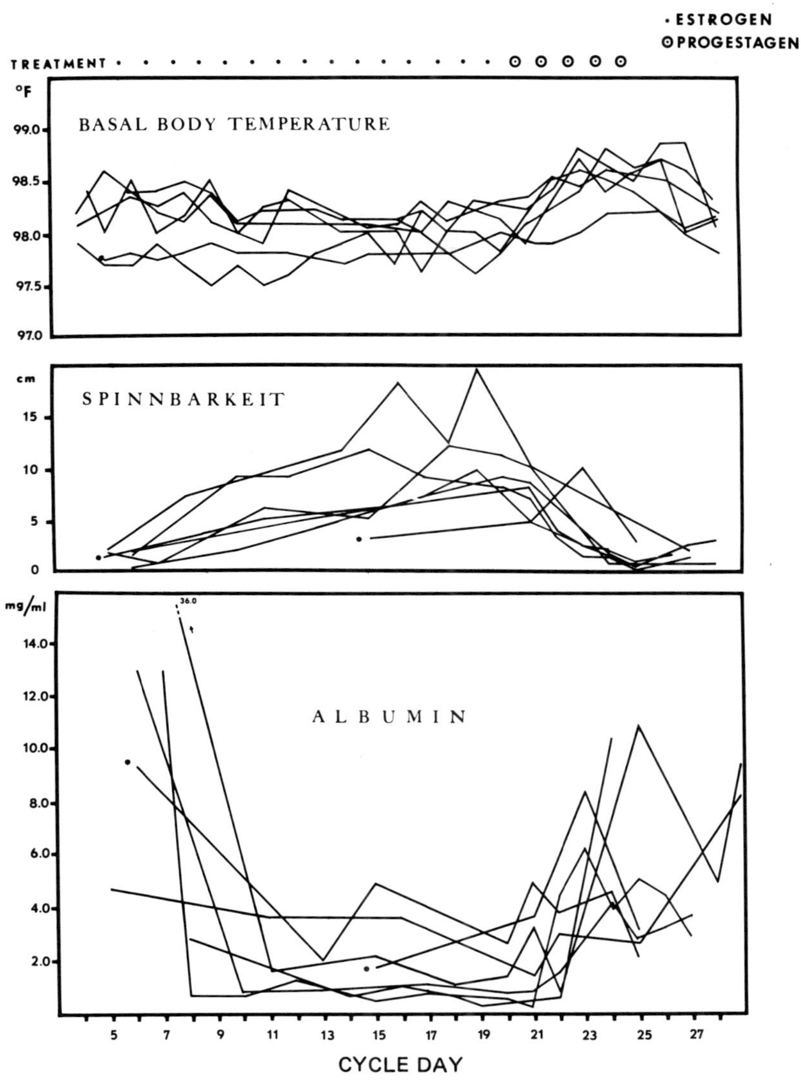

Fig. 9. Albumin in cervical mucus, spinnbarkeit and BBT during 7 cycles under the sequential regimen. This material is identical with the one in Fig. 7 and Fig. 11.

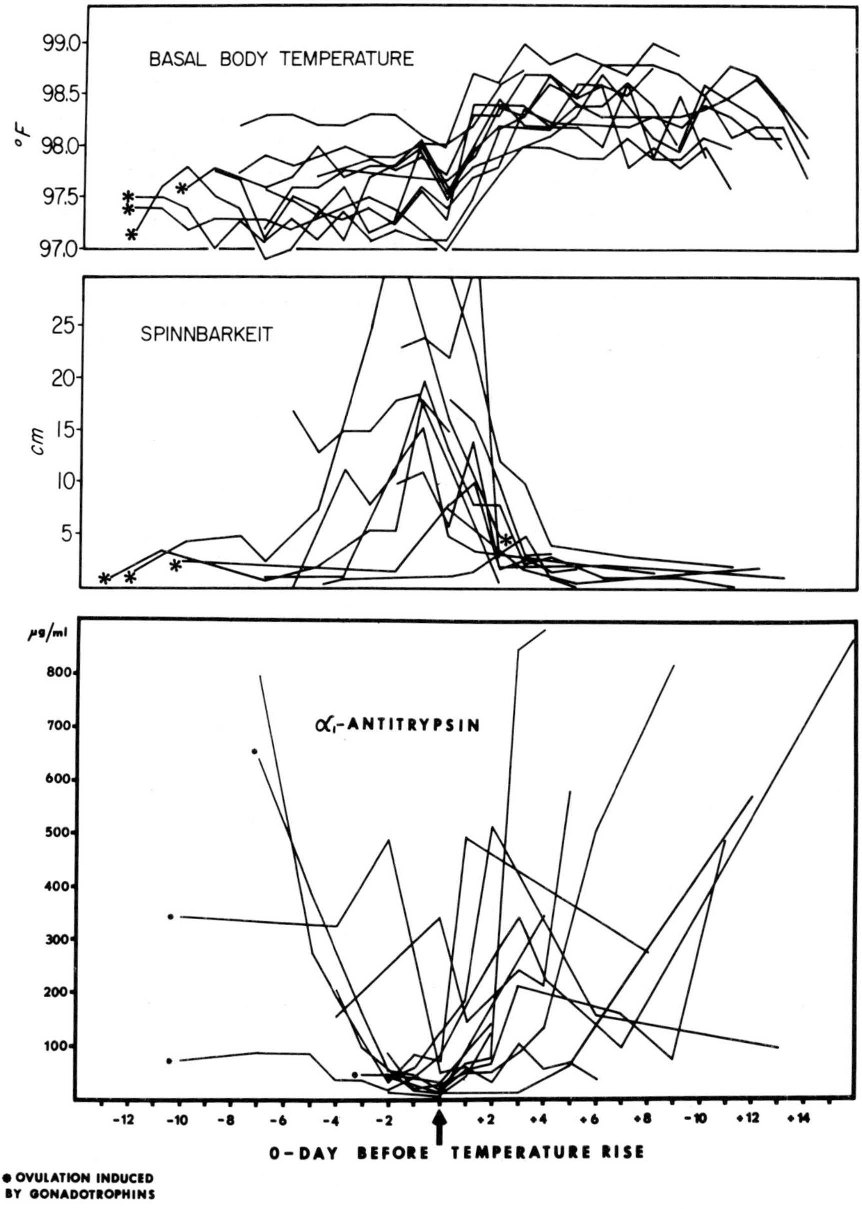

Fig. 10. α_1-Antitrypsin in cervical mucus, spinnbarkeit and BBT during 12 ovulatory cycles. This material is identical with the one in Fig. 6. The arrangement of the curves is the same.

Changes of α_1-antitrypsin levels in cervical mucus occur at a much lower level of concentration. However, the picture is essentially the same, as shown in Fig. 10 (12 cycles). In most instances the α_1-antitrypsin values decrease preovulatory to extremely low values of less than 30 μg/ml. They increase up to almost 20 times higher after day zero. Fig. 11 demonstrates the courses of α_1-antitrypsin in cervical mucus during 7 cycles under the sequential regimen. It shows, similar to the albumin, a decrease of α_1-antitrypsin from high postmenstrual values during the first days of estrogen intake. The values remain low under estrogen and increase again after administration of additional gestagens. However, the levels of α_1-antitrypsin (similar to the albumin levels) during the time of estrogen intake are not as low as preovulatory values during ovulatory cycles. Therefore, the difference appears to be not so pronounced.

A few other serum proteins have been determined in two of these cycles during the sequential regimen. The data are compiled in Fig. 12, illustrating the courses of albumin, prealbumin, α_1-Antitrypsin, transferrin, C_3-Complement (β_{1C}/β_{1A}-globulin), Immunoglobulin A (IgA) and Immunoglobulin G (IgG) in comparison with the locally produced lysozyme (muramidase) and spinnbarkeit and BBT. This figure demonstrates that other serum proteins present in cervical mucus, such as, prealbumin, transferrin, C_3-complement, IgA and IgG, follow a similar pattern under the influence of estrogen and gestagen. The increase in concentration of these constituents of cervical mucus occurs almost simultaneously.

Quantitative data on soluble macromolecular constituents of cervical mucus suggest, on the basis of present information, a strong influence of sex steroids on the concentration of a variety of components. Estrogens lower the concentration, whereas, gestagens seem to antagonize this effect. This might be partially a result of an increase or decrease in the water content of the mucus gel. However, the differences between minimum and maximum concentrations are considerably high, and the question as to why the mucus gel is taking up water or why it is losing water under the influence of these steroids still remains to be answered.

Acknowledgement: Most of the Ovulatory Cycles have been studied in close cooperation with Dr. *M. J. Pearl,* Associate Professor of Obstetrics and Gynecology at the Albany Medical College, Albany, N. Y. The skillful technical assistance of Miss *Margarita Feifel* and Mr. *Bruno Dragoje* is herewith gratefully acknowledged.
Supported by USPHS-Grant D 2682 and D 03696.

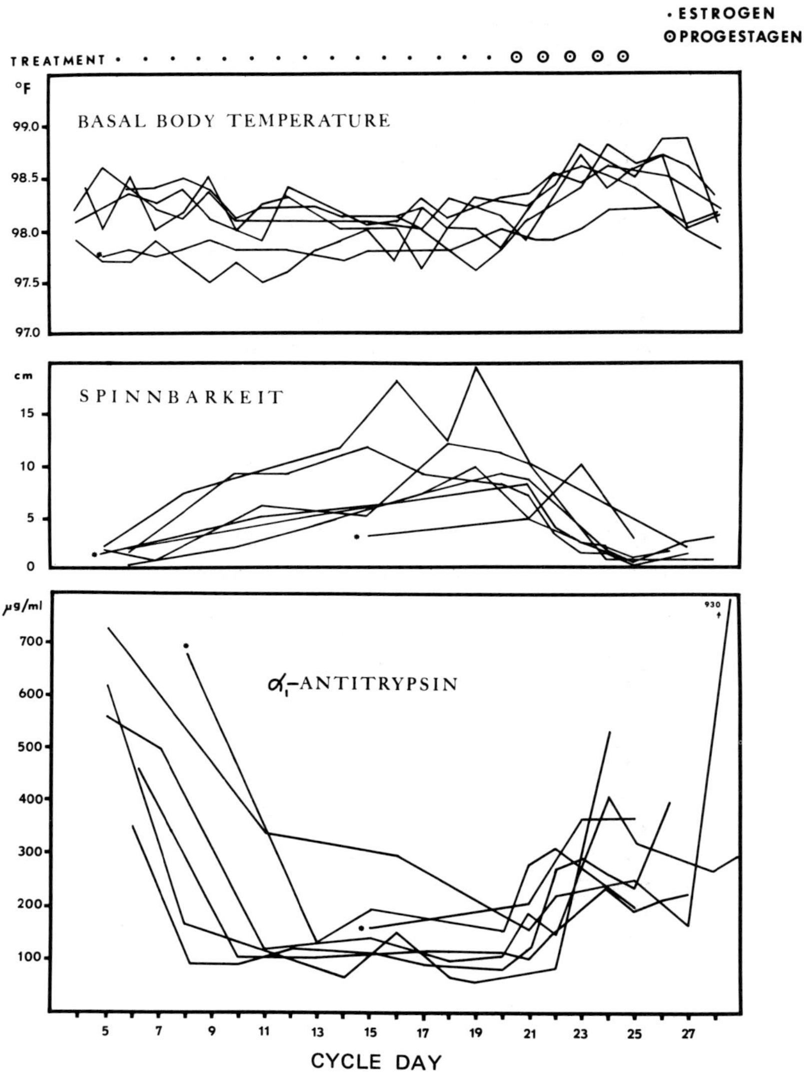

Fig. 11. α_1-Antitrypsin cervical mucus, spinnbarkeit and BBT during 7 cycles under the sequential regimen. This material is identical with the one in Fig. 7 and Fig. 9.

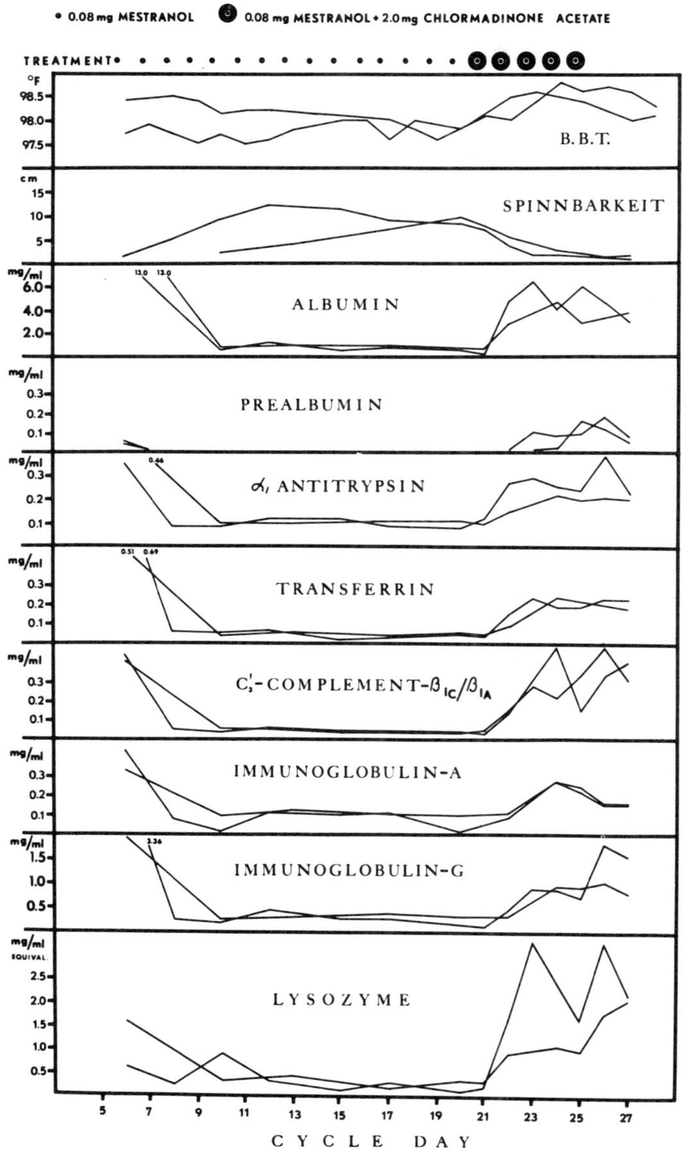

Fig. 12. Albumin, prealbumin, a_1-Antitrypsin, transferrin, β_{1A}/β_{1C}-globulin (C_3-complement), immunoglobulin A (IgA) and immunoglobulin G (IgG) and the locally produced lysozyme (muramidase) in cervical secretions together with spinnbarkeit and BBT curves during two cycles under the sequential regimen (C-Quens, *E. Lilly*). The respective observations on these two cycles are included in Fig. 7, 9 and 11.

References

[1] *Bane, A.:* Veterinary medical considerations on the cervix as a fertility factor. Acta Obstet. Gynec. Scand. **37**, Suppl. **1**, 59 (1958).

[2] *Baumstark, J. S.:* Studies on the elastase-serum protein interaction. I. Molecular identity of the inhibitors in human serum and direct demonstration of inhibitor-elastase complexes by zone and immunoelectrophoresis. Arch. Biochem. Biophys. **118**, 619 (1967).

[3] *Beller, F. K. & H. Vogler:* Die Bestimmung des Ovulationstermines bzw. der praeovulatorischen Phase unter besonderer Berücksichtigung der physikalischen Eigenschaften des Zervixschleimes. Z. Geburtsh. Gynak. **158**, 58 (1962).

[4] *Beller, F. K. & G. Weiss:* The fibrinolytic enzyme system in cervical mucus. Fertil. Steril **17**, 654 (1966).

[5] *Beller, F. K.:* Personal communication.

[6] *Bennett, J. P., D. K. Vallance & B. H. Vickery:* Inhibition of gamete migration and fertilization in the rabbit. Postgrad. Med. J. **39**, 39 (1967).

[7] *Blair, G. W. S., S. J. Folley, F. H. Malpress & F. W. V. Coppen:* Variations in certain properties of bovine cervical mucus during the estrous cycle. Biochem. J. **35**, 1039 (1941).

[8] *Blair, G. W. S.:* Ein Microviscosimeter für Nicht-Newtonsche Flüssigkeiten. Kolloid Z. **78**, 19 (1937).

[9] *Clift, A. F., F. A. Glover & G. W. S. Blair:* Rheology of human cervical secretions; effects of menstrual cycle and pregnancy. Lancet **I**, 1154 (1950).

[10] *Clift, A. F.:* Observations on certain rheological properties of human cervical secretions. Proc. Roy. Soc. Med. **39**, 1 (1945).

[11] *Cohen, M. R.:* Methods of determination of ovulation. Lying-in: The Journal of Reproductive Medicine **1**, 181 (1968).

[12] *Cohen, M. R., I. F. Stein & B. M. Kaye:* Spinnbarkeit: a characteristic of cervical mucus. Fertil. Steril. **3**, 201 (1952).

[13] *Erbring, H.:* Untersuchungen über die Spinnbarkeit flüssiger Systeme. Kolloidbeihefte **44**, 171 (1936).

[14] *Gibbons, R. A. & P. Mattner:* Some aspects of the chemistry of cervical mucus. a) Internat. J. Fertil. **11**, 366 (1966). b) in *B. Westin* and *N. Wiquist* "Fertility and Sterility", page 695. Excerpta Medica Foundation, Amsterdam 1967.

[15] *Gibbons, R. A.:* On polydispersity. Nature **200**, 665 (1963).

[16] *Gibbons, R. A. & G. P. Roberts:* Some aspects of the structure of macromolecular constituents of epithelial mucus. Ann. N. Y. Acad. Sci. **100**, Art. 2, 157 (1963).

[17] *Hafez, E. S. E.:* "Uterine cervix". Summary of the North American Conference on Fertility and Sterility in Ocho Rios/Jamaica, January 1969. Science (in press).

[18] *Hafez, E. S. E. & D. L. Black:* The mammalian uterotubal junction in *E. S. E. Hafez* and *R. J. Blandau*. "The mammalian oviduct" p. 85. The University of Chicago Press, Chicago, 1968.

[19] *Heimburger, N. & H. Haupt:* Charakterisierung von α_1-X-Glycoprotein als Chymotrypsin-Inhibitor des Humanplasmas Clin. Chim. Acta **12,** 116 (1965).

[20] *Heinen, G., H.-J. Keutel & H. Ott:* Immunelektrophoretische Untersuchungen des menschlichen Scheideninhaltes unter physiologischen und pathologischen Verhältnissen. Arch. Gynäk. **200,** 259 (1965).

[21] *Herzberg, M., C. A. Jöel & A. Katchalsky:* The cyclic variation of sodium chloride content in the mucus of the cervix uteri. Fertil. Steril. **15,** 684 (1964).

[22] *Hummel, K. P., F. L. Richardson & E. Fekete:* Chapter 13 "Anatomy" p. 247 in *E. L. Green* "Biology of the Laboratory Mouse". 2nd Ed. McGraw-Hill Book Co., New York 1966.

[23] *Jollès, P.:* Recent developments in the study of lysozymes. Angew. Chem. (Internat. Ed.) **3,** 28 (1964).

[24] *Jurow, H. N.:* Cyclic variations in the cervix of the guiena pig. Amer. J. Obstet. and Gyn. **45,** 762 (1943).

[25] *Kuettner, K. E., H. L. Guenther, R. D. Ray & G. F. B. Schumacher:* Lysozyme in proesseous cartilage. Calc. Tiss. Res. **1,** 298 (1968).

[26] *Macdonald, R. R.:* Cervical mucus and the management of abortion. J. Obstet. Gynecol. Brit. Comm. **70,** 580 (1968).

[27] *Mancini, G., A. O. Carbonara & J. F. Heremans:* Immunochemical quantitation of antigens by single radial immunodiffusion. Immunochem. **2,** 235 (1965).

[28] *Moghissi, K. S., D. Dabich, J. Levine & O. W. Neuhaus:* Mechanism of sperm migration. Fertil. Steril. **15,** 15 (1964).

[29] *Moghissi, K. S. & O. W. Neuhaus:* Composition and properties of human cervical mucus. II. Immunoelectrophoretic studies of the proteins. Amer. J. Obstet. Gynec. **83,** 149 (1962).

[30] *Moghissi, K. S.:* Cyclic changes of cervical mucus in normal and progestin-treated women. Fertil. Steril. **17,** 663 (1966).

[31] *Neuhaus, O. W. & K. S. Moghissi:* Composition and properties of human cervical mucus. III. A preliminary study of the mucoid component. Fertil. Steril. **13,** 550 (1962).

[32] *Odeblad, E.:* The functional structure of human cervical mucus. Acta Obstet. Gynec. Scand. **47,** Suppl. 1, 57 (1968).

[33] *Odeblad, E.:* The physics of the cervical mucus. Acta Obstet. Gynec. Scand. **38,** Suppl. 1, 44 (1958).

[34] *Odeblad, E.:* Research in Obstetrics and Gynecology with nuclear magnetic resonance. Acta Obstet. Gynec. Scand. **38,** 599 (1959).

[35] *Odeblad, E.:* Undulations of macromolecules in cervical mucus. Int. J. Fertil. **7,** 313 (1962).

[36] *Odeblad, E.:* Micro-NMR in high permanent magnetic fields: theoretical and experimental investigations with an application to the secretions from single glandular units in the human uterine cervix. Acta Obstet. Gynec. Scand. **45,** Suppl. 2, (1966).

[37] *Ossermann, E. F. & D. P. Lawlor:* Serum and urinary lysozyme (muramidase) in monocytic and monomyelocytic leukemia. J. Exp. Med. **124,** 921 (1966).

[38] *Platt, H. A.:* Mucus flow effects of hormones in the intact animal. Proc. 5th World Congr. Fertil. Steril., Stockholm, 1966, Int. Congr. Ser. No. 133, pp. 726–728. Excerpta Medica Found. Amsterdam-New York (1967).

[39] *Platt, H. A.:* The application of biophysical chemistry to the study of human cervical mucus. Ann. N. Y. Acad. Sci. **130** (Art. 3), 925 (1966).

[40] *Pommerenke, W. T.:* Cyclic changes in the physical and chemical properties of cervical mucus. Amer. J. Obstet. Gynec. **52,** 1023 (1946).

[41] *Roland, M.:* The Fern test, a critical analysis. Obstet. Gynecol. **11,** 30 (1958).

[42] *Schmidt-Matthiessen, H.:* Das normale menschliche Endometrium. G. Thieme Verlag, Stuttgart (1963).

[43] *Schultze, H. E. & J. F. Heremans:* Molecular biology of human proteins. Vol. I, Nature and metabolism of extracellular proteins. Amsterdam, Elsevier Pub. Co., 1966.

[44] *Schultze, H. E., K. Heide & H. Haupt:* α_1-Antitrypsin aus Humanserum. Klin. Wschr. **40,** 427 (1962).

[45] *Schumacher, G. F. B., E. K. Strauss & G. L. Wied:* Serumproteins in cervical mucus. Amer. J. Obstet. Gynec. **91,** 1035 (1965).

[46] *Schumacher, G. F. B. & G. L. Wied:* Semiquantitative microanalysis of proteins in cervical mucus. Proc. 5th World Congr. Fertil. Steril., Stockholm, 1966, Int. Cong. Ser. No. 133, pp. 713–722. Excerpta Med. Found., Amsterdam-New York (1967).

[47] *Schumacher, G. F. B.:* Protein analysis of secretions of the female genital tract. Lying-in: The Journal of Reproductive Medicine **1,** 61 (1968).

[48] *Schumacher, G. F. B. & M. J. Pearl:* α_1-Antitrypsin in cervical mucus. Fertil. Steril. **19,** 91 (1968).

[49] *Schumacher, G. F. B. & M. J. Pearl:* Muramidase (lysozyme) in cervical secretions. in *H. Peeters* "Protides of biological fluids". Proceedings of the XVIth Meeting. Vol. 16, p. 525. Pergamon Press, Oxford, 1969.

[50] *Schumacher, G. F. B. & M. J. Pearl:* Cyclic changes of muramidase (lysozyme) in cervical mucus. North American Conference on Fertility and Sterility, Ocho Rios/ Jamaica, 1969 (Report).

[51] *Schumacher, G. F. B.:* Unpublished observations.

[52] *Schumacher, H.:* Neue Erkenntnisse über die Biochemie des Lysozyms und ihre klinische Bedeutung für die Augenheilkunde. Fortschr. Augenheilk. **8,** 142 (1958).

[53] *Schwick, H. G.:* Quantitative Immunpräcipitation zur Bestimmung einzelner Plasmaproteine. 15th Colloq. Ges. physiol. Chem., Mosbach pp. 55–67. Ed.: *O. Westphal,* Springer Verlag Berlin, Heidelberg, New York 1964.

[54] *Schwick, G. & K. Störiko:* Qualitative and quantitative determination of plasma proteins by immunoprecipitation. Lab Synopsis **1,** 1 (1965). Lloyd Brothers, Inc., Cincinnati, Ohio (now Certified Blood Donor Service, Woodbury, N. Y.).

[55] *Scott-Blair, G. W. & F. A. Glover:* More early pregnancy tests from studies of bovine cervical mucus. Brit. Vet. J. **113,** 417 (1957).

[56] *Shettles, L. B.:* The polysaccharide composition of human cervical mucus. Fertil. Steril. **2**, 261 (1961).

[57] *Skerlavay, M., J. A. Epstein & A. J. Sobrero:* Cervical Mucus Amylase levels in normal menstrual cycles. Fertil. Steril. **19**, 726 (1968).

[58] *Smith, D. C., W. Hunter, L. R. Spadoni & W. Herrmann:* Control of alkaline phosphatase activity in the cervical mucus. 16th Ann. Meeting Pac. Coast Fertil. Soc. Palm Springs, Calif. 1968 (Report).

[59] *Snell, G. D., K. P. Hummel & W. H. Abelmann:* A technique for the artificial insemination of mice. Anat. Rec. **90**, 243 (1944).

[60] *Störiko, K.:* Normal values for 23 different human plasma proteins determined by single radial immunodiffusion. Blut **16**, 200 (1968).

[61] *Stambough, R. & J. Buckley:* Zona pellucida dissolution enzymes of the rabbit sperm head. Science **161**, 585 (1968).

[62] *Van Klingenberg, G. A.:* Extremely high alkaline phosphatase activity in the vaginal mucus of the cow. Nature **172**, 397 (1953).

[63] *Vickery, B. H. & J. P. Bennett:* The cervix and its secretions in mammals. Physiol. Rev. **48**, 135 (1968).

[64] *Werner, I.:* The chemistry of cervical mucus. Acta Obstet. Gynec. Scand. **38**, Suppl. 1, 39 (1959).

[65] *Wheeler, J. D. & S. Danziger:* A histochemical study of the cervix uteri. Obstet. Gynec. **5**, 739 (1955).

[66] *Zachariae, F.:* The acid mucopolysaccharide of the cervical mucus. Acta Obstet. Gynec. Scand. **38**, Suppl. 1, 86 (1959).

Discussion

H. G. Goslar: Have you found any correlation, any close relation, between the isoelectric point of the mucus on one side and the estrogen level on the other; and between the viscosity and the isoeletric point? I ask you because you have a conglomerate between neutral and acid mucopolysaccharides in the mucus, and perhaps you may have a change in the correlation of the components.

G. F. B. Schumacher: First of all I have to ask you a question back: isoelectric point of which substances, of the mucoid, the gelforming glycoproteins? I must confess I did not study the isoelectric points of cervical mucus constituents. Actually, I cannot remember anything in the literature which gives sufficient information to answer your question.

E. Johansson: May I just show Fig. 13 concerning the reference point in the cycle? This figure shows a composite of 16 ovulatory cycles in which LH and progesterone have been measured around the time of ovulation in daily samples. The study was made in collaboration with Dr. *Leif Wide.* The day zero is the first significant rise in the LH concentration. The day +1 is the most likely day for ovulation – for that we have a lot of evidence. The nadir in the temperature curve coincides with the first significant rise of LH. At the same time we found a small pre-ovulatory increase of plasma progesterone. Between the first and the second day there is a rapid increase in the plasma levels of progesterone indicating the formation of a corpus luteum. The basal body temperature was clearly elevated two days after the first rise of plasma LH. Thus, when you use the nadir in the basal body temperature curve as the day of ovulation you are likely to be 1 to 1 1/2 day before the actual ovulation.

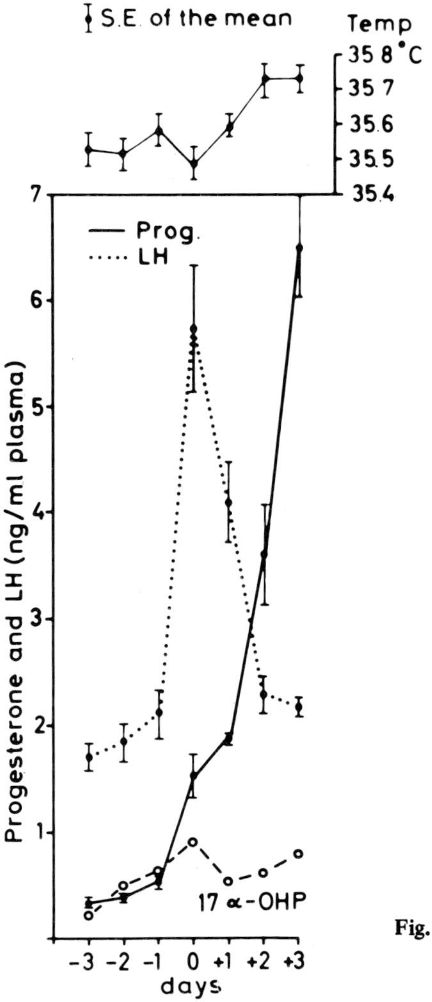

Fig. 13

G. F. B. Schumacher: I think this is a very interesting observation showing more accurately the relationship between serum progesteron and LH level and basal body temperature rise. However, we are still in trouble in regard to the actual time of follicle rupture. We would like to relate some biological and biochemical findings to it. Pregnenediol excretion or serum progesterone does not give enough information in regard to time of ovulation as far as hours are concerned. As you probably recognize from our curves, particularly the ones of the albumin, the minimum appears in some cases even before the last day of low basal body temperature. Now the question is: are these changes going on before the actual time of follicle rupture or not? After I saw your slide, I am much more inclined to believe that these changes do occur one to two days before ovulation.

Duration of Transport of Spermatozoa through the Ductular System

C. G. Heller, F. Teshima and M. J. Rowley

Pacific Northwest Research Foundation, Seattle, Washington, USA

Summary: Testes of normal men were exposed to single doses of ionizing radiation or injected with tritiated thymidine. The selected X-ray dose was sufficient to cause degeneration of spermatogonia while allowing each of the other germinal cells to mature normally. The pre-leptotene spermatocyte (the least advanced unaffected cell) is released into the tubular lumen 46 days following irradiation. Since spermatozoa were absent from the seminal fluid 67 days after irradiation, the longest travel time of a sperm discharged into the tubular lumen is three weeks (67–46=21 days). A second group of men was given intratesticular injections of thymidine. Seminal fluid for autoradiography was collected every third day. The most advanced cell labeled within one hour is the preleptotene spermatocyte. This is released into the lumen of the tubule as a mature sperm 46 days after labeling. Counts of labeled spermatozoa indicate that the earliest sperm appeared in the ejaculate 46 or 47 days after injection. The greatest number appeared in the ejaculate in 54 to 56 days. Combining irradiation and thymidine data we conclude that the duration of transport of spermatozoa in man varies from one to 21 days with the majority of the spermatozoa appearing in the ejaculate 10 to 14 days after release from the Sertoli cell cytoplasm.

Introduction

Little or no exact data is available on the transit time of human spermatozoa. *Freund* [7] postulated 16 to 18 days and in 1969 *Freund & Davis* [8] determined the disappearance of spermatozoa following vasectomy to be a median of 15 days. *Edwards* [6] stated 19–23 days, *Brown* [2] stated 24 days, *MacLeod & Gold* [16] in 1952 stated 35 days and in 1956 stated "2–3 weeks", while *Czerniak* [4] reported 5–12 days.

We proposed, therefore, to establish the time of transit of human spermatozoa from the time of their release from the Sertoli cell cytoplasm until their appearance in the ejaculate. Two methods were applied. One was to label the testicular spermatozoa with tritiated thymidine and note the time of appearance of labelled spermatozoa in the ejaculate. The other was to stop the replication of spermatozoa by

Manuscript received: 15 March 1969

administering X-ray irradiation to the testes and follow the depletion process by observing the diminishing number of spermatozoa in the ejaculate until no further spermatozoa were observed.

Material and methods

Adult normal male volunteers were divided into two groups. A group of five subjects was given intratesticular injections of thymidine-H^3 at four sites in each testis for a total of 80 microcuries. The second group of nine men received testicular X-ray irradiation of 100 or 300 roentgens. Two of these subjects were first injected with thymidine and then immediately irradiated with 100r. Each subject was vasectomized at the termination of the investigation.

Thymidine-H^3

Seminal fluid was collected once weekly until 30 days post injection, from which time we requested that seminal fluid be collected as often as every third day if possible. Due to personal differences collections were at slightly different intervals. Each specimen was counted by two technicians using the haemocytometer method.

Radioautographs were made from smears of each seminal fluid specimen. The smears were coated with Diaphane fixed in 95 % alcohol, and stained with Casarett's stain (1962). Radioautography was done with Kodak NTB-3 Nuclear Track Emulsion using a "loop" method. Exposures were from 21 to 280 days with subsequent development in a two-stage developer [17].

Counts of labelled spermatozoa were made on each slide and reported as cells labelled per 1000 counted. Only those spermatozoa having four or more grains per nucleus were included. All slides were coded for counting.

X-ray irradation

X-ray irradiation was administered from a unit featuring twin X-ray tubes within a lead lined box, especially designed for this purpose. The patient lay prone over the unit with his testes suspended in a plexiglass cup filled with water at scrotal temperature. The remainder of the body was fully protected by the lead shielding. Dosimetry revealed uniform delivery within ± 0.5r at any point within the plexiglass box. Those subjects undergoing testicular X-ray irradiation collected seminal fluid once weekly throughout the experiment.

Results

Thymidine-H^3

Thymidine-H^3 is incorporated into the DNA of the nuclei of germinal cells. In man the most advanced labelled cell in the maturation process is the pre-leptotene spermatocyte [9, 10]. It takes 46 days from the time of labelling of the pre-leptotene cell until it becomes a spermatozoa ready for release. During this time it undergoes development as a spermatocyte, maturation-division, and further development as a spermatid. Such labelled spermatozoa about to be released are seen 46 days after labelling in Fig. 1, along with the residual bodies of companion spermatozoa that have recently left the Sertoli-cell cytoplasm. This, then, becomes day zero in the timing of the transit of the spermatozoa through the ductular system.

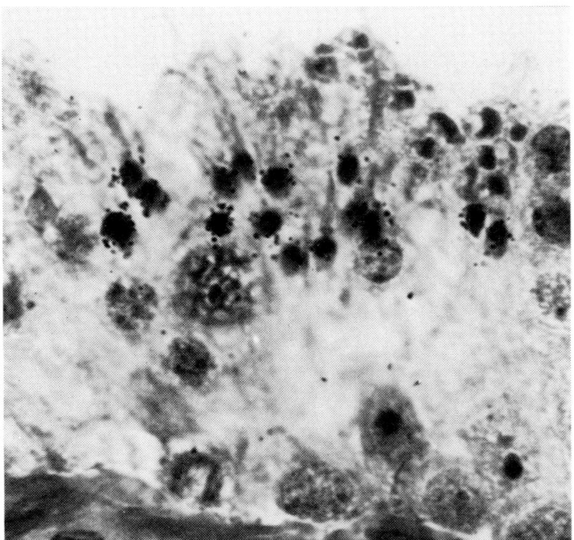

Fig. 1
X 1200. Labelled spermatozoa found in Cell Association II 46 days after Thymidine-H^3 injection. Note residual bodies along edge of lumen.

X-ray irradiation

X-ray irradiation, applying 100r to 300r, causes discernable damage to the spermatogonia in man within 4 to 6 hours [12]. The spermatogonia all fail to undergo mitosis to become pre-leptotene spermatocytes. At this X-ray dosage none of the pre-leptotene spermatocytes or any of the cells more advanced in their maturation are affected. They proceed through the maturation process normally in both numbers and rate. Spermatozoa are seen as the last remaining germinal cells leaving the Sertoli-cell cytoplasm 46 days following irradiation. Fig. 2 reveals a few remaining spermatozoa on day 46 just emerging from the cytoplasm along with the residual bodies of the companion spermatozoa which have just preceded them into the tubular lumen. Day 46 following irradiation thus becomes day zero in terms of determining the transit time consumed during residence in the tubular system.

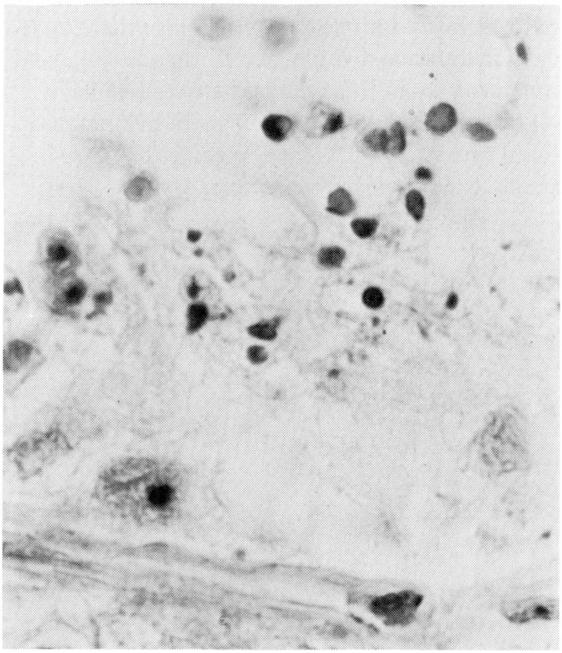

Fig. 2
X 1200. Spermatozoa and residual bodies found in Cell Association II 46 days after X-ray irradiation with 100r. Only other cell present within the tubule is a Sertoli cell.

The most rapid transit time

Each of the men injected with tritiated thymidine revealed labelled spermatozoa by day one or in the first ejaculated specimen collected after day one (Fig. 3). No subject had labelled spermatozoa in the ejaculate during the five days prior to day zero or before. In one subject labelled spermatozoa appeared on day zero. Of four collecting on day 1, three revealed labelled spermatozoa and one failed to do so. In the latter, the first labelled spermatozoa were observed in the next specimen collected (day 15). Thus we observed labelled spermatozoa by day 1 in four subjects. In the fifth subject no ejaculate was obtained between day 1 and day 15. The curves of labelling are shown for two men in Fig. 4.

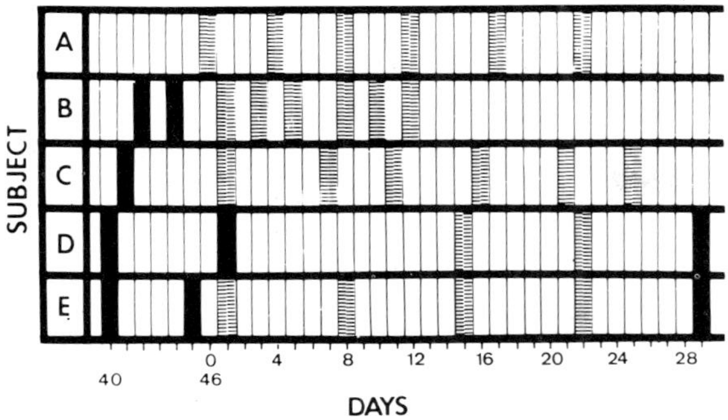

Fig. 3. The times of semen collection in five subjects receiving thymidine-H^3. Day 0 is day 46 after thymidine-H^3 injection and/or X-ray irradiation. The black bars indicate no labelled spermatozoa; striped bars labelled spermatozoa and white bars show the days that no collections were made. The specimens containing the largest number of labelled sperm per 1000 for subjects A, B, C, D and E were collected on 12, 12, 11, 15 and 8 days respectively. Subjects D and E also received X-ray irradiation. Note absence of labelled spermatozoa in semen collected on day 29 in these two men.

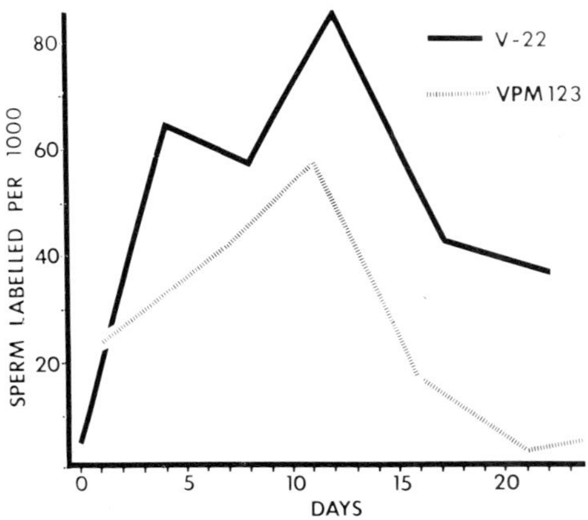

Fig. 4
The number of labelled spermatozoa per 1000 in two subjects injected with thymidine-H^3. Day 0 is day 46 after thymidine-H^3 injection.

The mode and median transit time

The mode and median transit time was determined from the time that the maximal number of labelled spermatozoa appeared in the smears. The mode as well as the mean were found by statistical analysis to be 12 days for each. The maximal labelling occurred on days 58, 58 and 57 for those receiving thymidine-H^3 alone. The two subjects who received X-ray irradiation at the same time revealed maximal labelling at 54 and 61 days (the midpoint being 57 or 58days also), no samples having been collected between 54 and 61 days in either subject (Fig. 3, subjects D and E).

Maximal transit time

The maximal transit time cannot be determined with the use of thymidine labelling alone because the spermatogonia are also labelled with the isotope. The resulting labelled spermatozoa continue to be excreted for a long period after the spermatozoa resulting from pre-leptotene spermatocytes are gone. Therefore the irradiation study was necessary to establish the maximal travel time. The testicular irradiation of nine men (including the two receiving thymidine) resulted in a maintenance of sperm count at normal pre-radiation levels for the first 46 days until day 1 transit time. At this time sperm count began to drop, and continued to decrease until only an occasional spermatozoon was found in the ejaculate on day 20–22. Allowances must be made for the fact that the days of collection varied. This depletion is illustrated in Fig. 5 in two men. It should be noted that the last labelled spermatozoa

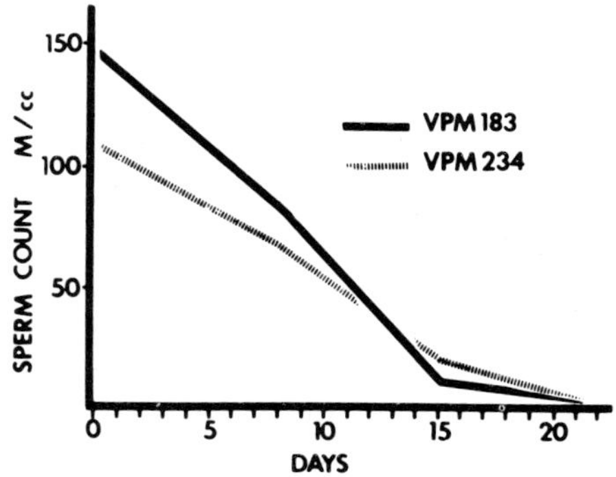

Fig. 5
The depletion of sperm count in two men irradiated with 100r (VPM 234) and 200r (VPM 183) of X-rays. Day 0 is day 46 after irradiation.

in the seminal fluid in the two subjects injected with thymidine-H^3 and then irradiated were observed on day 22 for each. There has been almost no individual variation that we can report in these men. All showed labelling on day 0, 1 or the first day a sample was available (Fig. 3), and all were depleted of sperm following X-ray irradiation by 21 ± 1 days.

Discussion

The precision of this means of determining the most rapid, the modal and the maximal transit time of spermatozoa traveling through the ductular system depends not only upon the data and their analyses as presented in the results, but depends upon the assumption that day 46 of spermatogenesis (the time of preleptotene maturation into spermatozoa) is also really day zero of transit time.

Examination of testicular biopsies (untreated controls, X-ray irradiated and thymidine-H^3 labelled) reveals that no spermatozoa are ever found in Cell Associations I and III which precede and follow Cell Association II [10]. Thus all spermatozoa are released from the Sertoli-cell cytoplasm during Cell Association II, which has a duration of 2.7 days [11]. It is not apparent from examining Cell Association II at what exact time within the Cell Association release occurs, but it is hypothesized that the majority are released on day 46. Neither the beginning nor the end of release has been further determined.

It has been assumed, for the purpose of clarity and practicability in this dissertation, that the release occurs only during a one-day period. We already know that release is not an instantaneous or simultaneous process. This is demonstrated in Fig. 1 and 2 where spermatozoa still held within the Sertoli-cell cytoplasm are associated with the residual bodies of spermatozoa already released.

It may be that some spermatozoa are in fact released before day zero. This cannot occur any sooner than one day before day zero and probably involves only a minority of the spermatozoa. Hence the early release will have no bearing on the modal or maximal transit time but may cause an error in the most rapid transit time, so that it becomes one day more than assumed in the results.

Czerniak [4] used tracer doses of radioactive phosphorus intravenously to try to determine the timing of human spermatogenesis and spermatozoon travel time. Unfortunately his measurements were not histological in regard to the timing of spermatogenesis. He has subsequently been proven incorrect by the thymidine-H^3 work reported by *Heller & Clermont* [9, 10]. The combination of the times that he gives for spermatogenesis and transport (2 weeks) probably reflects transport time alone. Although it is known that ^{32}P can label spermatogonia [5], *Czerniak* did not continue his experiments for a long enough time to see the excretion of cells labelled when they were spermatogonia or even spermatocytes.

The use of tritiated thymidine as a tracer to measure transit time is a recently developed technique. *Swierstra & Foote* [21], *Hunt & Foote* [13], *Amann, Koefoed-Johnsen & Levi* [1] and *Orgebin-Crist* [20] have studied the transit time of rabbit spermatozoa using thymidine-H^3. Their results set the travel time for the rabbit at 4–10 days. The travel time for bull spermatozoa has also been determined (6–11 days) using thymidine-H^3 [14, 19].

Only one study has been done utilizing X-ray irradiation to determine transit time. *Oakberg & DiMinno* [18] determined the average transport time in the mouse to be 8 days using this technique.

The combination of X-ray irradiation and thymidine-H^3 labelling has never been used before. We feel that it is a particularly useful combination to bracket the travel time with precision.

Summary and conclusion

The duration of transit of human spermatozoa through the ductular system, beginning with the release from the Sertoli-cell cytoplasm to their appearance in the ejaculate, has been determined using two methods of dating. These two methods were mutually confirmatory. The shortest transit time was one day, the median transit time 12 days and the maximal transit time 21 days.

Tritiated thymidine labelled sperm are released from the Sertoli-cell cytoplasm into the seminiferous tubule lumen 46 days after intratesticular injection of the isotope. A few labelled sperm appear in the ejaculate on day 46. Maximal labelling is observed between day 54 and day 61 following isotope injection. Thus the most rapid transit time is one day or less and the median and the modal transit time is 12 days as determined by thymidine-H^3 labelling.

Applying 100r or 300r directly to the testes alone caused immediate spermatogonial degeneration while allowing normal maturation of all more mature cells of the germinal series. The last of these normal cells were released from the Sertoli-cell cytoplasm 46 days following irradiation. The ejaculated seminal fluid was essentially azoospermic by day 67 after irradiation. Therefore the maximal transit time is 21 days ± 1.

Acknowledgement: The authors wish to acknowledge the statistical help of Dr. *Kathleen O'Keefe* in the evaluation of the data.

This investigation has been made possible through grants-in-aid from the Atomic Energy Commission (AT [45–1] - 1780) and the National Institutes of Health (HD00804).

References

[1] *Aman, R. P., Koefoed-Johnsen, H. H. & Levi, H.,* Excretion pattern of labelled spermatozoa and the timing of spermatozoa formation and epididymal transit in rabbits injected with thymidine-H^3. J. Reprod. Fertil., **10,** 169–183, 1965.

[2] *Brown, R. L.,* Spermia transport in man. J. Urol., **50,** 786–788, 1943.

[3] *Casarett, G. W.,* One solution stain for spermatozoa. Stain Techn., **28,** 125–127, 1953.

[4] *Czerniak, P.,* Duration of spermatogenesis in man as studied by P^{32}. Sym. on the Use of Radioisotopes in Animal Biology and the Medical Sciences, Academic Press, pp. 305–320, 1961.

[5] *Dawson, R. M. C.,* The labelling of ram semen *in vivo* with radioactive phosphate and (carboxy C^{14}) stearic acid. Biochem. J., **68,** 512–519, 1958.

[6] *Edwards, J.,* Effect of unilateral castration on spermatogenesis. Proc. Roy. Soc. London, **128,** 407–421, 1940.

[7] *Freund, M.,* Semen output and daily sperm production in man. J. Reprod. Fertil., **6,** 269–286, 1963.

[8] *Freund, M. & Davis, J. E.,* Disappearance rate of spermatozoa from the ejaculate following vasectomy. Fertil. Steril., **20,** 163–170, 1969.

[9] *Heller, C. G. & Clermont, Y.,* Spermatogenesis in man: an estimate of its duration. Science, **140,** 184–186, 1963.

[10] *Heller, C. G. & Clermont, Y.,* Kinetics of the germinal epithelium in man. Rec. Prog. in Hormone Res., **20,** 545–575, 1964.

[11] *Heller, C. G., Heller, G. V. & Rowley, M. J.,* Human spermatogenesis: an estimate of the duration of each cell association and of each cell type. III Internat. Cong. of Endo., Internat. Cong. Series, 1969 (in press).

[12] *Heller, C. G., Wooten, P., Rowley, M. J., Lalli, M. F. & Brusca, D. R.,* Action of radiation upon human spermatogenesis. Excerpta Medica Found., Proc. of VI Internat. Cong. of Endocrinol., pp. 172–174, 1966.

[13] *Hunt, W. L. & Foote, R. H.,* Efficiency of liquid scintillation counting and autoradiography for detecting tritium in spermatozoa. Rad. Res., **31,** 63–74, 1967.

[14] *Koefoed-Johnsen, H. H.,* Influence of ejaculation frequency on the time required for sperm formation and epididymal passage in the bull. Nature, London, **185,** 49–50, 1960.

[15] *MacLeod, J.,* Current reviews: human semen. Fertil. Steril. **7,** 368–386, 1956.

[16] *MacLeod, J. & Gold, R. Z.,* The kinetics of human spermatogenesis as revealed by changes in the ejaculate. Biol. of the Testes, N. Y. Acad. Sci., **55,** 707–725, 1952.

[17] *Needham, G.,* The Practical Use of the Microscope, Charles C. Thomas, 1958.

[18] *Oakberg, E. F. & DiMinno, R. L.,* X-ray sensitivity of primary spermatocytes of the mouse. Int. J. Radiat. Biol., **2,** 196–209, 1960.

[19] *Orgebin-Crist, M. C.,* Recherches expérimentales sur la duree de passage des spermatozoides dans l'épididyme du taureau. Ann. Biol. anim. Bioch. Biophys., **2,** 51–108, 1962.

[20] *Orgebin-Crist, M. C.,* Passage of spermatozoa labelled with thymidine-H^3 through the ductus epididymus of the rabbit. J. Reprod. Fertil., **10,** 241–251, 1965.

[21] *Swierstra, E. E. & Foote, R. H.,* Duration of spermatogenesis and spermatozoon transport in the rabbit based on cytological changes, DNA synthesis and labelling with tritiated thymidine. Am. J. Anat., **116,** 401–411, 1965.

Discussion

J. M. Bedford: Have you any evidence at all relating to the effect of frequency of ejaculation upon this movement?

C. G. Heller: Nothing that you would not already know: the more frequent the ejaculation, the lower the total count.

J. M. Bedford: I just wondered whether there is any effect of frequency of ejaculation on the speed of transport?

C. G. Heller: That would be very interesting, but you recognize, to conduct such experiments is very difficult. One cannot have all these refinements.

R. J. Ericsson: One comment on this – it has been shown in the rat that transport time changes in mated versus non-mated rats. The difference is a matter of days. 12 to 15 in the non-mated and 7 to 10 in the mated.

J. M. Bedford: In fact Orgebin-Crist has shown in the bull that speed of transport changes a little, by about a 20 %, with frequent ejaculation in the bull.

C. G. Heller: Yes, they have done a nice piece of work in the bull.

R. G. Edwards: I am actually surprised to hear Dr. *Heller's* results on the transport to the vas after one day from shedding. In animals, the period of transport is very constant and the peak of labelled spermatozoa were ejaculated within a very particular time. Now in the mouse and the rat, which I have worked with, it was 7 or 14 days, but we never got the kind of variation that you have reported here.

C. G. Heller: Yes, but this is a minimal number, this is not the maximum.

R. G. Edwards: But it still means you are getting some mixing.

C. G. Heller: Yes, we are getting a great deal.

R. G. Edwards: Some must be moving through to the vas very quickly. Is there any reason why the human should differ from animals in this respect?

C. G. Heller: I see no reason, no.

Un known speaker: Has age in the human any influence, for instance in puberty, on the transit time? Do you have any data concerning this?

C. G. Heller: No, our men were very close in age; I think the youngest was about 28, the oldest about 38. We have no further information.

Un known speaker: What do you think?

C. G. Heller: You remember my lecture began by criticizing people that had no data? I have no data.

M. C. Chang: How many days from the testis to the tail of the epididymis?

C. G. Heller: I can only make a very vague guess, and the guess would be based on the exhaustion following vasectomy in men, this is computed at 14 days. If this is correct, then it would take roughly 7 days for the normal sperm or average sperm to travel through the epididymis. But this is a guess on my part.

R. J. Ericsson: As I interpret Dr. *Heller,* he is saying that as early as one day, he does find that he gets labelled sperm from the testis through the epididymis into the ejaculate: as early as one day and as late as 21 days – an average of approximately 12. I donot think this gives you any information on transport time. All you are seeing is a mixing – some very early and some very late.

R. G. Edwards: These early sperms – were they labelled in DNA? Did you do controls to see that the DNA of the sperms was labelled, or whether a different pattern of labelling existed in those ejaculated after one day compared to those ejaculated after 12 to 21 days? Were you sure that you were labelling the DNA in every case?

C. G. Heller: We regarded a spermatozoon as having been labelled if it had at least 4 granules directly over the nucleus. If it had 3 or less we didnot count it.

R. G. Edwards: In that case, you actually have statistical distribution of the number of grains in your one-day sample and in your twelve-day sample. You now have a graincount for the twelve-day group, compared to the one-day group.

C. G. Heller: We spent months doing grain-counts, and they yielded no useful information.

R. G. Edwards: Would you say that the sperms on day one were as heavily-labelled, and in exactly the same place, as those found on day twelve.

C. G. Heller: Yes.

Mechanisms Concerned in Ovum Transport

K. W. Humphrey

Department of Veterinary Physiology, University of Sydney, Sydney, Australia

Summary: The forces involved in ovum transport in the oviduct are:

1) Contraction of the musculature.
2) Ciliary action.
3) Oviducal fluid which suspends the ova and acts as a stimulus for peristaltic contractions.

The ampulla-isthmus and isthmic-uterine junctions are most important in regulating the passage of ova. Failure of either of the sphincters to close normally results in very rapid transport of ova to the uterus. The oviducal musculature, particularly that of the sphincter has a rich non-adrenergic nerve supply, but little is known of the importance of the nervous control of the tube. A β-mimetic drug C_c-25 reduces tubal opening pressure in women and may correct infertility due to tubal spasm.

The oviducal musculature is the most important factor in ovum transport – estrogens cause closure of the ampulla-isthmus and perhaps the isthmic-uterine junction, and increase the contractility of the muscular walls. Exogenous estrogens cause both tube-locking in the ampulla and accelerated transport to the uterus depending on the species, the time of treatment and perhaps the dose and potency of the estrogen.

Normal tubal transport depends on the balance between estrogen and progesterone. Pregnancy can be prevented by excessive doses of either hormone or perhaps by anti-estrogens or anti-progestational agents.

Disturbances of the mechanisms that move ova from the follicles to the site of implantation are the major causes of early embryonic mortality in mammalian species. We would like to know more of these mechanisms and derangements so that we can minimize infertility at this stage and, conversely, to obtain contraception.

I would like to describe my observations on ovum transport in the mouse oviduct *in vitro* and give a review of tubal transport in other species.

My technique is similar to that of *Burdick, Whitney* and *Emerson* [1] – the excised oviduct was placed in a cavity slide in buffer solution kept at 37 °C. The tube was straightened by cutting the mesosalpinx. The movement of ova was readily seen with a binocular microscope and sub-stage lighting; it continued for up to 30 minutes, but stopped at room temperature or after squashing of the tube.

Manuscript received: 16 May 1969

Fig. 1 shows that the anatomy is basically similar to that in other species – the ampulla has a wide lumen and is thin-walled, and communicates with the peritoneal cavity by the infundibulum surrounded by the ciliated fimbriae. In the mouse, the ampulla forms the first 2 of 8 loops. Loop 1 is not distensible while loop 2 distends greatly with fluid at estrus. The ampulla-isthmus junction is obvious in the mouse, rat, rabbit and most other species.

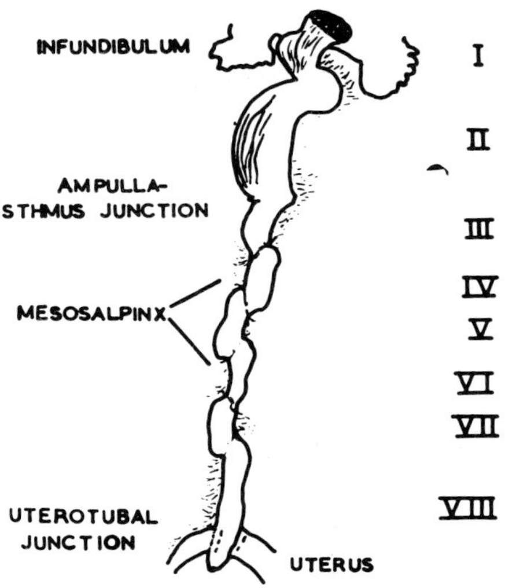

Fig. 1
Diagram of mouse oviduct.

The isthmus has a narrow lumen in all species except the opossum [2]. The walls have well developed musculature which increases towards the uterus. In the mouse, loops 2 to 7 are greatly coiled – ova are readily seen following straightening. Loop 8 is slightly arched; it terminates as the intramural section and intra-uterine papilla to form the isthmus-uterine junction. Ova in this section are only visible after compression under a cover-slip. All photos shown are of compressed, straightened oviducts.

The loops are defined by the attachment of the mesosalpinx, which spirals around the straightened oviduct and crosses over or under the tube at the end of each loop. This spiralling is obviously an artifact of straightening – it is a constant feature of the preparation and reflects some feature of the anatomy of the tube.

Fig. 2 and 3 show that ova are readily visible in these preparations; they can be counted, and the location and movement of the ova determined. Fig. 3 shows 2-cell ova in the ampulla.

Using this technique, it was possible to watch ovum transport in various parts of the tube on different days of pregnancy.

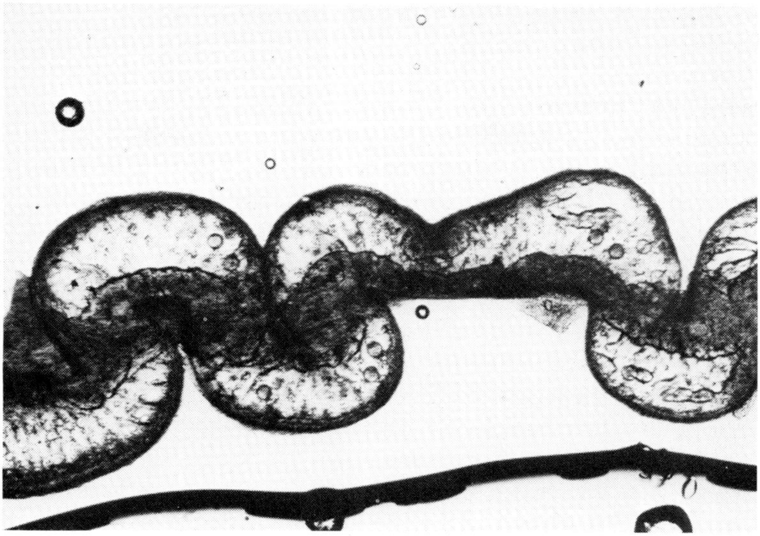

Fig. 2. Isthmus of the mouse oviduct, showing the spiral attachment of the mesosalpinx (X 33).

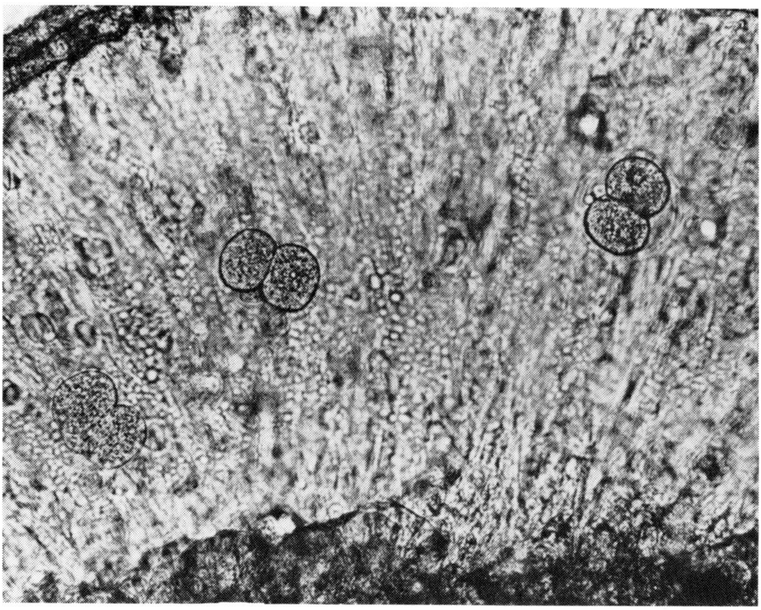

Fig. 3. Ampulla of the mouse oviduct on day 1, showing three 2-cell ova (X 83).

Briefly, the mechanisms of transport of ova through the oviducts are:
1) Muscular forces
2) Action of cilia
3) Fluid secreted by the oviduct.

The importance of the fluid in which ova are suspended cannot be overlooked. Fluid produced by secretory cells, principally in the ampulla, is the natural medium for capacitation of sperm, loss of the corona radiata, fertilization of the ova and cleavage of the egg, and, from my observations, the stimulus for peristaltic contraction of the oviducal musculature. The relative importance of these forces varies with different animals and in different sections of the tube at different times of pregnancy.

The term 'ovum reception' or 'pick-up' refers to the movement of newly ovulated eggs from the follicle through the infundibulum of the oviduct. The mechanism varies with the anatomy of the region. In the mouse, rat, hamster and vole, the ovary and fimbria of the infundibulum are completely enclosed by the ovarian bursa, with only a small opening to the peritoneal cavity. At estrus, under the influence of estrogens, there is increased secretion by the ampulla, and perhaps the bursa, and closure of the ampulla-isthmus junction, at least in the rat and mouse [3, 4], which causes marked distension with fluid of the bursa and the final loop of the ampulla. The cumulus cells surrounding the newly ovulated ova fuse together, and this large mass is moved through the infundibulum by the ciliary current working against the flow of fluid.

In other species – rabbits, guinea-pigs, sheep and primates – where the bursa is open, the ovary is approximated to the infundibulum by the contraction of the smooth muscle in the suspensory apparatus of the ovary and tube and engorgement and movement of the fimbriae [5]. Direct observations at laparotomy or culdoscopy in rabbits and primates [6–9] have shown that the fimbria become engorged and erected with blood and actively embrace the ovary, while there is contraction of the musculature of the meso-ovarium, meso-tubarium and tubo-ovarian ligaments. In the mouse and rat, similar muscular contractions cause churning movements of the fluid and ova in the bursa (*Humphrey*, unpublished observations). *Blandau* [9] used cinematography and recorded that in the rabbit the eggs adhere to the follicles for some time, until picked up by the sweeping action of the engorged fimbria. Again, the cumulus cells greatly increase the efficiency of ovum reception in both mono- and multi-ovular species. The movement of ova from the fimbria into the ampulla is apparently due to the action of cilia [3]. The importance of "suction" in the ampulla, due to muscular activity in the ampulla and mesosalpinx [10] or by rapid absorption of fluid by the lymphatics of the ampulla [11] has been largely discounted, following the findings of *Clewe* and *Mastroianni* [12] that ovum reception in rats and rabbits was not prevented by ligatures close to the infundibulum. In fact, *Strange* [5] has postulated that the vascular musculature at the base of the infundibulum in the human oviduct forms a functional sphincter.

Ciliated cells are very numerous on the folds of the fimbria and ampulla in the mouse, rat, rabbit and other species and produce a current that carries dye particles and ova in cumulus and fluid from the bursa into the ampulla [3, 4, 13]. In the tortoise and turtle, these appear to be both ascending and descending ciliary currents [3].

The efficiency of the cilia depends on the size and mass of the object to be moved [14]. This is confirmed by the observation that resin spheres were transported through the rabbit infundibulum only when the spheres were encased in cumulus cells [15].

The number and activity of the ciliated cells depends on the ovarian cycle, but these are not converted to secretory cells. In the rabbit, ciliary activity is increased 20 % at the time of ovulation [16] while estrogens accelerate ovum movement in the ampulla [17]. The number of cilia on the oviducal fimbriae of the Rhesus monkey reaches a maximum in the middle of the menstrual cycle, while deciliation occurs at the end of the cycle or following ovariectomy [18]. The ciliated cells in the ampulla do not degenerate, but do show cyclical changes in height. Deciliation does not occur in the rabbit following ovariectomy.

Movement of ova from the infundibulum to the ampulla-isthmus junction appears to be extremely rapid in all species studied, taking only 5 minutes in the rabbit [15, 19]. In his observations and cinematography, *Harper* found that in the first centimetre of the ampulla movement of ova in cumulus was regular, and due to ciliary action, while progression through the remainder of the ampulla was irregular, and coincided with definite segmental muscular contractions.

Similarly, in mice killed 1 hour after the expected time of ovulation, ova in cumulus were found in the dilated section of the ampulla, close to the ampulla-isthmus junction (*Humphrey,* unpublished observations). The cumulus cells had disappeared in 75 % of mice by 18 hours.

Although cilia on the folds of the ampulla could be seen beating, causing slight movement or vibration of the eggs in cumulus, rotation of the ova occurred only when the walls of the ampulla contracted [4].

Fertilization occurs in the final loop of the ampulla in nearly all species – shrews, tenrecs and ferrets are notable exceptions. The fertilized ova are retained by the ampulla-isthmus junction for some considerable time – 24 hours in the hamster and mouse [4, 20], 30 hours in the guinea-pig [2], 36–48 hours in the rabbit [22], over 36 hours in the pig [23, 24] and 36–48 hours in the ewe [25]. The ampulla-isthmus junction in the rat and mouse is formed by an external constriction of the wall with local proliferation of blood vessels and internal folds of mucous membrane [4] (see Fig. 4 and 5). *Greenwald* [22] states that, in the rabbit, no sphincter is visible histologically, but *Sjöberg* [26] described a sudden localized increase in adrenergic muscular innervation at the ampulla-isthmus junction in the prominent circular muscle layer. This also occurs in the cat and the human.

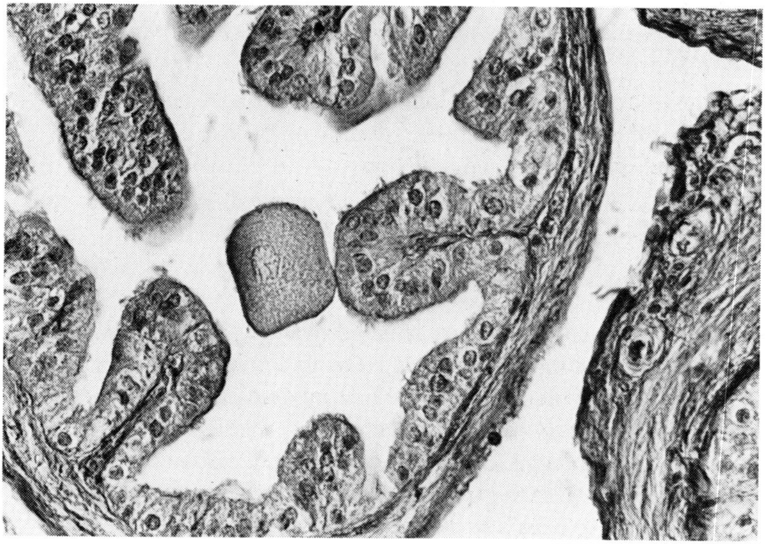

Fig. 4
Transverse section of the ampulla on day 1, showing prominent longitudinal folds bearing many ciliated cells (× 130).

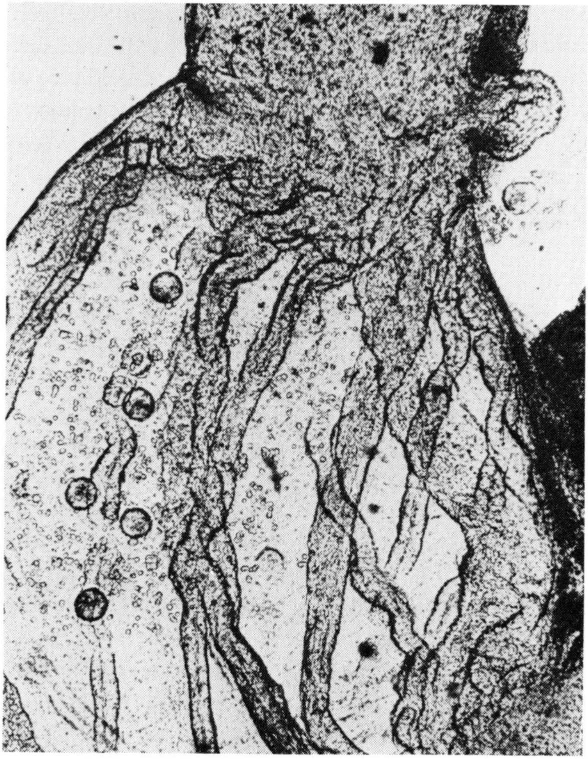

Fig. 5
Ampulla-isthmus junction on day 1, showing five 1-cell ova in loop 2 of the ampulla (× 33).

Ova then pass through the ampulla-isthmus junction, apparently due to relaxation of the ampulla-isthmus junction, and peristaltic contractions of the ampulla. Exogenous estrogens cause prolonged closure of the ampulla-isthmus junction in mice, rats, rabbits and guinea pigs. Progress through the isthmus is rapid in all species: roughly 12 hours in the rabbit, ewe and sow [22, 23, 24, 25]. Cilia are present in the isthmus in some species, but muscular contractions seem to be the most important factor in the progression of ova [4, 25, 27].

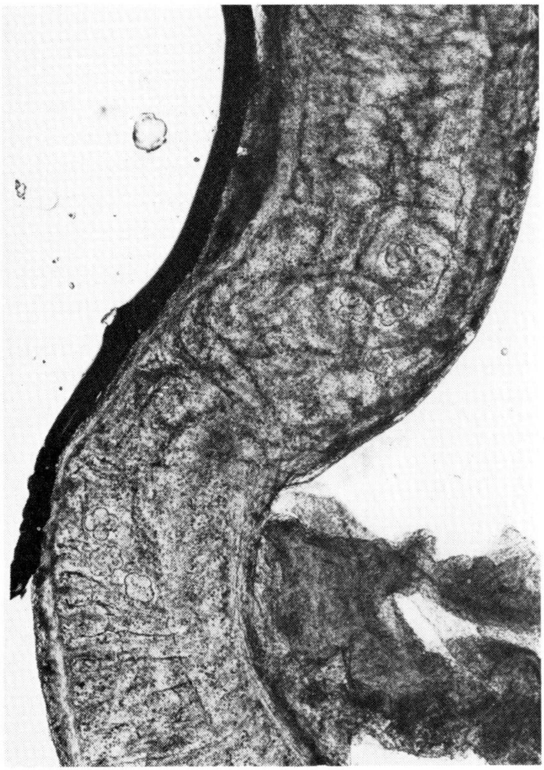

Fig. 6
Isthmus of the mouse oviduct (X 33).

The thick muscular wall of the isthmus is composed of two layers of spirally orientated fibres (Fig. 6) – a thin outer longitudinal layer, and a thick inner circular layer [27, 28, 29, 30]. In my observations, ova and fluid were present in only one or two dilated, adjoining loops of the isthmus, and were churned back and forth by segmental peristaltic-antiperistaltic contraction of the walls. All other loops were not dilated and were quiescent. Progression of ova to other loops was prevented by folds of mucous membrane at the end of the loop. Occasionally ova were forced into a quiescent loop, which then dilated and commenced segmental contractions. The existence, and the importance, of such folds in the coiled tube *in vivo* is not known. In the rabbit oviduct, there are kinks or elbows which are functional blocks to insufflation and perhaps to passage of ova [31].

Strong peristaltic contractions in the mouse oviduct were only found in those dilated loops containing ova, and in loop 3 near the ampulla-isthmus junction. The importance of this latter contraction is unknown – it may produce a blocking effect at the ampulla-isthmus junction; it may prevent adovarian movement of ova; or act as a pacemaker. Similar contractions occur in the rabbit, sheep and human oviduct *in vitro* [29, 32, 33, 34]. The productive contractions of the isthmus appear to be a typical smooth muscle response to distension similar to the small intestines – it continues in preparation *in vitro* with no nerve or blood supply. Similarly, *Black* and *Asdell* [27] attribute ovum transport in the rabbit isthmus to segmental contractions of the circular muscle layer. This muscle layer has abundant non-adrenergic nerve terminals [26, 34, 35], it is stimulated by nor-adrenalin, adrenalin or perivascular nerve stimulation [36, 37], and its contractility is increased by estrogens [29].

Thus the tone and the contractility of the oviduct in rabbits, sheep and the human depends on the hormonal background and the autonomic innervation [25, 26, 27, 29, 34, 36]. To-and-fro movement of ova certainly also occurs in the guinea pig, rabbit and sheep *in vitro* due to segmental contraction of the tubal musculature – possibly in these and other species, as in the mouse, the local distension of the tube by ova and fluid is an important stimulus for ovum transport.

The utero-tubal junction is a functional sphincter and delays passage of ova to the uterus for about 24 hours in the rabbit, pig and mouse [4]. In different species, the mechanism of retention varies with the anatomy (see Figs. 7–10) – in some species there are complex folds of the lining mucosa which form a valve (rabbit, guinea pig), or flexures in the lumen (sheep, cow, guinea pig), edema of the submucosa or folds of the intramural section (sheep, cattle, pigs, mice and rats), or a muscular sphincter with a vascular ring (primates) [2, 4, 30, 38, 39].

In the rabbit and mouse [4, 33], the section of the isthmus adjacent to the intramural section forms part of the sphincter mechanism, since ova do not enter this terminal section of the isthmus until the time of entry to the uterus. *Black* and *Asdell* [33] believe that rabbit ova stay in the isthmus until the utero-tubal junction opens and the ova are carried through in a wave of fluid following a peristaltic rush. Estrogens appear to be the dominant factor controlling the action of this sphincter, and cause closure of the utero-tubal junction in the ewe, rabbit, mouse and woman. In the sow, rabbit and also woman, using the Rubin test, the pressures needed to force open the utero-tubal junction are greatest at estrus or following estrogen treatment [2, 31, 33, 40]. These latter workers found that the fluctuations in pressure recorded during insufflation of the rabbit oviduct were caused by kinks in the oviduct.

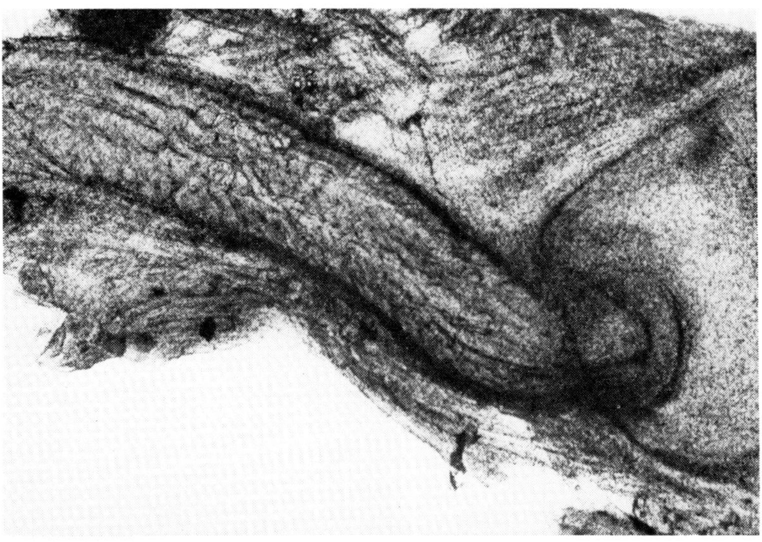

Fig. 7
Mouse utero-tubal junction. Several ova are in loop 8 of the isthmus (X 33).

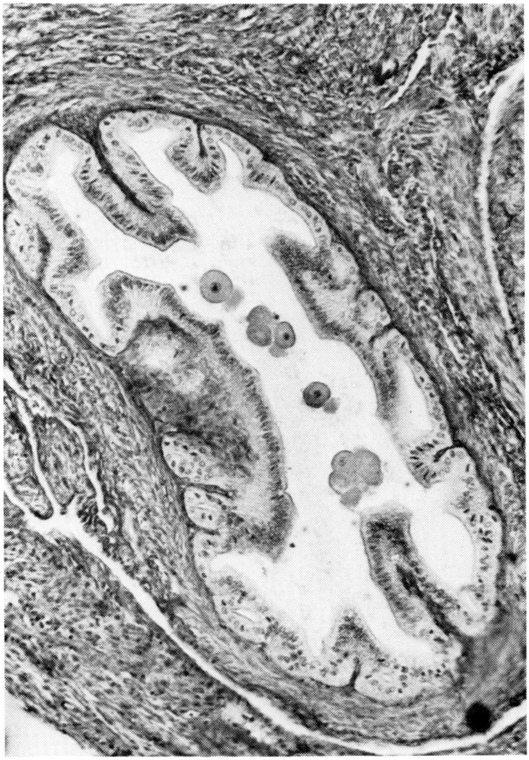

Fig. 8
Histology of the mouse utero-tubal junction (X 130).

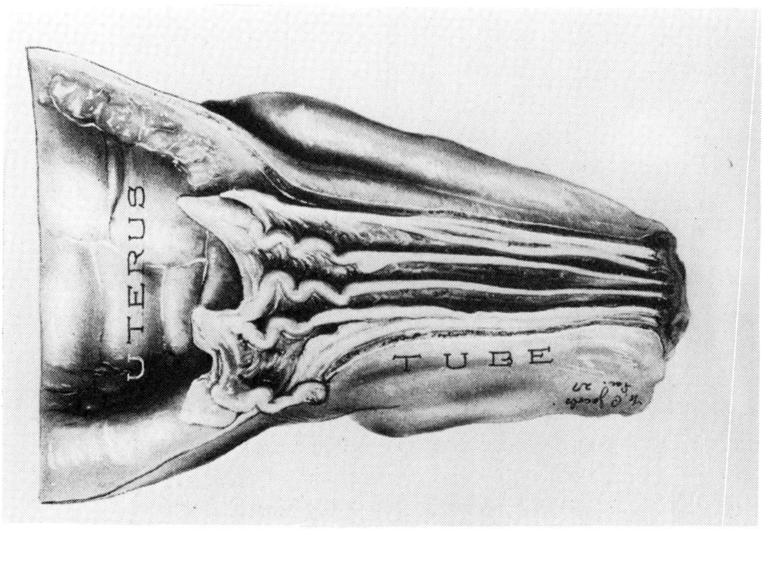

Fig. 9
Rabbit utero-tubal junction (after *Andersen* (1928)).

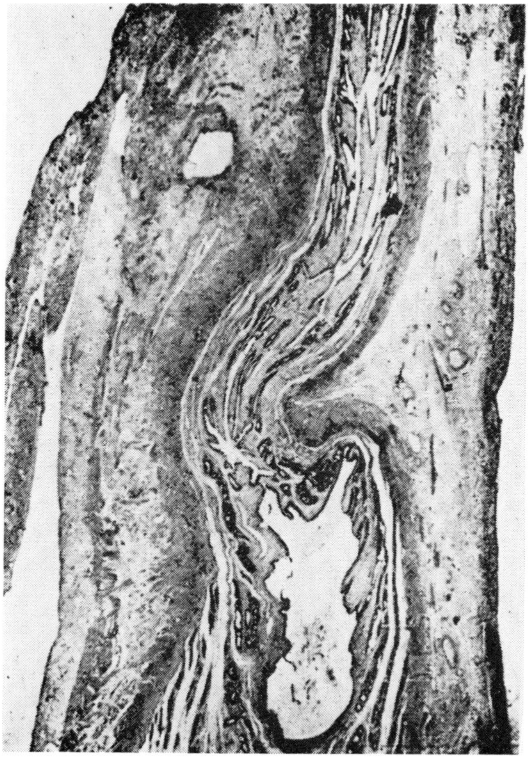

Fig. 10
Sheep utero-tubal junction (after *Edgar* & *Asdell* (1960)).

The role of the autonomic nervous system in the control of the utero-tubal junction is uncertain. There is a marked increase in the content of non-adrenergic nerve terminals in the isthmus and particularly at the utero-tubal junction in the rabbit and cat [26, 41], and the pressure of gas needed to open the utero-tubal junction and isthmus in women is reduced [42] by treatment with a β-mimetic drug C_c-25 (*p*-hydroxyphenyl-isopropylarterenol).

Stimulation of the hypogastric nerves causes closure of the isthmis lumen in the rabbit [26] and the uterus of estrogen-primed, ovariectomized rabbits; [43] – this is abolished by treatment with progesterone. *Greenwald* [29] and *Nakanishi* and *Wood* [44] believe that the entire isthmus contracts and acts as a sphincter in the rabbit and human female. However, there are distinct ampulla-isthmus and utero-tubal junctions in these species, and probably the isthmus is quiescent when ova are not present.

The nervous control of the oviducal musculature is gradually being elucidated. The parasympathetic supply is uncertain; there is no evidence of cholinergic ganglia in the human oviduct. Acetylcholine antagonizes the action of adrenaline on the rabbit oviduct and generally inhibits contractility. The sympathetic supply is partly derived from post-ganglionic fibres in hypogastric nerves and from lumbo-sacral ganglia and so-called "short" adrenergic fibres arising from the peripheral ganglia at the utero-vaginal junction [26, 45]. The number of adrenergic nerve terminals in the ampulla is much less than in the isthmus in the rabbit and human oviducts [41, 46]. The adrenergic terminals in the ampulla are almost all found around the blood vessels, but in the isthmus many terminate in the thick, well-developed layer of circular muscle fibres. The oviduct seems to have typical smooth muscle adrenergic innervation (α-adrenergic, stimulatory; β-adrenergic, inhibitory).

The ampulla-isthmus and isthmus-uterine junctions have particularly dense innervation. Non-adrenaline causes sustained contraction of the oviduct in man and the rabbit, and this effect was abolished by the α-blocking agent, phenoxybenzamine. The β-mimetic agent C_c-25, reduces tubal opening pressure in women, and may be of use if tubal spasm occurs and is the cause of infertility [42].

Secretions

This will be discussed by Dr. *Koester* and other speakers in this session.

Volumetric collection of oviduct fluid from the rabbit and sheep was performed by *Clewe* and *Mastroianni* [47, 48]. Estrogens increase fluid secretion, which is at a peak at estrus in the rabbit and sheep [48, 49]. Most of the secretory cells, and hence fluid production, are in the ampulla [48, 50]. As previously mentioned, the fluid is important in the capacitation of sperm (as outlined by other speakers yesterday), the dispersal of the cumulus and corona radiata cells [51, 52] fertilization *in vitro* [53] and presumably *in vivo* and possibly blastulation.

Mouse ova retained in the oviduct by estrogen injection show delayed development of the blastula [54]. *Kirby* [55, 56] showed that mouse ova transferred to extra-uterine sites develop embryonic tissue only if the ova are recovered from the uterus. *Daniel* and *Krishnan* [57] have isolated a uterine fluid which promotes blastocyst formation *in vitro* in the rabbit – this fluid may be formed in the oviduct.

The effect of seminal components on the oviduct cannot be overlooked. Prostaglandin E or F respectively inhibit and decrease tone of the rabbit oviduct, while Prostaglandin E increases contractility of the isthmus of the human oviduct [58]. The endocrine control of tubal transport is still imperfectly understood, but the papers by Drs. *Koester, Bennett* and *Blye* should shed some light on this problem. I would like to briefly summarize my results and interpretation of this problem.

The effects of ovarian hormones on tubal transport depend on:

1) Dose and nature of the hormone – potency, duration of action and route of administration;
2) Time of treatment – i.e., the location of the ova at this time.

The effects of estrogens on tubal transport are extremely variable, differing markedly in different animals, and even between oviducts in the same animal [54, 59]. The extreme variability often makes it difficult to generalize, particularly in regard to doses, especially when extremely high doses of a long-acting estrogen are used (e. g., by *Greenwald* [60] in the rabbit and mouse).

In ovariectomized rabbits and sheep, the oviducal musculature is atrophic and atonic [61, 25]. Estrogens increase the contractility of the uterus and oviduct in these and other species [17, 25, 59, 61].

Ova transferred to the oviducts of untreated, spayed rabbits are quickly lost to the uterus. Physiological doses of estradiol benzoate (1–4 μg) cause retention of some ova in the oviduct, and expulsion of uterine ova to the vagina. In mice, estradiol on day 1 causes tube-locking of most ova in the ampulla until day 4; treatment on day 2 or 3, when most are near the uterotubal junction, reduces the recovery of ova by more than 50 %. Ligation of various parts of the tract and ovum transfers have shown that this low recovery is the result of accelerated movement to the uterus and vagina.

Thus estrogens increase oviducal contractility resulting in accelerated transport; and cause closure of the ampulla-isthmus junction and to a lesser degree the utero-tubal junction resulting in tube-locking of ova in the ampulla at the time of treatment [54, 62, 64, 63].

For disbelievers in tube-locking, Fig. 11 shows well-developed blastocysts in the ampulla on day 4, following estradiol treatment on days 1–3.

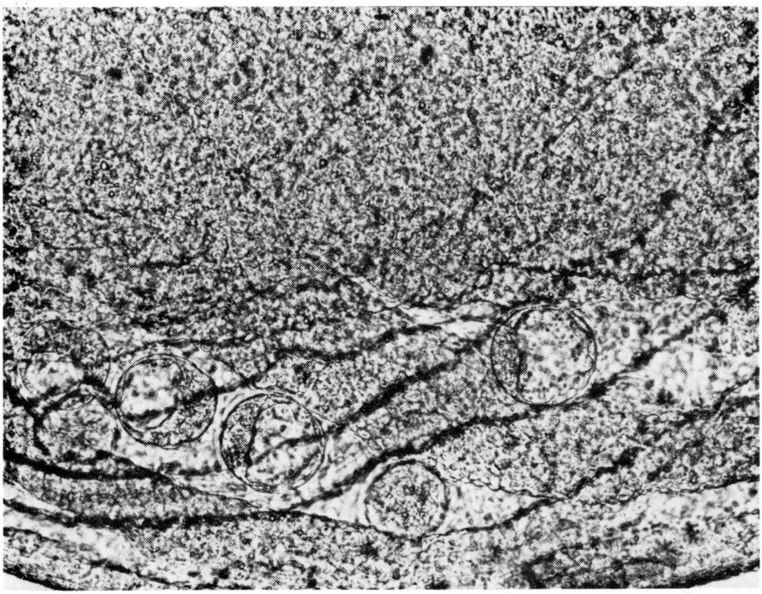

Fig. 11
Ampulla of oviduct of mouse killed on day 4 after treatment with oestradiol on days 1–3. Large blastocysts are seen in the ampulla (X 83).

Progesterone after ovulation has little or no effect on tubal transport. However, progesterone before ovulation causes marked acceleration of the ova in rabbits [65], and interferes with sperm transport and capacitation, as pointed out by the speakers yesterday.

The ovarian hormones modify tubal transport by an effect on the oviducal and uterine musculature.

In conclusion, much more is now known of ovarian control of oviducal secretions and of autonomic influences on the oviduct *in vitro.* Little is known of the importance of the autonomic nervous system and the ovarian hormones. The most important mechanism affecting ovum transport in the oviduct and uterus is the musculature. Prevention of tubal spasm in women may correct somes times of infertility, while derangement of ovum transport is a useful contraceptive technique. Accelerated ovum transport is preferable to tube-locking as a means of contraception, since ectopic pregnancies may occur with the latter, or the ova may escape from the oviduct and implant belatedly in the uterus.

References

[1] *Burdick, H. W., Whitney, R. & Emerson, B. B.* (1942). Endocrinology, **31**, 10.

[2] *Andersen, D. H.* (1928). Amer. J. Anat. **42**, 155.

[3] *Blandau, R. J.* (1961). In "Sex and Internal Secretions", vol. II, p. 797–882. Ed. *W. C. Young:* Williams & Wilkins Co., Baltimore.

[4] *Humphrey, K. W.* (1968a). J. Endocrin. **40**, 267.

[5] *Strange, H. H.* (1952). Arch. für Gynäk. **182**, 77.

[6] *Westman, A.* (1937). J. Obstets. Gynec. Brit. Emp. **44**, 821.

[7] *Decker, A.* (1951). Fert. & Steril. **2**, 253.

[8] *Doyle, J. B.* (1956). Obstets. Gynec. **8**, 686.

[9] *Blandau, R. J.* (1967). Procs. Symp. Oviduct, Pullman University, Washington. Eds. *E. S. E. Hafez & R. J. Blandau.*

[10] *Westman, A.* (1952). Proc. Conf. Studies on Testis, Ovary, Eggs and Sperm.Ed. *E. T. Engle;* p. 163–175. C.C. Thomas, Springfield, Illinois.

[11] *Dubreuil, G.* (1944–1945). Gynec. et Obstets. **44**, 397.

[12] *Clewe, T. H. & Mastroianni, L. Jr.,* (1958). Fert. & Steril. **9**, 13.

[13] *Alden, R. H.* (1942). Anat. Rec. **84**, 137.

[14] *Kneer, M. & Cless, H.* (1951). Gebwitsh. u. Fraucnh. **11**, 233.

[15] *Harper, M. J. K.* (1961). Proc. IVth Internat. Congr. Anim. Reprod., The Hague, p. 375.

[16] *Borell, U., Nilsson, O. & Westman, A.* (1957). Acta obst. et gynec. Scand. **36**, 22.

[17] *Harper, M. J. K.* (1966). Endocrinology, **78**, 568.

[18] *Brenner, R. M.* (1967). J. Cell Biol. **35**, Abstr. 16A.

[19] *Harper, M. J. K.* (1965). Endocrinology, **77**, 114.

[20] *Strauss, F.* (1956). J. Embryol. exp. Morph. **4**, 42.

[21] *Squier, R. R.* (1932). Contrib. Embryol. Carnegie Inst., Washington, **23**, 223.

[22] *Greenwald, G. S.* (1961). Fert. & Steril. **12**, 80.

[23] *Andersen, D. H.* (1927). Amer. J. Physiol. **82**, 557.

[24] *Oxenreider, S. L. & Day, B. N.* (1965). J. Anim. Sci. **24**, 413.

[25] *Wintenberger-Torres, S.* (1961). Ann. Biol. Anim. Biochem. Biophys. **1**, 121.

[26] *Sjöberg, N. O.* (1967). Acta physiol. Scand., Suppl. 305.

[27] *Black, D. L. & Asdell, S. A.* (1958). Amer. J. Physiol. **192**, 63.

[28] *Mastroianni, L. Jr.,* (1962). Clin. Obstes. Gynec. **5**, 781.

[29] *Greenwald, G. S.* (1963). Fert. & Steril. **14**, 666.

[30] *David, A.* (1968). Amer. J. Obstets. Gynec. **101**, 417.

[31] *Stavorski, J. & Hartman, C.* (1958). Obstets. Gynec. **11**, 622.

[32] *Breipohl, W.* (1938). Ztschr. für Geburtsh. u. Gynak. **118**, 1.

[33] *Black, D. L. & Asdell, S. A.* (1959). Amer. J. Physiol. **197**, 1275.

[34] *Brundin, J.* (1964). Acta physiol. Scand. **61**, suppl. 219.

[35] *Brundin, J.* (1965). Acta physiol. Scand. **66**, suppl. 259.

[36] *Nakanishi, H., Wansbrough, H. & Wood, C.* (1967). Amer. J. Physiol. **213**, 613.

[37] *Longley, W. J., Black. D. L. & Currie, G. N.* (1968). J. Reprod. Fertil. **17**, 95.
[38] *Vasen. L. C.* (1959). Internat. J. Fertil. **4**, 309.
[39] *Edgar, D. G. & Asdell, S. A.* (1962). J. Endocrin. **21**, 315.
[40] *Bernstein, F. & Feresten, M.* (1940). Endocrinology **26**, 946.
[41] *Brundin, J. & Wirsen, C.* (1964a). Acta physiol. Scand. **61**, 203.
[42] *Leeton, J. F., Wansbrough, H., Walker, D. W., Wood, C. & Nakanishi, H.* (1968). J. Obstets. Gynec. Brit. Comm. **75**, 577.
[43] *Miller, M. D. & Marshall, J.* (1965). Amer. J. Physiol. **209**, 859.
[44] *Nakanishi, H. & Wood, C.* (1968). J. Reprod. Fertil. **16**, 21.
[45] *Rosengren, E. & Sjöberg, N. O.* (1968). Acta physiol. Scand. **72.**
[46] *Brundin, J. & Wirsen, C.* (1964b). Acta physiol. Scand. **61**, 505.
[47] *Clewe, T. H. & Mastroianni, L., Jr.* (1960). J. Reprod. Fertil. **1**, 46.
[48] *Restall, B. J.* (1966). Aust. J. Biol. Sci. **19**, 187.
[49] *Mastroianni, L., Jr. , Beer, F., Shah, U. & Clewe, T. H.* (1961). Endocrinology, **68**, 92.
[50] *Hadek, R. D.* (1955). Anat. Rec. **121**, 187.
[51] *Mansani, F. E. & Mazzella, A.* (1955). Rivista di Obstet. e Gin. **10**, 653.
[52] *Zamboni, L., Hongsanand, C. & Mastroianni, L., Jr.* (1965). Fert. & Steril. **16**, 177.
[53] *Suzuki, S. & Mastroianni, L., Jr.* (1965). Amer. J. Obstets.Gnyec. **93**, 465.
[54] *Humphrey, K. W.* (1968b). J. Endocrin. **42**, 17.
[55] *Kirby, D. R. S.* (1963). J. Reprod. Fertil. **5**, 1.
[56] *Kirby, D. R. S.* (1965). J. Reprod. Fertil. **10**, 403.
[57] *Daniel, J. C., Jr. & Krishnan, R. S.* (1967). J. Cell. Biol. **35**, 30 A.
[58] *Sandberg, F., Ingel-Sunberg, A. & Ryden, G.* (1963). Acta obst. et gynec. Scand. **42**, 269.
[59] *Noyes, R. W., Adams, C. E. & Walton, A.* (1959). J. Endocrin. **18**, 108.
[60] *Greenwald, G. S.* (1967). Anat. Rec. **157**, 163.
[61] *Noyes, R. W.* (1959). Fert. & Steril. **10**, 480.
[62] *Humphrey, K. W. & Martin, L.* (1968a). J. Reprod. Fertil. **15**, 191.
[63] *Humphrey, K. W. & Martin, L.* (1968b). Aust. J. Biol. Sci., **21**, 1239.
[64] *Yanagimachi, R. & Sato, A.* (1968). Fert. & Steril. **19**, 787.
[65] *Chang, M. C.* (1966). Endocrinology **79**, 939.

Discussion

R. I. Dorfman: You alluded in your talk to the possibility of different kinds of effects due to different kinds of estrogens. I wonder if you would indicate the different types of estrogens that you mean.

K. W. Humphrey: I found with estradiol that a single dose on day one caused tube-locking; treatment on day two caused acceleration. With treatment on days one to three you get a combination of both effects. But treatment with anti-estrogens and weak estrogens such as dimethylstilbestrol (DMS), MRL 37, MER 25, U11, 100 A & U11, 555 A; most of these compounds, given on days one to three of pregnancy, cause "tube-locking" predominantly without any acceleration at all. And again, this "tube-locking" is in the ampulla.

R. I. Dorfman: What results do you get with estriol?

K. W. Humphrey: We have not looked at estrone or estriol effects in the mouse.

R. I. Dorfman: Studies with estriol might be particularly interesting because it is a weak estrogen and naturally occurring.

K. W. Humphrey: We have thought of this. We have looked at the effect of estriol on delayed implantation but, unfortunately, not on tubal transport.

A. Psychoyos: I have some information concerning the effect of estriol on the tubal transport and also its action in inducing implantation. The striking thing is that the same dose (30 to 40 μg causing a rapid expulsion of the eggs from the oviduct when given on the first or the second day of pregnancy, is the dose needed to induce implantation in animals in delay.

For estradiol only 0.05 μg – 0.1 μg induce implantation whereas are needed 5–10 μg to induce expulsion. So, though there is a big difference between these two estrogens as to their potency on implantation there is almost none for the tubal transport. I think this is important, indicating for these estrogens the existence of at least two different sites of action.

K. W. Humphrey: What species was that please? The rat. That is interesting, because estriol is much less potent than estradiol in other parameters of reproduction, of course.

D. R. S. Kirby: Dr. *Humphrey,* there is in reviews of this subject an oft-quoted reference to *Alden* (J. exp. Zool. **90,** 159–169, 1942), where it is stated he ovariectomized the pregnant rat and gave it progesterone. This it was claimed accelerated the transport of the eggs so that they arrived in the uterus earlier than usual. In fact, if you look at this reference *Alden* did not administer progesterone – it is a mistake and has been perpetuated by review after review. We looked at it in the mouse, and found that if you ovariectomized the mouse early in repgnancy, say on day 1 or day 2 after the sperm plug, and give daily progesterone (2 mg/day) it delays the rate of tubal transport by a day. Instead of arriving early on day 3, the blastocysts do not arrive until early on day 4. If you give progesterone and estrogen, say 2 mg of progesterone and 0.04 μg of estradiol, at the appropriate times, the rate of egg transport through the oviducts is normal. So it is not the trauma of ovariectomy that causes delay but the balance between progesterone and estrogen.

K. W. Humphrey: This is most interesting. You certainly do require a balance of progesterone and estrogen for reasonably normal tubal transport in ovariectomized rabbits and mice, from your work. Estrogens have the most dramatic effects on tubal transport and progesterone tends to be overlooked, but certainly a balance is required. Now, I ovariectomized mice on day two, that is the second day after finding the plug, and I find the recovery of eggs (giving the animals progesterone) on day four was extremely low. I never followed this up, but this would tie in with your results that they probably are still in the oviduct at this time.

Egg-Uterus Interrelationships

C. E. Adams

Agricultural Research Council, Unit of Reproductive Physiology and Biochemistry,
University of Cambridge, England

Summary: The first part of the paper reviews egg-uterus interrelationships with particular reference to the activity of the corpus luteum. Then follows an account of synchronisation requirements, including recent experiments involving asynchronous egg transfer either to estrogen or progesterone dominated uteri.

In estrous does eggs that were transferred to the uterus were soon expelled into the vagina, and the proportion recovered from the uterus fell from 69 % to 0 % from $\frac{1}{2}$ to 4 hours after transfer. The fact that 17–36 % of the eggs were unaccounted for could mean that some also degenerated *in utero*. The significance of myometrial activity is discussed relative to fertility.

In luteal phase uteri (recipients 9 days pseudopregnant) 60 hr. morula quickly showed degenerative changes so that none underwent blastulation. Under these conditions it appeared that the uterus was hostile rather than merely lacking in factors conducive to embryonic development.

In the majority of Eutherian mammals so far investigated, the eggs normally enter the uterus 3 to 4 days after ovulation, by which time they vary in development from the 4 cell stage in the pig to the early blastocyst in the rabbit. It is surprising to find that according to Marshall's Physiology of Reproduction and 'Sex and Internal Secretions', the above information is available for not more than ten species. The timing of the egg's entry into the uterus is quite critical for its future wellbeing. This fact has been clearly established by means of the egg transfer technique, which permits one to vary at will the stage of egg development relative to the uterine environment.

During the egg's sojourn in the oviduct, the uterus under the influence of progesterone undergoes profound changes, which affect not only the endometrial lining with which the eggs come into direct contact but also the outer muscular layer, the myometrium. *Corner* [13] already expressed the opinion that one of the most important functions of the oviduct is to hold back the eggs until the uterus is ready for them. Nowadays the existence of a tubal blocking mechanism (at the level of

Manuscript received: 13 March 1969

the ampullary-isthmic junction rather than at the utero-tubal junction) is well established; *Greenwald* [14] regards it as the most important factor in regulating egg transport. Certainly, if no such blocking mechanism existed and the rapidity of transport characteristic of the ampulla [15] was maintained throughout the isthmus the eggs could reach the uterus within the first hour of ovulation. The eggs would then perish not only because the endometrium was unprepared to receive them but also on account of myometrial activity. Even though *Corner's* concept appears to have been derived primarily from a consideration of endometrial function, it is still equally valid.

The present discussion will be limited to the preimplantation period. The first part will deal with observations on egg-uterine interactions and the corpus luteum, and the shedding of the zona pellucida. The second part is concerned with the fate of eggs when they are transferred out of synchrony with the uterine environment. The rabbit was chosen as the experimental animal for the following reasons:

1) No seasonal anoestrum.
2) Ovulation is easily timed; it can be induced either by mating or by the injection of luteinising hormone.
3) All preimplantation stages are well documented.
4) The genital tract is relatively large; the various regions are easily distinguishable; the uterus is bicornuate and intrauterine migration of ova cannot occur.
5) Egg transfer and recovery are easy to perform and efficient.
6) Ligation of any part of the tract is easy to perform.

The effect on the corpus luteum

Perhaps the most obvious and significant effect of the presence of an embryo within the uterus is the prolongation of the life span of the corpus luteum and, in some species, e.g. guinea pig, rat and mouse, though not in others, e.g. pig and sheep, the stimulation of luteal growth and secretory function. The effect of the embryo on corpus luteum function has been comprehensively reviewed recently [20].

Concerning the influence of the embryo on the endometrium, there are two questions of obvious relevance to the present discussion. First, the earliest time of action and second the mode of action. On the former aspect, information is beginning to accumulate both on the basis of analyses of the secretory function of the corpora lutea and on egg transfer, whereby the embryo is transferred at progressively later stages. In the rat, Moor cites evidence, based on secretory function, that the embryo first affects the corpora lutea by the 4th or 5th days of pregnancy, and in the pig before the 14th day, and possibly as early as the 7th day of pregnancy. It is notable that this action is taking place whilst the embryo is still lying free in the uterus, i.e. before attachment or implantation. In the rabbit, *Beier* [6] found that the protein concentration of endometrial secretion on the 6th day of pregnancy was much higher than in the corresponding stage of pseudopregnancy (8.0 mg/uterus v. 3.2 mg), and the relative proportions of albumin, utero-globin and β globulins were also changed. In an intensive study of embryo-uterine relationships in the sheep, employing the egg transfer technique [21, 22], it was found that the presence of an embryo first influences the life span of the corpus luteum between the 12th and 13th days after oestrus. In the pig, this effect may occur even earlier, even though the estrous cycle is 5 days longer than in the sheep.

Regarding the mechanism of action, there is still speculation. *Moor* [20] thinks that the effect of the early embryo is "concerned with preventing the uterus from acquiring its lytic properties". This would apply especially to species like the sheep and pig; in the latter, it appears that more than 4 embryos are required to overcome the lytic effect [26]. Where there is a pseudopregnancy, e.g. as in the rabbit, the conceptus need not influence the life span of the corpora lutea until a later stage.

Shedding of the zona pellucida

As a rule the zona pellucida is shed from the blastocyst *in utero* before implantation. In the hamster the process is strongly progesterone dependent [24], whereas in the rat [18] and possibly also in the mouse [25] it is estrogen dependent. Some workers [23] have concluded that shedding depends upon the maturity of the ovum rather than on a specific uterine environment. However, *Orsini & McLaren* [25], on the basis of their own observations on the mouse, sought to modify that conclusion, suggesting that 'the attainment by the ovum of a particular developmental stage is a necessary but not a sufficient condition for loss of the zona'. They imply that the shedding is normally dependent on an 'appropriate' (uterine) environment.

Residence within the uterus is by no means a pre-requisite for shedding since it may occur, though delayed, in the ligated oviduct of the same species [24, 25], or even of a different species (rat → mouse) [29], in some cases (sheep → rabbit) under estrogen dominance [3], and also *in vitro* [9].

Zona shedding may occur in more than one way as suggested by *Mintz* [19] and reiterated by *Orsini & McLaren* [25]. In some species it seems likely that a uterine factor could enhance an endogenous mechanism thereby hastening the process in comparison with the extra-uterine sites.

The rabbit blastocyst is exceptional in that it does not shed the zona pellucida before implantation; in fact, the zona plays an essential role in the economy of the blastocyst, notably in its expansion which is marked in this species. When rabbit morula are cultured *in vitro* blastulation will occur but expansion is very limited and the zona often ruptures [4, 17]. However, if such ruptured blastocysts are transferred to the uterus of synchronised recipients expansion will occur. This indicates that the zona had not sustained irreversible damage *in vitro,* and that a factor, deficient *in vitro,* could be supplied by an appropriate uterine environment.

Egg transfer studies: the need for synchronisation of the developing egg and the uterine environment

Chang [10], working with rabbits, firmly established the importance of synchronising the stage of egg development with the luteal phase of the recipient. He concluded "that there is little or no chance for a transferred ovum or blastocyst to develop into young when it is in the tract one or two days before or after the corresponding corpus luteal stage". For example, the transfer of '2 day' ova, recovered from the tubes, yielded young (approx. 30 %) only in combination with a 2 or 3 day uterus; with a more advanced endometrium (4 or 5 days p.c.) no young were born. At examination two days after the transfer of 2 day eggs to 4 day uteri, *Chang* found "normal late morulae or blastocysts, almost resembling 4 day blastocysts", whilst at 6 or 7 days degenerating large blastocysts were recovered. No swellings indicative of viable implantations were found.

In the first three species to which egg transfer was extensively applied, namely rabbit, rat and mouse, it so happens that implantation occurs relatively early (days 5–8) and only during a limited period of time. Consequently, it was thought that the implantation requirements were dictating the need for the synchronisation of the donor and recipient animals. However, it is now well recognised that asynchronous transfers may fail even before implantation. Moreover, it is noteworthy that in sheep, where there is a relatively long estrous cycle and a functional corpus luteum, synchronisation is still necessary [21, 22] even though eggs may be transferred over a much longer period than in Rodents. In cattle, too, preliminary observations point to the need for synchronisation [28]. Very recently, the need for 'proper synchrony' between embryonic and uterine development has also been demonstrated in the pig [5]. Also of interest is their finding that there was a significant difference in survival rate between native (recipient already pregnant) and transferred eggs, if the latter were 12 or 24 hr less developed.

Fate of eggs in estrogen dominated uteri

What happens to eggs if they should arrive prematurely in the uterus? *Chang* [11] reported on the transfer of 82 recently ovulated eggs to the uteri of 6 does that had been mated 12 hr. earlier. Twenty four hours after transfer he was able to recover only 11 eggs (13 %), all of which were degenerate, including 9 that had lost the zona pellucida. In view of the fact that unfertilized eggs can be recovered from the uterus 6 days after ovulation, *Chang* inferred that the eggs had degenerated *in utero.* However, the analogy may not be entirely appropriate because unfertilized eggs do not normally enter the uterus until about 60 hr. after ovulation, by which time the environment has changed considerably from that prevailing at estrus. The presence or absence of a mucin layer also constitutes another difference.

The experiment was therefore repeated using as recipients estrous does in which the vagina was ligated in order to determine the proportion of eggs expelled, if any. With one exception, freshly ovulated eggs from which the cumulus had been removed with hyaluronidase were transferred to the uteri near the tubo-uterine junction.

Autopsy took place 0.5, 1, 2 and 4 hr. after transfer to permit examination of the uteri and vagina for eggs. The results are shown in Fig. 1, and illustrated in Fig. 2. It will be seen that eggs soon appeared in the vagina with the result that the proportion recoverable from the uterus fell from 69 % to zero over the 4 hr. period.

Fig. 1. Fate of eggs transferred to uterus of estrous does.

Interval from egg transfer to recovery (hr.)	No. of does	No. of eggs transferred		No. of eggs recovered					Eggs missing	
				uterus		%	vagina	%		
		L	R	L	R				No.	%
0.5	2	32	26	22	18	69	8	14	10	17
1.0	2	28	31	8	15	39	17	29	19	32
2.0	2	34	23	7	7	25	23	39	20	36
4.0†	1	26		0	0	not searched			26	100
4.0*	1	49•	17	16	9	38			41	62

† vagina not ligated, * uterine horns ligated, • morulae.

Of the eggs that were unaccounted for (17–36 %), it seems likely that some degenerated *in utero* since others were recovered showing damage to the zona pellucida (Fig. 3). Further, in one recipient in which both uterine horns were ligated near the cervices only 33 % of the morulae and 53 % newly ovulated eggs transferred were recovered. It therefore appears as if the morulae were equally susceptible to damage *in utero.*

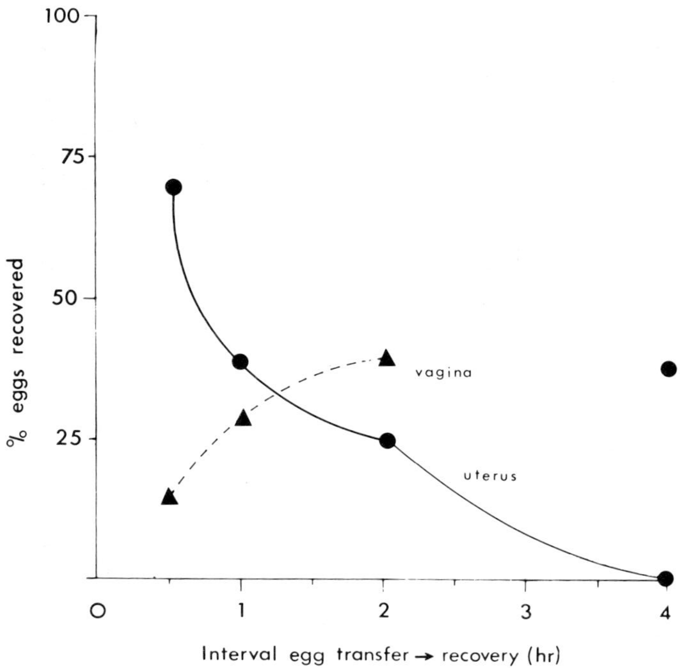

Fig. 2. Fate of newly ovulated eggs transferred to the uterus of estrous does, based on Fig. 1.

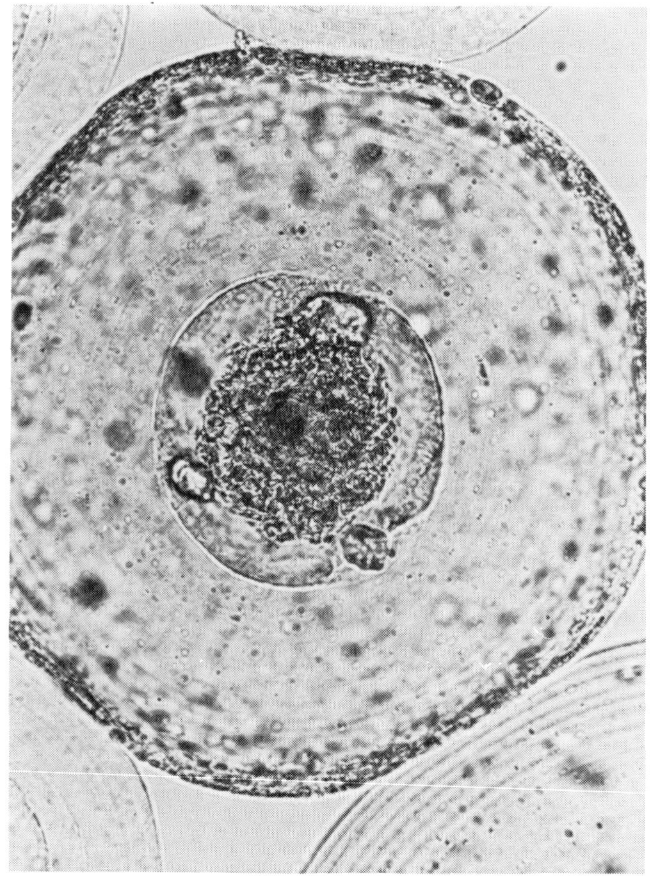

Fig. 3. A rabbit egg recovered 6 1/2 h after transfer to the uterus of an estrous doe, showing degeneration of the cytoplasm and damage to the zona pellucida.

Hitherto, myometrial activity has attracted most attention from the point of view of parturition mechanisms; in other words, in the terminal stages of pregnancy. So far as the establishment of pregnancy is concerned its role is less obvious. At estrus, myometrial activity is at a peak under the influence of estrogen, and this appears to be advantageous for sperm transport. However, before the egg reaches the uterus it is necessary for myometrial activity to decline in order to be compatible with the eggs retention – this normally occurs as estrogen dominance gives way to progesterone.

A relationship between luteal function and retention of eggs within the uterus was demonstrated by *Adams* [1], who used a surgical approach to reduce the numbers of corpora lutea. It was found that to prevent eggs being expelled nearly the full complement of corpora lutea was required. If all of the corpora lutea were removed shortly after ovulation, eggs that were transferred to the uterus 45 hr. later were subsequently not recoverable from that site. In further experiments in which the vagina was ligated it was established that the eggs were being expelled from the uterus.

Whereas almost maximal numbers of copora lutea are needed to render myometrial function compatible with pregnancy, only a minimal amount of luteal tissue is required to induce endometrial proliferation (for references see *Adams* [1]. Qualitatively, such an endometrium is satisfactory in so far as it will sustain blastocyst development. This implies that in the rabbit at least the endometrium and myometrium have widely differing thresholds for progesterone – a divergence that may have clinical significance. For example, we may visualise a situation in which the endometrial condition is satisfactory and yet the pregnancy could have failed already due to the egg having been expelled on account of inadequate luteal function.

Fate of eggs in progesterone dominated uterus

Adams [2] reported the results of transferring 60 hr. morulae, also recovered from the tubes, to the uteri of recipients whose luteal stage was either synchronous with, or 24 or 48 hr. in advance of that of the donors. The findings at autopsy made 72 hr. after transfer are shown in Fig. 4. It will be seen that with increasing asynchrony

1) the proportion of transferred eggs recovered fell from 83 % to 36 % and
2) the proportion of normal blastocysts fell even more sharply from 89 % to 5 %, with a corresponding increase in the porportion of morulae and degenerate blastocysts.

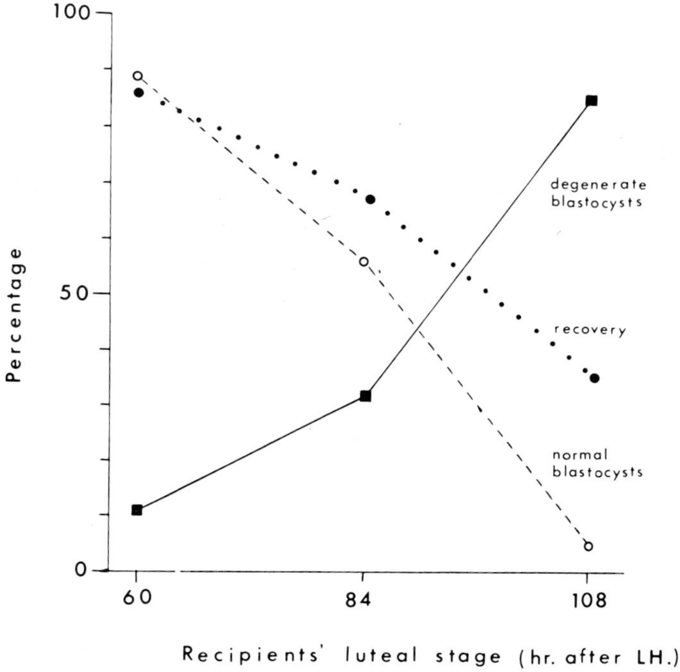

Fig. 4. Fate of 60 h morula transferred to the uterus of recipients given LH, 60, 84 or 108 h before transfer.

A further indication of the unfavourableness of the asynchronous uterus was the decreasing size of those recovered blastocysts whose mean surface area was 18.2 mm^2 (60 → 60), 9.3 mm^2 (60 → 84) and 4.3 mm^2 (60–108 hr.).

In a further experiment 65 similar morulae were transferred to the uteri of 7 does on the 8th or 9th days of pseudopregnancy. At autopsy 3 or 4 days after transfer there were 77 corpora lutea, and a total of 73 eggs including 'natives' and 'aliens' was recovered. With certain exceptions, it proved difficult to identify with certainty the native and alien eggs, of which the latter were mostly degenerate late morulae. Only five of the transferred morulae had reached an early blastocyst stage (cf. 108 hr. p.c.) before degenerating Fig. 5. In a second group of recipients examined earlier, at 6 $\frac{1}{2}$, 24, or 48 hr. after transfer, it proved easier to identify the transferred eggs, though by 48 hr. the picture was already confusing. Out of 71 morulae transferred to 5 does (see Fig. 6), 61 were recovered but not a single one showed any signs of blastulation. In fact, the morula recovered only 6 $\frac{1}{2}$hr. after transfer alrea-

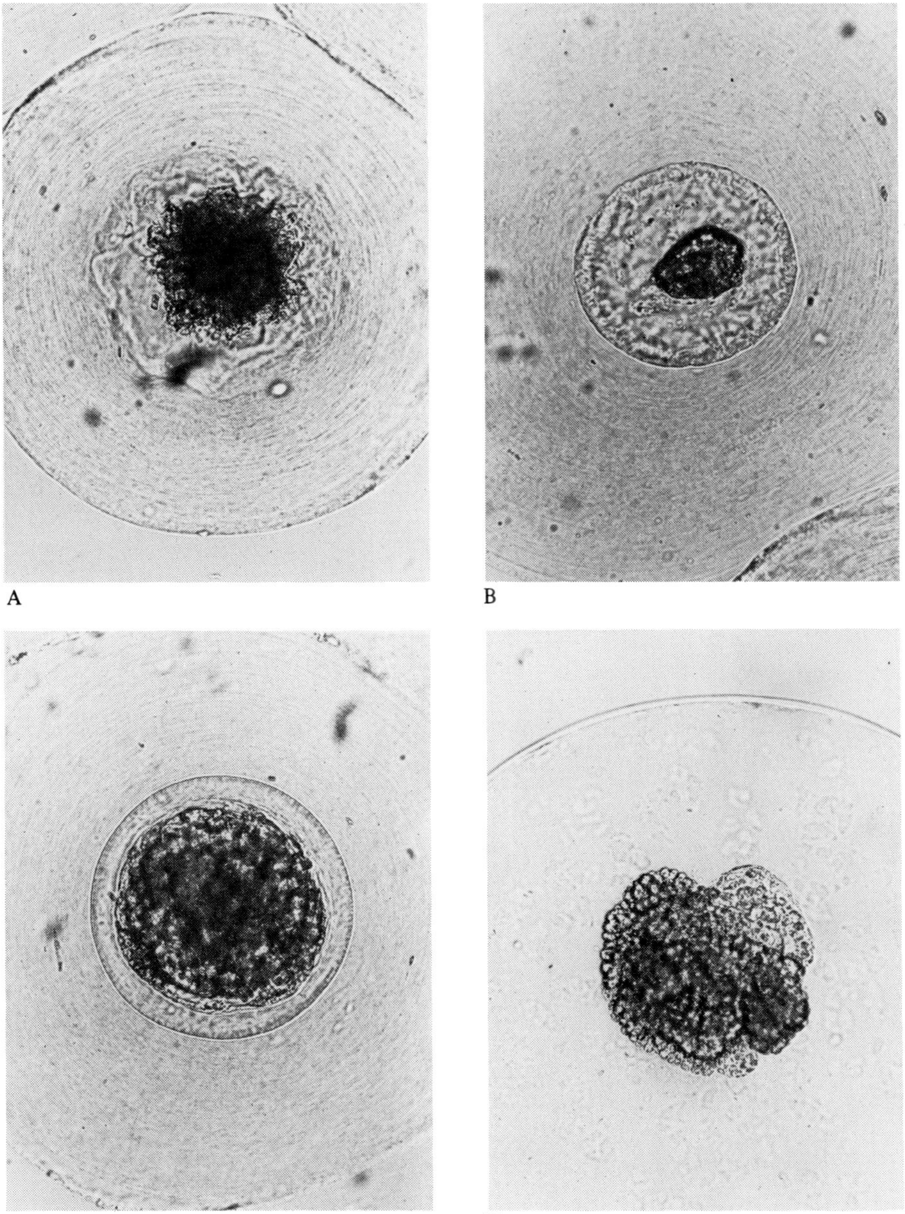

A B

C D

Fig. 5. The fate of 60 h rabbit morula transferred to uterus of recipients on the 8 th day of pseudopregnancy.

A and B. Degenerate late morulae – early blastocyst stages.

C. Degenerate blastocyst. D. Native egg.

Fig. 6. Fate of 60 hr. morula transferred to the uterus of recipients on the 9th day of pseudopregnancy.

Interval egg transfer →recovery (hr.)	No. recipients	No. eggs		No. of		Appearance of morula at recovery
		transferred	recovered	corpora lutea	native eggs recovered	
6.5	1	14	12	11	7	Easily identifiable, showing first signs of degeneration
24	2	28	24	22	14	Most morula markedly degenerate
48	2	29	25	16	3	All morula markedly degenerate

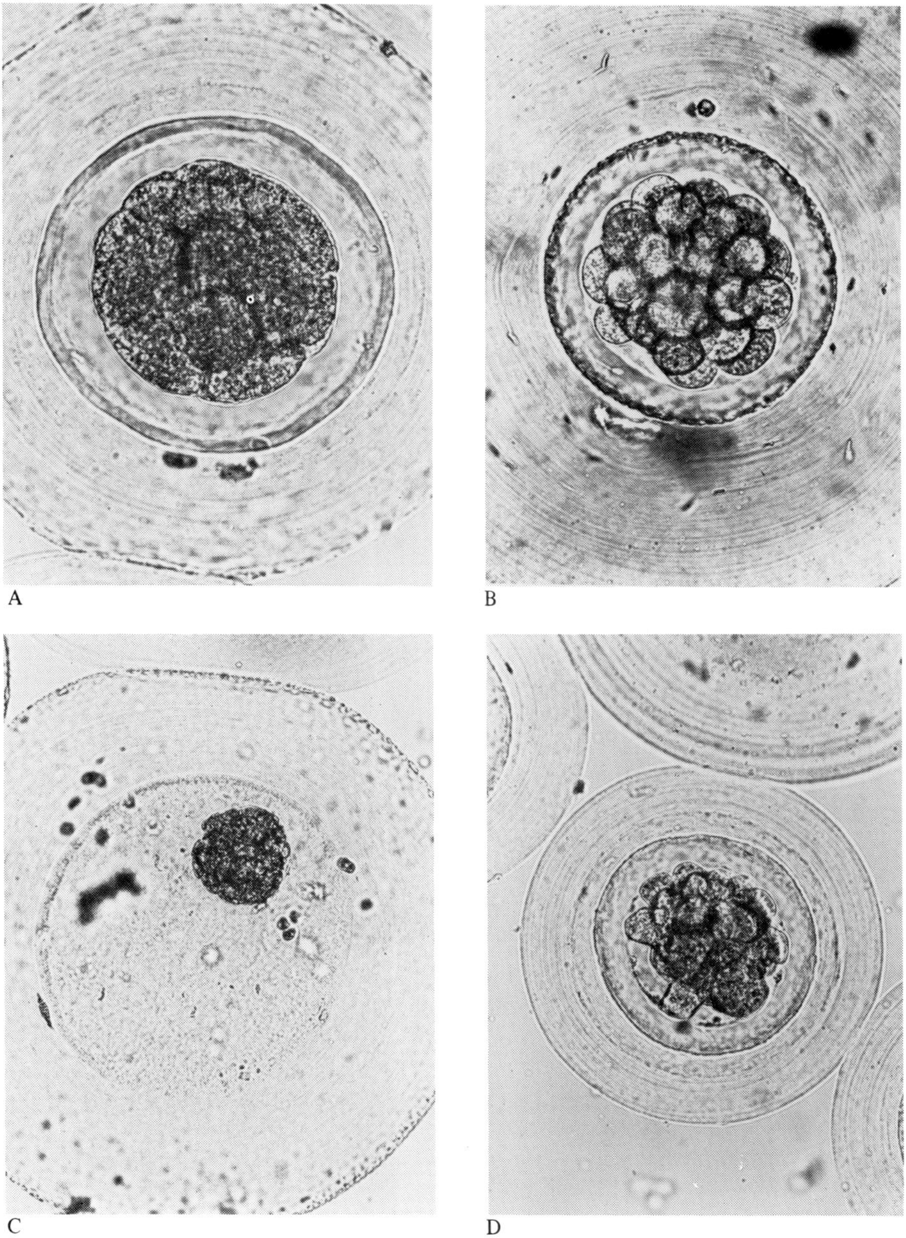

Fig. 7. The fate of 60 h rabbit morulae transferred to the uterus of recipients on the 9 th day of pseudopregnancy.

A. Normal morula before transfer.

B. Degenerating morula recovered 6 1/2 h after transfer.

C. 24 h after transfer and D. 48 h after transfer.

dy showed some deterioration of the blastomeres. At 24 hr. the blastomeres had shrunk away from the zona pellucida, which itself appeared thin and abnormal. After 48 hr. the eggs appeared very degenerate (Fig. 7).

The fact that the eggs underwent degenerative changes comparatively quickly as opposed to merely failing to develop suggests that the uterine environment was not only lacking in some essential factor(s) but rather hostile as well. The present findings provide indirect evidence that fractions, such as blastokin [16] or uteroglobin [6] exist in the uterus for only a limited period of time, and incidentally draw attention to the fact that the composition of the uterine secretions may change quite significantly whilst the endometrium shows comparatively little morphological change.

It is suggested that the sensitivity of the transitional morula – blastocyst stage (cf. 60–108 hr. p.c.) could be expolited in the screening of compounds that might be developed to interfere with uterine secretion in the pre-implantation stage.

Addendum

On page 156, it was stated that morulae transferred to the uterus of recipients on the 9th day of pseudopregnancy show some deterioration of the blastomeres after only 6 $\frac{1}{2}$ hr. By re-transferring the morulae to synchronised recipients, it has now been established, that they are still capable of developing to term. It appears, therefore, that more than 6 hr. exposure is required to produce an irreversibly damaging effect. The experiments are being continued in order to determine more precisely the temporal relationships between morphological changes in the egg and its death.

References

[1] *Adams, C. E.* (1965). Influence of the number of corpora lutea on endometrial proliferation and embryo development in the rabbit. J. Endocrin. **31,** XXIX–XXX.

[2] *Adams, C. E.* (1968). Ovarian control of early embryonic development within the uterus. Reproduction in the female mammal, pp. 532–548. Ed. *Lamming, G. E. & Amoroso, E. C.,* Butterworths, London.

[3] *Adams, C. E., Moor, R. M. & Rowson, L. E. A.* (1968). Survival of cow and sheep eggs in the rabbit oviduct. Proc. VIth Int. Congr. Anim. Reprod. Paris.

[4] *Adams, C. E.* (1969). The development of rabbit eggs after culture *in vitro* for 1–4 days. J. Embryol. exp. Morph. *(in press)*

[5] *Bazer, F. W., Clawson, A. J., Robison, O. W. & Ulberg, L. C.* (1969). Uterine capacity in gilts. J. Reprod. Fert. **18,** 121–124.

[6] *Beier, H. M.* (1968). Biochemisch-Entwicklungsphysiologie-Untersuchungen am Proteinmilieu für die Blastozystenentwicklung des Kaninchens (Oryctolagus cuniculus). Zool. Jb. Anat. **85,** 72–90.

[7] *Blandau, R. J.* (1961). Sex and Internal Secretions. 3rd Ed. Ed. *W. C. Young.* Ch. 14, p. 846. The Williams & Wilkins Co., Baltimore.

[8] *Boyd, J. D. & Hamilton, W. J.* (1952). Ch. **14,** p. 37. Marshall's Physiology of Reproduction, vol. 2. 3rd Ed. Ed. *A. S. Parkes,* Longmans, Green & Co., London.

[9] *Brinster, R. L.* (1963). A method for *in vitro* cultivation of mouse ova from two cell to blastocyst. Exp. Cell Res. **32,** 205–208.

[10] *Chang, M. C.* (1950). Development and fate of transferred rabbit ova or blastocyst in relation to the ovulation time of recipients. J. exp. Zool. **14,** 197–216.

[11] *Chang, M. C.* (1955). Développement de la capacité fertilisatrice des spermatozoides du lapin à l'intérieure du tractus génital femelle et fecondabilité des oeufs de lapine. La fonction tubaire et ses troubles. Masson. pp. 40–52.

[12] *Chang, M. C.* 1968). The reaction of the uterine endometrium on spermatozoa and eggs. pp. 459–477. Reproduction in the female mammal. Eds. *Lamming, G. E. & Amoroso, E. C.,* Butterworths, London.

[13] *Corner, G. W.* (1947). The hormones in human reproduction. Princeton.

[14] *Greenwald, G. S.* (1968). Hormonal regulation of egg transport through the mammalian oviduct. Progress in Infertility. Ed. *Behrmann & Kistner;* Little, Brown & Co. p. 157–179.

[15] *Harper, M. J. K.* (1962). Egg movement through the ampullary region of the Fallopian tube of the rabbit. Proc. IVth. Int. Congr. Anim. Reprod. The Hague. **2,** 375–380.

[16] *Krishnan, R. S. & Daniel, J. C.* (1967). 'Blastokinin': Inducer and regulator of blastocyst development in the rabbit uterus. Science **158,** 490–492.

[17] *Lewis, W. M. & Gregory, P. W.* (1929). Cinematographs of living developing rabbit eggs. Science **69,** 226–229.

[18] *Meyer, R. K.* (1966). Discussion. In Egg Implantation, pp. 28–29. Eds. *G. E. W. Wolstenholme & M. O'Connor.* London. J. & A. Churchill Ltd.

[19] *Mintz, B.* (1965). In Preimplantation stages of Pregnancy, p. 167. Eds. *G. E. W. Wolstenholme & M. O'Connor,* Churchill, London.

[20] *Moor, R. M.* (1968). Effect of embryo on corpus luteum fraction. J. anim. Sci. Suppl. **1,** 97–118.

[21] *Moor, R. M. & Rowson, L. E. A.* (1964). Influence of the embryo and uterus on luteal function in the sheep. Nature (London) **201,** 522–523.

[22] *Moor, R. M. & Rowson, L. E. A.* (1966). The corpus luteum of the sheep: functional relationship between the embryo and the corpus luteum. J. Endocr. **34,** 233–239.

[23] *Noyes, R. W., Dickmann, Z., Doyle, L. L. & Gates, A. M.* (1963). Ovum transfers, Synchronous and Asynchronous, in the study of implantation. In Delayed Implantation, Ed. *Enders, A. C.* 213–232. Univ. of Chicago Press, Chicago.

[24] *Orsini, M. W.* (1965). Factors affecting loss of the zona pellucida in the hamster. J. Anat. **99,** 922.

[25] *Orsini, M. W. & McLaren, A.* (1967). Loss of the zona pellucida in mice, and the effect of tubal ligation and ovariectomy. J. Reprod. Fert. **13,** 485–499.

[26] *Polge, E. J. C., Rowson, L. E. A. & Chang, M. C.* (1966). The effect of reducing the number of embryos during early stages of gestation on the maintenance of the pregnancy in the pig. J. Reprod. Fert. **12,** 395–397.

[27] *Pincus, G.* (1930). Observations on the living eggs of the rabbit. Proc. Roy. Soc. **B 107,** 132–167.

[28] *Rowson, L. E. A. & Moor, R. M.* (1966). Non surgical transfer of cow eggs. J. Reprod. Fert. **11,** 311–312.

[29] *Tarkowski, A. K.* (1962). Interspecific transfers of eggs between rat and mouse. J. Embryol. Exp. Morph. **10,** 476–495.

Discussion

J. M. Bedford: You quite rightly emphasized the importance of the synchrony of the egg in the uterus in all these species. I wonder if you or anyone else would like to make a comment on the situation in man. You know this Estes operation of repositioning the ovaries. Well it is held that pregnancies are sometimes established following this procedure.

C. E. Adams: Yes, I do know of Este's operation but unforunately I have no direct information on its effectiveness. As yet, egg transfer has not been performed in man. However, in every other species so far investigated the exacting requirements for synchrony have been found to exist. By analogy one would expect that the same conditions would apply in man. On this basis, I question the value of Este's operation. I would like to see some actual results. The need for synchrony does not apply to the tube. For example, in inter-species transfer (sheep → rabbit) the eggs survive so well that asynchronous transfer has become a routine procedure in our laboratory.

J. M. Bedford: I wonder if anyone else in the audience knows?

E. J. Plotz: It certainly works sometimes, yes.

M. Chang: I would like to congratulate you on your presentation. Not only because your English is good, but also because your work is quite good. At the same time I would like to apologize for my English, not only because I speak poorly but because I wrote rather badly in 1950. It always gives me great pleasure if I can repeat my experiment; it gives me even more pleasure if other people can repeat my experiment. Now we appreciate the importance of synchronization more as we know more about development of eggs in relation to the development of endometrium. You have demonstrated that very nicely in your recent experiment. There are two things I would like to mention: firstly, I think Dr. *Bedford* also did some work on the transfer of unfertilized eggs into the uterus: I would like him to speak a few words. Now the second point is about the possibility of fertilization in the human uterus. I have argued with Dr. *Morris* of Yale many times. Based on the study of laboratory animals, I do not think that the unfertilized eggs or newly fertilized eggs can survive in the uterus but he insists that it could in the human, because the transplantation of the ovary into the human uterus was reported to be able to produce normal babies. I asked him if he did the work himself. No, he did not. As you know, it is possible that in such preparations the tube may be still intact and function inside the uterus, and the eggs may have been fertilized in the tube and properly transported into the uterus at the right time. I think the medical profession ought to take this problem very seriously, to determine whether with ovary alone transplanted into the uterus, the eggs can get fertilized and survive.

K. W. Humphrey: I agree with you.

J. M. Bedford: The only comment is, that rabbit eggs can be normally fertilized in the uterus, whether or not there are any Fallopian tubes – in other words, a uterus which is separated from the Fallopian tubes will sustain complete capacitation and eggs in the uterine lumen can be fertilized. But, one can leave the eggs in the uterus only about six to seven hours. If one leaves them longer, the zona and the ooplasm start to degenerate. Incidentally, there is a great difference between the uterus of the ovariectomized animal, as you stress, and the estrous uterus. The ovariectomized uterus does not seem particularly inimical to the egg; the corona cells and some of the cumulus will stay around the egg, and the egg remains farily normal for a much longer period in this estrogen deficient milieu.

C. E. Adams: Yes, as a matter of fact, in some cases we recovered the eggs very soon after transfer and found motile sperm within the eggs. Some of these eggs were then re-transferred to the tubes of other unmated recipients, which went on to produce young. However, the proportion of eggs developing to term was rather low (38 %).

K. W. Humphrey: One point; I have been transferring day four blastocysts to ovariectomized mice, and I find that if we prime them with estrogen and transfer them on the second day of progesterone treatment they would not implant. If we do not prime them, they do implant in a certain percentage of the mice, so that estrogen pre-treatment definitely does reduce the survival of the egg.

C. E. Adams: Yes, I agree.

A. Psychoyos: We have shown in the Rat (Proceedings of the VI. Inter. Cong. Anim. Reprod. p 463, 1968) that if you transfer two cell eggs into the uterus of ovariectomized animals these eggs under appropriate hormonal treatment develop normally and in a big percentage, give normal embryos. When we transfered one-cell eggs, out of 500 transferred, only 2 developped. I think that even if fertilisation can take place into the uterus, there will be always the problem of the first segmentation. The problem is the same for *in vitro* development.

C. E. Adams: I think it is important to stress this difference between the animal with ovaries and the one that has been spayed. For instance, the uterus of rabbits that had been spayed three weeks before egg transfer was able to 'support' development from the morula to the blactocyst stage. Development may also occur in the vagina (*Adams,* (1965) (Preimplantation stages of pregnancy, Ed. *G. E. W. Wolstenholme* & *M. O'Connor,* Churchill, London).

So it would seem that the secretions of the uterus are very important when under ovarian influence. I visualize the asynchronous uterus as being a hostile rather than just a neutral situation.

R. Eliasson: Referring to Dr. *Bedford's* question about the human. The sensitivity of the human myometrium is dependent on the ovarian hormone balance. For example it is much less sensitive to prostaglandins in the secretory phase than in the ovulatory phase. When we do *in vivo* recordings of the motility and add prostaglandins into the vagina, the motility is not changed during the proliferative phase but in most cases a marked effect is noted at the time of ovulation.

The Effect of Drugs on Egg Transport

J. P. Bennett

Institute of Hormone Biology, Syntex Research, Palo Alto, California, USA

Summary: Movement of eggs in the oviduct and the actions of the isthmo-ampullar and utero-tubal junctions are under delicate hormone control and can be affected by ovariectomy, exogenous hormones and a number of synthetic and natural substances. The actions upon egg transport of synthetic steroidal estrogens, prostaglandins and other drugs with effects upon smooth muscle (e.g., reserpine) are discussed. The actions of some of these compounds produce acceleration of the eggs down the oviduct where premature arrival in the uterus results in egg degeneration. Alternatively, the compounds may produce an arrest of the rate of egg transport termed "tube-locking", usually occurring at the isthmo-ampullar junction. The function of a particular compound in producing arrest or acceleration of the rate of egg transport is dependent on the dose level administered and the species used.

The mechanisms of action of the described compounds are discussed, with the emphasis on how such knowledge may be helpful in the design of unique contraceptives.

The greatly increased interest in the effect of various drugs on mammalian egg transport stems partly from the possible application of such studies in controlling fertility. An ability to interfere with the delicate hormone balance required to produce normal ovum pick-up, and ovum transport through the oviduct offers exciting possibilities for the development of new methods of human contraception.

With few exceptions, the duration of egg transport in the oviducts of most mammalian species occupies 3 to 4 days (Fig. 1). The eggs pass rapidly down the thin-walled, ciliated ampulla after ovulation, remain for a prolonged period at the isthmo-ampullar junction where they are fertilized, and subsequently proceed slowly down the muscular walled isthmus through the utero-tubal junction and into the uterus. This movement of the eggs in the oviduct and the actions of the isthmo-ampullar and utero-tubal junctions are under delicate endocrine control and can be affected by ovariectomy, exogenous hormones and a number of synthetic and natural substances (Fig. 2). An arrested or accelerated rate of egg transport down the oviduct can be produced by such interference, which may also result in eventual egg degeneration.

Manuscript received: 21 April 1969

Fig. 1. The duration of egg transport in oviducts of some common mammals.

Animal	Approximate Time of Egg entry into Uterus after Ovulation (hrs)
Dog	168
Cat	148
Horse	98
Rat	96
Rhesus monkey	96
Cow	90
Guinea pig	82
Sheep	72
Mouse	72
Rabbit	60
Pig	50

Fig. 2. Some compounds affecting the rate of egg transport in the oviduct of common laboratory mammals.

Compound	Function on Egg Transport	Species	Authors
Synthetic and Steroidal estrogens	Accelerate or arrest	Rabbit	*Greenwald* [21, 23, 25, 26]
		Mouse	*Burdick & Pincus* [8]
		Guinea pig	*Deanesly* [13]
			Greenwald [25]
		Hamster	*Greenwald* [21, 25]
	Accelerate only	Rat	*Greenwald* [21]
			Gardner, Gnoj, Watnick & Gibson [20]
			Bennett, Vallance & Vickery [2]
			Bennett, Vickery & Dorfman [5]
Synthetic Non-steroidal estrogens	Accelerate	Rat	*Duncan, Stucki, Lyster & Lednicer* [14]
			Greenwald [24]
			Harper & Walpole [31]
			Jensen [34]
Synthetic progestins	Accelerate	Rabbit	*Chang & Bedford* [11]
			Chang [9, 10]
			Bennett, Vickery & Dorfman [5]
		Rat	*Davis* [12]
			Wakeling [46]
Gonadotropins	Accelerate	Rabbit	*Wislocki & Snyder* [46]
		Mouse	*Harrington* [32]
		Monkey	*Bennett* [3]
Prostaglandins E_1	Accelerate	Rat	*Duncan* [16]
Reserpine, Chlorpromazine, Tetrabenazine	Arrest	Mouse	*Bennett & Kendle* [4]
			Kendle & Bennett [37]

Estrogens

The effect of exogenous estrogens upon egg transport varies with the species (Fig. 3). In the mouse and rat the administration of relatively high doses of estrogen accelerates the rate of egg transport down the oviduct. The eggs arrive prematurely in the uterus where they degenerate (Fig. 4a, b).

Fig. 3. The effect of single injections of estradiol cyclopentylpropionate given to animals shortly after mating, upon the rate of ovum transport in the Fallopian tube (after [25]).

Species	Dose (μg) Accelerating Rate of Ovum Transport	Dose (μg) "Tube-locking" Ova at Isthmo-Ampulla Junction	Dose Interrupting Pregnancy in + 80 % Animals
Mouse	> 1	1	1
Rabbit	25	100	50
Rat	>10	None	10
Guinea pig	50–100	250	10
Hamster	100	250	25

Studies of *Bennett, Kendle, Vallance* and *Vickery* [2], using uterine ligation to collect eggs prematurely ejected from the oviduct showed that the optimum dose response of estrogens in accelerating the rate of egg transport in rats was Day 3 of pregnancy, Day 1 corresponding to the observation of spermatozoa in the vagina. This coincides with the passage of the eggs down the muscular isthmus of the oviduct.

High, relatively unphysiological doses of estrogen in the rabbit, guinea pig or hamster produce an arrest of egg transport in the oviduct termed "tube-locking", while lower doses of estrogen are required for acceleration. The eggs of the rat cannot be "tube-locked" with estrogens during tubal transport, but this action is possible in the mouse, using lower doses than are required to accelerate the rate of egg transport. According to *Greenwald* [26] the site of regulation of arrested egg transport in the above-mentioned species is the isthmo-ampullar junction (Fig. 5). The position of this junction in the oviduct varies with the species.

Greenwald [26] has used experiments in the rabbit to explain the mechanism of action of estrogens on tubal egg transport. He observed that retention of eggs occurred in the oviduct at the isthmo-ampullar junction after four days of estrogen treatment (10 μg estradiol benzoate by injection) despite the increasing levels of progesterone after ovulation, or massive doses of 17α-hydroxyprogesterone caproate (100 mg for four days). This led him to postulate that regulation of egg transport in this region was controlled more by the amount of estrogen present than progestin.

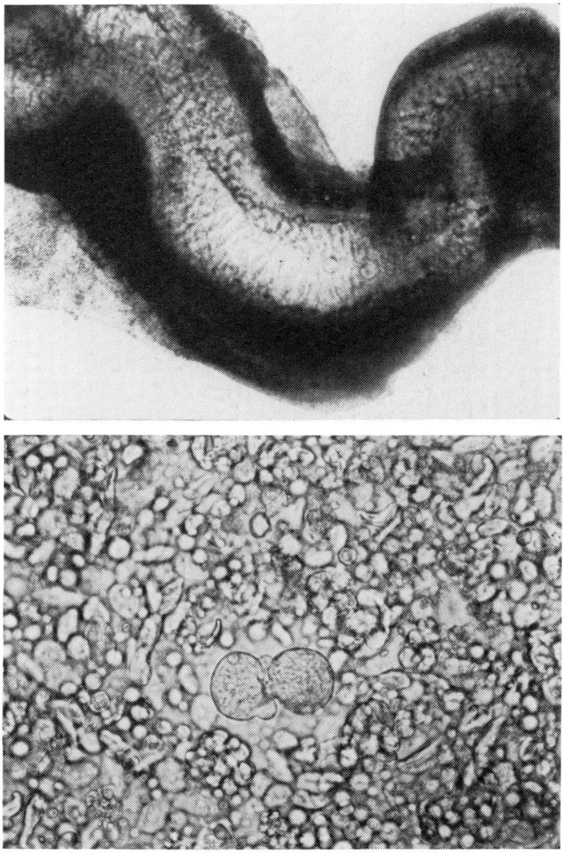

Fig. 4
Eggs prematurely traversing the isthmus
a) and degenerating in the uterus
b) after a single dose of 400 μg/kg mestranol given on Day 1 of pregnancy (day of sperm in vaginal lavage) in the rat.
(Magnification a) = X 10.5; b) = X 67)

In further studies he noted that although 25 μg and 250 μg of estradiol cyclopentyl propionate produce diametrically opposed effects on the rate of rabbit egg transport: changes in tubal motility as measured by alterations in the intraluminal pressure were not necessarily associated with the rate of egg migration. In addition, rapid transport of eggs occurred through an oviduct of 28-day castrate rabbits which showed little spontaneous muscular activity [23]. Thus it was concluded that a premature or delayed relaxation of the tubal egg locking mechanism was more important for regulation of egg transport than changes in tubal motility.

Various explanations have been given to explain tube locking of eggs, most referring to effects on the utero-tubal junction. Among such explanations are included closure of the tubal lumen due to edema of the subserosa and muscle layers [17]; partial or complete absence of circular muscle activity at the site of the block [7] and/or variation in the rate and amplitude of muscle activity depending on balance of estrogen

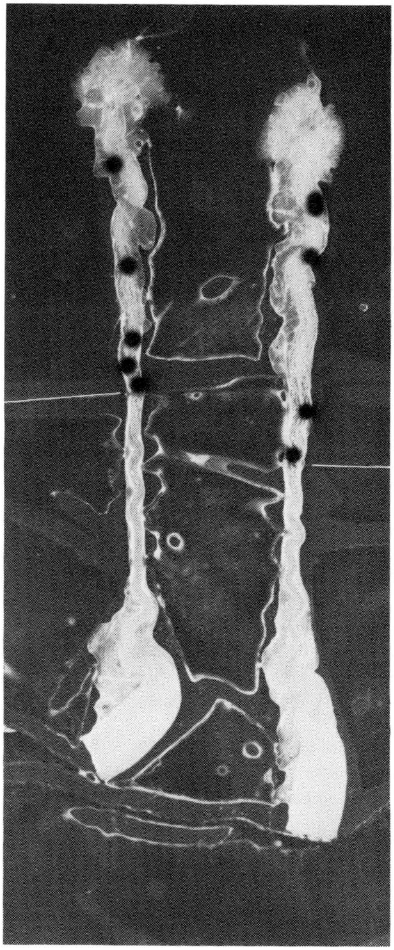

Fig. 5
Autoradiograph of radioactive ^{198}Au artificial eggs at the isthmo-ampullar junction of the rabbit 8 hours after transfers made 10 3/4 hours after ovulation (From [28]).

and progesterone [33]. All of these explanations could equally apply to a functional block at the isthmo-ampullar junction. Further, the work of *Greenwald* [21, 23, 26] suggests that the isthmo-ampullar junction is generally the site of egg transport regulation. Although the level of estrogen is critical for the rapid transport of eggs through the ampulla, a critical balance of estrogen and progesterone is responsible for retention of the eggs at the isthmo-ampullar junction until the uterus is ready to receive them. The nature of the blocking mechanism is unknown, but *Greenwald* [25] suggests a sphincter-like activity as probable, due to contraction of layers of smooth muscle locally at the isthmo-ampullar junction or including the isthmus. This would seem possible, since fluid retention and some ballooning of the ampulla occurs at the junction during treatment with estrogens.

After estrogen dosing in the rat, *Bennett, Vickery & Dorfman* [5] observed an increased volume of tubal secretions in the ampulla (Fig. 6) while the eggs were briefly resident at the isthmo-ampullar junction. The motility of the tubal smooth muscle was, however, extremely active compared to controls and the eggs commenced rapid passage through the junction only eleven hours after a single oral dose of 400 μg/kg mestranol. The lack of the "tube-locking" phenomenon in the rat may well be due to this very high muscular activity produced by exogenous estrogens.

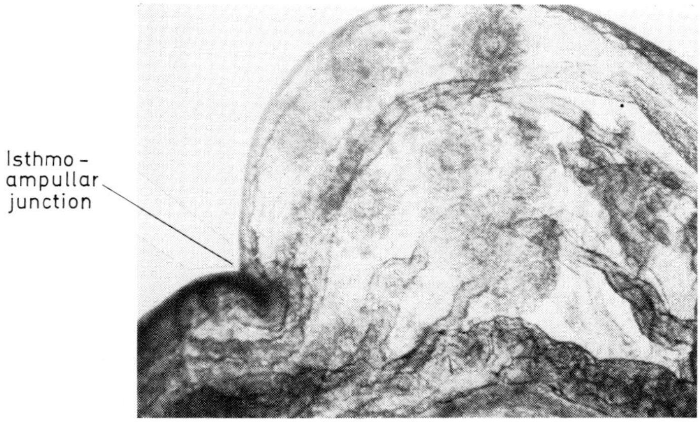

Fig. 6. Ova in the ampulla of the rat at the isthmo-ampullar junction, illustrating 'ballooning' of the oviduct from increased tubal fluid 8 hours after dosing with 400 μg/kg mestranol on Day 1 of pregnancy (8 hours after ovulation).

Although degeneration of eggs occurs upon administration of estrogens during the transport phase, the successful transfer of eggs from estrogen-treated rabbits [22, 38] and rats [43] to undosed animals suggests that estrogens do not directly affect egg viability. The eventual degeneration of eggs after accelerated transport is probably due to the premature stage of development of the egg with respect to that of the uterus [41].

A number of synthetic estrogens have been tested during the early stages of pregnancy in laboratory animals with the aim of separating the estrogenic activity and the antifertility effects on egg transport to develop a unique postcoital human contraceptive. *Gardner, Gnoj, Watnick & Gibson* [20] reported such a separation, but compared antifertility in the rat with estrogenicity in the mouse. Ethinyl estradiol and a number of halogenated 17-ethinyl steroids were compared in the rat and mouse by *Bennett, Kendle, Vallance & Vickery* [2] and observed to be active in accelerating the rate of egg transport. The trifluoro-ethynyl-3,17β-dihydroxy estra-1,3,5(10)-triene derivative (BDH 6146) exhibited a ratio of postcoital antifertility to estrogenic activity in the rat of 30 : 1, respectively. The postcoital antifertility activity was found to be due to premature ejection of the eggs into the uterus.

The estrogens have previously received the most intense study for effects on egg transport but the compounds described below suggest the diversity of chemical substances now being utilized.

Synthetic non-steroidal estrogens

Duncan, Lyster & Wright [15] worked with several non-steroidal antifertility agents and found them to act upon egg transport. They showed that U 10293, [2,8-dichloro-6,12-diphenylbenzo (b,f) (1,5) diazocine] given at oral or subcutaneous doses of 0.5 mg daily for 7 days to rats, from the time of mating was an effective antifertility agent. The compound was estrogenic and the antifertility data suggested loss of eggs when the compound was given during the early stages of pregnancy.

The dihydronaphthalene derivatives are orally effective non-steroidal compounds which interrupt pregnancy in rats, rabbits and guinea pigs. U 11 100A (1- 2[p-(3,4-dihydro-6-methoxy-2-phenyl-1-naphthyl)phenoxy] ethyl pyrrolidine hydrochloride) is active as a single dose during the egg transport phase of pregnancy. *Greenwald* [24] showed that a single dose of 80 μg per rat of U 11 100A on Day 1 of pregnancy caused acceleration of egg transport beginning on Day 3 of pregnancy. Although a supposed estrogenic activity of the compound was observed in several tests (e.g., inhibition of decidual reaction) the vaginal cytology showed no evidence of estrogenic stimulation. This compound and others of its type may, in fact, be inhibiting estrogen synthesis in the ovary or be acting as an anti-estrogen and thereby be interfering with the delicate hormone balance required in the regulation of the isthmo-ampullar junction.

Other orally active non-steroidal compounds which have been shown to possess estrogenic activity and to act as effective antifertility agents during the days of egg transport in the rat are shown in Fig. 7 and 8.

Progestational compounds

Studies of progestins upon egg transport are few, in comparison with those on estrogens. *Chang* and *Bedford* [11] and *Chang* [9] showed that progesterone or medroxyprogesterone acetate, given for a few days before ovulation produces accelerated transport of eggs in the rabbit. *Chang* suggested that exogenous progestins

Fig. 7. Some non-steroidal estrogens effective as antifertility agents during the period of egg transport.

Type	Compound (Type Formulae)	Animal Tested	Effective Antifertility dose/day (mg/kg)	Days of Pregnancy when Dosed	Authors
Triphenyl Ethylenes	ICI 4647*	Rat	0.1 – 0.4	1 to 4	*Harper & Walpole* [31]
Triphenyl Ethylenes	U 1155A*	Rat	0.5	–1 to 6	*Duncan, Stucki, Lyster & Lednicer* [14]
3,4 Diphenyl Chromenes	Compound VII*	Rat	0.02	1 to 5	*Bencze, Carney, Barsky, Renzi, Dorfman & Stevens* [1]
2,3 Dephenyl Benzofuranes	Compound IV*	Rat	4 to 16	1 to 5	*Grover, Chawla, Anand, Kamboj & Kar* [27]
3,4 Diphenyl Coumarins	Compound XIII*	Rat	0.4	–1 to 6	*Lednicer, Lyster & Duncan* [30]
Cyclohexylidene Methane	F 6103*	Rat	4 to 8	1 to 4	*Jensen* [34]
Cyclohexane Carbonic Acid	ORF 3858*	Monkey	0.5–2.0	1 to 6	*Morris, Van Wagenen, McCann & Jacob* [40]

* see Fig. 8

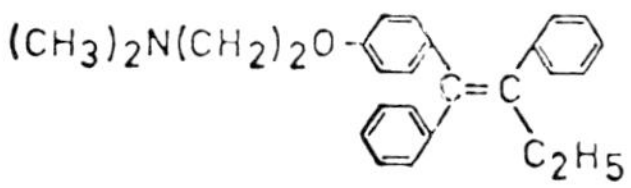

ICI 46 474

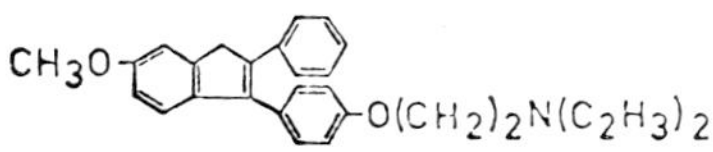

U-11 555 A

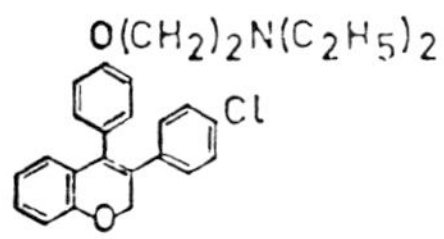

"Compound VII"

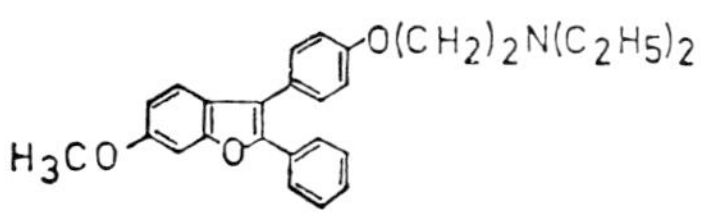

"Compound IV"

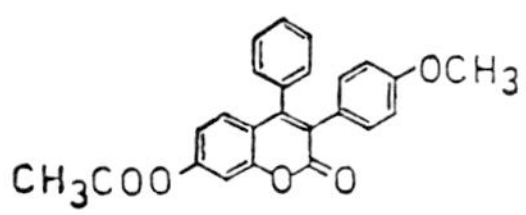

"Compound XIII"

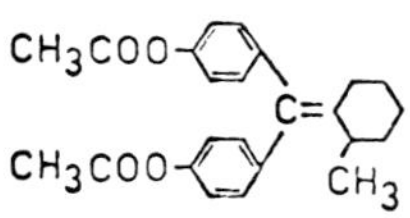

F6103

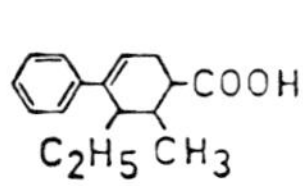

ORF 3858

Fig. 8

given before ovulation produce an accelerated rate of egg transport by inducing a rebound estrogen effect after ovulation. However, recent work of *Bennett, Vickery & Dorfman* [5] shows that if progestin dosing is initiated before ovulation in the rabbit and continued after ovulation for a further 8 days, an accelerated rate of egg transport still occurs. Thus progestins can act directly on the rate of egg transport in the rabbit.

The 19-norprogestins also act on egg transport, as a result of conversion to estrogens [42]. *Davis* [12] produced sterility in rats with a subcutaneous dose of 0.9 mg norethynodrel given on Day 2 postcoitum. *Wakeling* [46] in the rat, and *Chang* [9, 10] in the rabbit showed that norethynodrel can accelerate the rate of egg transport. *Kar, Setty & Kamboj* [35] discovered that topical administration of representatives of the above-mentioned steroidal and non-steroidal estrogens and some progestins were active postcoitally in the rat. The antifertility effect is almost certainly one of effect upon egg transport. Such a finding, if confirmed in man, offers scope for development of unique forms of contraception (e.g., an "after the fact" perfume!).

Other agents with effects upon the rate of egg transport

Among other agents observed to produce alteration in the rate of egg transport are gonadotropins, some prostaglandins and certain drugs acting on the central nervous system, e.g., reserpine.

Superovulation with gonadotropins accelerates egg transport in mice [32], rabbits [47], sheep [44], cattle [45] and monkey [3] by elevating the release and/or synthesis of ovarian estrogens and hence the systemic levels of these hormones. The alteration of endocrine balance results in premature ejection of the eggs from the oviduct into the uterus. Here they either degenerate or are further ejected via the vagina.

It is surprising that so few studies have been reported of direct effects upon the rate of egg transport with prostaglandins. These are naturally occurring hydroxy acidic lipids [19], with particular effects on smooth muscles found in high concentrations in human and ram semen [18, 19]. Uterine and oviduct smooth muscles are contracted or relaxed by prostaglandins, the effect depending upon the species and type of prostaglandin used [18], but they possess a somewhat short biological half-life.

Duncan [16] has initiated some studies of antifertility with the prostaglandins. Using 2 μg/kg/min of P.E. infused into the heart of rats from proestrus, subsequent mating produced incomplete inhibition of pregnancy when examined on Day 8 of pregnancy. With the production of synthetic prostaglandins it may be possible to enhance the direct effects of these compounds upon the muscle tone of the oviduct and tubal junctions and thus regulate egg transport.

Among other drugs active on smooth muscle which have been found to alter the rate of egg transport are reserpine, chlorpromazine and tetrabenazine [4, 36, 37]. A reversible arrest of egg transport and development in the mouse was produced by relatively high doses of these compounds given intraperitoneally (Fig. 9). The role of these compounds in central and peripheral depletion of tissue amines and consequent reduction in smooth muscle activity was not directly implicated in their effects upon egg transport.

Fig. 9. The effect of some smooth muscle stimulants on the rate of egg transport in mice compiled from [4, 36, 37].

Compound	Dose mg/kg	Day of Pregnancy when Dosed	Mean % Oviduct Traversed (± SE) on Day 4 of Pregnancy	Mean No. Eggs Recovered	Mean Esophageal Temperature °C ± SE
Reserpine	10	1	27.8 [1])	8.6	–
Dose Vehicle	–	1	100 [2])	8.0	–
Tetrabenazine	200	1–3	43.2 ± 4.13	8.0	28.6 ± 1.28
Chlorpromazine	20	1–3	37.4 ± 3.13	8.2	31.9 ± 2.75
Dose Vehicle	–	1–3	100 [2])	9.6	39.0 ± 0.02

[1]) Measured center position of egg group.
[2]) Position of eggs recovered from uterus.

Studies of the mechanism of action of the compounds showed that all produced a marked hypothermia which was implicated in the arrested egg transport. Hypothermia produced in mice without drugs but in environmental cages was effective in preventing egg transport, while prevention of the fall in body temperature completely inhibited the drug-induced effects on egg transport.

The above review of the progress made in our knowledge of the control of egg transport with drugs is encouraging. Many different compounds have been shown effective in regulating egg transport and they have each contributed to our understanding of the complex mechanisms underlying such transport and the species differences that exist. Hopefully, this information will assist in the development of new drugs effective in problems of infertility and especially in the initiation of unique methods of contraception.

References

[1] *Bencze, W. L., Carney, R. W. J., Barsky, L. I., Renzi, A. A., Dorfman, L. & de Stevens, G.* (1965), Synthetic Estrogens and Implantation Inhibitors, Experientia **21**, 261.

[2] *Bennett, J. P., Kendle, K. E., Vallance, D. K. & Vickery, B. H.* (1966), A Comparison of the Antifertility and Hormonal Activities of Some New Synthetic Estrogens, Acta Endocrinol. **53, 433.**

[3] *Bennett, J. P.* (1967), Artificial Insemination of the Squirrel Monkey, J. Endocrinol. **37**, 473.

[4] *Bennett, J. P. & Kendle, K. E.* (1967), The Effect of Reserpine Upon the Rate of Egg Transport in the Fallopian Tube of the Mouse, J. Reprod. Fertil. **13**, 345.

[5] *Bennett, J. P., Vickery, B. H. & Dorfman, R. I.* (1969), Some Effects of Progestins on Pregnancy in the Rabbit, Proc. N. Amer. Conf. Fertil. & Steril., Jamaica, West Indies.

[6] *Bennett, J. P., Vickery, B. H. & Dorfman, R. I.* (1969), The Effect of Estrogens Upon Egg Transport (in prep.).

[7] *Black, D. L. & Asdell, S. A.* (1958), Transport Through the Rabbit Oviduct, Amer. J. Physiol. **192**, 63.

[8] *Burdick, H. O. & Pincus, G.* (1935), The Effect of Estrin Injections Upon the Developing Ova of Mice and Rabbits, Amer. J. Physiol. **111**, 201.

[9] *Chang, M. C.* (1964), Effects of Certain Antifertility Agents on the Development of Rabbit Ova, Fertil. & Steril. **15**, 97.

[10] *Chang, M. C.* (1966), Effects of Oral Administration of Medroxyprogesterone Acetate and Ethynyl Estradiol on the Transportation and Development of Rabbit Eggs, Endocrinology **79**, 939.

[11] *Chang, M. C. & Bedford, J. M.* (1966), The Effects of Various Hormones on the Transportation of Gametes and Fertilization in the Rabbit, Proc. IV International Congress on Animal Reproduction, The Hague, p. 1.

[12] *Davis, B. K.* (1963), Termination of Pregnancy in the Rat with Norethynodrel, Nature (London) **197**, 308.

[13] *Deanesly, R.* (1961), The Effect of Estrogen on Tubal Transport and Ovo-Implantation in the Guinea Pig, Proc. IV International Congress on Animal Reproduction, The Hague, p. 371.

[14] *Duncan, G. W., Stucki, J. C., Lyster, S. C. & Lednicer, D.* (1962), An Orally Effective Mammalian Antifertility Agent, Proc. Soc. Exptl. Biol. Med. **109**, 163.

[15] *Duncan, G. W., Lyster, S. C. & Wright, J. B.* (1965), Reproductive Mechanisms Influenced by a Diazocine, Proc. Soc. Exptl. Biol. Med. **120**, 725.

[16] *Duncan, G. W.* (1967), Inhibition of Deciduomata Development with Non-Steroidal Antifertility Agents, Proc. Society for Reproduction, Exeter, England.

[17] *Edgar, D. G. & Asdell, S. A.* (1960), The Valve-Like Action of the Utero-Tubal Junction of the Ewe, J. Endocrinol. **21**, 315.

[18] *Eliasson, R.* (1959), Studies on Prostaglandin, Acta Physiol. Scand. **46,** Suppl. 158.

[19] *Euler von, U. S.* (1936), On the Specific Vaso-Dilating and Plain Muscle Stimulating Substances from Accessory Genital Glands in Man and Certain Animals, (Prostaglandin and Vesiglandin), J. Physiol. **88**, 213.

[20] *Gardner, J. N., Gnoj, O., Watnick, A. S. & Gibson, J.* (1964), New Contraceptive Agents with High Activity in Rats, Steroids **4**, 801.

[21] *Greenwald, G. S.* (1961), A Study of the Transport of Ova Through the Rabbit Oviduct, Fertil. & Steril. **12**, 80.

[22] *Greenwald, G. S.* (1962), The Role of the Mucin Layer in Development of the Rabbit Blastocyst, Anat. Record **142**, 407.

[23] *Greenwald, G. S.* (1963), *In vivo* Recordings of Intraluminal Pressure Changes in the Rabbit Oviduct, Fertil. & Steril. **14**, 666.

[24] *Greenwald, G. S.* (1965), Effects of a Dihydronaphthalene Derivative (U-11100A) on Pregnancy in the Rat, Fertil. & Steril. **16**, 185.

[25] *Greenwald, G. S.* (1967), Species Differences in Egg Transport in Response to Exogenous Estrogen, Anat. Record **157**, 163.

[26] *Greenwald, G. S.* (1968), Hormonal Regulation of Egg Transport Through the Mammalian Oviduct. In: Progress in Infertility, *Behrman & Kistner* (Ed.), Little, Brown & Co.

[27] *Grover, P. K., Chawla, H. P. C., Anand, N., Kamboj, V. P. & Kar, A. B.* (1965), New Antifertility Agents 2,3-Diphenylbenzofurans, J. Med. Chem. **8**, 720.

[28] *Harper, M. J. K., Bennett, J. P., Boursnell, J. C. & Rowson, L. E. A.* (1960), An Autoradiographic Method for the Study of Egg Transport in the Rabbit Fallopian Tube, J. Reprod. Fertil. **1**, 249.

[29] *Harper, M. J. K.* (1966), Contrasting Endocrine Activities of *cis* and *trans* Isomers in a Series of Substituted Triphenyl-ethylenes, Nature (London) **212**, 87.

[30] *Harper, M. J. K.* (1967), A New Derivative of Triphenylethylene: Effect on Implantation and Mode of Action in Rats, J. Reprod. Fertil. **13**, 101.

[31] *Harper, M. J. K. & Walpole, A. L.* (1967). Mode of action of I.C.I. 46, 474 in preventing implantation in rats, J. Endocrinol., **37, 83.**

[32] *Harrington, F. E.* (1965), Transportation of Ova and Zygotes Through the Genital Tract of Immature Mice Treated with Gonadotrophins, Endocrinology **77**, 635.

[33] *Hartman, C. G.* (1939), Ovulation Fertilization and the Transport and Viability of Eggs and Spermatozoa. In: Sex and Internal Secretions, 2nd Edition; *E. Allen* (Ed.), Williams & Wilkins, p. 630.

[34] *Jensen, E. N.* (1968), Antifertility Properties of Two Diphenylethenes. Thesis – Department of Pharmacology, Royal Veterinary College, Sweden.

[35] *Kar, A. B., Setty, B. S. & Kamboj, V. P.* (1968), Postcoital Contraception by Topical Application of Some Steroidal and Non-Steroidal Agents, Amer. J. Obstet. & Gynecol. **102,** 306.

[36] *Kendle, K. E. & Bennett,J. P.* (1969), Studies Upon the Mechanism of Reserpine Induced Arrest of Egg Transport in the Mouse Oviduct I. Effect of Hormone Replacement, J. Reprod. Fertil. (in press).

[37] *Kendle, K. E. & Bennett, J. P.* (1969), Studies on the Mechanism of Reserpine Induced Arrest of Egg Transport in the Mouse Oviduct II. Comparative Effects of Some Agents with Actions on Smooth Muscle and Tissue Amines, J. Reprod. Fertil. (in press).

[38] *Ketchel, M. M. & Pincus, G.* (1964), *In vitro* Exposure of Rabbit Ova to Estrogens, Proc. Soc. Exptl. Biol. Med. **115**, 419.

[39] *Lednicer, D., Lyster, S. C. & Duncan, G. W.* (1965), Mammalian Antifertility Agents II. Basic Ethers of 3,4-Diphenylcoumarins, J. Med. Chem. **8**, 725.

[40] *Morris, J. McL., Wagenen, van, G., McCann, T. & Jacob, D.* (1967), Compounds Interfering with Ovum Implantation and Development II. Synthetic Estrogens and Antiestrogens, Fertil. & Steril. **18**, 18.

[41] *Noyes, R. W. & Dickmann, Z.* (1960), Relationship of Ovular Age to Endometrial Development, J. Reprod. Fertil. **1**, 186.

[42] *Petrow, V.* (1966), Steroidal Oral Contraceptive Agents, Essays in Biochem., **2, 117.**

[43] *Pincus, G., Banik, U. K. & Jacques, J.* (1964), Further Studies on Implantation Inhibitors, Steroids **4**, 657.

[44] *Robinson, T. J.* (1951), The Control of Fertility in Sheep II., J. Agricul. Sci. (Cambridge), **41**, 6.

[45] *Rowson, L. E. A.* (1951), Methods of Inducing Multiple Ovulations in Cattle, J. Endocrinol. **7**, 260.

[46] *Wakeling, A.* (1965), Effect of Norethynodrel on Spontaneous Ovulation and Fertilization in Adult Female Rats, J. Reprod. Fertil. **10, 1.**

[47] *Wislocki, G. & Snyder, F.* (1933), The Experimental Acceleration of the Rate of Transport of Ova Through the Fallopian Tube, Bull. Hopkins Hosp. **52,** 379.

Discussion

R. J. Ericsson: If you give estrogens to the male and then mate, you will get an alteration in ovum transport and subsequently an anti-fertility effect; the male then serves as an unique applicator for the anti-fertility. It is also not unreasonable to expect estrogens to do this – nafoxidine hydrochloride (U-11,100A) will also do the same thing.

J. P. Bennett: You refer to giving the estrogen to the male and then mating. The problem there, I think, is that you are probably cross-contaminating the females, if you are keeping them in cohabitation. They are coprophagic animals and they can pick up the estrogen from the male. So that you are, in effect, indirectly dosing the female.

R. J. Ericsson: Yes, we realize the problem exists ...

J. P. Bennett: We had the same problem when we were working in the male and we did discover that there is cross-contamination.

R. J. Ericsson: If you put the female in for only one night and you put them in a clean cage, and then use an orally-active estrogen, I think you eliminate the problem.

J. P. Bennett: Well, not totally.
All I can say is that in the tests we have done in the male with estrogens, we have been able to show cross-contamination from the feces. We have also tried both situations; putting the male in for only a certain time ... this helps, but I do not think it eradicates the problem completely.

R. P. Blye: I believe that some of Dr. *Ericsson's* comments refer to his studies with carbestrol, which is a cyclohexenecarboxylic acid. This compound is metabolized in the rat to a very active estrogen, and some of it is excreted unchanged. I quite agree that the females are picking up this drug or its metabolites from the males either via the feces or via the urine. We must remember also that the semen is contaminated with urine during its passage through part of the reproductive tract. The zygotes are very, very sensitive to small doses of these potent orally-active estrogens, so that I am sure that exposure of the female for only one evening to urine or feces of a male treated with the compound would be adequate to interrupt pregnancy.

J. P. Bennett: We can even carry this point a stage further. In some cases where we have been dosing animals with very potent estrogens, other animals in the same cage quarters, or in the same room will show estrogenic effects due to airborne contamination.

R. J. Ericsson: The work we did with U-11,100A was never published – we used ^{3}H-U-11,100A and studied radioactivity of rabbit sperm. As I recall, we calculated that somewhere between 5 and 6 million molecules did bind to sperm in vivo. Not only is it in the seminal secretions, but also bonded directly to the sperm, and therefore capable of transfer.

J. P. Bennett: That is interesting.

R. J. Ericsson: This is also true of the estrogens (*Ericsson* et al. Acta endocrin. **56**, 424, 1967).

J. P. Bennett: I do not know.

C. E. Adams: May I comment on your reference to superovulation leading to acceleration of egg transport, particularly with reference to the rabbit, because in this species it is rare to find accelerated transport, except possibly in some immature animals. Accelerated transport and failure of fertilization seem to go together. In superovulated ewes egg transport is not normally accelerated, unless the number of ovulations exceeds 20.
It is also interesting that in the sheep one does not normally see accelerated transport, but if the number of ovulations exceeds 20, then accelerated transport is common.

J. P. Bennett: This suggests progestins are effective in the sheep.

C. E. Adams: It used to be supposed that increased progesterone levels were responsible for accelerated egg transport in both the sheep and cow. However, I think one must also pay attention to the degree of follicular development that exists before ovulation and the residual influence of estrogen. Hitherto, this aspect has been neglected.

R. G. Edwards: I would like to take up this point that Dr. *Adams* raised, because I am very surprised to hear you say that in the mouse super-ovulation causes loss of eggs.

J. P. Bennett: I was, in fact, quoting someone else's work –

R. G. Edwards: If mice are super-ovulated, many of them carry large numbers of fetuses to full term, and it looked in preliminary work as if tubal transport was normal, as well as implantation and fetal development. I think it is a question of using large doses of gonadotropins.

J. P. Bennett: This is possible. In the monkey there is definite acceleration with gonadotropins. When the dose is brought down until there is only one egg arising from the ovary, the egg still arrives in the uterus ahead of schedule. This causes complications if you are trying to do artificial insemination and development in this species.

P. Eckstein: In connection with the last bit you said, how do you know when ovulation had occurred, and how certain can you be that transport had in fact been accelerated ?

J. P. Bennett: This is a good point. It is a problem that has come up in *Marston* and *Kelly's* work and others working in the primate. I can tell you that we did work through a large number of squirrel monkeys to a point where we felt fairly sure we were recovering the egg, from the top of the fimbria. It was also possible to see that the stigma on the ovary was ruptured. Now there is a possibility from that of maybe 12 hours out, I agree, but I do not think much more than that.

P. Eckstein: You mean you did two laparotomies, you looked ...

J. P. Bennett: We examined about 30 to 40 monkeys, and traced the ovulation back, through several laparotomies.

P. Eckstein: Not in the same animals?

J. P. Bennett: No, step-wise in different animals.

P. Eckstein: I did not think this can be done. The other point I wanted to raise was this – in your original figure you gave tubal transport speeds for a number of species. The one animal 'out' seemed to be the pig: 50 hours compared with 70, 80 or 90 hours and even longer for almost all other species. Is there any explanation for that?

J. P. Bennett: I wish Dr. *Polge* was here from Cambridge to answer your question; he is the expert in reproduction in the pig.

P. Eckstein: The pig is certainly an exception. It is an established fact; *Pomeroy* found this in the late 50's. It might be even less – 40 hours, under normal circumstances. Can you give any explanation for this?

J. P. Bennett: Maybe the endocrine system is different in the ovary of the pig at this stage.

P. Eckstein: It does not seem to be due to excessive ovulations, because the pig does not produce a large amount of progesterone.

J. P. Bennett: The anatomy is, of course, different, so this may be enough in itself to produce the result. There may be an increased suction from the longer uterus for example.

K. W. Humphrey: There is not much retention at the utero-tubal junction in the pig, but they are retained at the ampullo-isthmic junction for 24–48 hours, and then there is fast transport through the isthmus to the uterus.

A. A. Haspels: You showed an interesting figure about stilbestrol. In 45 females we used 25 mg stilbestrol, or 2 mg of ethinyl-estradiol for 5 days as a morning after pill. When coitus had taken place less than 48 hours before, we saw no pregnancies. And we would like to ask if this is the right dose? Have others seen pregnancies, if stilbestrol was started more than 48 hours post-coital?

J. P. Bennett: How many cycles was that in treatment?

A. A. Haspels: For 5 days, continuously.

J. P. Bennett: But how many cycles of treatment? I mean, you gave it for 5 days after coitus to women, but to how many women and how many cycles were studied?

A. A. Haspels: 45 women. I gave it less than 48 hours after coitus. I gave it for 5 days, 25 mg of stilbestrol a day. Is this a high dose?

Tubal Secretion and Egg Development

H. Koester

Universitäts-Frauenklinik, Giessen/Lahn, Germany

Summary: The tubal secretions, its hormonal regulation and its influence on migrating eggs within different parts of the Fallopian tubes were investigated by using radioactive sulfur as a tracer substance. In rabbits ^{35}S was applied intravenously and excreted by the secretory cells of the tubal mucosa as shown in autoradiographic studies. Since there was no radioactivity found in other tissues, it could be taken as a parameter for the secretory activity. By measuring the radioactivity of the different segments of the Fallopian tubes it could be shown that the main activity is found within the mucosa of the isthmus. Since the utero-tubal junction is closed during the first three days of pregnancy tubal secretions flow from the isthmus towards the infundibulum with different speeds within different segments of the tube. By these different speeds of the tubal flow an explanation can be given for the so-called "tube locking" of the eggs at the ampullary-isthmic junction. At this segment of the tube a balance exists between the cilia, which beat in direction of the uterus, and the tubal secretions, which flow towards the ovarian end of the tubes. This balance can be altered by ovarian hormones.

Sulfur uptake out of the tubal secretions during the first cleavage stages of tubal eggs could be shown by autoradiography in eggs flushed out of the tubes and in eggs, which were sectioned within different parts of the tubes. Sulfur is taken up by the zona pellucida, the cell plasma and by the cell membranes. The possible meaning of the observation is discussed.

If one closes both sides of the rabbit oviduct there is an accumulation of liquid in the Fallopian tube [20]. In 1956 *Bishop* was able to prove that this was due to a genuine secretion of the epithelium in the tube. *Bishop* [2], as well as *Mastrioanni* and his group [15] were able to show that the secretion was subject to cyclic variations. As was proven in the case of the sheep [4], the monkey [14] and in other animals, there occurs at the time of ovulation a sudden sharp rise, and with the development of the corpus luteum, a decrease in the amount of secretion. In cycles where ovulation did not take place the rise in secretion did not occur, and in castrated animals the amount of secretion was reduced to a third of its previous value [13]. The observations suggest, that the tubal secretion is controlled by hormones. It is stimulated by estrogens and inhibited by progesterone.

Manuscript received: 12 May 1969

Cyclic changes were also observable in the metabolic processes in the epithelium and the secretions of the oviduct. We shall not examine the diversity of these processes in this paper, except to say that the cyclic changes were also apparently dependent on hormones. Thus, in each phase of the cycle differing concentrations of lipids, glycogen and glucose, alkaline phosphatase, unspecific esterase, dehydrogenase, peptidase and other enzymes were found. It was the same in the case of anorganic and organic substances, i.e. amino acids. Almost all shared the common characteristic of a change in concentration in connection with the ovulation and continued during the passage of the egg through the tube. These substances were apparently of importance to the ovum.

An observation made by *Friz* [7] seemed interesting to us for this reason. He thought the could prove that the secretion of fluid in the different segments of the oviduct varied with the movement of the ova, and was always strongest at the location where the ova were at that moment. The strongest secretory activity, then, was thought to follow the ova wavelike in its passage from the ovarian end of the Fallopian tube to the uterine end. To investigate this interesting problem was the object of our own experiments performed on rabbits.

In order to ascertain the change in secretory type, the Fallopian tubes were investigated 14 hours and 62 hours after ovulation; that is, on the first and third day of the three-day passage of the ova through the tube. To measure the intensity of the secretion in the various segments of the tube we took advantage methodologically of the fact that parenterally applied anorganic radioactive sulphur is incorporated quite selectively in the mucopolysaccharids, which make up the largest part of the tubal secretions. A few hours after application of sulphur the radiograph showed activity only in the secretory cells of the epithelium, not, however, in the other tissues. The amount of radioactivity could thus be used as a parameter for the secretory activity.

The Fallopian tubes were divided into five equal segments seven hours after the addition of $Na_2{}^{35}SO_4$. After the segments were reduced to ash the radioactivity was measured with a standardized and automatized measuring system.

In order to enable a comparison in activity between the single tube segments, the radioactivity measured for each segment was set in relation to the median activity of the whole tube. The resulting "relative segment activity" was statistically certified. It was possible, using this method, to compare a sufficient number of tube segments with one another.

During evaluation of the experiments we found that a wave-type run off of the secretion from the ampullary to the uterine end of the tube does not exist. The secretory activity was always strongest in the isthmus and weakest in the ampulla. The reason for this is as follows: the epithelium of the isthmus (Fig. 1) contains

Fig. 1. Isthmical segment (PAS'reaction).

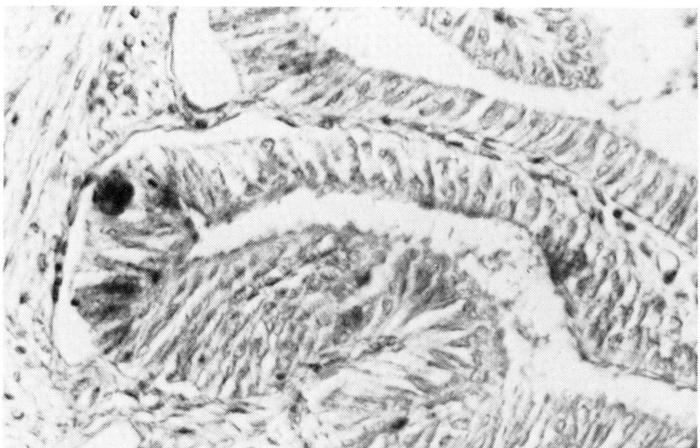

Fig. 2. Ampullary segment (PAS'reaction).

mainly secretory cells and only widely scattered ciliar cells; the epithelium of the ampulla (Fig. 2) contains mostly ciliar cells and only a very few secretory cells.

On the first day after ovulation (Fig. 3) the radiograph of the isthmus shows a corresponding intensive blackening of the film along the entire epithelium. In the auto-

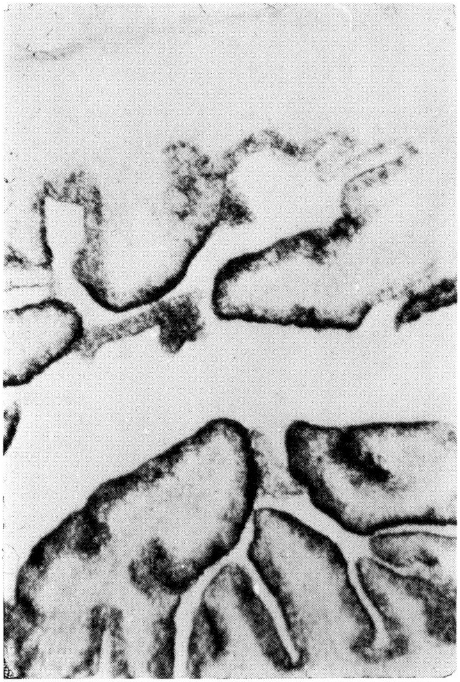

Fig. 3
Autoradiograph of isthmical segment at first postovulatory day.

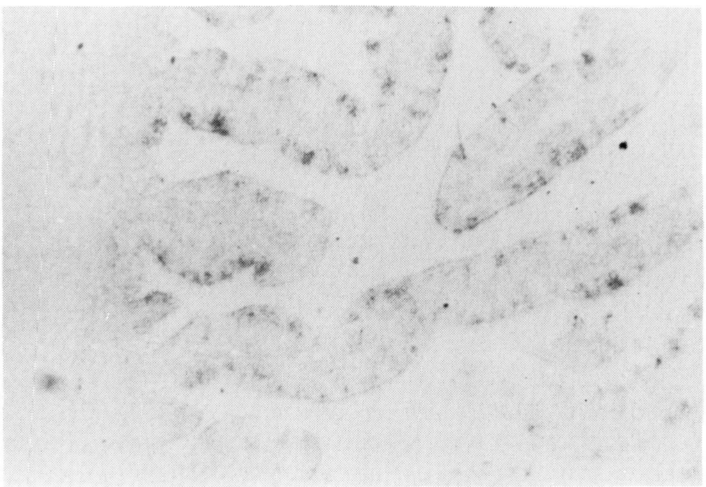

Fig. 4. Autoradiograph of ampullary segment at first postovulatory day.

radiograph of the ampulla (Fig. 4) only a very few active secretory cells can be detected. The measurement curves (Fig. 5) show the same finding. The "relative segment activity" of the isthmical segments 4 and 5 is significantly higher than that of the ampullary segments 1 and 2.

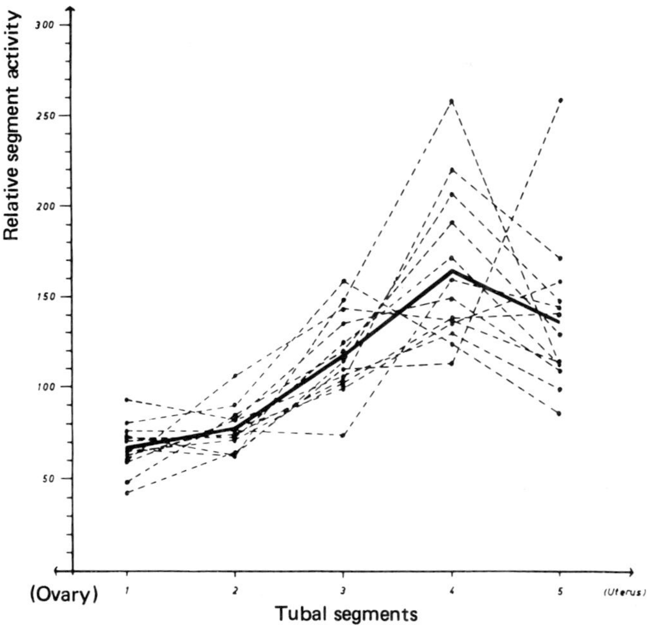

Fig. 5. Relative segment activity at first postovulatory day.

On the third day after ovulation, the last day of the passage of the egg through the tube, the exhaustion of the secretory activity is discernible, and is quite clearly indicated in the radiograph as well. Secretory activity in the isthmus (Fig. 6) is scarcely noticeable. Secretion in the ampulla has also ceased. In the measurement curve (Fig. 7) the "relative activities" of the isthmic and ampullary segments correspond much more than they did before.

The difference between the first and third days is clear when one compares the summal curves (Fig. 8). There is a difference between the first and third days insofar as the "relative secretory activities" of each tubal segment at the ends of the passage approximate one another. Due to the strong concentration of secretory cells in the isthmus the decrease in secretion on the third day after ovulation as described by

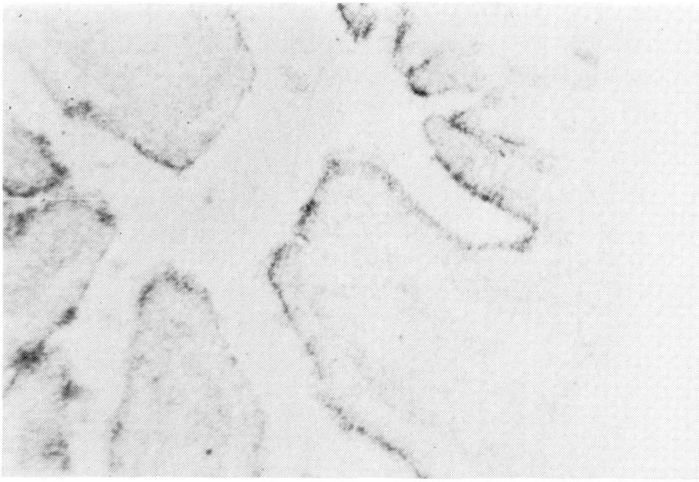

Fig. 6. Autoradiograph of isthmical segment at third postovulatory day.

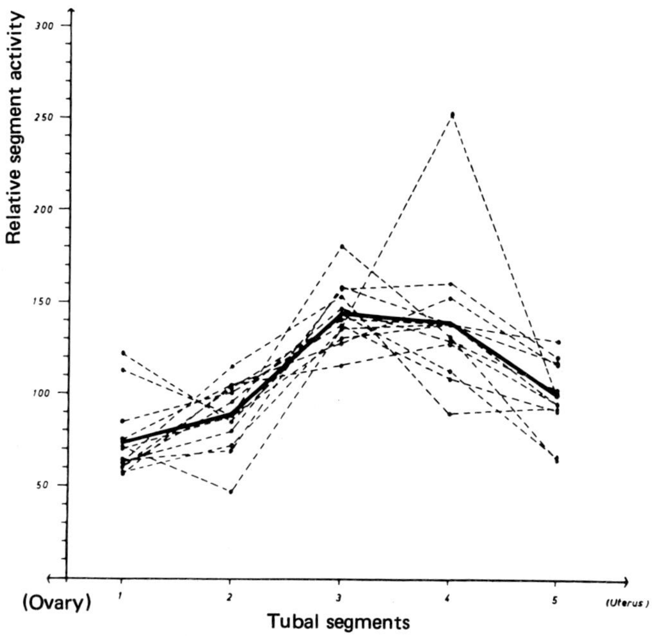

Fig. 7. Relative segment activity at third postovulatory day.

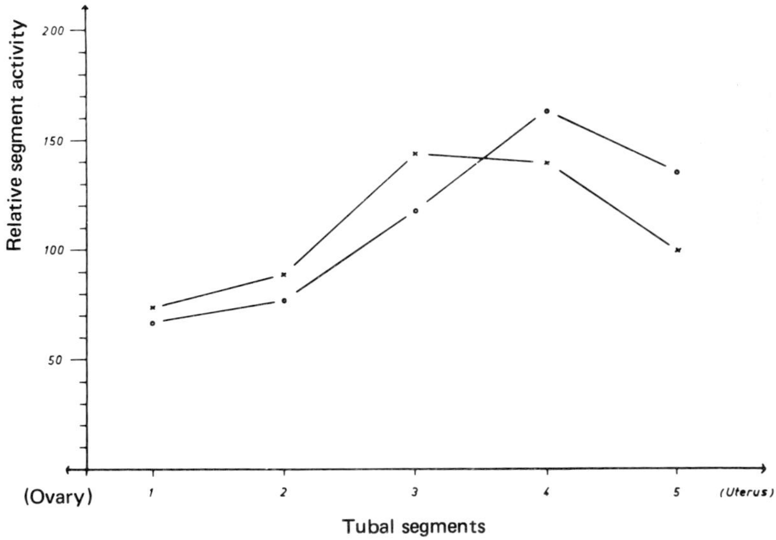

Fig. 8. Comparison of the relative segment activities of first (o).and third (x) postovulatory day.

Bishop and others is especially noticeable, and is indicated by a corresponding decrease in radioactivity. Thus the relation of the activity of the single segments in the different parts of the tube undergoes a change as a result of the decrease in total secretion. This finding is noteworthy, and I will return to it later.

The hormonal dependency of these changes can be demonstrated if an ovariectomy is performed on the animal after ovulation and hormone treatment is undertaken. With a daily dose of 100 γ 17β-estradiol, a consistently strong secretion is observed, even on the third day (Fig. 9). It can be demonstrated by a comparison of the summal curves (Fig. 10) that by administration of an estrogen the type of secretion present on the first day can be sustained.

When 10 mg progesterone (Proluton ® Schering) was administered daily after ovariectomy the autoradiograph of the isthmus (Fig. 11) shows almost no secretion. This is indicated in the measurement curve (Fig. 12) by a correspondence in values between the isthmus and ampullary segments.

This effect is also caused by hormone compounds used to inhibit ovulation (Fig. 13) Rabbits treated after ovariectomy with a combination of 5 γ ethinyl-estradiol and 50 γ Norgestrel (Eugynon® Schering) show a typical progesterone curve on the third day, in spite of the estrogen in the hormone compound. Here, also, the production of secretion exhausts itself and the "relative activities" of isthmus and ampulla reach almost a common level.

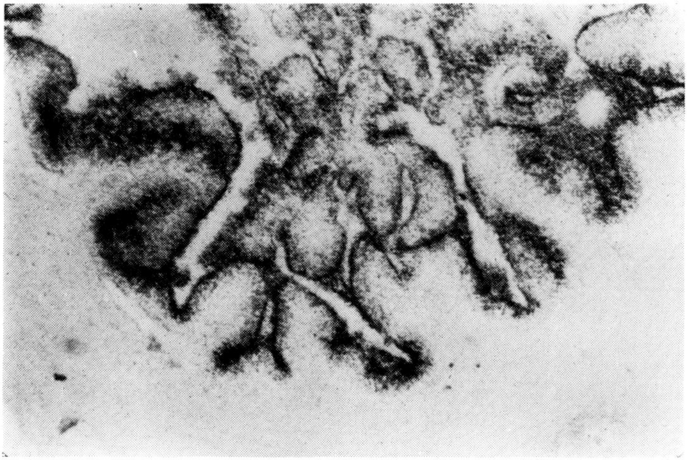

Fig. 9. Autoradiograph of isthmical segment at third day after ovariectomy and estrogen treatment.

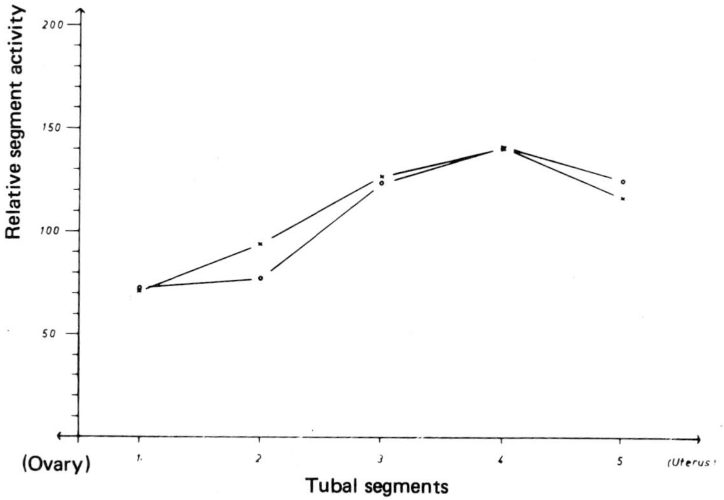

Fig. 10. Comparison of the relative segment activities of first (o) and third (x) day after ovariectomy and estrogen treatment.

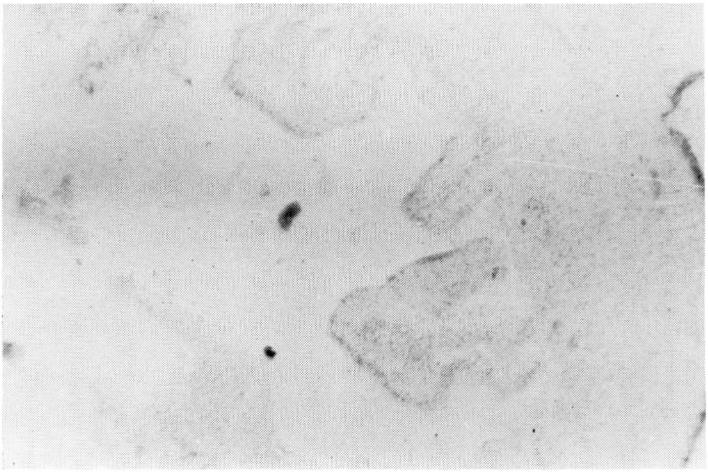

Fig. 11. Autoradiograph of isthmical segment at third day after ovariectomy and progesterone treatment.

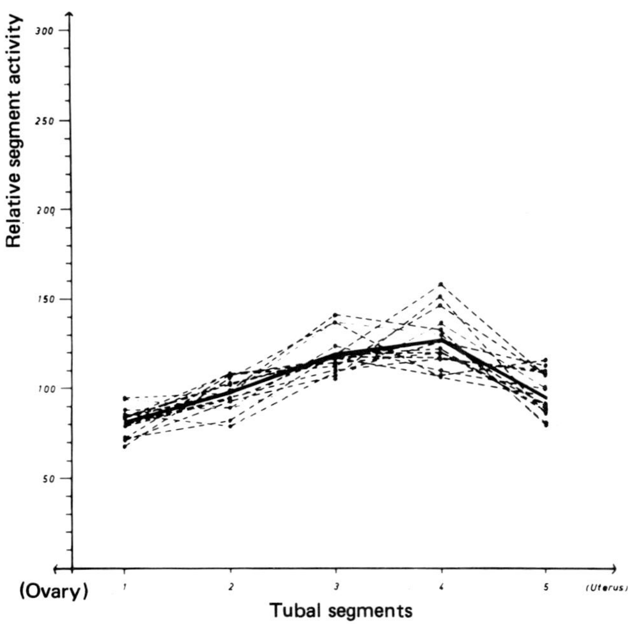

Fig. 12. Relative segment activity at third day after ovariectomy and progesterone treatment.

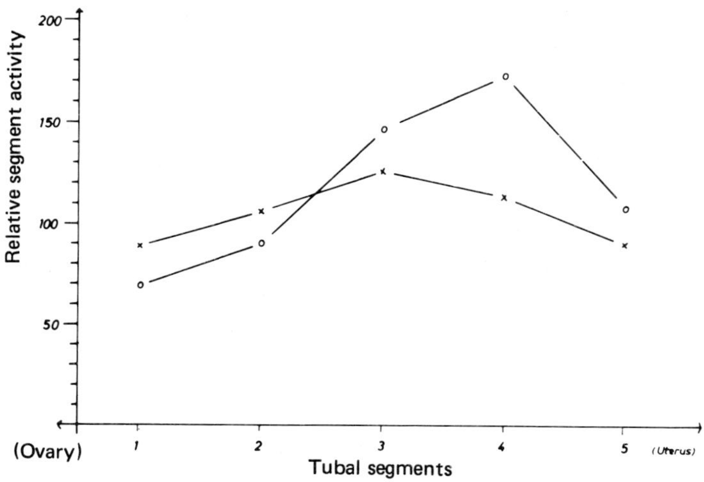

Fig. 13. Comparison of the relative segment activities at third day after ovariectomy and treatment with ethinyl-estradiol (o) and a compound (x) of ethinyl-estradiol and norgestrel.

What is the significance of these observations for the ova present in the oviduct during these three days? The secretion pattern discussed here seems to offer an explanation for the problem of the "tube-locking" at the isthmo-ampullary junction, a problem that has not yet been solved completely satisfactorily. An explanation is made possible when one looks at the secretory activities of the separate segments in relation to their inside diameter.

The measurements of the inside diameter of the segments nearest to the isthmo-ampullary junction yield the following values: ampulla, 1.72 mm average diameter; isthmus, only 0.79 mm average diameter. Calculation of the inside sectional area gives these values: ampulla, 2.32 mm^2; isthmus, 0.49 mm^2. The sectional area of the ampulla is thus 4.73 times larger than that of the isthmus.

Our observations showed that the stronger secretory activity of the tube is in the isthmus; that in the ampulla is much weaker. The preponderant secretion is thus produced in the narrowest part of the tube. We know from the work of *Black* and *Asdell* [3] that the utero-tubal junction is closed during the three-day passage of the eggs through the tube. Thus the tubal secretion cannot flow into the uterus, but must flow from the isthmus into the ampulla, and from there into the abdominal cavity. This results in the formation of definite flow patterns inside the tube.

The differing sizes of the tube caliber create differing secretion flow velocities. These are defined by the law of continuity:

$$\text{constant} = q_1 \cdot v_1 = q_2 \cdot v_2 \ .$$

This equation states that the flow velocity (v) is dependent upon the sectional area of the tube (q). In other words the same amount of liquid flows faster in a small tube than in a large one.

Assuming the tube would have a smooth inner surface calculations using the sectional areas of the isthmus and the ampulla would yield that the flow velocity of the secretion produced in the isthmus upon its arrival in the ampulla is only one fifth as great as in the isthmic segment 4. (Fig. 14)

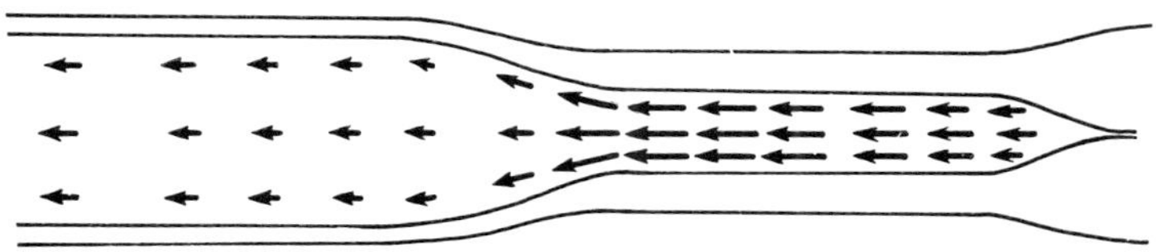

Fig. 14. Direction and amount of tubal flow.

Although the absolute amount of secretion produced on each day is known [2, 15] the velocity cannot be precisely calculated because the tube bore is not smooth. It is made narrower by the villi of the mucous membrane. Taking into consideration the fact that the "continuity equation" can be applied only "in approximation", its principle though holds true for the explanation of the flow patterns as Professor *Scharmann* of the Department of experimental physics of our university assured me.

Once again, the secretion has a high velocity in the isthmus and a lower velocity in the ampulla, at least in the segments bordering on the isthmo-ampullary junction. If one relates this to the ciliar activity, the factor most important in egg-transfer [5, 10, 12], it becomes possible to explain the differing speeds of egg migration in the differant segments and the phenomenon of "tube-locking" at the isthmo-ampullary junction.

The epithelium of the ampulla consists primarily of ciliar cells and only a few widely scattered secretory cells. The stroking movement of the cillia propels the ova quickly along in this part of the tube in the direction of the uterus [10, 11]. This takes place against the flow of secretory fluid. However, the velocity of the fluid is so low that the propelling force of the cilia is apparently stronger. The tube bore becomes narrower at the isthmo-ampullary junction and the fluid velocity rises sharply. At the same time the number of ciliar cells in relation to the number of secretory cells becomes much smaller. The balance between the propelling force of the cilia and the opposing secretory flow changes in favor of the latter. This process takes place under the influence of the estrogens, which are predominant during the estrus and on the first day after ovulation.

As the progesterone effect increases towards the end of the three-day egg migration, the frequency of the ciliar stroke increases by 20% with their highest speed 48 hours after ovulation [5], while the secretion in the isthmus decreases very sharply. An imbalance in forces occurs, which leads to the entrance of the ova into the isthmus.

This mechanism is assisted by a morphological change in the epithelium. Under the influence of estrogens the secretory cells arch far over the ciliar cells. This is of no importance in the ampulla (Fig. 15) due to the small number of secretory cells present. In the isthmus (Fig. 16), however the effect is a mechanical one, for here the small number of ciliar cells are set deeply among the high secretory cells. The tips of the cilia barely reach into the tube bore and cannot exercise an effect.

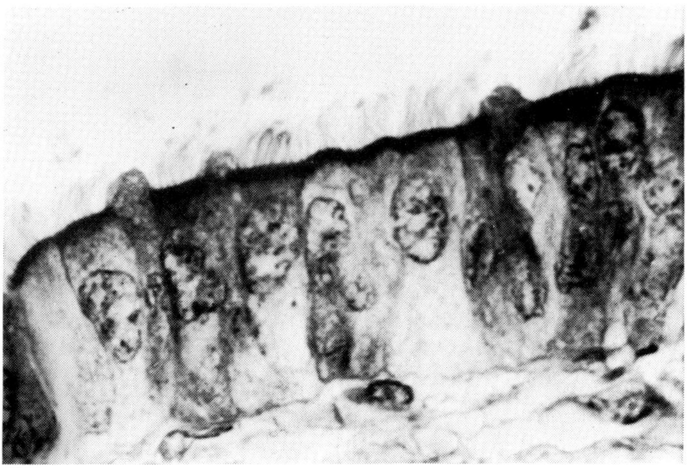

Fig. 15. Epithelium of the ampulla at first postovulatory day (*Masson* trichrome staining).

In the second phase of the egg passage the secretory cells grow shorter due to the influence of progesterone, feeing the ciliar cells (Fig. 17). The cilia now project far into the tube bore, even in the isthmus, and can exercise their forward driving force once again.

The assumption of *Greenwald* [8], that not the muscular tube motility but a blockade mechanism in the isthmo-ampullary junction regulates the egg transport, can be proven correct, if one determines the secretion-type of the tube and the influence of

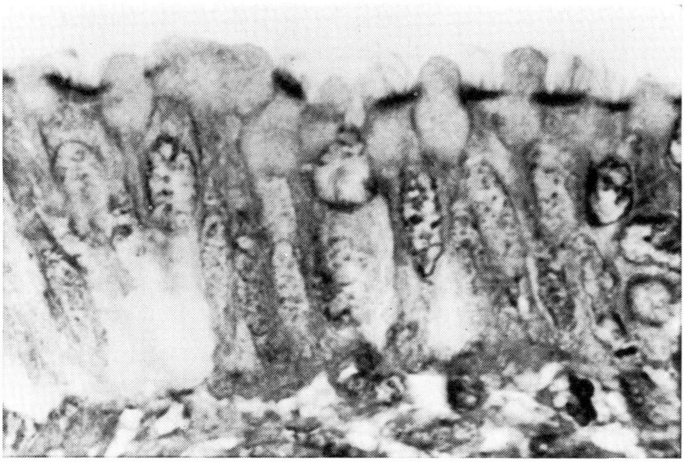

Fig. 16. Epithelium of the isthmus at first postovulatory day (*Masson* trichrome staining).

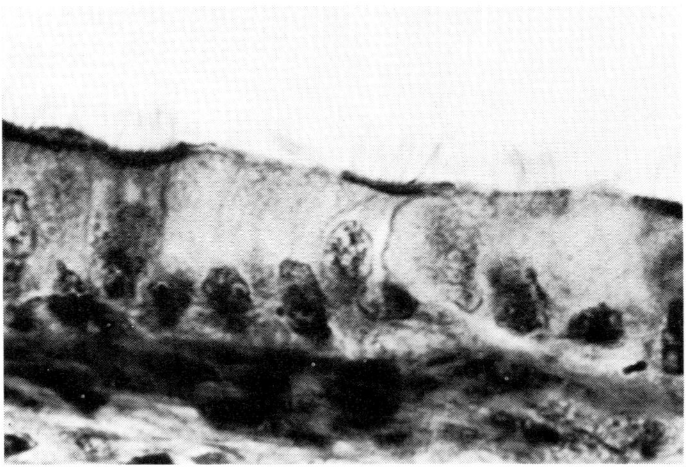

Fig. 17. Epithelium of the isthmus at third postovulatory day (*Masson* trichrome staining).

particular hormones upon it. This blockade is not, however, the result of some sort of closure mechanism usually assumed to act at this point, but is apparently caused by a balancing-out of ciliar stroke and opposing secretion flow, whose velocity is dependent upon the changing width of the tube bore and the hormonally influenced secretion production in the isthmus. This mechanism regulating the egg migration is in its turn dependent upon a particular estrogen-progesterone relationship [9] which effects the closure of the utero-tubal junction [3]. The converging relation between the two hormones which occurs after ovulation appears to be the most important factor regulating the egg migration, for the egg transport through the tube is normal only for the first three to four days after ovulation, as the formation of progesterone is rapidly increasing.

The question whether the tubal secretion is important only for the transport of the ova or whether it also plays a role in the egg development also should be mentioned shortly. It was once thought that the ova itself possessed enough nutrition for the entire migration period.

Shettles' assumption [17] that the granulosa cells still attached to the ova after ovulation supplied its nutrition could not be verified by electron-microscopy [18, 19]. In this connection we examined rabbit ova after i.v. application of radioactive sulphur. We observed them autoradiographically 14 hours after ovulation, at which point the cleavage of the fertilized ova begins and ova in one-, two-, and four-cell stages are to be found in the tubes [6].

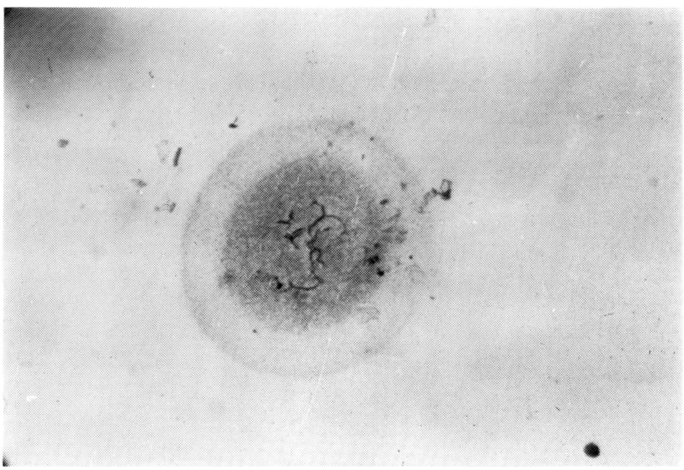

Fig. 18. Autoradiograph of tubal egg flushed out of the tube.

In the autoradiograph (Fig. 18) an egg, obtained by flushing the Fallopian tube, shows a distinctly stronger blackening of the film in the center, which is the area of the zona pellucida, than around the circumference, indicating that the mucin layer contains less sulphur.

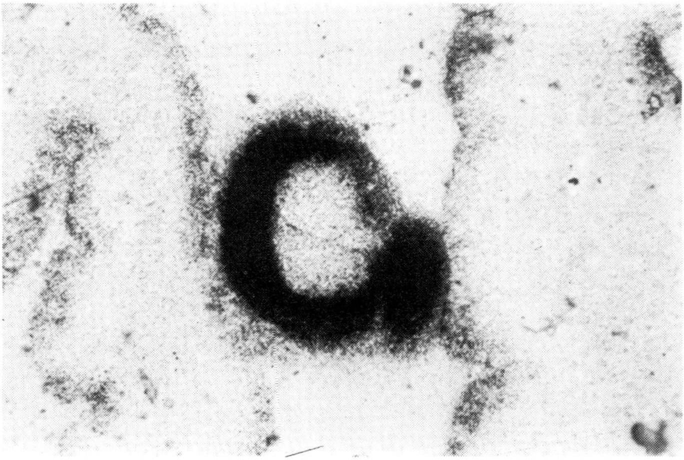

Fig. 19. Autoradiograph of tubal egg in 2-cell-stage.

This is clearer in the autoradiograph (Fig. 19) of a 5 μ thick histological, uncolored section of a tubal ovum in the 2-cell-stage. The zona shows an unusually strong incorporation of sulphur. However, the blastomeres already show an early sulphur incorporation, much earlier in time than the 36-hour development stage described by *Friz* [7] and the uterine stage described by *Moricard* and *Gothié* [16].

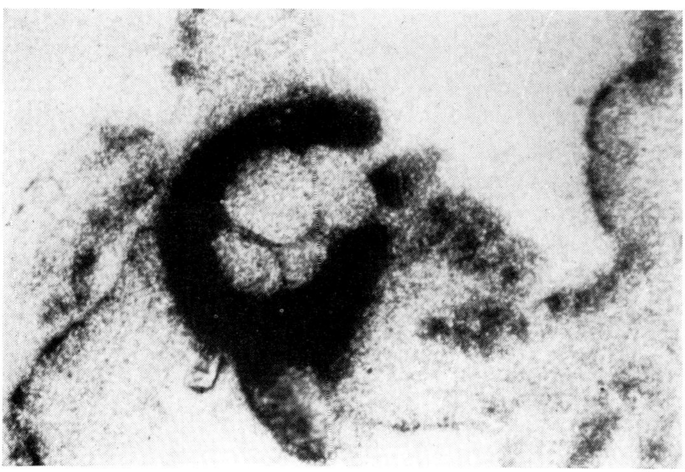

Fig. 20. Autoradiograph of tubal egg in 4-cell-stage.

Even in an 14-hour old ovum (Fig. 20), which has, however, already entered the four-cell stage, there is incorporation of S^{35} not only into the nuclear substance but even more into the nuclear membranes of the blastomeres.

The same distribution of radioactive sulphur is manifest (Fig. 21) in a 14-hour old egg which has not yet undergone cleavage.

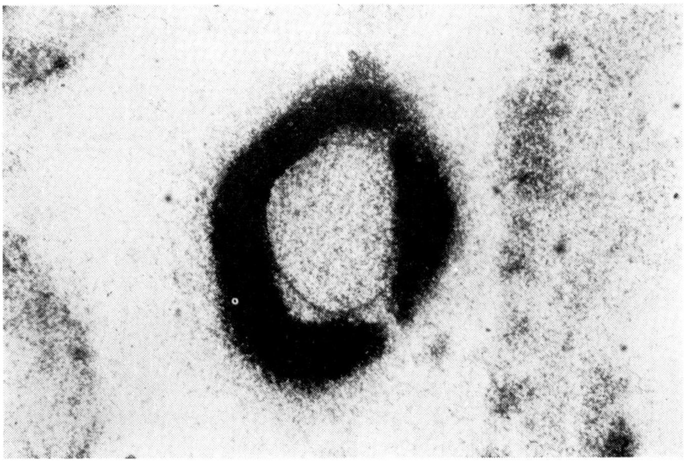

Fig. 21. Autoradiograph of uncleared tubal egg.

The autoradiographs show that the ovum does incorporate a sulphur-containing substance from the secretion at an early point in time. It is not only incorporated in large degree into the zona pellucida, but also into the blastomeres and above all into the nuclear membranes.

I cannot say yet which chemical form of S^{35} is absorbed by the ovum in such differing concentrations. The functional meaning of the unusually strong incorporation into the zona pellucida is not clear either, and is the object of further experiments. It probably has to do with a process connected with a change in construction of the zona pellucida and the maintenance of energy for the ovum. The observations of *Moricard* and *Gothié* [16] and *Ahlgren* [1] support this assumption. One can conclude from the observations, however, that the ovum is not autonomous during its migration through the tube, but already possesses a close relationship to the mothers organism. The tubal secretion is thus of primary importance for the transport and development of the ova.

Acknowledgement: This presentation is an excerpt from "Bibliotheca gynaecologica" **57**, 1970, Karger Basel - New York and printed with the friendly permission of the Karger Verlag.

References

[1] *Ahlgren, M.,* Development of fertilized tubal rabbit eggs in diffusion chambers in vitro. J. Repord. Fertil. **2** (1966), 145.

[2] *Bishop, D. W.,* Active secretion in the rabbit oviduct. Am. J. Physiol. **187** (1956), 347.

[3] *Black, D. L. & Asdell, S. A.,* Transport through the rabbit oviduct. Am. J. Physiol. **192**, (1958), 63.

[4] *Black, D. L., Duby, R. T. & Riesen, J.,* Apparatus for the continuous collection of sheep oviduct fluid. J. Reprod. Fertil. **6** (1963), 257.

[5] *Borell, U., Nilsson, O. & Westman, A.,* Ciliary activity in the rabbit Fallopian tube during estrus and after copulation. Acta obst. et gynec. scandinav. **36** (1957), 22.

[6] *Chang, M. C.,* Fertilization and normal development of follicular oocytes in the rabbit. Science **121** (1955), 867.

[7] *Friz, M.,* Experimenteller Beitrag zur Frage der Milieubeziehungen frühester Entwicklungsstadien des Säugereies. Gynaecologia **148**, (1959), 215.

[8] *Greenwald, G. S., In vivo* recording of intramural pressure changes in the rabbit oviduct. Fertil. Steril. **14** (1963), 666.

[9] *Hafer, E. S. E.,* Endocrine control of reception, transport, development, and loss of rabbit ova. J. Reprod. Fertil. **3** (1962), 14.

[10] *Harper, M. J. K.,* The mechanism involved in the movement of newly ovulated eggs through the ampulla of the rabbit Fallopian tube. J. Reprod. Fertil. **2** (1961) 522.

[11] *Harper, M. J. K.,* The effects of constant doeses of estrogen and progesterone on the transport of artificial eggs through the reproductive tract of ovariectomized rabbits. J. Endocrin. **30** (1964), 1.

[12] *Kneer, M. & Cless, H.,* Flimmerung und Strömung im menschlichen Eileiter. Geburtsh. Frauenhk. **11** (1951), 233.

[13] *Mastroianni, L., Beer, F., Shah, U. & Clewe, T. H.,* Endocrine regulations of oviduct secretions in the rabbit. Endocrinology **68** (1961), 92.

[14] *Mastroianni, L., Shah, U. & Abdul-Karim, R.,* Prolonged volumetric collection of oviduct fluid in the rhesus monkey. Fertil. Steril. **12** (1961), 417.

[15] *Mastroianni, L. & Wallach, R. C.,* Effect of ovulation and early gestation on the oviduct secretion in the rabbit. Am. J. Physiol. **200** (1961), 815.

[16] *Moricard, R. & Gothié, S.,* Etudes sur la fonction de reproduction au moyen du ^{35}S. Méiose – Fécondation – Segmentation. IIIe Int. Congrés on Animal Reproduction, Cambridge, Physiology, Sect. I, 1956, p. 31.

[17] *Shettles, L. B.,* Ovum humanum. Wachstum, Reifung, Ernährung, Befruchtung und frühe Entwicklung. Urban & Schwarzenberg, München und Berlin 1960.

[18] *Sotelo, J. R. & Porter, K. R.,* An electron microscope study of the rat ovum. J. Biophys. & Biochem. Cytol. **5** (1959), 327.

[19] *Stegner, H. E.* Über Reservesubstanzen und nutritive Beziehungen der Säugereizelle während der Blastogenese. Arch. Gynäk. **202** (1965), 255.

[20] *Woskressensky, M. A.,* Experimentelle Untersuchungen über die Pyo- und Hydrosalpinxbildung bei den Tieren. Zbl. Gynäk. **15** (1891), 849.

Discussion

G. A. Overbeck: You rightly pointed out the mechanical aspects of the speed of transport, but you mentioned very little about the possible effect of the viscosity of the secretions, and this, I think, would also have a marked effect on the speed. It could be influenced by the estrogens and progestrogens, and would nicely bring in line the problems of the sperm in the cervix with the problem of the egg in the oviduct.

H. Koester: I do not have experiences in the viscosity of the tubal secretions. So I cannot answer this question.

K. W. Humphrey: Dr. *Koester,* the volume of fluid secreted by the oviduct in the rabbit is in the region of a couple of milliliters a day, is that correct ?

H. Koester: It is up to about 3 ml. This is the highest observation.

K. W. Humphrey: So the actual speed of flow through the oviduct must be quite small.

H. Koester: Well, it is quite small; but the diameter of the isthmus probably is ever smaller than measured in the isolated organs of the experiment, because under the influence of the estrogens, the prostaglandins and so forth, the tubular muscles *in vivo* are more contracted.

K. W. Humphrey: You know, I feel the speed of flow is so low, that the rate of secretion and the viscosity does not come into it at all. Since *Black* and *Asdell* and *Greenwald* and also *Harper* have shown that the segmental contractions of the circular muscle, both in the ampulla and the isthmus is responsible for the movement of ova – I think your observations on secretions are coincidental.

H. Koester: I do not think so. There are other observations, also. *Harper* (1961) could show that, if the muscles of the tubes were paralysed, the speed of the tubal transport increased, because the ciliary activity was free now. *Greenwald* (1963) showed that he could speed up the transport of the ova with a low dose of estrogen, and could 'lock' them with high doses. Since both doses increased the muscular activity, the muscular activity could not be the reason for the rapid transport, or the 'tube locking' of the ova.

K. W. Humphrey: Certainly you have got the two factors working – muscular and ciliary – but I still believe the ciliary is less important, compared to the muscular activity. If you decrease the contractility of the rabbit isthmus, for example, in an ovariectomized animal, you get rapid transport through the oviduct, but this is not solely due to cilia. There is a lack of contraction of the isthmus and ova, with none of this segmental to- and-from movement, just passe straight through. This has been shown many times in sheep also.

H. Koester: This certainly is interesting, but how do you explain the rapid movement of the ova when you paralyse the muscles, so that you do not have any contractions at all, or when you increase the muscular activity with a low dose of estrogen? How do you explain it then?

In vitro Cultivation of Mammalian Ova

R. L. Brinster

Laboratory of Reproductive Physiology, Department of Animal Biology, School of Veterinary Medicine, University of Pennsylvania, Philadelphia, Pennsylvania, USA

Summary: Although mammalian ova have been cultivated *in vitro* with varying degrees of success for approximately 50 years, no completely successful method exists to cultivate embryos of most species from the time of ovulation or fertilization up to implantation. The greatest success has been obtained with embryos of the mouse and rabbit. *In vitro* development will occur, in the case of the mouse, from the one-cell stage to implantation in a few hybrid strains, and, in the case of the rabbit, from the late one-cell stage up to implantation. When the embryos of other species are placed *in vitro* only a few cleavages generally occur.

However, studies made on the preimplantation developmental stages have elucidated a number of important environmental conditions for culture of the embryos. In general, it has been found that culture media which give good results consist of the following:

1) a salt composition similar to blood plasma;
2) a protein concentration of 1 to 10 mg per ml;
3) a glucose concentration of 1 mg per ml; and
4) a pyruvate concentration of about 5×10^{-4} M.

The embryos seem to develop best when placed in drops of medium under liquid paraffin oil in an atmosphere of 5 % CO_2 in air. Embryonic stages after the morula (and perhaps including the norula) seem to be benefited by the addition of serum (usually 10 %) and/or a complement of amino acids and vitamins. It seems certain that much more work will need to be done before we completely understand the environmental and nutritional requirements necessary for successful *in vitro* development of mammalian ova.

Introduction

Attempts to cultivate mammalian embryos *in vitro* date back almost a hundred years. The first report of an attempt to cultivate mammalian ova *in vitro* was that of *Schenk* [72]. He reported observing the first cleavage division in the rabbit and guinea pig after inseminating ovarian ova of the rabbit and guinea pig. A few years later, *Onanoff* [66] claimed *in vitro* fertilization and subsequent cleavage of ova of the rabbit and the guinea pig. He said these embryos developed to the eight-cell stage *in vitro* and that subsequent development occurred when they were placed in the abdominal cavity of females or males. However, most of the work on the cultivation of mammalian ova has occurred during the last thirty-five or forty years,

Manuscript received: 15 May 1969

and the interest in the culture of mammalian ova has increased considerably since then. During the last ten years there has been a tremendous increase in the number of studies in this area, largely stimulated by two desires.

1) the desire to better understand the development of mammals in general and man in particular and
2) the desire to regulate fertility.

Despite the fact that the embryos of many species have been subject to study, substantial success has only been obtained with the embryos of the mouse and the rabbit. When the embryos of such other species as the rat, guinea pig, sheep, goat, cow, monkey, and man are placed *in vitro,* only a few cleavage divisions generally occur [25]. However, it seems only a matter of time until the routine manipulation and development of preimplantation stages of all mammals is possible. This is both because of the tremendous importance of this field and because of the tremendous amount of effort now being devoted to the study of these developmental stages. First, I would like to discuss the general methods that have been employed, pointing out those which seem most useful, and then to discuss culture media in more detail.

General methods

Handling embryos

Some of the information about culturing embryos has recently been reviewed [25], and only those aspects which seem particularly important will be mentioned. As mentioned above, most information is available about the mouse and rabbit embryo. Therefore, what will be said here applies primarily to those species. The first problem one encounters in studying the development of these embryos is the matter of obtaining the experimental material, that is, the embryos.

Ovarian ova or oocytes can be obtained by removing the ovaries from any particular species and dissecting the oocytes from the maturing follicles. Two methods have generally been used for this. In the first method, the ovary is shredded into fine pieces and the oocytes are then picked out of the debris. In general, dissecting microscopes are used in the handling of mammalian ova and a magnification of 12 × to 100 × is employed. The disadvantage of the first method is that the oocytes are scattered among a large amount of debris, and it is time consuming to find them. In addition, oocytes from all stages of follicular development are liberated. The advantage is that liberation of the follicles from the ovaries is rapid. In the second method, the ovary is examined under a dissecting microscope, and only large Graafian follicles are ruptured with a fine probe. In some cases, it may be desirable to flush the larger follicles using a hypodermic needle and syringe. The disadvantage of this method is that searching for the follicles in the smaller mammals, can be time consuming. The advantage is that only ova from large follicles are liberated, and they are relatively easy to find because the amount of debris is less than by the first method. After the oocytes are picked up, they are handled the same as unfertilized ova are handled, as will be described later.

To obtain post-ovulation ova or embryos, one must first be sure that the embryos have been ovulated from the ovary. This is generally done by following the estrus cycle of the mammal or by controlling the estrus cycle with gonadotropins. The second method is used most often because it allows greater flexibility in experimental planning, and in many cases, the yield or the number of ova obtained per animal can be increased by increasing the dose of gonadotropins. In general, the gonadotropins used represent a mixture of ovarian tropic hormones, a preparation containing primarily a follicle stimulating hormone is given first, followed in several days by a luteinizing hormone preparation. In my laboratory, for ovulation in the mouse, we use two International Units of Equinix (Ayrst Pharmaceuticals, New York) followed 48 hours later by two International Units of luteinizing hormone (Pregnyl, Organon, West Orange, New Jersey, USA). Both hormones are injected intraperitoneally. Ovulation occurs approximately twelve hours after the second injection. If fertile ova are desired, the animals are placed with a male at the time of the second injection. For superovulation, five International Units of each preparation are injected intraperitoneally. Some variation among strains in their susceptability or in their response to the hormones is noted and doses must be adjusted to take this into account. In the rat, the doses have not been worked out quite so well, but we have found that 25 International Units of each preparation works fairly well when given to 28-day old rats. Again, it seems probable that each laboratory will have to make modifications of the dose to achieve the desired effect. In the New Zealand white rabbit, six months of age and older, we use 150 International Units of Equinix injected intermuscularly followed 72 hours later by 150 units of Pregnyl injected intravenously. The average yield from superovulated animals in all three species has been between 25 and 30 embryos per female. Timed ovulation and/or superovulation has been achieved in a variety of other mammalian species, including the hamster [45], pig [51] cow [5], Rhesus monkey [80], and human [76].

Embryos are removed from the reproductive tract by one of two methods:

1) by incising the wall of the oviduct or shredding the oviduct, or
2) by flushing an appropriate culture nedium through the reproductive tract.

Method two offers several advantages in that it is quicker, less damage is likely to be done to the embryos, and less debris is mixed with the embryos. One need not flush the entire reproductive tract since the embryos are generally located in a short section. Regardless of what method is used to recover the embryos, it is important that the procedure be executed as quickly as possible, since it has generally been found that there is always a certain amount of tissue, fluid, and debris with the embryos and that the embryos do very poorly in this mixture. Consequently, it is important to recover the embryos rapidly, remove them from the debris, and place them in fresh culture medium.

The embryos are generally picked up and manipulated during subsequent procedures in glass pipettes drawn to the proper diameter. Pasteur pipettes are a readily available source of glass tubing which can be easily drawn to the various bore diameters needed. The bore diameter at the tip of the pipette should be slightly larger than the diameter of the embryo, and the tip should be broken off smoothly and at right angles or with a slight bevel to the long axis of the tubing. It is important that the experimenter gain considerable experience in recovering the embryos, making pipettes, and manipulating the embryos before experiments are undertaken. Once this experience is obtained, it is possible to recover several hundred mouse or rabbit embryos and set them up under the desired experimental conditions within about two to three hours. If difficulty is encountered with embryos sticking to glass surfaces, such as occurs in protein free solutions, manipulations will be facilitated by coating the glassware with silicone. In general, for embryos measuring less than 1 mm in diameter, mouth suction on tubing attached to the Pasteur pipette is fast and easy to control during manipulations. On the other hand, for embryos over 1 mm in diamter, such as five day rabbit blastocysts, mechanical means of inducing suction are superior. The mechanical means which may be used are a rubber bulb, a glass syringe or a mechanical syringe. Fig. 1 shows the pipettes and apparatus used in my laboratory.

For some experiments, the external layers of the embryo must be removed. In some cases, these may be removed by exposure to the appropriate enzyme solutions. To remove the cumulus cells which surround one-cell embryos, hyaluronidase at a concentration of 300 units per ml in a phosphate-buffered salt solution can be used [17]. In addition to the cumulus cells, in the mouse and the rat, the solution will remove the corona radiata. However, in the rabbit, monkey, human, and some other species, hyaluronidase is not effective in removing the corona cells. To remove the corona cells in these species, one can draw the embryos back and forth in a narrow pipette with a bore diameter approximately the same size as the external diameter of the zona pellucida [21]. Alternatively, in the rabbit, the corona cells may be removed by shaking the embryos in 1 ml of culture medium in a serological test tube [6]. The zona pellucida of the mammalian embryo and the mucinous coat of the rabbit embryo may be removed with a solution of pronase at a concentration of 2.5 mg per ml, in phosphate buffered salt solution [17, 58]. It is desirable to watch the embryos and remove them from the solution as soon as the desired effect has been

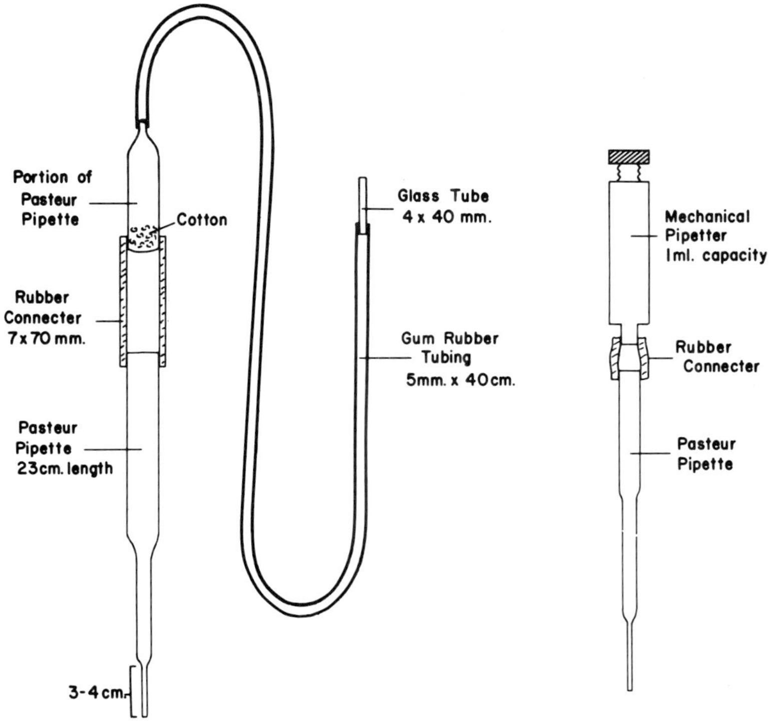

Fig. 1. Pipettes used to handle mammalian embryos. See text for further details.

achieved. This will prevent excessive damage to the embryos by the enzymes. This is particularly important in the case of pronase. It has been found that mouse embryos can be exposed to hyaluronidase at the above concentration for thirty minutes without harmful effect.

Culture vessels

The variety of vessel types and methods which may be used to grow mammalian embryos *in vitro* seems almost endless. Some of these are shown in Fig. 2.

Method A has been developed by *Brinster* [11] to cultivate mouse embryos during the preimplantation period. In this method, 10 ml of paraffin oil is placed in a disposable plastic tissue culture dish, 15 x 60 mm (Falcon Plastics, Inc.), and small drops of 50 to 100 μl of medium are placed under the oil. The oil prevents evaporation, aids in maintaining sterility of the culture medium, and may have some effect on gas diffusion. The culture dish is then placed in a chamber with a controlled temperature and atmosphere.

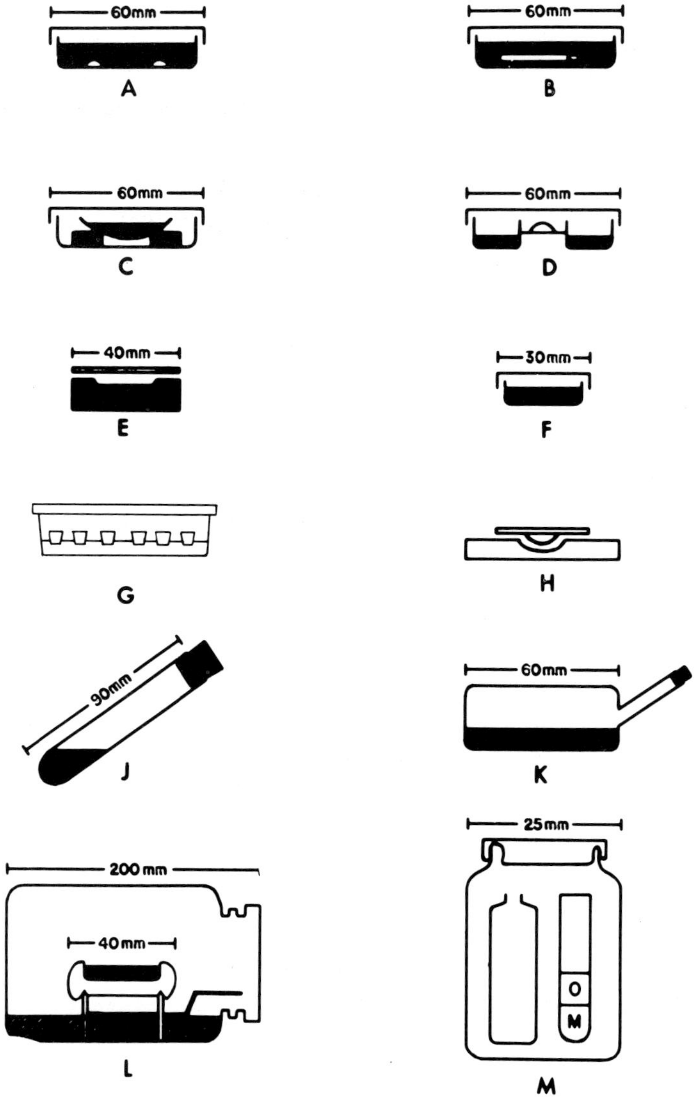

Fig. 2. Culture vessels used for *in vitro* studies on mammalian embryos. See text for further details.

In method B the Petri dish again contains 10 ml of sterile paraffin oil, and on the bottom of the dish are small capillary tubes with an internal diameter of 1 to 2 mm. The embryos are placed in culture medium at the bottom of the tube and a small air bubble is left at the top of the tube. *Mulnard* [62] has used this method extensively.

In method C, a 15 x 60 ml Petri dish serves as a container to hold a sponge or glass wool on the bottom. On top of this is placed a watch glass to hold the culture medium [33]. The glass wool or sponge is moistened and serves to saturate the atmosphere inside the dish with water vapor. This method has been used to culture fallopian tubes of the mouse and ovarian pieces [29].

Method D is similar to method C, but the entire dish with a sponge in the surrounding well is available as a disposable plastic dish from Falcon Plastics. The culture medium volume can be as small as 50 μl or as large as 1 ml.

Method E is an embryological watch glass with a culture medium volume of 1 to 4 ml. The dish may be left uncovered, covered with a glass slide, or the medium may be covered with paraffin oil. This method is particularly useful for culturing large numbers of embryos in single drops and gives excellent results when the medium is covered with paraffin oil.

Method F is a small plastic disposable culture dish, approximately 25 to 30 mm in diameter, containing 1 ml of culture medium. The culture medium may be covered with sterile paraffin oil or may be left exposed to the atmosphere.

In method G, a portion of a plastic disposable tissue culture plate (Falcon Plastics, Inc.) is used. The entire plate contains 60 small numbered wells, each with a volume of 10 μl, into which the medium is placed. The medium is then covered with sterile paraffin oil. A cover is placed on top of the plate. Since the bottom and the cover of the tissue culture plate are transparent, the embryos can be easily observed. This method has been used recently in my laboratory and has proven quite useful.

Method H is the standard hanging drop method for culturing small numbers of cells in a small volume.

Method J is a serological or small tube containing the culture medium. It has been used by *Hammond* [48] and *Whitten* [86]. The tube may be left opened and placed in a controlled atmosphere or gassed and stoppered.

Method K is a standard Carrel flask, as used by *Pincus* [68, 69], containing 2 to 10 ml of culture medium, and it can be gassed as in method J. The advantage is that the embryos can be more easily seen on a flat surface.

Method L has been used by *Mintz* [59] for cultivation of mouse embryos and consists of a large bottle which contains a bicarbonate-buffered balanced salt solution on the bottom to maintain humidity. Embryos are then grown in a small dish in a platform over the salt solution. The atmosphere in the bottle is controlled by gassing with the appropriate gas mixture.

Method M has been developed by *Brinster* [19] to measure carbon dioxide production from radioactively labeled substrates. It consists of a scintillation vial containing a 2 ml ampule with the top broken off and a serological test tube 7 mm by 45 mm. Fifty to 100 μl of culture medium are placed in a serological test tube under three drops of sterile paraffin oil. The ampule is used later in the experiment to contain Hyamine. The top of the scintillation vial is closed with a rubber serum stopper and the vial may be gassed by inserting two 20-gauge needles through the rubber stopper. One needle is connected to a tank containing the proper gas mixture and the other needle acts as an exhaust vent. After approximately 30 to 60 seconds of rapid gas flow, the needles are removed and the scintillation vial then acts as a single closed culture chamber.

Several of the methods shown in Fig. 2 (A, D, E, and J) have been examined for their relative ability to support development of the two-cell mouse embryo into a blastocyst. The results of these tests are shown in Fig. 3. Of the methods examined,

Fig. 3. Comparison of different culture vessels in Fig. 2 for *in vitro* development of mouse embryos.

Treatment Method	Response Replicate Number 1	2	3	4	Mean Angular Response	% Response
1. Petri dish – A	7	6	9	10	55.18	66.6
2. Tubes – J	8	5	7	6	47.43	54.2
3. Watch glass – E	8	3	1	0	27.45	25.0
4. Center-well dish – D	5	4	6	4	38.95	39.6

Response is the number of normal blastocysts from 12 two-cell embryos after three days cultivation *in vitro* in BMOC-2. Medium volume was 50 μl in 1, 2 ml in 2, 2 ml in 3, and 1 ml in 4. The mean angular response is the statistical transformation used in order to compare treatment effects. The standard error of the mean in this example is ± 6.14 angular degrees.

method A produced the greatest number of blastocysts and was by far the easiest to use. Method J produced the second largest number of blastocysts but was considerably more difficult technically. This arises primarily from the difficulty encountered in visualizing the embryos or in recovering the embryos. Further studies in our laboratory indicated that paraffin oil exercised a beneficial effect on the culture system. For instance, development in an ambryological watch glass, such as E, is very poor if the medium is exposed, even if the atmosphere is completely humidified and controlled. This is true even if the watch glass is covered with a glass slide and petroleum jelly seal. However, development with method E is not different than development with method A if paraffin oil is placed over the medium in E.

In fact, most of the vessels shown in Fig. 2 allow good development of embryos provided the surface of the medium is covered with paraffin oil. Unfortunately, in the case of studies on fat soluble compounds, it is not possible to cover the surface of the medium with paraffin oil since the substance under study would rapidly partition itself between the medium phase and the oil phase. In these cases, it is necessary to use a method such as C, D, J. K. or L. The volume of culture medium used in the various culture vessels does not seem to effect development [12] except when the medium is not covered with paraffin oil. In cases where there is no paraffin oil to prevent evaporation, large volumes of medium should be employed to buffer against evaporation and the resulting hyperosmolarity of the medium.

Incubation

One of the most important aspects of cultivating mammalian embryos is the control of the environment to which the culture vessels are exposed. The system used in my laboratory is shown in Fig. 4. The main characteristics of the environment controlled are:

1) temperature,
2) humidity,
3) flow of the gas, and
4) gas composition.

The effects of temperature on the development of mammalian embryos *in vitro* has been studied very little. *Alliston* [2] has shown that rabbit embryos cultivated at 40 °C for six hours do not develop as well as controls cultivated at 37 °C when transferred into foster mothers. Two-cell mouse embryos begin to show irreversible, morphological changes when the temperatures reach 42° to 45 °C (*Brinster,* unpublished) and development is poor at temperatures of 39° to 40 °C. Development of the mouse embryo seems to be slowed at temperatures between 35° and 37 °C. Several studies [31, 47] have found that rabbit embryos can be stored at low temperatures, but development appears to be arrested during storage. When these stored embryos are transferred to foster mothers subsequent development is not as good as in controls. On the basis of studies done on the cultivation of other types of mammalian cells *in vitro,* it seems best to use a temperature of 37° to 37.5 °C to maintain cultures of mammalian ova. Although there is no direct evidence that short periods of exposure to room temperature is harmful to the embryos, it is probably best to maintain the embryos as close as possible to 37 °C during the flushing and recovery period and during periods of observation.

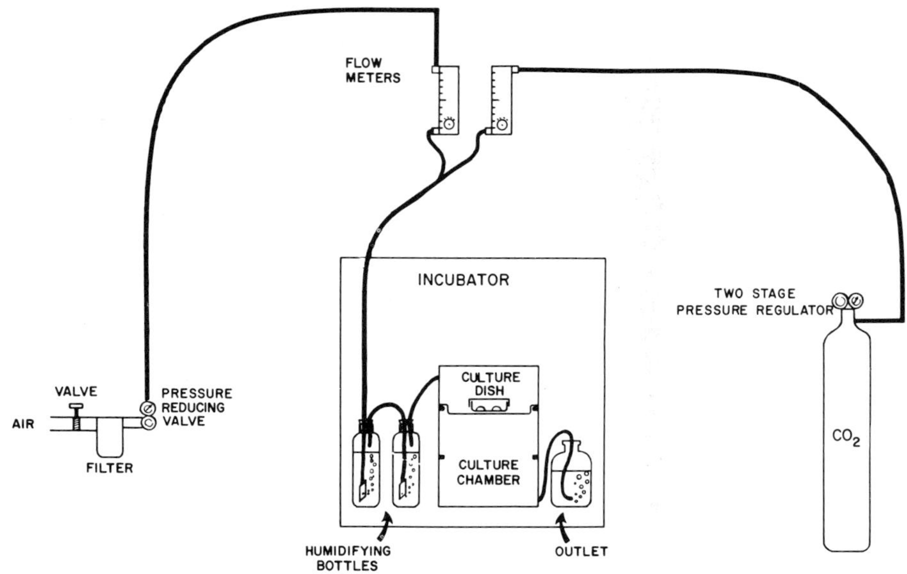

Fig. 4. Equilibration of a 25 liter volume with incoming gas at a flow rate of 1 liter/min. System used to control the environment of the embryo cultures (not drawn to scale). The air is supplied under pressure from a central source or aquarium pump. A valve is placed in the line to turn the flow on or off, and a filter is used to remove dirt and oil residues from the air. The filtration necessary depends on the quality of air supplied. A single-stage reducing valve regulates the pressure. Five to 7.5 pounds per square inch is desirable for flexible rubber hose lines. The capacity of the air flow meter is determined by the culture chamber size (see text). Carbon dioxide (100 %) is delivered from a commercial tank by means of a two-stage regulator at a pressure of 5 to 7.5 pounds per square inch. The capacity of the CO_2 flow meter is determined by the capacity of the air flow meter and the required composition of the final gas mixture. For 5 % CO_2 in air, the CO_2 meter is one-twentieth the capacity of the air meter. Mixing of the gas occurs in the delivery lines. Gas washing tubes in one liter bottles are used to humidify and warm the gas mixture. The culture chamber must be airtight, and the size is determined by experimental requirements. The outlet tube is submerged in water and serves to indicate continous flow [25].

When embryos are cultured in medium covered with mineral oil, the relative humidity of the atmosphere probably is not very important. However, when embryos are cultivated in medium exposed to the atmosphere it is extremely important that the atmosphere be saturated with water vapor, otherwise evaporation of the culture medium and subsequent changes in osmotic pressure of the medium will result. Development is decidedly poor when the osmotic pressure of the culture media is above that found in blood serum [16].

A desirable flow rate for the gas through the culture chamber depends on two things, the size of the chamber and the time one is willing to wait for equilibration. The formula for the change in gas composition within the culture chamber is a differential equation and can be solved for any of the desired values. For instance, the time necessary for the gas in the chamber to equilibrate can be obtained from the following formula:

$$V \frac{dC}{dt} = FC_x - FC,$$

where

- C the concentration in the chamber,
- C_x the concentration of the entering gas,
- V the volume of the culture chamber,
- F the flow rate in liters per minute,
- $\frac{dC}{dt}$ the change in concentration with time.

If the initial concentration in the chamber is zero, the above formula reduces to

$$C = [1 - e^{-(F/V)T}]C_x;$$

For C to be 99 % of C_x, $e^{-(F/V)T}$ must be

$$0.01, \text{ or } \frac{F}{V} T = 4.60$$

if F = 1 liter/min and V = 25 liter, F/V = 0.04/min; therefore,

$$T = \frac{4.60}{0.04} = 115 \text{ min for 99 \% equilibration.}$$

The composition of the gas in the chamber from 0 to 115 minutes is shown in Fig. 5. The time necessary for partial equilibration can be calculated for various chamber sizes and flow rates.

This is an important formula, since the composition of the gas within the culture medium is dependent upon the composition of the gas in the gas phase of the culture chamber. Equilibration of the culture medium can take place no faster than the equilibration of the culture chamber volume with the composition of the inflow of gas. In fact, equilibration of the culture medium will show a definite time lag behind the equilibration of the gas phase in the culture chamber. This time lag is dependent on the physical barriers to diffusion of the gas into the medium. Diffusion of the gas into the medium, or from the medium, is slowed by the layer of paraffin oil as in method A. This may account for part of the beneficial

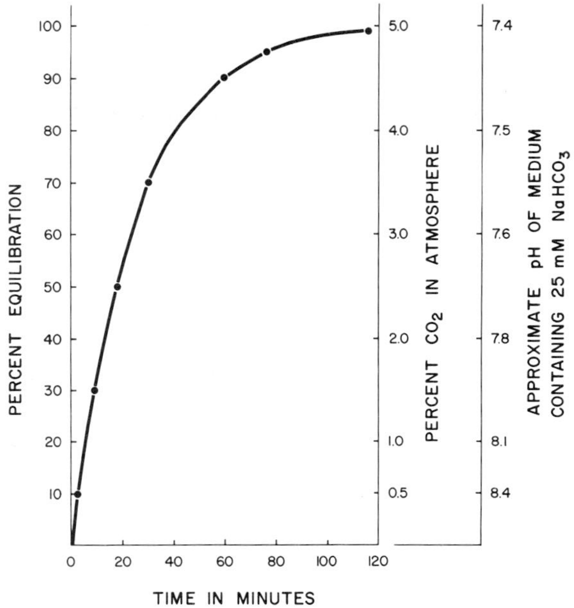

Fig. 5
Effect of osmolarity on the development of two-cell mouse ova. The left abscissa gives the percent equilibration of the culture chamber volume with the incoming gas. The inner right abscissa gives the composition of the gas in the chamber when the incoming gas is 5 % CO_2. The outer right abscissa gives the pH of a medium containing 25 mM $NaHCO_3$ if there was instanteneous equilibration of the gas with the medium. Effect of osmolarity on the development of two-cell mouse ova.

effects of the paraffin oil. This interference is due primarily to the interphase between the medium and the oil and between the oil and the atmosphere, and not primarily to a decrease in mobility of the gas molecules within the oil. This increase in diffusion time of the gas in those culture vessels where paraffin oil covers the culture medium is beneficial when the vessels are to be used for short periods of time outside the incubator. It takes several times as long for precipitation of the medium to occur in a vessel where paraffin oil covers the surface as it does in a vessel where the medium is exposed directly to the atmosphere.

The gas composition of the atmosphere in which the cultures of the embryos are maintained has received very little attention. Generally, there are three major components of the gas: nitrogen, oxygen and carbon dioxide. Nitrogen is considered inert and is varied depending on the amount of the other components desired. The oxygen tension of the culture medium has received some attention. *Whitten* [86] showed that the eight-cell mouse embryo would develop in an atmosphere of 5 % CO_2 in air or nitrogen, but not in 5 % CO_2 in oxygen. In our laboratory, using a continuous flow system we obtained slightly different results. Like *Whitten,* we found that two-cell embryos when grown in 5 % CO_2 and oxygen degenerated within 24 hours. However, in contrast to the findings of *Whitten,* we found that the two-cell mouse embryo had a definite and measurable requirement for atmospheric oxygen during the culture period. The embryos invariably degenerate in pure nitrogen without undergoing even one cleavage. At a concentration of 1.15 % oxygen, two-cell mouse embryos developed normally to blastocysts. At 0.56 % oxygen there is markedly reduced response, but a few embryos will develop to the blastocyst stage [3]. *Daniel* [38] has recently suggested that 10 % oxygen is the best concentration in which to culture 5 to 7 day rabbit embryos.

The amount of carbon dioxide which should be contained in the atmosphere is determined by whether the embryos require CO_2 in the medium and by the buffer system employed. Studies in our laboratory (*Thomson,* personal communication)

Fig. 6. Brinster's medium for ovum culture BMOC-2 (Source [16]).

	mM	Gm/l	ml of 0.154 M stock in 13 ml
NaCl	94.88	5.546	5.90
NaLactate	25.00	# 2.253	2.10
NaPyruvate	0.25	0.028	*2.10
KCl	4.78	0.356	0.40
$CaCl_2$	1.71	0.189	**0.20
KH_2PO_4	1.19	0.162	0.10
$MgSO_4 \cdot 7H_2O$	1.19	0.294	0.10
$NaHCO_3$	25.00	2.106	2.10
Pen Strep	100 U/ml of Pen, 50 µg/Ml of Strep		
Bovine Serum Albumin	1 mg/ml = 1 gm/l		

NaLac added as liquid prepared as follows:
Add 1.82 ml of concentrated lactic acid (85–90 %) to 200 ml of double distilled H_2O. Neutralize to pH 7.4 (about 15–20 ml of 1 N NaOH). This is enough lactic acid to make one liter of medium.

* NaPyr stock is o.00154 M Pyruvate in 0.154 NaCl (17 mg/100 ml).
** $CaCl_2$ stock is 0.11 M.
It is desirable to add 1 mg/ml glucose to this medium [27].

have shown that when noncarboxylic buffering compounds are substituted for bicarbonate in *Brinster's* medium for ovum culture (Fig. 6) development of the embryos is markedly reduced. For instance, when Tris or phosphate replaced bicarbonate in the medium, two-cell embryos underwent one, two, or three cleavages and then stopped dividing. However, the control embryos in the BMOC-2 with bicarbonate developed into blastocysts. The Tris and phosphate media were kept in air and the bicarbonate medium in 5 % CO_2. Phosphate buffer seems to be an acceptable substitute for bicarbonate when the medium is to be employed for short term experiments or during manipulation of embryos [57].

If a bicarbonate buffer system is employed, then close attention must be payed to both bicarbonate concentration and the gas composition of the atmosphere over the culture medium, since both of these affect pH. For example, with a bicarbonate concentration of 25 mM, the pH of the medium in Fig. 6 is 7.4 in 5 % CO_2. Lowering or raising the bicarbonate concentration tenfold, lowers or raises the pH one unit, respectively. If the bicarbonate concentration is held constant at 25 mM and CO_2 concentration in the atmosphere is raised from 5 % to 50 % then the pH falls approximately one unit from 7.4 to 6.4. On the other hand, if CO_2 concentration in the atmosphere falls from 5 % to 0.5 %, the pH of the culture medium rises approximately one pH unit, from 7.4 to 8.4. These changes in pH have a marked effect on the ability of the embryo to develop *in vitro* [13]. It has been my experience that problems associated with pH, bicarbonate concentration and CO_2 composition of the atmosphere are some of the most troublesome obstacles encounterd in the cultivation of mammalian embryos. It is quite common to discover some fundamental error relative to these physical principles when trying to ascertain why embryos are not growing in a newly established culture laboratory. When studying the development of mammalian ova *in vitro,* it is especially important to be able to assess accurately the effects of cultivation and various experimental procedures on the development of the embryos. There are a number of criteria which can be used for this assessment. Morphological criteria have been used to a considerable degree in the past. Among these are the development of the blastocyst, cleavage in experiments on fertilization, the number of cleavage divisions which occur, or the general morphological appearance of the cells. One of the best criteria of effects is the ability of the embryo to re-implant into a suitably prepared foster mother and go through normal subsequent development. Recently a number of quantitative biochemical parameters have been used to assess the development *in vitro.* These include such things as increases in incorporation of radioactive isotopes in RNA and protein of the embryo [27, 75] and increases in enzyme activity (*Auerbach,* personal communication). These parameters are direct and accurate indicators of development; they are quantitative and can be treated statistically. The assessment of development in a quantitative manner is very desirable in order to permit use of the information in statistical comparisons of different treatments of the embryo. The application of data transformations to the analysis of experimental results from experiments on preimplantation embryos has been discussed by *Biggers* and *Brinster* [7].

Culture medium

Introduction

A large variety of culture media have been employed in an attempt to cultivate the embryos of various mammalian species. These will not all be reviewed here and the reader is referred to recent reviews on the subject [25]. Classical studies on the early cleavage stages of the rabbit embryo were made in 1929 by *Lewis* and *Gregory* [53] and were later extended by *Pincus* in the thirties and *Chang* in the 1950's and 60's. The first success in cultivating mouse embryos was by *Hammond* in 1949 [48]. As mentioned before, attempts to cultivate the embryos of other mammalian species have met with limited success, and only a few cleavages have been obtained *in vitro*. In the last few years, with the intense interest in cultivation of mammalian ova *in vitro*, evidence is emerging for a beneficial effect of follicular fluid. *Thibault* [78] has cultivated cow embryos for periods up to 48 hours in the liquid of the large Graafian follicle, and obtained some success. Recently, *Edwards, Bavister* and *Steptoe* [40] have been able to fertilize the human oocyte in follicular fluid. With the increasing number of workers entering this field, and the considerable effort being expended in an effort to cultivate mammalian embryos *in vitro*, one should expect to be able to cultivate the embryos of most mammalian species *in vitro* in the next decade.

Ionic composition

The ionic composition of most culture media used to cultivate mammalian embryos *in vitro* has resembled closely the ionic composition of blood serum. Some work has been done to determine what ions are essential or beneficial to the embryo and most of this work has been done using mouse embryos. *Whitten* [86] found that the omission of calcium, magnesium, or potassium from the medium prevented growth, and that development was delayed when phosphate was absent from the medium. He also found that growth occurred when the osmolarity of the culture media was reduced to 0.09. *Brinster* [13] made a detailed study of the effect of osmolarity on the development of the early mammalian embryo. He showed that the maximum number of two-cell mouse embryos developed into blastocysts when the osmolarity of the culture media was 0.276, and that development occurred between 0.200 and 0.354 osmols. The effect of osmolarity on the development of the two-cell mouse embryo is shown graphically in Fig. 7. The normal osmolarity of tissue culture media and blood serum is approximately 0.308. The reason why mouse embryos should develop more readily *in vitro* in a hypo-osmolar culture medium is not known. However, this optimum at low osmolarity *in vitro* has been found for other body cells. For instance, *Eagle* [39] found an optimum osmolarity of 0.277 for He la cells and mouse fibroblasts grown *in vitro*.

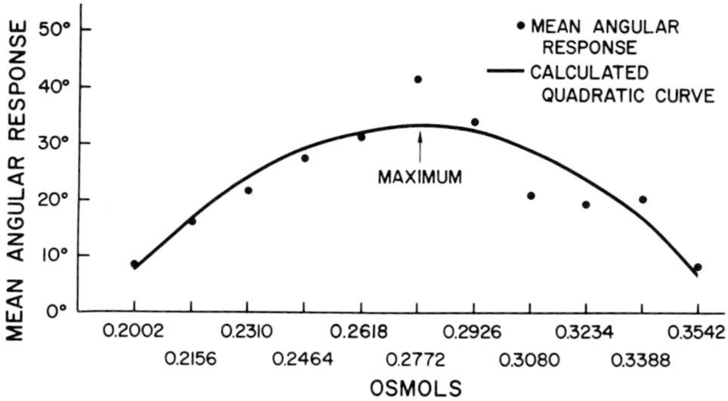

Fig. 7. See text for details.

Recently *Whitten* and *Biggers* [88] have succeeded in cultivating one-cell stage mouse ova to blastocysts completely *in vitro* in the F_1 Hybrid Strain C57 B1/10J x SJL/J. Unfortunately other species including the parent strains did not develop from the one-cell stage completely to the blastocyst. In this respect, the parent strains were similar to other strains of mice in that the one-cell stage would develop to the two-cell stage, or the two-cell stage would develop to the blastocyst stage, but entire development from the one-cell to the blastocyst stage did not take place *in vitro.* The culture medium they used was essentially the same as that shown in Fig. 6 with two exceptions. The first exception was that the concentration of crystalline bovine serum albumin was increased from 1 to 4 g per Liter, and the second exception was that the sodium chloride was reduced to 4 g per liter (giving the medium an osmolarity of 0.256).

Since it has been shown that raising the albumin concentration from 1 to 10 mg per ml has no effect on development [15], whereas osmolarity has a significant effect on development of the mouse embryo [13], it seems probable that the hypo-osmolar condition of the culture medium allows development of the hybrid species employed by *Whitten* and *Biggers.* The fact that the one-cell embryo of this hybrid mouse could develop to a blastocyst completely *in vitro* suggests that the block to development between first and second cleavages, which has been observed previously in the mouse embryo, may not be due to the lack of a specific molecule secreted by the fallopian tube. However, one wonders why the hypo-osmolar condition allows development in only selected hybrid species, and more important, what biochemical and physiological effect the hypo-osmolarity is having on the embryo to bring about the change which results in development. *Whittingham* (1968) used a culture medium quite similar to that described by *Whitten* and *Biggers* and was able to obtain fertilization of the mouse oocyte completely *in vitro.* Although oocytes used in the fertilization experiments were taken from Swiss mice as well as hybrid mice, the best results were obtained using oocytes from F_1 hybrids (C57 B1 x Balb/C). It is interesting that *Whitten* obtained his results, and *Whittingham* obtained his best results with a hybrid, and that in both cases, one parent of the hybrid was from the strain C57 B1. More work certainly needs to be done to clarify the effect of hypo-osmolarity on the mammalian embryo.

The importance of calcium and magnesium in the culture medium has been examined by *Brinster* [12] and it was shown that a reduction in calcium concentration resulted in significant decrease in the number of blastocysts which developed from two-cell mouse embryos *in vitro.* However, the reduction of magnesium to one-fourth the level contained in the culture medium shown in Fig. 6 did not have a significant effect on development. Furthermore, there was no interaction between the effects of calcium and magnesium in these studies.

The effects of ionic composition of the culture medium on rabbit embryos has not been examined. However, the composition of the oviductal fluid in the rabbit has received more attention than in any other species. Consequently, we have information about the ionic composition of culture media for the mouse embryo and we know little about the tubal fluid composition in this species, whereas in the rabbit embryo we know little about the culture ionic composition requirements and know a considerable amount about the ionic composition of the oviductal fluid. In general, it has been found that the ionic composition of oviductal fluid in the rabbit as well as in the sheep is very similar to that found in blood plasma and therefore very similar to that shown in Fig. 6.

One of the most important ions in the culture medium is the hydrogen ion. The level or concentration of this ion in the culture medium is regulated as mentioned above by two main factors, the amount of bicarbonate in the medium and the amount of CO_2 in the atmosphere. *Whitten* [86] found that the development of eight-cell mouse embryos would occur between a pH of 6.9 and 7.7. *Brinster* [13] found that the development of two-cell mouse embryos into blastocysts occurred between pH 5.87 and 7.78 in an atmosphere of 5 % CO_2 in air. It has proved quite difficult to determine an accurate pH optimum for the embryos. This is because the availability of the important energy sources for the embryo, pyruvate and lactate, is related to the hydrogen ion concentration of the medium. The lower the pH of the medium, the lower the concentration of the substrate necessary to obtain optimum development, suggesting that the uptake of these compounds is related to the amount of the compound in the acid form. If this is true, it suggests that the membrane of the developing embryo is able to show selective permeability to substrates necessary for its development, and that in some cases the selective permeability is dependent on hydrogen ion concentration in the medium. So far, comparable studies have not been made for the rabbit embryo or for the embryos of other species.

The pH of oviductal fluid has been determined or estimated for a number of species. *Blandau, Jensen* and *Rumery* [10] found that the pH of the reproductive fluid of the rat was significantly higher than the pH of the peritoneal fluid. *Vishwakarma* [81] found that the mean pH of the fluids from the ligated rabbit uterus was 7.86 and that of fluids from the ligated oviduct was 7.91. It seems doubtful that fluids collected from the ligated portions of the reproductive tract accurately represent the normal condition of secreted fluid. *Hamner* and *Williams* [49] found that the bicarbonate concentration of rabbit oviductal fluid was 1.76 mg per ml which is close to the concentration found in blood serum. Similar values have been found by *Restall* [71] for sheep oviductal fluid. If CO_2 tension in the oviductal fluid is close to that of blood, then the pH would be approximately 7.4 for these fluids.

Amino nitrogen composition

Most culture media used in the early studies to grow mammalian ova *in vitro* contained a variable percentage of serum (see [4]). Within the last few years attempts have been made to define, in more specific terms, the actual requirements of mammalian embryos for amino nitrogen. In 1956, *Whitten* [86] showed that eight-cell mouse embryos would develop into blastocysts when cultivated in Krebs-Ringer bicarbonate containing 1 mg per ml of glucose and bovine serum albumin at a concentration between 0.03 and 6 %. In 1957, he showed that the bovine serum albumin could be replaced with glycine or other amino acids and simple peptides. *Brinster* [15] has studied the effect of various exogenous amino nitrogen sources on the development of the two-cell mouse embryo into the blastocyst. He found that concentrations between 1 mg per ml and 10 mg per ml bovine serum albumin allowed maximum development. The requirement for protein can be supplied by mouse serum or bovine serum (*Brinster,* unpublished), as well as by the constituent amino acids of albumin. Unlike most other cells grown *in vitro,* the two-cell mouse embryo does not seem to have a requirement for the essential amino acids. The removal of any single amino acid from the culture medium does not prevent the development of the two-cell embryo into a blastocyst. Cystine is the only amino acid whose removal significantly decreased the number of blastocysts forming. However, the two-cell mouse embryo will not develop into a blastocyst when the culture medium contains only a single amino acid. On the other hand, a number of single amino acids will allow development of the eight-cell mouse embryo into a blastocyst [30, 87].

Recent studies have shown that glutathione, either oxidized or reduced, allows development of two-cell mouse embryos into blastocysts without any other exogenous amino nitrogen source [22]. In fact, the development of the embryos with culture media containing albumin is no better than in culture media containing glutathione. These studies suggest that the embryo can rely to a substantial degree on endogenous amino acid supplies but requires exogenous amino nitrogen to supplement its endogenous stores. Studies on the protein content of the mouse embryo lend further support to this hypothesis. *Brinster* [18] found that protein content in the developing mouse embryo decreased by 25 % during the first four days of development. Several studies suggest that blastocyst development is improved by the addition of serum and/or an amino acid complement. Thus, if the basic medium shown in Fig. 6 is supplemented with 10 % serum, the mouse blastocyst will attach to the culture dish surface and begin to spread. This has been demonstrated by a number of workers, all using medium which contained serum and/or amino acids [34, 46, 59].

The amount of work that has been done on specific amino nitrogen requirements of the rabbit embryo is not as extensive as for the mouse embryo. Most workers have employed complex protein components in the culture media for the rabbit embryo, and in fact, the embryos seem to do best when serum forms a substantial part of the culture medium. *Chang* [32] found that certain heterologous sera contained an ovicidal factor. He found that sera from fowl, goat, cattle, sheep, and man contained this factor whereas the sera of rabbit, horse, dog, guinea pig, and rat were free of the factor. The factor was undializable and heat labile and was lost after storage of the sera for 17 days at 3 °C. The characteristics of this ovicidal factor seem quite similar to such factors as complement which have been identified in sera.

Adams [1] found that a simple culture media of Krebs-Ringer bicarbonate and 0.2 % bovine plasma albumin, fraction 5, allowed cleavage of the rabbit embryo from the two-cell to the 16 cell stage. *Purshottam* and *Pincus* [70] suggest that the best growth of the preimplantation rabbit embryo is obtained in 100 % rabbit serum.

Daniel [35] studied the effect of single amino acids on the growth of five day rabbit blastocysts. He added the single amino acids to Hams F 10 medium plus 15 % rabbit serum and the criterion of development was a change in blastocyst volume determined at 4 hours and 24 hours after the beginning of culture. *Daniel* selected for study those amino acids found in the uterus in high concentration by *Gregoire, Gongsakdi,* and *Rakoff* [44]. In some cases during culture, four hour development was faster under *in vitro* conditions than would be expected *in vivo.* However, after 24 hours there was no case where blastocyst development *in vitro* was as good as would be expected *in vivo.* The concentrations of amino acids which *Daniel* recommends as optimum are shown in Fig. 8. *Staples* [73] has employed *Daniel's* medium plus 5 to 10% rabbit serum to grow five day blastocysts for 8, 16, and 24 hour periods, after which he transferred the blastocysts to foster mothers. Implantation of cultured blastocysts was significantly lower than controls for the 24 hour group but not for the 8 or 16 hour group.

Fig. 8. Optimum concentration of five amino acids for five-day rabbit blastocysts. (Source [35, 44]).

	Concentration in Uterine Fluid mg/ml	Concentration in F 10 as Modified by *Daniel* mg/ml
Alanine	0.064	0.1
Glycine	0.268	0.2
Glutamic acid	0.119	0.2
Threonine	0.040	0.1
Serine	0.104	0.1

Mauer, Hafez, Ehlers and *King* [56] were unable to demonstrate a need for glycine, serine, glutamine, methionine or tryptophane in the culture medium for preblastocyst rabbit embryos. Three of these were found beneficial to the rabbit blastocyst by *Daniel* [35]. It seems that there are changes in the amino nitrogen requirements of the preimplantation rabbit embryo as it undergoes development, just as in the case of the mouse embryo [15, 30].

Brinster [26] has studied the development of the one-cell and early two-cell rabbit embryo into the morula using the standard culture system [11]. He found that the rabbit embryo was able to develop from the two-cell stage to the morula in a medium containing an amino nitrogen source as the only metabolizable component. This amino nitrogen source could be bovine serum albumin, oxidized glutathione, or certain single amino acids such as glutamine, proline, or alanine. Although a requirement for a non-nitrogenous energy source cannot be demonstrated during this period, a beneficial effect of pyruvate and lactate was shown. It appears that an amino nitrogen source is essential for development of the early rabbit embryo. Other studies suggest that development is improved by the addition of whole serum to the culture medium.

Onuma, Maurer, and *Foote* [67] have been able to cultivate the two- and four-cell stages of the rabbit embryo through to the blastocyst and obtain hatched blastocyst stages *in vitro* for the first time. For these studies, they used the culture method of *Brinster* [11] as shown in Fig. 1 A. The culture medium consisted of rabbit or bovine serum to which was added 0.05 % glucose. They found that treating the ova with 1 % pronase solution before placing them *in vitro* significantly increased the number of completely hatched blastocysts. More recently, *Maurer* (personal communication) has eliminated the pronase treatment of the ova and has obtained a high percentage of hatched blastocysts from two- and four-cell rabbit ova placed *in vitro* in a culture medium consisting primarily of treated rabbit or bovine serum. From the evidence available so far, it seems that serum has an important role in the culture medium for rabbit embryos, particularly after blastocyst formation.

Energy source composition

In 1949, *Hammond* [48] showed that eight-cell and some late four-cell embryos could develop into blastocysts when the medium contained glucose as an energy source, but no two-cell embryos developed into blastocysts in this medium containing glucose as an energy source. *Whitten* [86, 87] confirmed his findings and demonstrated that a number of other compounds such as mannose, lactate, pyruvate, and malate could provide energy for development of the eight-cell embryo. However, compounds such as fructose, lactose, maltose, galactose, acetate, propionate, citrate, glycerol and glycine could not provide the energy for development. He further demonstrated that some two-cell mouse embryos could develop into blastocysts when the medium contained calcium lactate or isotonic lactic acid. *Brinster* [14, 16] extended these studies on energy source for the developing mouse embryo and determined the optimum concentration of those compounds which would allow development (Fig. 9). He found that the compunds which will allow development of the two-cell embryo form a group with pyruvate as the central compound.

Fig. 9. Possible energy sources for the developing two-cell mouse embryo. (Source [14])

I. Compounds which will not support development

Malate*	Glucose*
Fumarate	Fructose*
Succinate	Ribose
Iso-citrate	D-glyceraldehyde
Citrate*	Glucose-6-phosphate
Acetate*	Fructose-1,6-diphosphate
Cis-aconitate	Bovine serum albumin*
α-ketoglutarate	

(All compounds tested at 10^{-2}, 10^{-3}, 10^{-4}, 10^{-5} molar except glucose and fructose, which were tested at 2.78×10^{-2}, 5.56×10^{-3}, and 2.78×10^{-3} molar; Bovine serum albumin tested at 1 mg/ml).

II. Compounds which will support development and their optimum concentration.

Pyruvate	(5×10^{-4} M) #
Lactate DL	(5×10^{-2} M)
Oxaloacetate	(5×10^{-4} M) #
Phosphoenolpyruvate	(1×10^{-2} M)

* Compounds which will support development of the eight-cell but not the two-cell embryo. Fumarate, Iso-citrate, and Cis-asconitate were not tried with eight-cell embryos [30].

\# Compounds which will support development of the one-cell mouse ovum and oocyte [8].

Cole and *Paul* [34] used Waymouth's medium, supplemented with nucleosides, ATP and serum, to culture one-cell, two-cell, and four-cell mouse embryos. This medium worked best when employed with a feeder layer of HeLa cells, and the authors considered that one of the contributions of the feeder layer was to supply lactate and pyruvate. *Biggers, Whittingham,* and *Donahue* [8] have shown that pyruvate and oxaloacetate, but not lactate or phosphoenolpyruvate, will allow cleavage of the one-cell embryo into the two-cell embryo and maturation of the ovarian oocyte. The addition of cumulus cells to the culture medium, containing lactate, phosphoenolpyruvate or glucose, will allow cleavage of the one-cell embryo into the two-cell embryo and oocyte maturation. Development of the eight-cell embryo into a blastocyst occurs when the culture medium contains singly a number of energy sources which do not allow development of two-cell embryos [30]. These findings of different energy source requirements for *in vitro* development at different developmental stages suggest strongly that there are marked changes in the metabolic capability of the mouse embryo as it undergoes development from the oocyte to the implanting blastocyst.

Of considerable interest in respect to changes in metabolic capability of the mouse embryo, is the fact that glucose and malate will allow development of eight-cell embryos, but not two-cell embryos. Studies in our laboratory have shown that the difference, using glucose as an energy source, is not due to changes in permeability of the embryo to glucose. *Wales* and *Brinster* [84] have shown that the uptake of glucose by the eight-cell embryo is not significantly greater than that of the two-cell embryo. Therefore, it appears that the ability of the eight-cell embryo to survive on glucose depends on the development of the necessary enzyme systems, as has been suggested by *Brinster* [16]. In fact, it has now been shown that one of the key enzymes, hexokinase, which is a regulatory enzyme for glucose utilization, shows a dramatic 7.5 fold increase in activity from the one-cell stage to the blastocyst in a mouse embryo [23]. Other studies in our laboratory by *Wales* and *Biggers* [83] have shown that in the case of malate, there is a marked difference in uptake between the two-cell embryo and the eight-cell embryo. The eight-cell embryo is approximately seven times as permeable to malate as the two-cell embryo. This may account for the ability of the eight-cell embryo to develop on malate. However, the amount of substrate taken up by the embryos was too small to allow identification of the intracellular form. Therefore, differences in uptake could reflect differences in permeability, or differences in utilization within the cell. However, the large uptake of malate by the eight-cell embryo must account for the ability of the eight-cell embryo to develop on malate.

Wales and *Whittingham* [85] have also shown that the uptake of pyruvate and lactate are the same in fertilized one-cell and two-cell embryos, and a difference in uptake does not account for the difference in the ability of the two compounds to support *in vitro* development. However, utilization of pyruvate for both stages, as shown by CO_2 formation (Fig. 10) is twice as great as lactate [20, 85]. The greater utilization of pyruvate probably accounts for its ability to support the development of the one-cell stage at a time when lactate will not support development. The unfertilized mouse ovum also produces twice as much CO_2 from pyruvate as from lactate (Fig. 10) and this probably accounts for the ability of pyruvate, but not lactate, to support maturation of the oocyte [20].

Fig. 10. Comparison of carbon dioxide production from glucose, pyruvate and lactate for preimplantation mouse and rabbit embryos.

Stage of Development	Glucose		Pyruvate		Lactate	
	Mouse	Rabbit	Mouse	Rabbit	Mouse	Rabbit
Unfertilized	0.13	0.61	7.24	16.68	3.09	14.71
Fertilized	0.68	0.81	6.95	14.47	3.31	11.28
Day 2	1.19	5.98	6.03	13.42	2.77	22.11
Day 3	2.16	13.71	7.25	22.31	4.54	23.06
Day 4	8.84	50.33	11.96	201.2	11.20	178.2
Day 5	14.69	1335.15	15.73	2032.9	15.06	655.2
Day 6	–	4904.98	–	9882.3	–	3844.3

All values are $\mu\mu$moles of CO_2 produced/embryo/hr.
The substrate concentration was for glucose, 1 mg/ml; for pyruvate, 5×10^{-4} M; and for lactate, 5×10^{-2} M. The values for the mouse are from *Brinster* [19, 20] and the values for the rabbit are from *Brinster* [24, 28]

The reason that oxaloacetate, although similar in structure to malate, allows development of the two-cell embryo, one-cell embryo, and maturation of the oocyte, is that the oxaloacetate is decarboxylated to pyruvate in the culture medium. This occurs under normal conditions when oxaloacetate is in solution and probably accounts for the inhibitory interaction when pyrarate and oxaloacetate are employed together in the culture medium [16]. The rate of decarboxylation appears to be increased in the presence of the embryo, thus suggesting that the embryo contains the enzyme to facilitate this conversion (*Brinster,* unpublished).

Recently a simple system has been developed by *Brinster* [19] for determining carbon dioxide production from labeled substrates in the culture medium. Fig. 2 M shows the culture vessel used for this procedure. Carbon dioxide production from glucose, pyruvate and lactate have now been determined for all stages of the mouse and rabbit embryo from ovulation to implantation (Fig. 10). Pyruvate appears to be the best energy source for the early stages of development in both the mouse and the rabbit. The blastocyst, prior to implantation, appears to oxidize pyruvate, glucose, and lactate equally well. Lactate formation from glucose by the mouse embryo has recently been measured using column chromatography [82]. There was approximately a 200-fold increase in lactate formation during the preimplantation period. It seems that the major increases in lactate formation as well as in CO_2 formation from glucose occur at the same time in development, that is, fertilization and blastocyst formation.

Fig. 11. Molar ratio of radioactive carbon dioxide production compared to oxygen uptake in the preimplantation mouse and rabbit embryo.

Stage of Development	Glucose		Pyruvate		Lactate	
	Mouse	Rabbit	Mouse	Rabbit	Mouse	Rabbit
Unfertilized	0.02	–	1.04	–	0.45	–
Fertilized	0.10	0.03	0.99	0.51	0.47	0.39
Day 2	0.18	0.22	0.90	0.50	0.41	0.83
Day 3	0.25	0.37	0.85	0.59	0.53	0.61
Day 4	0.48	0.21	0.66	0.83	0.62	0.73
Day 5	0.62	0.56	0.66	0.86	0.63	0.28
Day 6	–	0.51	–	1.03	–	0.40

Substrate utilization is from Fig. 10.
Oxygen consumption for the mouse was taken from *Mills & Brinster* [57], and for the rabbit from *Fridhandler* [41].

The carbon dioxide formed (Fig. 10) from various substrates by mouse and rabbit embryos cultivated *in vitro* compared to total oxygen consumption by the embryo is shown in Fig. 11. For the mouse embryo, only about 2 % of the oxygen consumption can be accounted for by CO_2 formed from glucose in the unfertilized ovum. On the other hand, 65 % of the total oxygen consumption in the blastocyst stage can be accounted for by exogenous glucose. In the unfertilized ovum, 100 % of the oxygen consumption can be accounted for by CO_2 from pyruvate, and this value drops to approximately 65 % at the blastocyst stage. The oxidation of pyruvate, glucose, and lactate by the rabbit embryo appears to be very similar to that found in the mouse embryo. It would appear that a general characteristic of the preimplantation stages of the mammalian embryo is the requirement for pyruvate for the first two to three days of development. In later stages, either pyruvate or glucose can be used.

Part of the reason for the apparently constant 65 % of total oxygen consumption being accounted for by exogenous substrate at the blastocyst stage of the mouse embryo probably results from utilization of endogenous glycogen by the blastocyst. Using histochemical techniques, *Thomson* and *Brinster* [79] have shown that the mouse embryo contains a large store of glycogen in the early stages of preimplantation development. This store of glycogen is not markedly reduced until blastocyst formation begins, then the glycogen is rapidly utilized during the last day and a half of preimplantation development. It appears that 35 % of oxygen consumption which cannot be accounted for by exogenous substrate in the late blastocyst could be accounted for by oxidation of the endogenous glycogen. It has further been demonstrated by *Thomson* and *Brinster* [79] that this endogenous store of glycogen is unavailable even to embryos which are maintained in culture media with suboptimal energy source concentration. This suggests that the necessary enzymes are not available for utilization of the stored glycogen during the early preimplantation period. Recently, *Stern* and *Biggers* [74] using a biochemical technique have demonstrated the formation of glycogen between the one-cell and eight-cell stage of development. They show that most of the glycogen contained in the mouse embryo is formed during this period and that only a small part of this is utilized between the eight-cell stage and the blastocyst stage. Further studies are needed to resolve the differences between the findings using the histochemical technique and the findings using the biochemical technique.

Brinster [27] has studied the incorporation of carbon from glucose and pyruvate into the mouse embryo over relatively long periods of time (20 to 24 hours). These studies show that incorporation of glucose carbon is greater than pyruvate carbon for all the developmental stages (Fig. 12). This is in marked contrast to substrate oxidation where pyruvate carbon is oxidized much better than glucose carbon up to the blastocyst stage, after which both substrates seem to be oxidized equally well. Only a small amount of labeled carbon is incorporated into glycogen (the ethanol precipitated fraction) before the eight-cell stage of development. The rate of accumulation of glucose and pyruvate carbon into the glycogen fraction represents less than 2 % of the rate necessary to account for the glycogen formation rate shown by *Biggers* and *Stern* [74]. Other possible sources of carbon, for the glycogen synthesis they found, might be:

1) pre-formed glycogen precursors measured by the histochemical staining technique but not by the biochemical technique;
2) endogenous protein, the carbon from which is approximately three times as great as the carbon used to form glycogen (*Brinster* 1967a, 1969c);
3) exogenous glycogen precursors such as protein.

It would also be interesting to determine whether glycogen synthesis as determined biochemically occurs during *in vitro* development. Histochemically, there appeared to be little difference between embryos grown *in vivo* and those kept *in vitro* [30]. Culture medium energy sources have not been studied in as much detail for the rabbit embryo as they have for the mouse embryo. The critical phase for rabbit

Fig. 12. Carbon accumulated from radioactive pyruvate and glucose by mouse embryos (Source [27]).

Growth Period	Energy Source in Medium							
	P*G				PG*			
	Activity in Embryo				Activity in Embryo			
	Protein (TCA ppt.)	Glycogen (ETOH ppt.)	ETOH Supernatent	Total	Protein (TCA ppt.)	Glycogen (ETOH ppt.)	ETOH Supernatent	Total
1-cell to 2-cell	0.279	0.004	0.479	0.761	1.114	0.018	0.527	1.648
2-cell to 4-cell	0.258	0.004	0.425	0.686	1.072	0.020	0.628	1.719
8-cell to Morula	0.665	0.006	0.573	1.244	2.657	2.912	1.443	7.012
Morula to Blastocyst	0.869	0.010	0.914	1.793	3.372	14.020	2.617	20.008

Activity is in $\mu\mu$ atoms of carbon per embryo per hr, and each value is the mean of 4 determinations. Embryos cultured approximately 24 hours in medium containing both pyruvate (5×10^{-4} M) and glucose (1 mg/ml).

P*G = $^{14}C(U)$ pyruvate and cold glucose.

PG* = Cold pyruvate and $^{14}C(U)$ glucose.

embryo development has been the transition from morula to blastocyst and the subsequent expansion of the blastocyst. This block has now, at least partially, been overcome in studies by *Onuma, Maurer* and *Foote* [67] and *Krishnan* and *Daniel* [52]. *Maurer* and colleagues have succeeded in obtaining development of the two-cell rabbit embryo into a blastocyst and expansion of the blastocyst in a culture medium containing predominantly bovine or rabbit serum. *Krishnan* and *Daniel* [52] have isolated from the six-day rabbit uterus a substance which stimulates expansion of the blastocyst *in vitro.* Clearly more studies are needed in this area to determine if the expansion obtained by *Maurer* and colleagues *in vitro* is comparable to that found *in vivo,* and whether the substance isolated by *Daniel* and colleagues is an essential part of the natural process of blastocyst expansion.

Early studies by *Fridhandler, Hafez,* and *Pincus* [42] and *Fridhandler* [41] on the energy metabolism of the rabbit embryo using Cartesian diver techniques, indicated little evidence for glycolytic activity in the premorula rabbit embryo. The blastocyst, however, did show glycolytic activity. Actually at the time of blastocyst formation, both glycolytic activity and Krebs cycle activity is markedly increased in the rabbit embryo. These findings have been confirmed and extended using radioactive substrates (see Fig. 10 and 11). *Fridhandler* [41] and *Brinster* [24] found that the C_1 to C_6 ratio was high in the rabbit embryo before blastocyst formation, indicating an active pentose shunt. After blastocyst formation the C_1 to C_6 ratio was approximately one indicating the pentose shunt had become less important in relation to the Krebs cycle. In mouse embryos [19] the C_1 to C_6 ratio remains at about 1.6 throughout the preimplantation period, indicating the pentose shunt is not very important relative to the Krebs cycle in this species. Thus although both embryos oxidize very little glucose during the first day or two of development, the rabbit oxidizes its glucose by the pentose shunt predominantly, and the mouse oxidizes its glucose predominantly through the Embden-Meyerhof cycle and the Krebs cycle.

Mounib and *Chang* [61] cultured six-day rabbit blastocysts in Warburg vessels in Krebs-Ringer phosphate (pH 7.4) in the presence of labeled glucose, fructose, and pyruvate. They found that the glucose was oxidized to a greater degree than fructose at low concentrations, but not at high concentrations. Pyruvate C_1 oxidation was greater than pyruvate C_2 oxidation, perhaps indicating that acetate was used for synthetic processes.

Daniel [35] examined the effect of pyruvate, lactate, and glucose on the development of the five-day rabbit blastocyst. All three compounds improved development for a four hour period, but the *in vitro* blastocysts at 24 hours were smaller than equivalent age *in vivo* blastocysts. On the basis of his work, *Daniel* suggests that the concentration of pyruvate be raised to 1 mg per ml and the concentration of glucose be reduced from 1 to 0.2 mg per ml in the F 10 medium, in order to obtain optimum growth of the five-day rabbit blastocyst. He also suggests 0.8 mg per ml of sodium lactate and 0.1 mg per ml of glycogen be added to this medium.

Recently, *Daniel* [36] studied the effect of energy sources when employed singly on the development of various preimplantation stages of the rabbit embryo. Their ability to grow *in vitro* for 12 to 24 hours was considered evidence that they were utilizing a particular energy source. The data from these studies are shown in Fig. 13, and it can be seen that most of the intermediates will support growth of blastocysts, but the initial cleavage stages will only grow in the presence of pyruvate, phosphoenolpyruvate and lactate. These results are very similar to those found for the

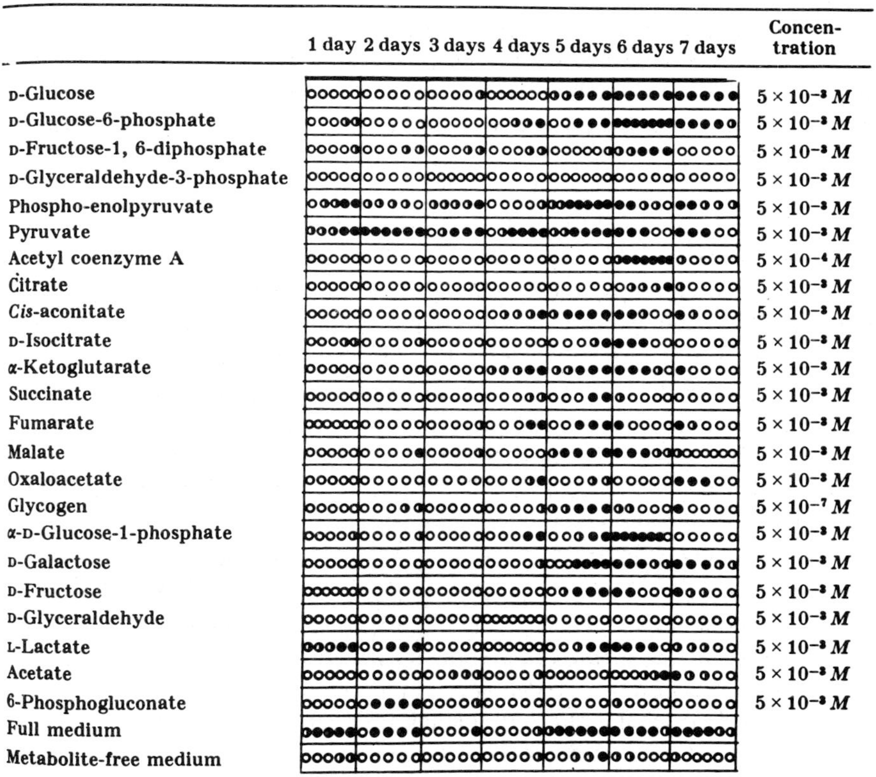

Fig. 13. The effect of energy sources on the development of preimplanted rabbit embryos. Each circle is a separate embryo; open if there was no growth, solid if growth continued for 24 h, half-open if some growth was observed by 12 h but did not continue for 24 h. Basic medium was F 10. From *Daniel* [36].

preimplantation stages of the mouse embryo. However, *Brinster* [26] found a marked difference in the minimum requirements of the preimplantation stages of the mouse and rabbit embryo. He found that the mouse embryo required both an energy source and an amino nitrogen source, whereas the rabbit embryo would develop in a medium containing only an amino nitrogen source. Nevertheless, *Brinster* was able to show a beneficial effect of pyruvate and lactate on the development of early stages of the rabbit embryo. Not surprisingly, the optimum concentration for pyruvate was approximately 10^{-4} M and for lactate 1 to 3 x 10^{-2} M; not too different from the optimum concentration for these compounds for early stages of the mouse embryo.

How closely the energy source requirement *in vitro* resembles the energy sources available *in vivo* is not completely known. However, it has been shown [9, 50, 55] that the glucose concentration is very low in the rabbit oviduct, whereas the lactate and pyruvate concentrations are high Similar obserbations have been made on the fallopian tube fluid of the sheep [71]. *Lutwack-Mann* [54] also has shown that in the uterus of the rabbit the glucose concentration is low and the lactate concentration is high.

Additional factors

Cole and *Paul* [34] reported that they found a beneficial effect of nucleic acid precursors when employed in the culture medium for early cleavage stage mouse embryos. However, recently, *TenBroeck* [77] has shown that the presence of nucleic acid precursors in the culture environment, either singly or in combination, does not enhance development of the preimplantation mouse embryo. Since experiments with labeled nucleic acids have shown that these compounds are incorporated by the early mouse embryo [60] the lack of a beneficial effect cannot be due to impermeability of the embryos to the compounds. It therefore appears likely that the early embryo either contains an adequate endogenous supply or is able to synthesize the components required for the formation of RNA and DNA.

Daniel [37] found that the five day rabbit blastocyst had a requirement for certain vitamins and growth factors. He found that development was improved by any of the following: thiamine, riboflavin, niacin, pyridoxine, folic acid, inositol, or hypoxanthine. *Brinster* and *Brunton* (unpublished) using the techniques described by *Brinster* [11, 13] were able to determine no statistical difference between development of embryos grown in the medium shown in Fig. 6 or the same medium supplemented at a concentration of 10^{-3}, 10^{-4}, 10^{-5}, 10^{-6}, or 10^{-7} molar with any of the following vitamins: thiamine, riboflavin, niacin, pyridoxine, pantothenic acid, biotin, choline, inositol, B_{12}, or ascorbic acid. Each was employed singly in the medium.

The unfertilized mouse ovum can be fertilized and will undergo subsequent cleavage to the blastocyst in the explanted fallopian tube [29]. The oviducts, or the ampulla of the oviduct, is placed on tea-bag paper over a wire grid in a culture dish similar to Fig. 1 C. This method provides one means by which to study tubal influences on embryos.

Implantation of the embryo *in vitro* has also been studied to some extent. *Glenister* [43] cultured rabbit endometrium and blastocysts in apposition. The embryo is removed from the mother in the late blastocyst stage and the external coverings of the embryo are removed mechanically. The embryo is then placed on endometrial strips which are supported on top of agar medium. Some invasion of the endometrium by the trophoblasts took place.

A number of attempts have been made in the past to culture postimplantation mammalian embryos. These have met for the most part with limited success, but recently *New* and *Stein* [64, 65] have obtained relatively good development of postimplantation embryos in the early somite stages of mouse and rat. These embryos have been cultured on plasma clots within watch glasses enclosed in a petri dish such as is shown in Fig. 2 C. The cultures are incubated in 60 % oxygen and 3 to 5 % CO_2. Explanted mouse embryos will grow in culture for up to 40 hours and develop a functioning blood circulation in the embryonic yolk sac and often in the alantoid placenta. Embryos are usually harvested from the animal for explantation at eight to nine days in the mouse and 10 to 11 days in the rat. Development of the rat embryo in culture is very similar to that of mouse embryos except that the explants grow somewhat larger, and the most advanced stage of embryonic development that can be obtained is about 28 to 30 somites with only the anterior limb buds prominent. *New* [63] has found that the rat embryo will grow just as well in rat serum as on plasma clots. Serum may be taken from male or female pregnant or nonpregnant (rats). Development of rat embryos in heterologous serum is often poor.

Acknowledgement: During the past years, research in my laboratory has received financial support from the National Institutes of Health (HD 03071, HD 00239, HD 02315); National Science Foundation (GB 4465); Pennsylvania Department of Agriculture; Population Council; and the King Ranch, Inc.

References

[1] *Adams, C. E.* 1956. Egg transfer and fertility in the rabbit. Proceedings of the Third Int. Cong. Anim. Reprod. (Cambridge), Section **3**, 5.

[2] *Alliston, C.* 1965. Embryonic mortality following culture *in vitro* of one- and two-cell rabbit eggs at elevated temperatures. J. Reprod. Fertil. **9**, 337.

[3] *Auerbach, S. & Brinster, R. L.* 1968. Effect of oxygen concentration on the development of two-cell mouse embryos. Nature, **217**, 465.

[4] *Austin, C. R.* 1961. The Mammalian Egg. lst Ed., Springfield, Illinois: Charles C. Thomas.

[5] *Avery, T.* 1962. Investigations associated with the transplantation of bovine ova. II. Superovulation. J. Reprod. Fertil **3**, 212.

[6] *Bedford, J. M. & Chang, M. C.* 1962. Fertilization of rabbit ova *in vitro.* Nature, **193**, 898.

[7] *Biggers, J. D. & Brinster, R. L.* 1965. Biometrical problems in the study of early mammalian embryos in vitro. J. Exp. Zool. **158**, 39.

[8] *Biggers, J. D., Whittingham, D. G. & Donahue, R. P.* 1967. The pattern of energy metabolism in the mouse oocyte and zygote. Proc. Nat. Acad. Sci. **58**, 560.

[9] *Bishop, D.* 1956. Metabolic conditions within the oviduct of the rabbit. Inter. J. Fertil. **2**, 11.

[10] *Blandau, R., Jensen, L. & Rummery, R.* 1958. Determination of the pH values of the reproductive tract fluids of the rat during heat. Fertil. Steril. **9**, 207.

[11] *Brinster, R. L.* 1963. A method for *in vitro* cultivation of mouse ova from two-cell to blastocyst. Exp. Cell Res. **32**, 205.

[12] *Brinster, R. L.* 1964. Studies on the development of mouse embryos *in vitro.* Ph. D. thesis. University of Pennsylvania, Philadelphia, Pennsylvania.

[13] *Brinster, R. L.* 1965. Studies on the development of mouse embryos *in vitro.* I. The effect of osmolarity and hydrogen ion concentration. J. Exp. Zool. **158**, 49.

[14] *Brinster, R. L.* 1965. Studies on the development of mouse embryos *in vitro.* II. The effect of energy source. J. Exp. Zool. **158**, 59.

[15] *Brinster, R. L.* 1965. Studies on the development of mouse embryos *in vitro.* III. The effect of fixed-nitrogen source. J. Exp. Zool. **158**, 69.

[16] *Brinster, R. L.* 1965. Studies on the development of mouse embryos *in vitro.* IV. Interaction of energy sources. J. Reprod. Fertil. **10**, 227.

[17] *Brinster, R. L.* 1965. Lactic dehydrogenase activity in the preimplanted mouse embryo. Biochim. Biophys. Acta. **10**, 439.

[18] *Brinster, R. L.* 1967. Protein content of the mouse embryo during the first five days of development. J. Reprod. Fertil. **13**, 413.

[19] *Brinster, R. L.* 1967. Carbon dioxide production from glucose by the preimplantation mouse embryo. Exp. Cell Res. **47**, 271.

[20] *Brinster, R. L.* 1967. Carbon dioxide production from lactate and pyruvate by the preimplantation mouse embryo. Exp. Cell Res. **47**, 634.

[21] *Brinster, R. L.* 1968. Lactate dehydrogenase activity in the oocytes of mammals. J. Reprod. Fertil. **17**, 139.

[22] *Brinster, R. L.* 1968. Effect of glutathione on the development of two-cell mouse embryos *in vitro.* J. Reprod. Fertil. **17**, 521.

[23] *Brinster, R. L.* 1968. Hexokinase activity in the preimplantation mouse embryo. Enzymologia, **34,** 304.

[24] *Brinster, R. L.* 1968. Carbon dioxide production from glucose by the preimplantation rabbit embryo. Exp. Cell Res. **51,** 330.

[25] *Brinster, R. L.* 1969. Mammalian embryo culture. In The Mammalian Oviduct, ed. *E. S. E. Hafez* and *R. Blandau,* Chapter 18, New York: University of Chicago Press.

[26] *Brinster, R. L.* 1969. Culture of two-cell rabbit embryos to morulae. J. Reprod. Fertil. In press.

[27] *Brinster, R. L.* 1969. The incorporation of carbon from glucose and pyruvate into protein and glycogen of the preimplantation mouse embryo. In prep.

[28] *Brinster, R. L.* 1969. Radioactive carbon dioxide production from pyruvate and lactate by the preimplantation rabbit embryo. Exp. Cell Res. In press.

[29] *Brinster, R. L. & Biggers, J. D.* 1965. *In vitro* fertilization of mouse ova within the explanted fallopian tube. J. Reprod. Fertil. **10,** 277.

[30] *Brinster, R. L. & Thomson, J. L.* 1966. *In vitro* culture requirements of the eight-cell mouse ovum. Exp. Cell Res. **42,** 308.

[31] *Chang, M. C.* 1948. The effects of low temperature on fertilized rabbit ova *in vitro* and the normal development of ova kept at low temperature for several days. J. Gen. Physiol. **31,** 385.

[32] *Chang, M. C.* 1949. Effects of heterologous sera on fertilized rabbit ova. J. Gen. Physiol. **32,** 291.

[33] *Chen, J. M.* 1954. The cultivation in fluid medium of organized liver, pancreas, and other tissues of foetal rats. Exp. Cell Res. **7,** 518.

[34] *Cole, R. J. & Paul, J.* 1965. Properties of cultured preimplantation mouse and rabbit embryos, and cell strains derived from then. In Preimplantation Stages of Pregnancy, ed. *G. E. W. Wolstenholme* and *M. O'Connor,* London: J. and A. Churchill Ltd.

[35] *Daniel, J.* 1965. Studies on the growth of five-day old rabbit blastocysts *in vitro.* J. Embryol. Morph. **13,** 83.

[36] *Daniel, J. C.,* Jr. 1967. The pattern of utilization of respiratory metabolic intermediates by preimplantation rabbit embryos *in vitro.* Exper. Cell Res. **47,** 619.

[37] *Daniel, J. C.,* Jr. 1967. Vitamins and growth factors in the nutrition of rabbit blastocysts *in vitro.* Growth, **31,** 71.

[38] *Daniel, J. C.,* Jr. 1968. Oxygen concentrations for culture of rabbit blastocysts. J. Reprod. Fertil. **17,** 187.

[39] *Eagle, H.* 1956. The salt requirements of mammalian cells in tissue culture. Arch. Biochem. Biophys. **61,** 356.

[40] *Edwards, R. G., Bavister, B. D. & Steptoe, P. C.* 1969. Early stages of fertilization *in vitro* of human oocytes matured *in vitro.* Nature, **221,** 632.

[41] *Fridhandler, L.* 1961. Pathways of glucose metabolism in fertilized rabbit ova at various preimplantation stages. Exp. Cell Res. **22,** 303.

[42] *Fridhandler, L., Hafez, E. S. E. & Pincus, G.* 1957. Developmental changes in the respiratory activity of rabbit ova. Exp. Cell. Res. **13,** 132.

[43] *Glenister, T. W.* 1961. Organ culture as a new method for studying the implantation of mammalian blastocysts. Proc. Roy. Soc. B. **154,** 428.

[44] *Gregoire, A. T., Gongsakdi, D. & Rakoff, A.* 1961. The free amino acid content of the female rabbit genital tract. Fertil. Steril. **12,** 322.

[45] *Greenwald, G.* 1962. Analysis of superovulation in the adult hamster. Endocrinology, **71,** 378.

[46] *Gwatkin, R. B. L.* 1966. Amino acid requirement for attachment and outgrowth of mouse blastocyst *in vitro.* J. Cell Physiol. **68,** 335.

[47] *Hafez, E.* 1963. Storage of fertilized ova. Inter. J. Fertil. **8,** 459.

[48] *Hammond, J.,* Jr. 1949. Recovery and culture of tubal mouse ova. Nature, **163,** 28.

[49] *Hamner, C. E. & Williams, W. L.* 1965. Composition of rabbit oviduct secretions. Fertil. Steril. **16,** 170.

[50] *Holmdahl, T. H. S. & Mastroianni, L.,* Jr. 1965. Continuous collection of rabbit oviduct secretions at low temperature. Fertil. Steril. **16,** 587.

[51] *Hunter, R.* 1964. Superovulation and fertility in the pig. Animal Prod. **6,** 189.

[52] *Krishnan, R. S. & Daniel, J. C.,* Jr. 1967. Blastokinin: Inducer and regulator of blastocyst development in the rabbit uterus. Science, **158,** 490.

[53] *Lewis, W. H. & Gregory, P. W.* 1929. Cinematographs of living developing rabbit eggs. Science, **69,** 226.

[54] *Lutwack-Mann, C.* 1962. Glucose, lactic acid and bicarbonate in rabbit blastocyst fluid. Nature, **193,** 653.

[55] *Mastroianni, L.,* Jr. & *Wallach, R. C.* 1961. Effect of ovulation and early gestation on oviduct secretions in the rabbit. Amer. J. Physiol. **200,** 815.

[56] *Mauer, R. E., Hafez, E. S. E., Ehlers, M. H. & King, J. R.* 1968. Culture of two-cell rabbit eggs in chemically defined media. Exp. Cell Res. **52,** 293.

[57] *Mills, R. M.,* Jr. & *Brinster, R. L.* 1967. Oxygen consumption of preimplantation mouse embryos. Exp. Cell Res. **47,** 337.

[58] *Mintz, B.* 1962. Experimental study of developing mammalian eggs. Removal of the zona pellucida (mice). Science, **138,** 594.

[59] *Mintz, B.* 1964. Formation of genetically mosaic mouse embryos, and early development of lethal (t^{12}/t^{12}) – normal mosaics. J. Exp. Zool. **157,** 273.

[60] *Mintz, B.* 1964. Synthetic processes and early development in the mammalian egg. J. Exp. Zool. **157,** 85.

[61] *Mounib, S. M. & Chang, M. C.* 1965. Metabolism of glucose, fructose and pyruvate in the six-day rabbit blastocyst. Exp. Cell Res. **38,** 201.

[62] *Mulnard, J. G.* 1965. Studies of regulation of mouse ova *in vitro.* In Preimplantation Stages of Pregnancy, ed. *G. E. W. Wolstenholme* and *M. O'Connor,* London: J. and A. Churchill Ltd.

[63] *New, D. A. T.* 1966. Development of rat embryos cultured in blood sera. J. Reprod. Fertil. **12,** 509.

[64] *New, D. A. T. & Stein, K. F.* 1963. Cultivation of mouse embryos *in vitro.* Nature, **199,** 297.

[65] *New, D. A. T. & Stein, K. F.* 1964. Cultivation of postimplantation mouse and rab embryos on plasma clots. J. Embryol. Exp. Morph. **12,** 101.

[66] *Onanoff, J.* 1893. Recherches sur la fecondation et la gestation des mammifers. C. R. Soc. Biol. (Paris) **45,** 719.

[67] *Onuma, H. Maurer, R. R. & Foote, R. H.* 1968. *In vitro* culture of rabbit ova from early cleavage stages to the blastocyst stage. J. Reprod. Fertil. **16,** 491.

[68] *Pincus, G.* 1941. The control of ovum growth. Science, **93,** 438.

[69] *Pincus, G.* 1941. Factors controlling the growth of rabbit blastocysts. Amer. J. Physiol. **133,** 412.

[70] *Purshottam, N. & Pincus, G.* 1961. *In vitro* cultivation of mammalian eggs. Anat. Rec. **140,** 51.

[71] *Restall, B. J.* 1966. The fallopian tube of the sheep III, the chemical composition of the fluid from the fallopian tube. Australian J. Biol. Sci. **19,** 687.

[72] *Schenk, S. L.* 1880. Das Saugethierei kunstlick befruchtet außerhalb des mutterthieres. Mitt. Embr. Inst. K. K. Univ. Wien. **1,** 107.

[73] *Staples, R. E.* 1966. Development of five-day rabbit blastocysts after culture at 37 °C. J. Reprod. Fertil **13,** 369.

[74] *Stern, S. & Biggers, J. D.* 1968. Enzymatic estimation of glycogen in the cleaving mouse embryo. J. Exp. Zool. **168,** 61.

[75] *Taska, R. J.* 1968. Interrelationships between RNA and protein synthesis in preimplantation stage mouse embryos. Ph. D. thesis, Temple University, Philadelphia, Pennsylvania.

[76] *Taymor, M. L.* 1966. Induction of ovulation with human postmenopausal gonadotropin I. Fertil. Steril. **17,** 731.

[77] *TenBroeck, J. T.* 1968. Effect of nucleosides and nucleoside bases on the development of preimplantation mouse embryos *in vitro.* J. Reprod. Fertil. **17,** 571.

[78] *Thibault, C.* 1966. La culture in vitro de l'oeuf de vache. Ann. Biol. Anim. Biochim. Biophys. **6,** 159.

[79] *Thomson, J. L. & Brinster, R. L.* 1966. Glycogen content of preimplantation mouse embryos. Anat. Rec. **155,** 97.

[80] *Van Wagenen, G.* 1968. Induction of ovulation in Macaca mulatta. Fertil. Steril. **19,** 15.

[81] *Vishwakarma, P.* 1962. The pH and bicarbonate-ion content of the oviduct and uterine fluids. Fertil. Steril. **13,** 481.

[82] *Wales, R. G.* 1969. Production of carboxylic acids by the preimplantation mouse embryo. Submitted to: J. Reprod. Fertil.

[83] *Wales, R. G. & Biggers, J. D.* 1968. The permeability of two-and eight-cell mouse embryos to L-malic acid. J. Reprod. Fertil. **15,** 103.

[84] *Wales, R. G. & Brinster, R. L.* 1968. The uptake of hexoses by mouse embryos. J. Reprod. Fertil. **15,** 415.

[85] *Wales, R. G. & Whittingham, D. G.* 1967. A comparison of the uptake and utilization of lactate and pyruvate by one and two-cell mouse embryos. Biochim. Biophys. Acta, **158,** 703.

[86] *Whitten, W. K.* 1956. Culture of tubal mouse ova. Nature, **176,** 96.

[87] *Whitten, W. K.* 1957. Culture of tubal ova. Nature, **179,** 1081.

[88] *Whitten, W. K. & Biggers, J. D.* 1968. Complete development *in vitro* of the preimplantation stages of the mouse in a simple chemically defined medium. J. Reprod. Fertil. **17,** 399.

[89] *Whittingham, D. G.* 1968. Fertilization of mouse eggs *in vitro.* Nature, **220,** 592.

Discussion

C. R. Austin: Am I right that you can substitute protein with glutathione, is that correct?

R. L. Brinster: Yes.

C. R. Austin: You donot need a large molecule there at all?

R. L. Brinster: No. Well, in those experiments I had Ficoll but later found that it is not essential. The glutathione itself will exercise a protective effect, just like a large molecule, for some reason. Glutathione is 6-amino acids together, perhaps that is large enough. Glutathione is also known to have a very substantial protective effect on enzymes and in other systems where protein has been found to exercise protective action.

C. R. Austin: May I ask another question? Is the bicarbonate buffer system the only one that works at all well?

R. L. Brinster: Yes. Dr. *Thomson* has tried Tris and phosphate buffers and a number of other amino acid buffers. None of these will allow substantial development. 1, 2 or 3 cleavages is about all the development that can be obtained in the mouse *in vitro* system. They are adequate for short-term tests, but inadequate for complete development. I do not know why this is.

P. Soupart: It is interesting that your maximal jump in the utilisation of glucose in the rabbit embryo seems to appear between the 4 th and 5 th day, according to one of your figures. Because it is precisely at that time that the ultra-structural complexity of the mitochondria in the blastocyst reaches full development. And most probably becomes really functional.

R. L. Brinster: I think two things are involved here. One is that the increase in tissue mass is the largest during that period. The other one is the one you directly alluded to, the development of the mitochondrial oxidative system.

P. Soupart: It would be very interesting to demonstrate the appearance of mitochondrial, functional capability, for instance cytochrome oxidase activity, using cytochemical methods at that very time. It would correlate very nicely, I guess.

R. L. Brinster: Dr. *Chang,* did you do something on cytochrome oxidase in the rabbit at one time?

M. C. Chang: In the earlier time I played around, but I did not do it seriously. Did you try the rabbit serum and mouse serum?

R. L. Brinster: Yes, I have tried serums and they have worked well, but if you incorporate serum you have less idea of what the medium contains. Just for routine culturing serum is adequate for the rabbit.

D. W. Schomberg: Just a question for point of interest. Is glutathione broken down to its constituent amino acids, in a culture system?

R. L. Brinster: I do not know the answer to that question. I am interested in how glutathione is utilized because it is only three different amino acids and it could be studied quite easily.

Stages of the Development of the Human Egg

R. G. Edwards

Physiological Laboratory, University of Cambridge, Cambridge, England

Summary: The maturation of human oocytes *in vitro* has been used to study various aspects of early development.

1) Judged from the analysis of maturation of human oocytes *in vitro,* the interval between the LH surge and ovulation in man is 36 hours approximately.
2) The origin of Mongoloid children and other chromosomally unbalanced fetuses evidently occurs in the fetal ovary, i.e. many years before the egg is ovulated. Chromosomome pairing is faulty, leading to the formation of univalents in the fetal ovary. The eggs are then used in a regular order in the adult ovary, and become increasingly abnormal in later life.
3) Fertilization of human oocytes matured *in vitro* was achieved by the use of culture medium containing high quantitites of bicarbonate, and pyruvate. Follicular fluid was added to the medium in some cases. The movement of spermatozoa into the eggs and the formation of pronuclei were timed.
4) By use of laparoscopy, human eggs can be removed from Graffian follicles of women given gonadotropins, and replaced in the oviduct.
5) In the rabbit, blastocysts can be sexed with complete accuracy. Many of them will then develop to full term when tranferred into a recipient.

I wish to make today, very briefly, five points about early human and mammalian development. The first three involve the use of oocytes, maturing *in vitro,* for certain essential studies. As long ago as 1935, *Pincus* and *Enzmann* [1] showed that if a rabbit occyte is taken from its follicle and placed in a culture medium it would mature to the stage ready for fertilization. We have extended this work to many species, including man.

All of the eggs possessed by a female mammal are formed in the ovary before or shortly after birth; they also enter their meiotic prophase at this time. In the fetus therefore the eggs have reached and passed the stage of chiasma formation, and reached the stage of meiosis called diplotene. The eggs are all formed and are in that state at or shortly after birth. This stage is succeeded by a nucleated stage called dictyate (or diffuse diplotene), which, as is well known, persists into adult life until shortly before ovulation.

Manuscript received: 14 May 1969

Under the surge of LH in mid-cycle of the adult, the chromosomes of the oocyte reappear in diakinesis. The oocytes proceed in their development to the extrusion of the first polar body and to metaphase of the second meiotic division (metaphase-II). Ovulation occurs at this stage of development. Thus, if LH is injected into animals, the oocytes mature from dictyate to metaphase-II. Likewise, if ovarian oocytes in dictyate are placed in culture, they will proceed through the same stages. The important fact is that in every species we have examined, the oocytes proceed through these stages at the same rate after the injection of LH or when maturation of oocytes occurs *in vitro*.

1. Maturation of oocytes in vitro as a measure of the time of ovulation after LH release

Let us take one example from work on the pig oocyte, showing the work carried out with Dr. *Polge* some years ago. The timings of maturation in the pig are very similar to those occurring in the human oocyte. The relationship between oocyte maturation after the injection of LH into the pig and after placing oocytes in culture is shown in Fig. 1; the timings are identical. This holds for rabbits – Dr. *Chang* [2] showed this for rabbits in 1955 – and for every species we have since examined [3]. Now, we can reverse this argument. We know that the maturation of oocytes *in vitro* requires the same interval as oocytes maturing *in vivo*. If we can time these intervals during the culture of human oocytes, essential information on the interval between LH release and ovulation during the menstrual cycle can be obtained.

	Hours after release of oocytes from follicles, or injection of L. H. into mother.	
	Release	L. H.
Germinal vesicle	0–20	0–18
Metaphase-I	20–30	18–34
Anaphase / Telophase	30–43	34–37
Metaphase-II/ Polar Body	30–43	38

Fig. 1

Duration of meiosis in pig oocytes. I am indebted to Drs. *Polge* and *Hunter* for permission to quote their data on the injection of L. H.

	Hours after release of oocytes from follicles.
Germinal Vesicle	0–24
Diakinesis	25–28
Metaphase-I	26–35
Metaphase-II / Polar Body	36–43

Fig. 2

Duration of meiosis in human oocytes.

The rate of maturation of human eggs *in vitro* is shown in Fig. 2. For most oocytes, the timing to metaphase-II using a variety of culture media is 36 hours, or even an hour or two less [4], although a few oocytes mature more rapidly than this. So we predict that in man the interval between the LH surge and ovulation is 30–36 hours.

2. Analysis of chromosomes in diakinesis and metaphase-I

From the data given in Fig. 2, it is clear that the stages of meiosis called diakinesis and metaphase-I can be obtained quite easily. These are essentially the stages of meiosis where the chiasmata can be counted and analysed. Dr. *Henderson* and I believe that the analysis of the number of chiasmata and the examination of chromosome pairing at these stages is essential to our understanding of the cause of mongolism and, possibly, of the cause of infertility in older women [5].

So for, most of our analyses have been carried out on mouse oocytes. The egg in Fig. 3 came from a middle-aged female mouse, and was examined during diakinesis. Notice that most of the chromosomes have one chiasmata, others having two. But notice especially that all of the chromosomes are bivalents, i.e. they are associated with their homogue.

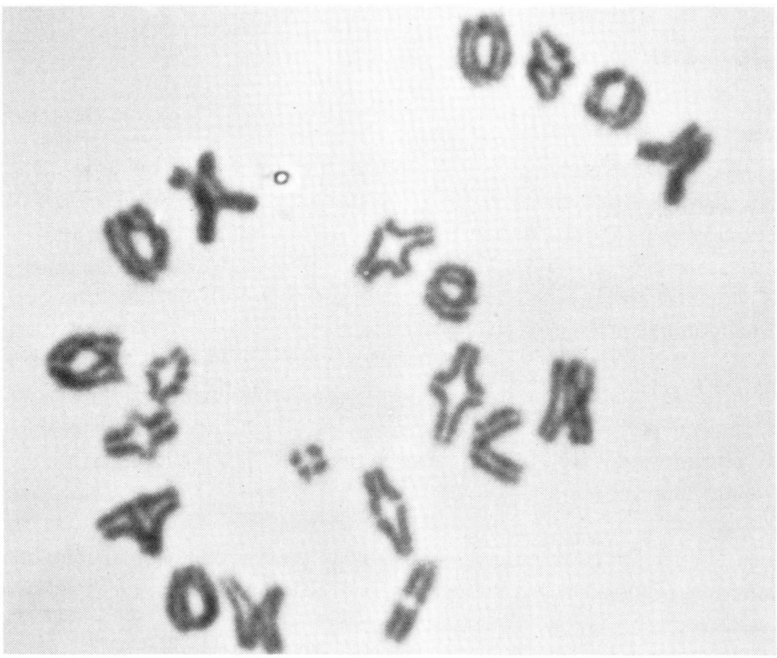

Fig. 3. Diakinesis/metaphase-I in an oocyte from a middle-aged mouse. All the bivalents are paired.

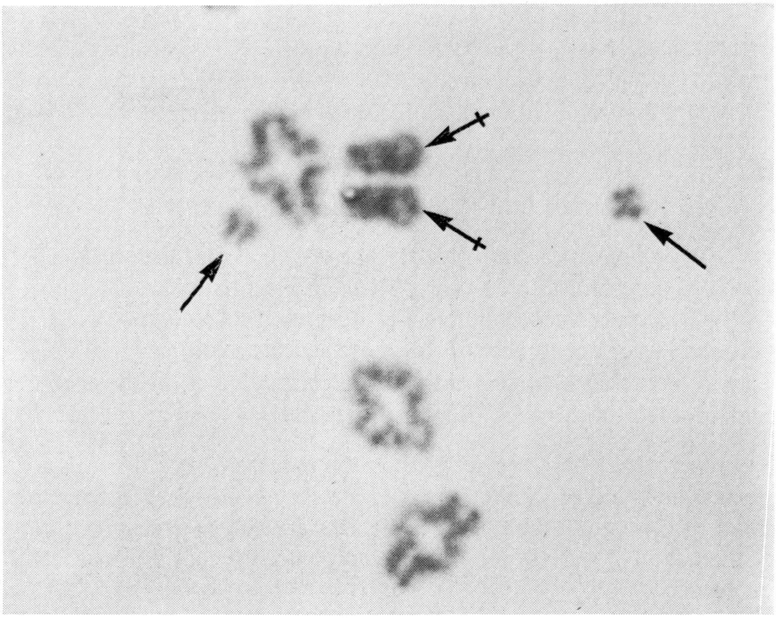

Fig. 4. Diakinesis/metaphase-I in an oocyte from an old mouse. Two pairs of univalents, one showing secondary pairing (‡), the other widely separated (↓).

An egg from an old mouse is shown in Fig. 4. Here, the two chromosomes in the centre are univalents with no chiasmata to associate them together, yet they are showing secondary association. Look at the two small chromosomes: they are univalents and are widely separated. This separation could lead to abnormal segregation of the two small chromosomes at anaphase-I and thus to the production of a chromosomally-unbalanced embryo.

We believe that the same applies in humans. Fig. 5 shows a human egg with two small separate univalents. We have labelled them as 22; they could be labelled 21 for these two chromosomes are in the same size range. This egg could therefore have been the precursor of a mongoloid embryo.

The fundamental question is why did the eggs display these chromosomal characteristics? We believe that the answer is a surprising one. If the number of chiasmata is counted in eggs taken from mice of increasing age and plotted graphically: the chiasma frequency decreases with maternal age (Fig. 6). There could be two major reasons for the decline in chiasma frequency with maternal age:

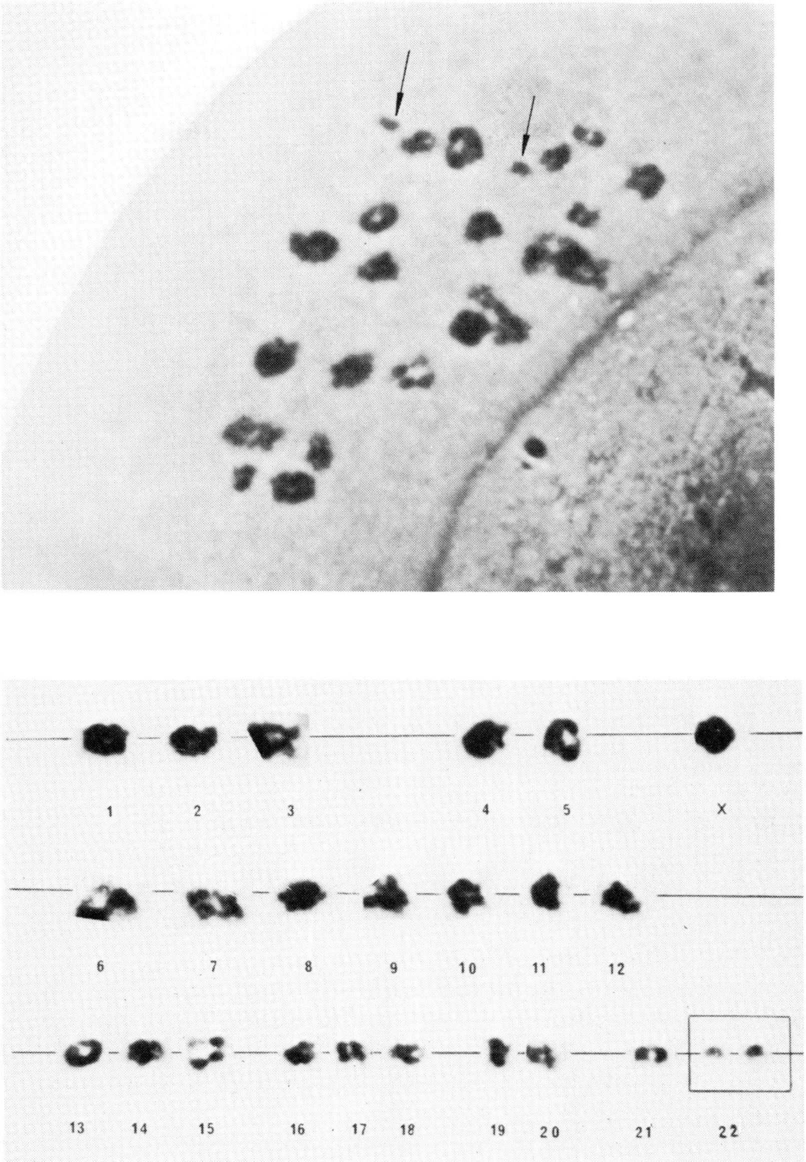

Fig. 5. Metaphase-I in a human oocyte. Two univalents are separate (↓ in 5A), and belong to group 21 or 22 (5B).

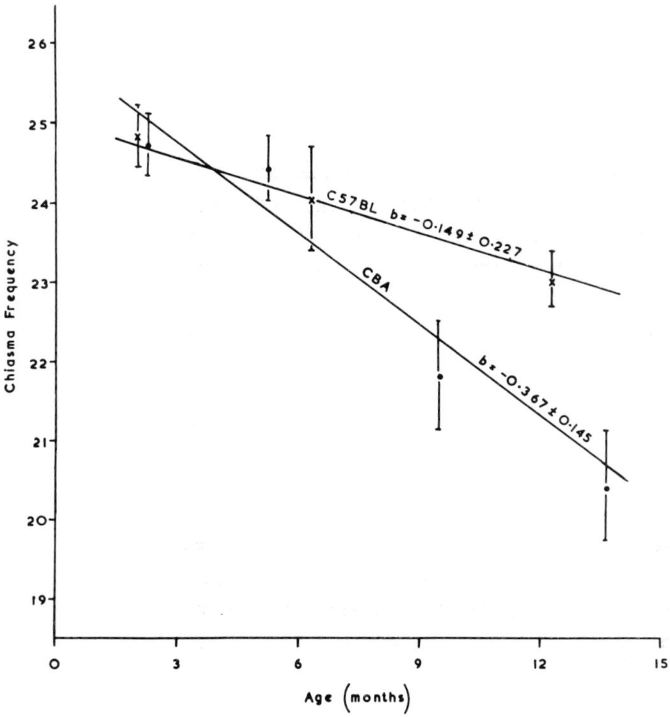

Fig. 6. Chiasma frequency and maternal age in two inbred mouse strains.

(i) the chromosomes are falling apart as the eggs grow older. This is the theory ("terminalisation of chiasmata") supported by many observations in classical cytogenetics.

(ii) The chromosomes are separate because they were originally formed this way in the oocyte during meiotic prophase in the fetus on eggs with low chiasma frequencies and univalents are then selectively ovulated at the end of reproductive life of the adult female. We believe that this is the correct interpretation.

Formal proof of one or the other of these alternatives should come from genetical studies. Suppose there are linked genes on these chromosomes (Fig. 7). Remember that chiasmata were formed in the fetal ovary, and therefore the recombination products were decided then. If the first reason is correct, i.e. terminalisation, then the recombination frequency of these genes in offspring of the female will not be affected as she grows older. In other words, the proportion of recombinants in offspring would be the same throughout maternal life. If the second theory is correct, i.e. that there are few chiasmata in oocytes formed late in the fetal ovary and these oocytes are ovulated late in maternal life, then recombination frequency will decline with advancing maternal age (Fig. 7).

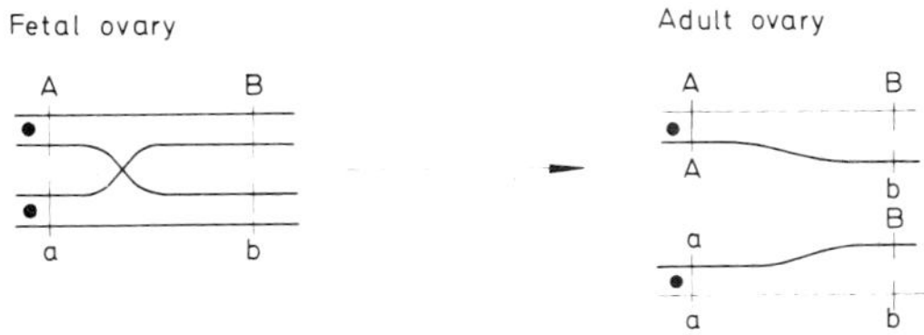

(A) Terminalisation: decline in chiasma frequency, no decline in recombination frequency.

(B) Failure of formation of chiasmata: decline in chiasma frequency, and decline in recombination frequency.

Fig. 7

Genetic situation in oocytes of older females resulting from (A) terminalisation of chiasmata (B) failure of chromosome pairing in the fetal ovary.

The available data shows that the recombination frequency declines with maternal age [6, 7]. The decline closely parallels the decline in chiasma frequency. This parallel indicates that the second suggestion is correct, i.e. oocytes with chiasmata are selectively ovulated late in life. Indeed, it seems that eggs ovulated throughout life are being selected one by one – those with high chiasmata frequencies early, and those with low chiasmata frequencies later. In other words, there seems to be a "production line" of eggs in the ovary, and there is some evidence that this production line is actually laid down in the fetus: perhaps even when the gonocytes enter the fetal ovary. Thus, we suggest that if a woman has 10^6 eggs, they each are numbered in the fetus, and during her adult life they are ovulated from no.1 in rotation to no. 10^6, sticking closely to a regular progression.

If our theory is correct, the outlook for averting the birth of Mongoloid babies is almost nil, since the original error occurs in the fetus. The same applies to other chromosomally unbalanced fetuses, for the larger chromosomes can form univalents too. In old mice, the great majority of oocytes carried at least one pair of univalents. One cause of infertility in older mice and women could thus be that the eggs are chromosomally abnormal, in addition to any uterine effects associated with age.

There has been criticism of our theory. If the theory is correct, eggs transferred from an old mouse into a young female would have a low chance of survival, whereas eggs transferred from a young mouse into a young female should live. There is one experiment in mice [8] which did not confirm this prediction. But this experiment is totally inconclusive: various strains of mice were indiscriminately mixed as donors and recipients. There is some work on the hamster which confirms our ideas completely: eggs from old hamsters placed into young hamsters do not live, whereas eggs from younger mice do survive [9]. The old eggs were clearly abnormal.

3. Fertilization in vitro

When eggs reach the stage of metaphase-II and the first polar body, they are ready for fertilization. We have had various trials and tribulations in our attempts to fertilise eggs *in vitro* over the last four or five years, including placing cow and human eggs into intrauterine chambers (where there was some evidence of fertilization). About three months ago *Bavister* [10], following a succession of studies on fertilization *in vitro* in the hamster, showed that pH control was critical. We therefore applied this system to the human, and spermatozoa were immediately found entering the eggs [11]. The medium for fertilization is a Tyrode's solution with pyruvate, albumin and extra bicarbonate, and the pH is 7.5 or 7.6. In our experiments we added follicular fluid to spermatozoa or eggs, since this was earlier considered essential for fertilization in hamsters.

In early experiments we examined the living eggs by phase-contrast microscopy, and then fixed them using acetic alcohol, which is needed to clear yolky eggs such as those of the cow and sheep. Later we fixed with formalin. We could identify all the stages of fertilization up to late pronuclear development.

The timings of human fertilization after insemination of the eggs are shown in Fig. 8. Examples of fertilization are shown in Fig. 9–11. In some eggs, spermatozoa failed to migrate completely through the zona pellucida and remained embedded

Fig. 8. Timings of human fertilisation *in vitro.* Details of the eggs.

Hours after insemination	Unpenetrated	Spermatozoa in zona	Spermatozoa in perivitelline space	Pronucleate
6 - $6\frac{1}{2}$	3			
7 - $10\frac{3}{4}$	5	1	4	
$11\frac{1}{2}$ and later	8	5	1	7
	16	6	5	7

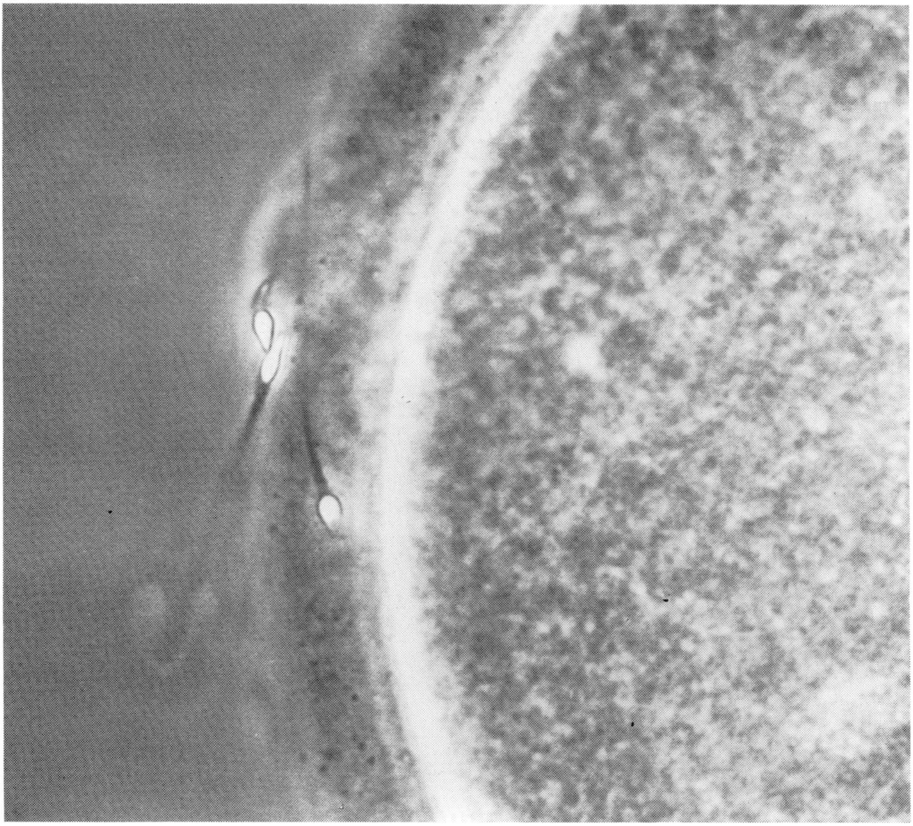

Fig. 9. Spermatozoon moving through the zona pellucida of a human egg; two other spermatozoa lie outside the zona.

deep in it. Other eggs had several pronuclei. No doubt some of these anomalies arose because the eggs were brought to Cambridge by car for a journey of some 160 miles, and this could obviously give rise to anomalies. Recently we have seen several examples of monospermic fertilization. All the stages of fertilization to late pronuclear formation have now been recorded.

Why do spermatozoa penetrate through the zona pellucida with medium of high pH, especially when ejaculated human spermatozoa – remember not capacitated spermatozoa – and eggs matured *in vitro* are mixed together? Perhaps the acrosome change so important in capacitation is related in some way to the steroids around the egg.

If different follicular fluids are used for fertilization, some give good and others poor rates of fertilization. In one experiment an efficient fluid was of a clear yellow colour and evidently came from a Graafian follicle, whereas an inefficient fluid was slightly pink, and probably came from a cystic follicle. In mares, fluid from preovulatory follicles have high levels of estradiol-17β, while cystic fluids do not [12].

Here, perhaps, is a clue – we have about 0.05 ml of the medium to repeat the use of these fluids in studies on fertilization and to analyse the fluids for their steroid content. We must also analyse the fluids immunologically to see if various immunoglobulins are present, especially that reacting with the acrosome [13]. This immuno-

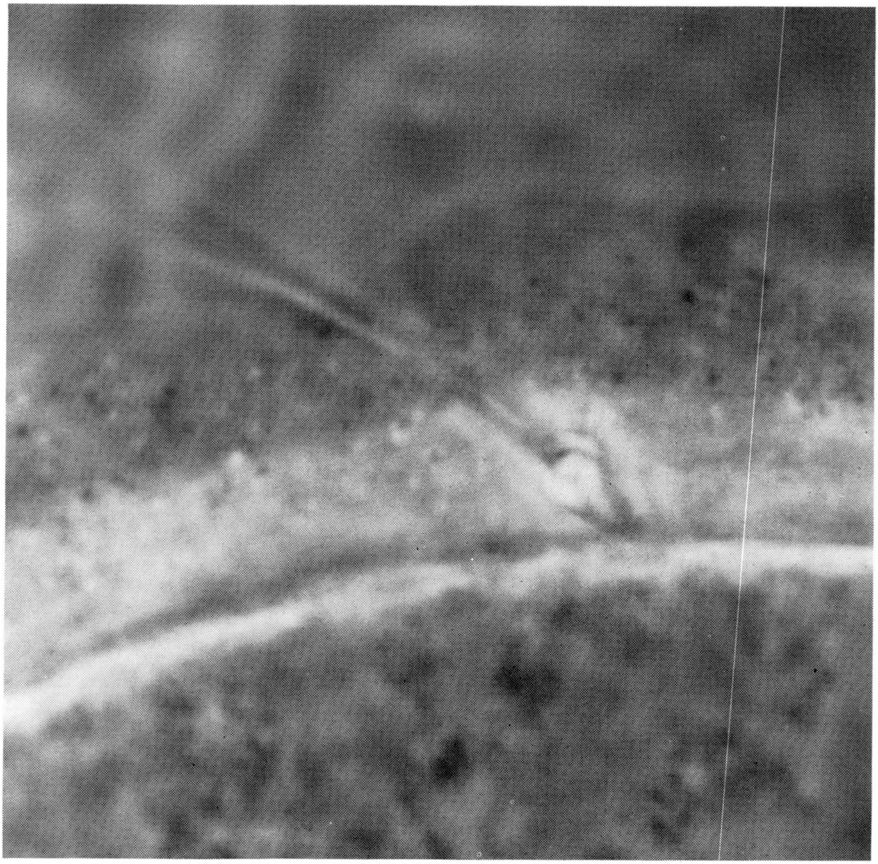

Fig. 10. Sperm-head and part of mid-piece have passed into the perivitelline space, the tail lies outside the egg. The sperm head is just touching the vitellus.

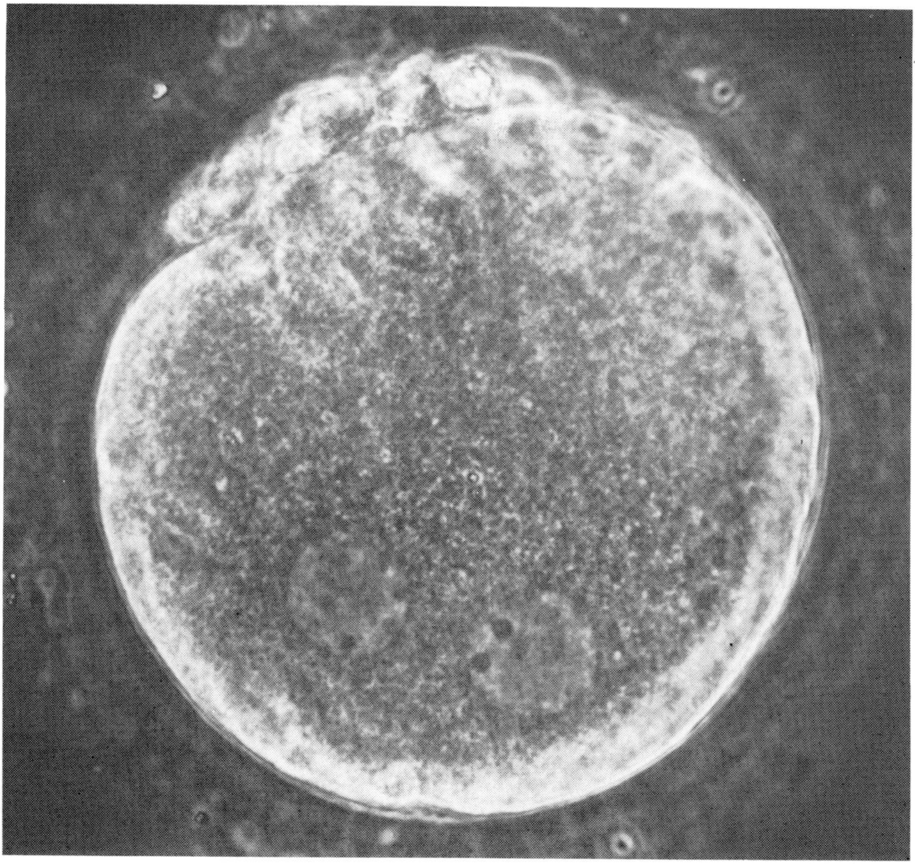

Fig. 11. Human egg with two pronuclei, and extruded bodies (including the polar bodies). The extruded bodies are an unhealthy sign.

globulin, which might be involved in capacitation, has been characterised as an IgG2 in the guinea pig [14]. If this immunoglobulin is present only in the fluids giving low rates of fertilization, it would imply that it is toxic to spermatozoa and is not involved in capacitation.

4. Clinical applications of human fertilization in vitro

Fertilised human eggs would be invaluable for curing certain forms of human infertility. But if we wish to produce human offspring, it is doubtful if oocytes maturing *in vitro* will be helpful, for we will probably have difficulty in obtaining cleaving ova. *Chang* [2] carried out some work with rabbit oocytes, and we have worked with rabbit, pig and cow oocytes (unpublished work with *C. Polge* and *L. E. Rowson*) to analyse the developmental capabilities after fertilization *in vivo* of oocytes matured *in vitro*. The results appear to be the same for all species, i.e. that development breaks down soon after fertilization. We simply do not know enough about the maturation of oocytes *in vitro* to produce embryos sufficiently viable to develop to full term. A two-cell pig egg that was matured *in vitro* and fertilized and grown *in vivo* is shown in Fig. 12. A sub-nucleus is present in one of the blastomeres.

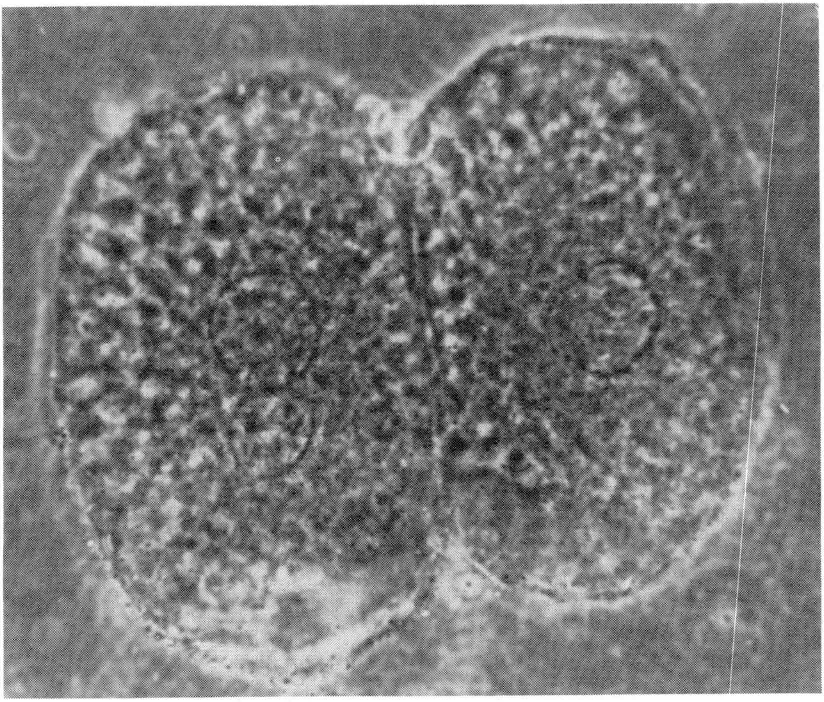

Fig. 12. A 2-celled pig egg. It was matured *in vitro*, and fertilised and cleaved *in vivo*. A sub-nucleus is present in one of the blastomeres. Stained with orcein. Courtesy of Dr. *C. Polge*.

There is one way of circumventing this difficulty. Again referring to *Chang* [2] and to unpublished work of ours, it is possible to obtain full-term rabbit fetuses if maturation is started *in vivo* by gonadotropin injections into the mother, and completed *in vitro* by removing the oocytes from the ovary after some hours. This means that as a clinical approach we will have to inject women with gonadotropins, and remove their eggs shortly before ovulation. *Patrick Steptoe* and I are attempting

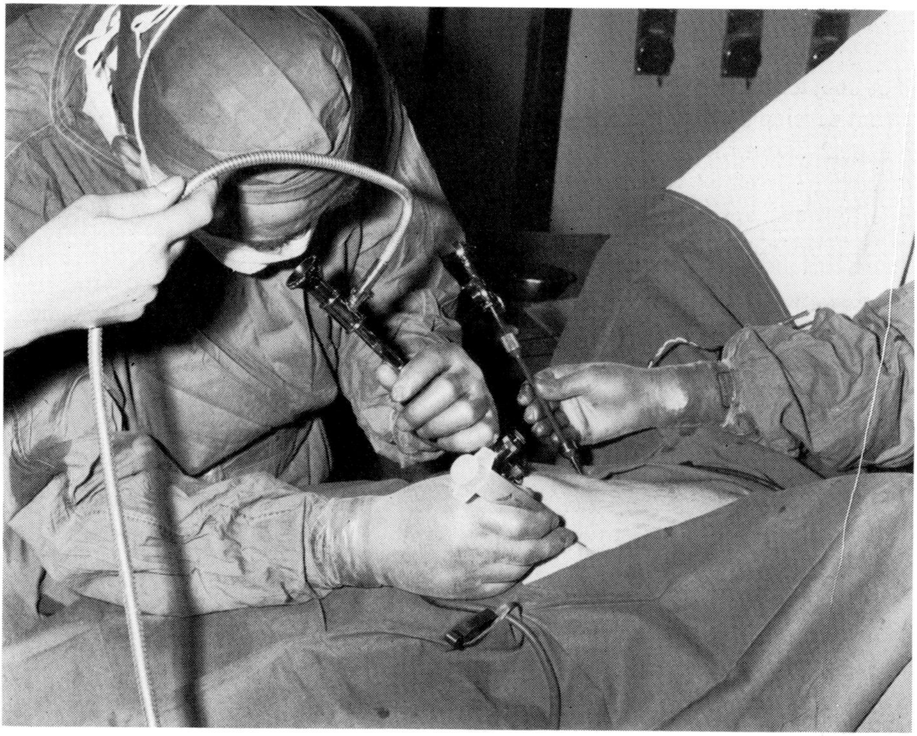

Fig. 13. Technique of laparoscopy. Syringe for collection of oocytes is being passed into the peritoneum. Courtesy of Mr. *P. C. Steptoe.*

to achieve this now, using a technique called laparoscopy. In laparoscopy, the necessary abdominal incisions are very small Fig. 13, and a telescope is used to look into the abdomen. A needle is passed into the abdominal cavity, in order to penetrate the follicles, and suck the eggs out. Our routine is thus to treat women with gonadotropins and use laparoscopy to withdraw the eggs from the follicle 30 hours after the injection of LH. We must make absolutely sure that the timing of ovulation we predicted earlier is right by examining the oocytes for their chromosomal state.

If we can fertilize oocytes freshly removed from the ovary, there should be fewer problems about further development. We might be able to use the rabbit as a host for these eggs for experimental purposes. Judged from work with unfertilized human eggs [4], the rabbit appeared to be an excellent host for the oocyte, although immunological abd other problems could obviously arise through the use of a foreign species as host.

5. Selection of embryos

If several eggs can be obtained from a woman and then fertilized, some choice over the type of embryo to be returned to her would become possible. In certain families, e.g. those where the mother is a carrier for hemophilia or other sex-linked diseases, the choice of girls would avert the birth of affected offspring.

Our first attempts to sex embryos were made on mice, by examining blastocysts for H2 and other transplantation antigens using a mixed cell antiglobulin test (Sell, Coombs + Edwards, unpublished). But we could not find these antigens on the embryos (Fig. 14).

Fig. 14. Antigens present on mouse Blastocysts.
Courtesy of Drs. *K. Sell & R. A. R. Coombs.*

H2-d	-
H2-k	-
B	+

This was a setback, because one of these antigens is sex-linked and might have shown the presence of a Y chromosome in the embryo. We therefore used a different approach in the rabbit, namely by studying the embryo for the presence or absence of sex chromatin [15]. The blastocyst was held in a large suction pipette and a small portion of trophoblast drawn into a fine suction pipette [16]; the small portion was then excised (Fig. 15). The small piece of excised tissue was then used for sexing the embryo by staining it and looking for the presence or absence of sex chromatin. The embryos were placed back in culture for six hours to allow the repair of the wound and recovery of the embryos, and then transferred to recipient hosts. Male embryos were transferred separately to females. At full term every fetus had been predicted correctly [16]. We would contrast these results with those obtained in attempts to control sex ratios by separating X and Y spermatozoa. After dozens of papers on the separation of spermatozoa, there still appears to be not the slightest hope of even inducing a slight change in the sex-ratio, using this method. On the other hand, our first attempts at sexing preimplantation embryos were totally correct.

It is doubtful that human embryos can be sexed in this way. But there are markers on the sex chromosomes, e.g. the gene for glucose-6-phosphate dehydrogenase, which could prove most useful. The successful utilization of this enzyme will probably depend on whether both chromosomes are genetically active in the female blastocyst, which could lead to differences in levels and activity of this enzyme in male and female embryos.

Acknowledgement: I wish to thank the Ford Foundation for their generous financial support.

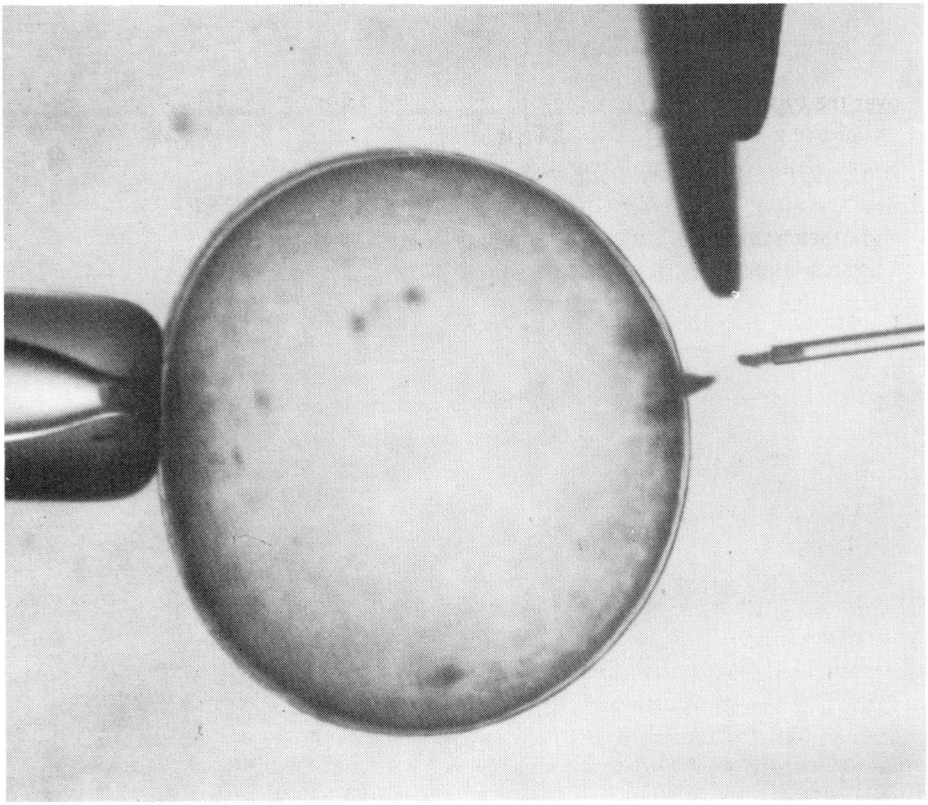

Fig. 15. Rabbit blastocyst held in a large suction pipette (left), with a small piece of trophoblast excised in a fine suction pipette (right). The scissors can be seen (top).

References

[1] *Pincus, G. & Enzmann, E. V.*, 1935. J. Exp. Med. **62,** 655.

[2] *Chang, M. C.*, 1955. J. Exp. Zool. **128,** 379.

[3] *Edwards, R. G.*, 1965. Nature, **208,** 349.

[4] *Edwards, R. G.*, 1965. Lancet, ii, 926.

[5] *Henderson, S. A. & Edwards, R. G.*, 1968, Nature, **218,** 22.

[6] *Bodmer, W. F.*, 1961. Nature, **190,** 1134.

[7] *Reid, D. H. & Parsons, P. A.*, 1963. Heredity, **18,** 107.

[8] *Talbert, G. B. & Krohn, P. L.*, 1965. J. Reprod. Fert. **11**, 399.
[9] *Blaha, G. C.*, 1964. Anat Rec. **150**, 413.
[10] *Bavister, B. D.*, 1969. J. Reprod. Fert (in press).
[11] *Edwards, R. G., Bavister, B. D. & Steptoe, P. C.*, 1969. Nature, **221**, 632.
[12] *Short, R. V.*, 1964. In: Recent progress in Hormone research., **20**, 303.
[13] *Beck, J. S., Edwards, R. G. & Young, M. R.*, 1963. J. Reprod. Fert. **4**, 103.
[14] *Johnson, M. H.*, 1968. J. Reprod. Fert. **16**, 503.
[15] *Edwards, R. G. & Gardner, R. L.*, 1967. Nature, **214**, 576.
[16] *Gardner, R. L. & Edwards, R. G.*, 1968. Nature, **218**, 346.

Discussion

J. M. Bedford: Just one small point, why do you wash the ova? Why do you wash your eggs before you analyse them?

R. G. Edwards: You remember you described yesterday the effect of serum on certain cells. Now we know that follicular fluid used in the medium for the fertilization of eggs contains most in all of the proteins in serum. We were concerned that if we put follicular fluid on the sperms, we would get an immunological effect. So we washed the eggs in order to avoid with this effect. When you get excessive contamination with follicular fluid, there is often agglutination of the sperms, especically in some animal species.

J. M. Bedford: Is this a prior thought or have you also got controls in which you have not washed the eggs and obtained poor results?

R. G. Edwards: It is difficult to wash eggs. We do a series of dilution, pipetting the eggs into relatively large amounts of washing medium. But much of the fluid seems to remain around the eggs.

P. Soupart: There is another reason for the washing of the eggs in the experimental procedure. And I would like to remind you of the theory which holds that there is a factor to be eliminated, the so-called "fertilizin". Preliminary washing of ova definitely has a bearing not only on the speed with which fertilization can be achieved in vitro but also upon the percentage of success. To emphasize the speed effect of washing, I would like to show a few figures. Fig. 16

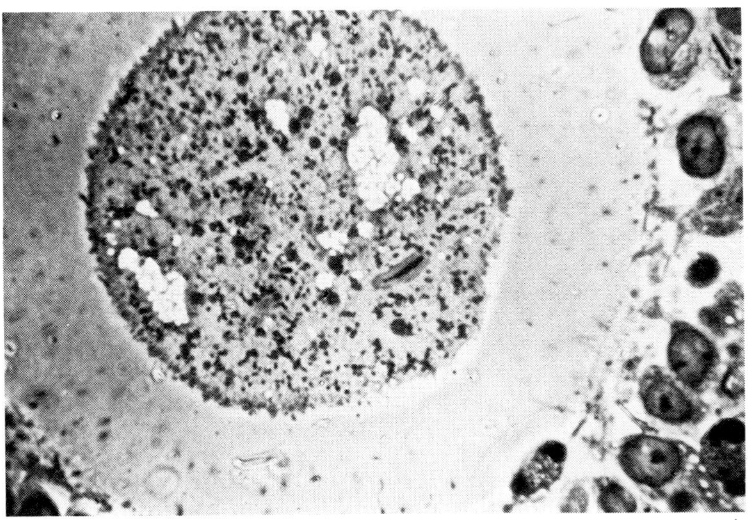

Fig. 16
In vitro fertilized rabbit ovum, fixed 2 h after mixing capacitated sperm and washed ova. Sperm head beginning to swell in vitellus.

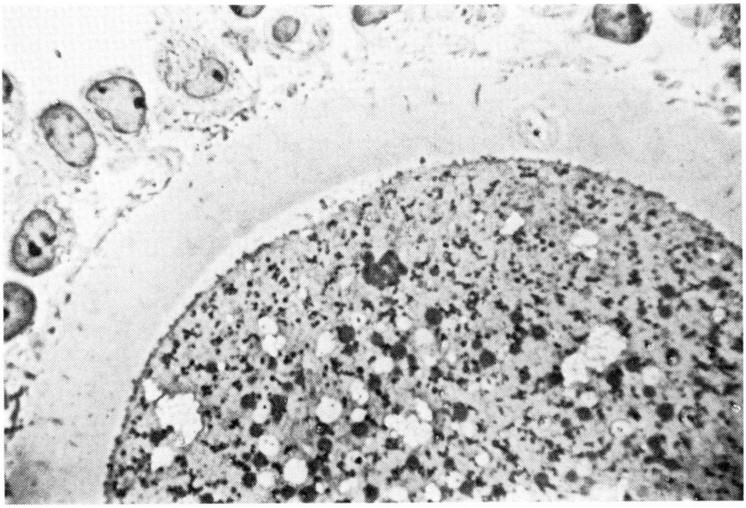

Fig. 17
Same ovum as Fig. 16. Serial section showing female chromosome after extrusion of second polar body.

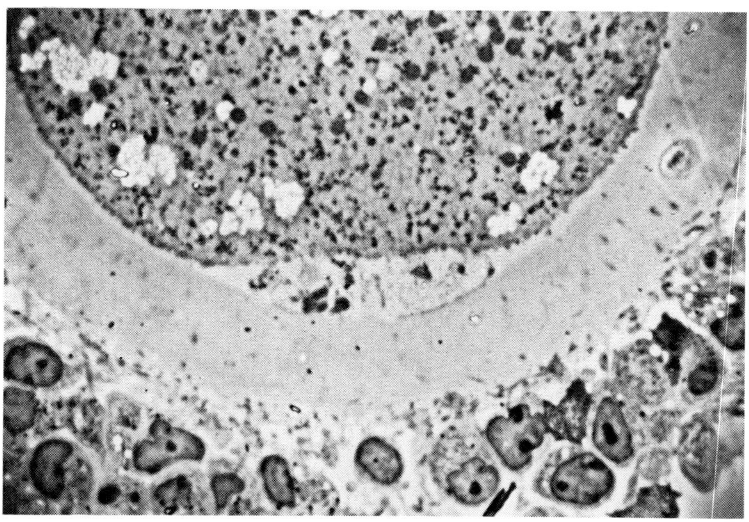

Fig. 18
Same ovum as Fig. 16. Serial section showing the two polar bodies in the perivitelline space.

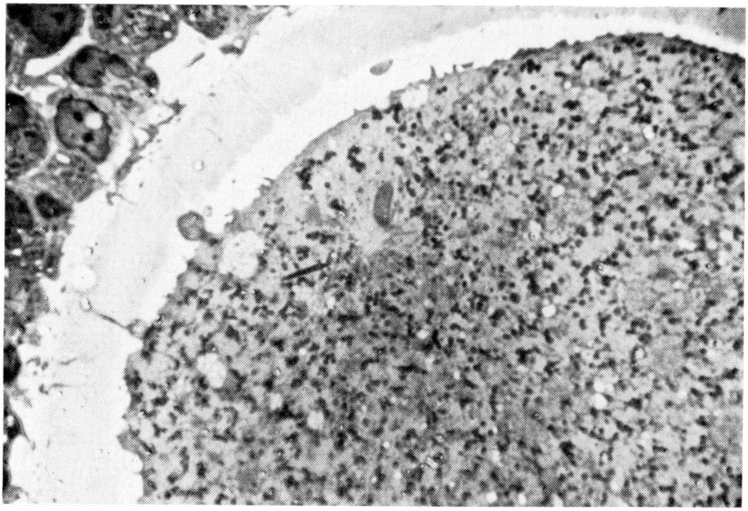

Fig. 19
Another example of rabbit ovum fertilized *in vitro* and fixed 2 h after mixing capacitated sperm and washed ova. Swelling of sperm head and sperm mid-piece.

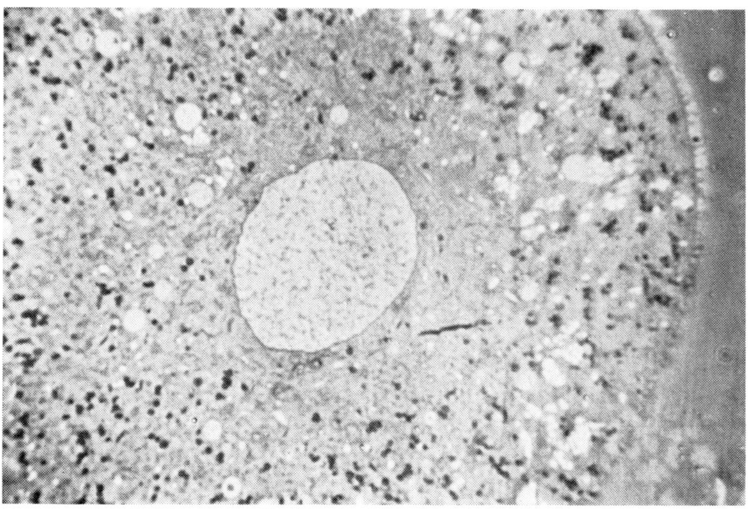

Fig. 20

Rabbit ovum fertilized *in vitro* and fixed 5 h after mixing capacitated sperm and washed ova. Section through male pronucleus and sperm mid-piece.

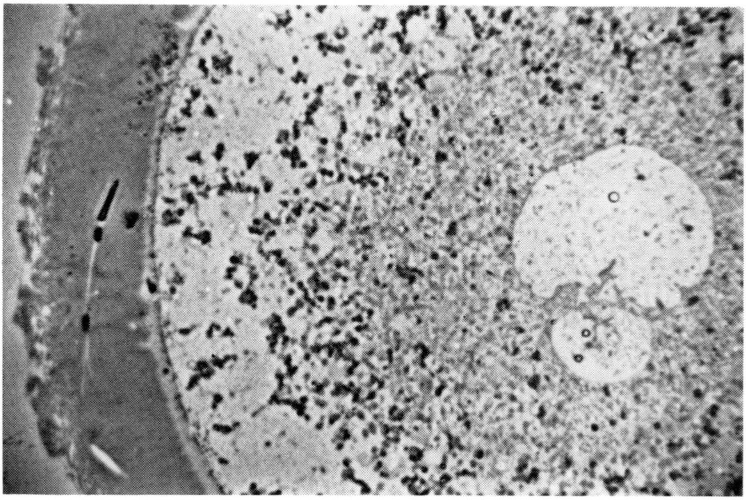

Fig. 21

Rabbit ovum fertilized *in vitro* and fixed 5 h after mixing capacitated sperm and washed ova. Section through male (large) and female (small) pronuclei. A sperm is visible in a penetration slit in the zona pellucida.

is a rabbit's egg, fertilized *in vitro,* which has been fixed *two hours exactly* after mixing sperm and washed eggs. As you can see by this time a sperm head is already there in the vitellus at around 4 o'clock. Fig. 17 will show you, in the some egg, the female chromosomes sitting at about 11 o'clock, and the beginning of the enlargement of the perivitelline space in which you will find the two polar bodies in adjacent section – which you will see on Fig. 18. Here are the two polar bodies in the perivitelline space. Fig. 19 is another example, showing a para-sagittal section of the sperm head, and part of the sperm mid-piece, sitting at about 8 o'clock, close to the edge of the egg. Fig. 20 this eggs was fixed about 5 hours after the mixing of sperm and washed egg. The male pronucleus and the mid-piece are very clearly showing. Fig. 21 is a serial section of the same egg showing a larger part of the tail. I think this is a very simple technique to appreciate exactly the details of *in vitro* fertilization. You simply fix the ova in osmium tetroxide and embed them in Araldite. For rapid analysis, you cut 0.5 μ serial section and stain a ribbon of 10 sections at a time with Toluidine blue, on a glass slide. In this way, an entire egg can be examined from pole to pole within 2 hours. If you want more detail at some level, you examine only 2 sections at a time and you can cut ultra thin sections for electron microscopic examination.

The Extra-Uterine Mouse Egg as an Experimental Model

D. R. S. Kirby

Department of Zoology, University of Oxford, Oxford, England

Summary: The mouse egg transplanted from the genital to an extra-uterine site has been used as an experimental model to investigate a range of mechanisms involved in conception. These include.

(1) The developmental capacity of the unimplanted mouse egg.

(2) The local effect of the uterus on the blastocyst during delayed implantation.

(3) The effect of the hormonal environment on the development of the extra-uterine conceptus.

(4) The endocrinological effect of an extra-uterine pregnancy on the host.

(5) Trophoblast invasion and its control.

(6) The immunological role of the uterus during development.

(7) The control of orientation of the implanting mouse egg.

The investigation of these topics will be reviewed briefly.

Introduction

The first report of an extra-uterine pregnancy was presented by *Albucasis,* a Spanish surgeon of the 10th century. He described an abdominal pregnancy in a woman which fistulated through the abdominal wall. Through the fistula were passed the macerated remains of the fetus. A remarkable episode in the history of the history of the extra-uterine pregnancy occured in the year 1500. *Jakob Nufer* a swine gelder of the Black Forest with considerable insight into obstetrics realised his wife had an extra-uterine pregnancy. At term he "slit her belly as he did the swine" and exposed the fetus. Both mother and infant, which lived for 77 years, made uneventful recoveries. The earliest account of an extra-uterine pregnancy in an animal was given by *Rommel* in 1680. He described a perfect embryo lying in the abdominal cavity of a hare.

Strangely enough the first account of transplantation of an extra-uterine pregnancy was in a woman. In 1917 *Wallace* [32] described how, upon exploring the abdomen at laparotomy, he found the conceptus developing in the oviduct. To set nature's mistakes aright he removed the conceptus and transplanted it at the lumen of the uterus, retaining it in place by two sutures. Pregnancy continued normally and a perfectly formed baby was born.

Manuscript received: 15 March 1969.

It was not until 1931 that *Nicholas,* [29] using the rat, made an experimental approach to the problem of extra-uterine pregnancy. *Fawcett, Wislocki* and *Waldo* [9] and *Fawcett* [8] investigated the developmental capacity of the mouse egg when transplanted to an extra-uterine site. The intention of this paper is to review those mechanisms involved in conception which have been investigated using the extra-uterine pregnancy as an experimental model.

The developmental capacity of the unimplanted mouse egg

During normal pregnancy the mouse oocyte is fertilized in the ampulla of the oviduct (day 0 of pregnancy). During the following 48 hours the "egg" undergoes cleavage divisions as it passes down the oviduct, to arrive in the uterus as a morula late on day 2 or early day 3 of pregnancy. Much attention is now being directed to these preimplantation stages of pregnancy.

The egg and the mother are not in direct physical contact but to what extent is the egg dependent on the oviducal or uterine environments for specific factors necessary for promoting normal development? This question can be investigated by removing eggs at different times from the genital tract and transplanting them to an extra-uterine site. The ability to develop normally in this site might indicate the essentiality or otherwise of the genital tract during the preimplantation stages.

An experiment was set up in which 1, 2 and 4 cell stage eggs and morula were transplanted from the oviduct to the kidney. The transplantation apparatus essentially consits of an Agla micrometer syringe (Burroughs & Wellcome Ltd., London) and an Oxford manipulator (Oxford Micro-Instruments, Ltd., Oxford) (Fig. 1).
Following transplantation only the extra-embryonic membranes and trophoblast developed (Fig. 2). Tissues of the embryo proper were not present. This finding has been confirmed (with very rare exception to the rule) by many workers (see [4, 15]). Blastocysts, on the other hand removed from the uterus are able to give rise, when transferred to the kidney, to embryos of normal appearance (Fig. 3). After further work [11] I postulated that the blastocyst must be exposed to the uterine environment to realise this totipotency. This assumption was based on the results of experiments in which blastocysts were prevented from entering the uterus by tying off the oviduct shortly after mating. On transfer to the kidney these tube-locked blastocysts failed to give rise to embryos. Recent work by *Billington, Graham* and *McLaren* [4], however, have shown that the uterine environment is not essential in this respect. They removed eggs from the oviduct in the 2, 4 or 8 cell stage of development and cultured them *in vitro* until blastocysts had formed. A proportion of these, when transferred to the kidney, developed into normal embryos.

Fig. 1
The Oxford Micro-manipulator holding a glass micropipette.
The polythene tube is connected to the "Agla"-microsyringe.

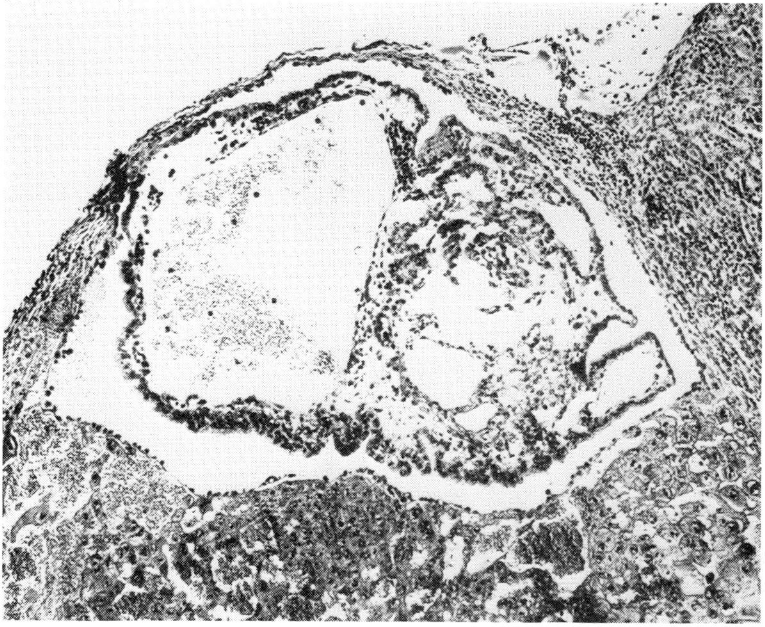

Fig. 2
Extra-embryonic membranes (yolk-sac and Reicherts) developing from a cleavage stage mouse egg transplanted to the kidney. Note the absence of the embryo.

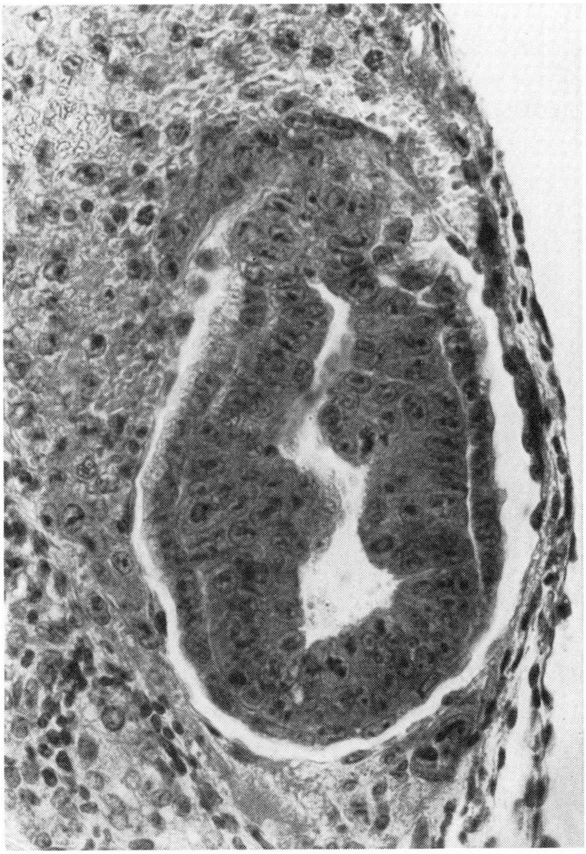

Fig. 3
Mouse embryo of normal appearance developing beneath the kidney capsule of an adult male host.

These results, then, show that the mouse eggs if transferred directly to the kidney in the early cleavage stages give rise (presumably) to blastocysts but these blastocysts produce only trophoblast and extra-embryonic membrane. Eggs allowed to develop into blastocysts either in the genital tract or *in vitro* give rise to embryos when transplanted to the kidney. These strange results clearly merit further investigation.

The endocrinological effects of an extra-uterine pregnancy

By mid-pregnancy the placenta of both rats [1, 2, 6] and mice [27] produces a luteotrophic hormone (L. T. H.). The placenta is a complex organ composed essentially of maternal (decidual) and fetal (trophoblastic) tissues. Either tissue (or both) could secrete the luteotrophic principal. Several workers [2, 7, 26] believe that the trophoblast is the major source. This conclusion was reached from experiments in which the fetal and maternal components of the mature placenta were separated surgically and assayed independently. However, the fusion and interdigitation of the two tissues renders complete separation by this method impossible.

This difficulty, however, can be overcome by transplanting the unimplanted blastocyst to an extra-uterine site. If the recipient is a virgin female any endocrinological changes not appearing in a sham operated control mouse may be safely attributed to the presence of trophoblast, uncontaminated with decidual tissue or products of decidual origin. Blastocysts transplanted in this way to the kidney can give rise to structures similar to the normal placenta (Fig. 4).

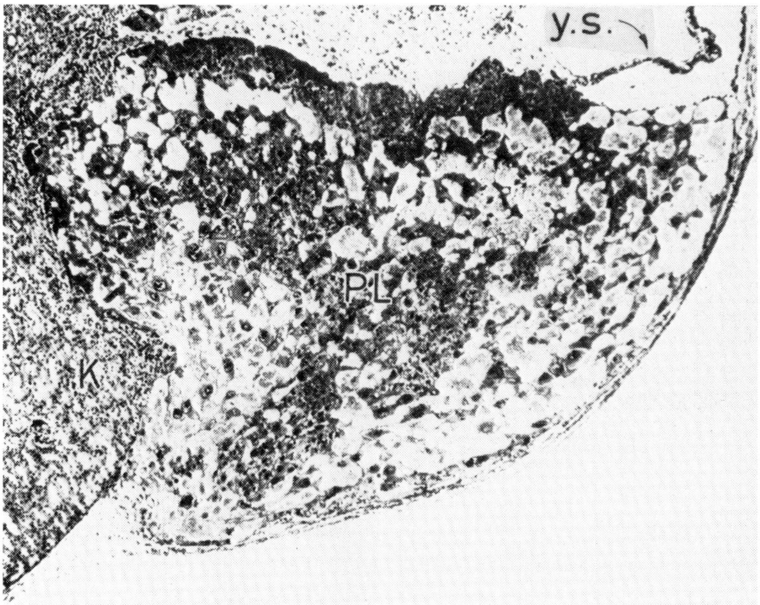

Fig. 4. 'Placenta' developing on kidney.
Y. S. Yolk sac
K. Kidney
P. L. Placental labyrinth.

Mice bearing such as extra-uterine pregnancy did not show the arrest of the estrus cycle, inhibition of mating with ensuing pregnancy, or stimulation of the corpora lutea sufficient to produce progestational changes in the genital tract. The extra-uterine trophoblast did however cause some changes characteristic of normal pregnancy. These were the involution of the adrenal X-zone (Fig. 5 a + b) and the stimulation

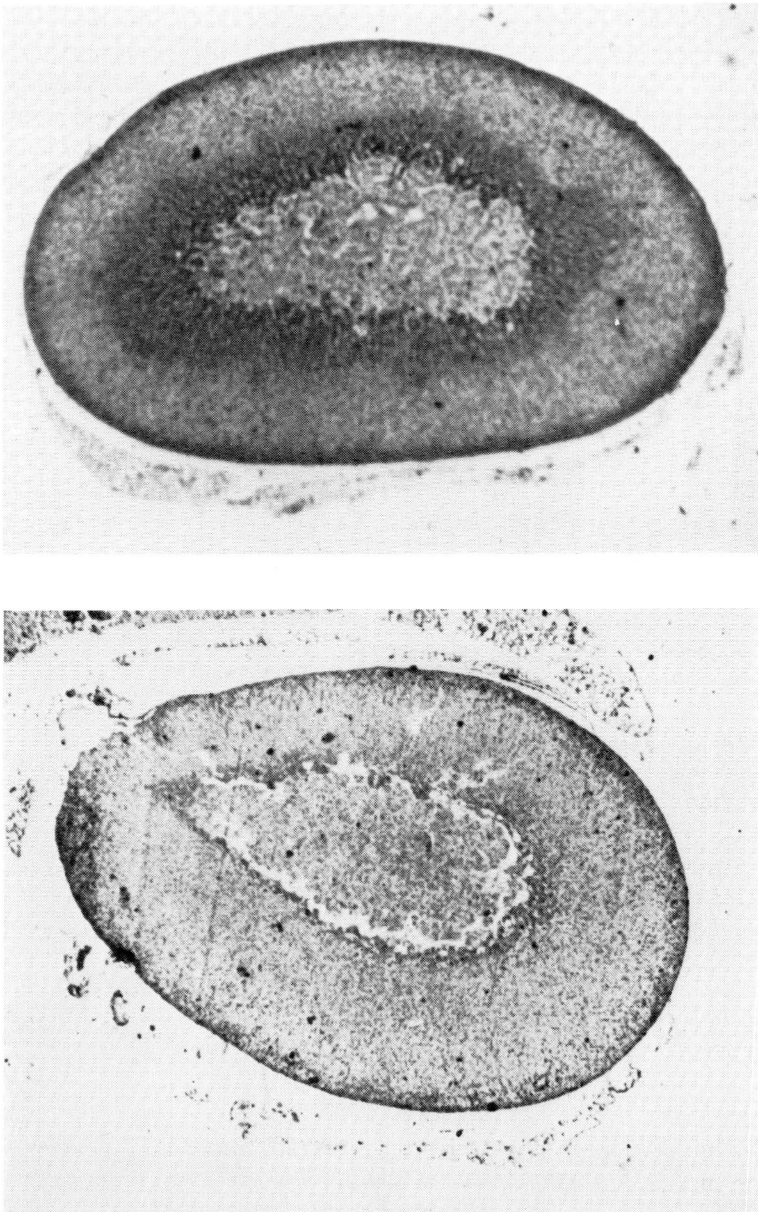

Fig. 5. Section on adrenal gland.
a) in virgin ♀, note the conspicuous, juxta-medullary X zone;
b) in mouse with an extrauterine pregnancy. X zone has involuted.

of the mammary glands [16]. These results can best be explained by assuming that the extra-uterine trophoblast secreted small quantities of L. T. H. and the mammary glands were more senditive to the hormone than were the corpora lutea. *Lyons, Li, Ahmad & Rice-Wray* [25] have confirmed that in the rat the mammary gland more readily responds to a hormone with luteotrophic activity than does the corpus luteum.

Recently *Zeilmaker* [33] has obtained results suggesting that extra-uterine trophoblast is more strongly luteotrophic than my experiments indicate.

The local effect of the uterus on the blastocyst during delayed implantation

During delayed implantation in the mouse, either naturally induced by lactation or experimentally by removal of the ovaries early in pregnancy, the embryos remain quiescent in the blastocyst stage of development. The blastocysts show no mitotic activity and DNA synthesis is inhibited [28]. The mechanism by which development is arrested is not understood. However, one may ask whether the mechanism is confirned to the uterus or generally effective throughout the body. This can be investigated by removing a blastocyst during delayed implantation from the uterus and transplanting it to an ectopic site in the same animal.

Mice were ovariectomised on day 1 or day 2 of pregnancy. On day 5 of pregnancy (i. e. 1 day beyond the normal time of implantation) blastocysts were removed from the left uterine horn by flushing out the contents through a uterine fistula. With the animal still under anesthetic one blastocyst was transferred beneath the kidney capsule. Four days later the mouse was killed. The right uterine horn was found to contain unimplanted blastocysts, whilst the blastocyst transferred to the kidney had "implanted" and developed into an embryo, the appearance of which was equivalent to a normal embryo of 8 days *post coitum* (Fig. 6).

This result shows that the factor (s) inhibiting development of the embryo during delayed implantation is probably localised to the uterine lumen [19].

Effect of the hormonal environment on the development of the extra-uterine mouse egg

Blastocysts give rise to embryos of normal appearance when transplanted to the kidney [10]. Spleen [13] (Fig. 7) and brain (*Kirby* unpublished) (Fig. 8), in both males, virgin and pregnant females. The sex and state of pregnancy of the host did not appear

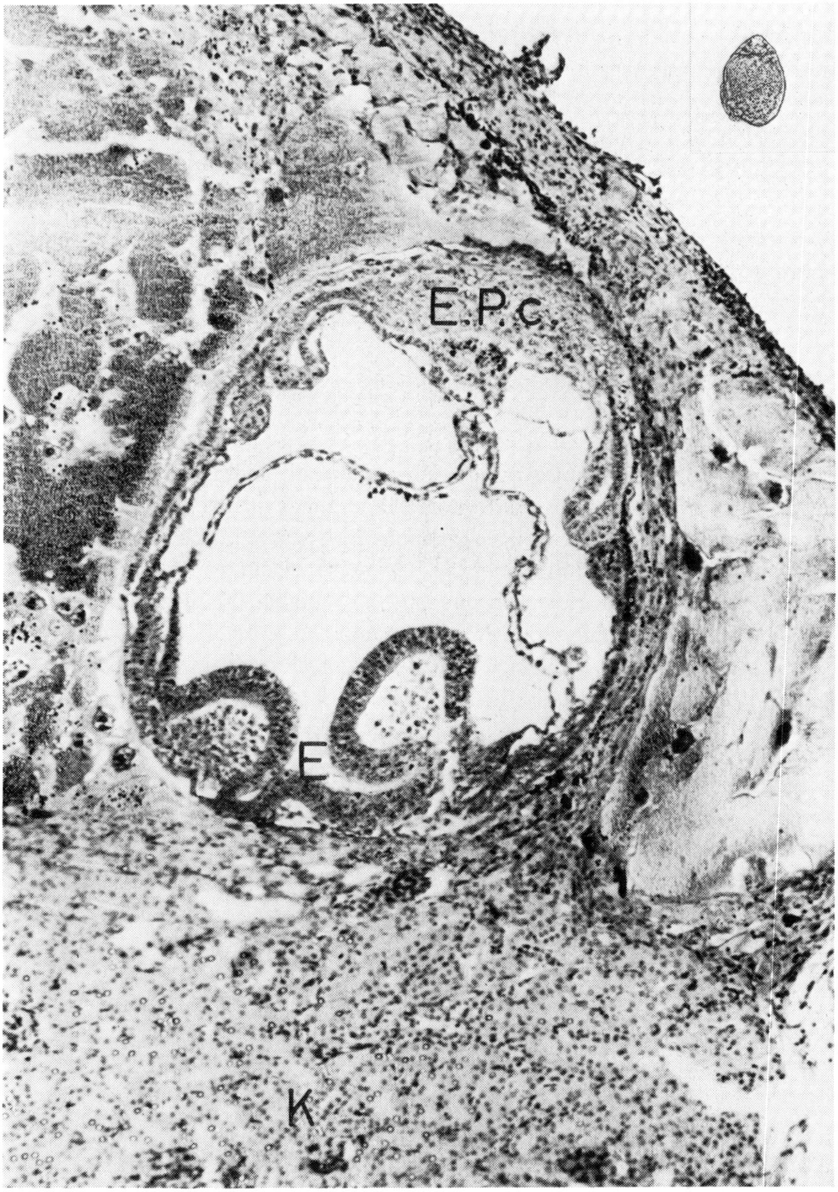

Fig. 6. Embryo developing on the kidney, and blastocyst from uterus (inset). Both removed from the same mouse 9 days *post coitum.* Blastocyst was transferred from uterus to the kidney 4 days before autopsy.

K. Kidney
E. P. C. Ectoplacental cone
E Embryo

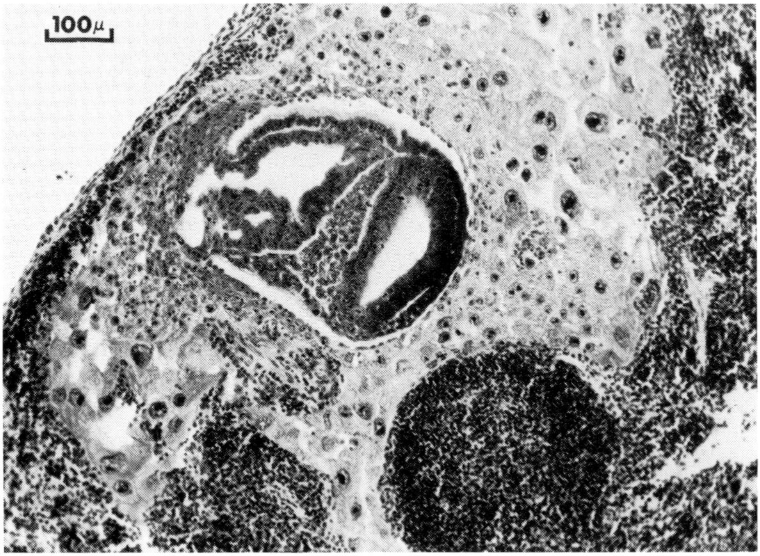

Fig. 7. Embryo developing in the spleen. Note the pattern of trophoblast invasion.

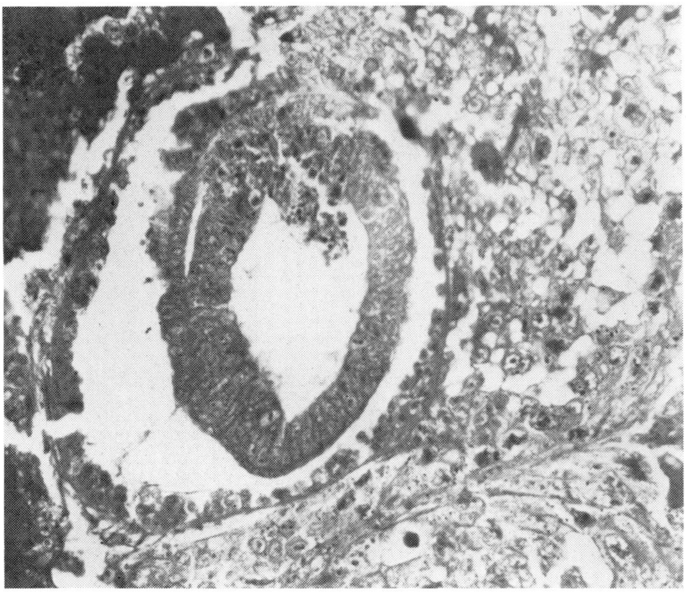

Fig. 8. Embryo developing in the brain.

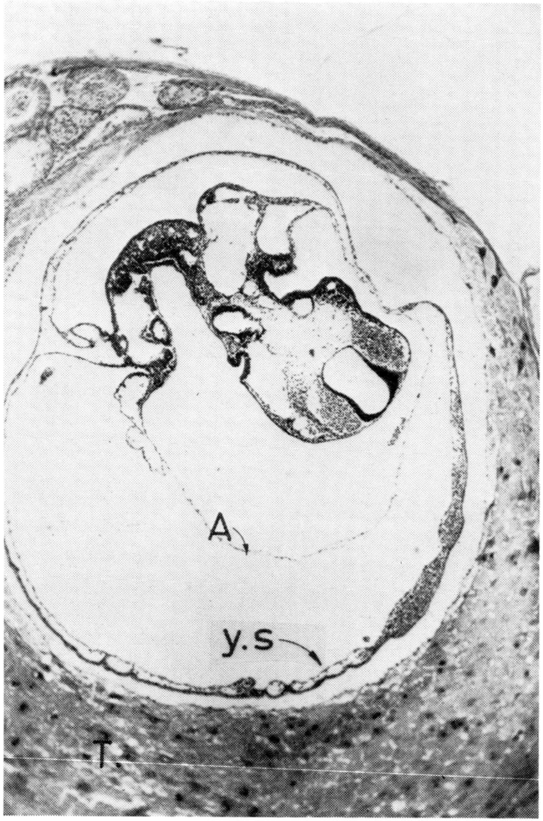

Fig. 9
Embryo developing in the scrotal testis.
S. T. seminiferous tubule, T trophoblast, A amnion, Y. S. Yolk-sac.

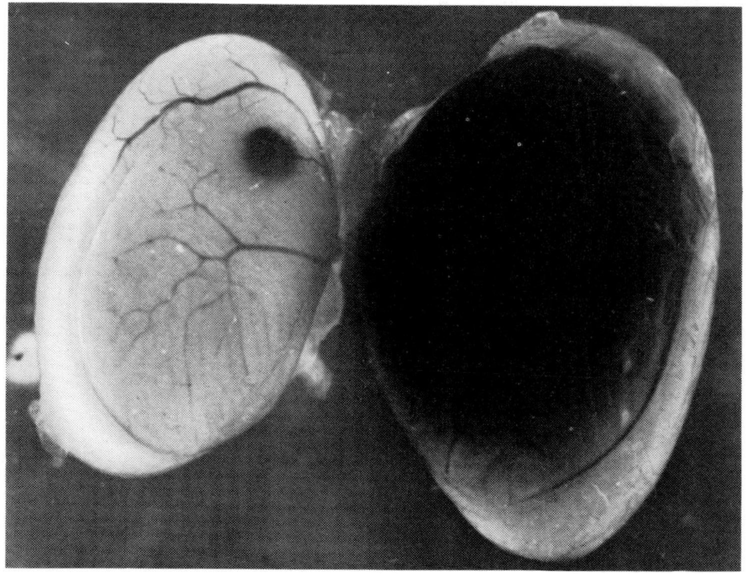

Fig. 10
Hemorrhagic destruction of testis caused by invasive trophoblast from transplanted blastocyst. Left scrotal and right cryptorchid testes of same mouse. Note in the latter case the trophoblast is very extensive.

to affect the success of the graft. The scrotal testis is a particularly favourable site for mouse embryo development (Fig. 9). Not so the cryptorchid testis (Fig. 10). In this organ the transplanted blastocyst gives rise only to trophoblast, development of the embryo proper seems to be totally inhibited. This does not seem to be due to any hormonal disturbance associated with cryptorchidy but probably due to the elevated temperature of this organ [14].

These results confirm those from many other sources that the precise hormonal requirements for normal pregnancy are necessary to prepare the uterus, not the embryo. Indeed mouse eggs transplanted to the unprepared uterus always fail to survive. Only when the hormonally inadequate uterus is damaged can implantation occur within its confines (*Cowell,* quoted by *Kirby* [18]).

The orientation of the implanting blastocyst

In the mouse the blastocyst invariably implants with the inner cell-mass directed away from the point of initial attachment between the trophoblast and the uterine epithelium (Fig. 12 a).

There seem to be six *a priori* ways how this orientation could occur. These are

1) The blastocyst develops at its prospective implantation site with the position of the inner cell mass determined in response to the underlying maternal tissue.

2) While the blastocyst is being manoeuvered by the myometrial contractions towards the antimesometrial side of the uterus its correct orientation is established as a result of differential compressibility of the abembryonic pole and the inner cell mass regions.

3) Uterine epithelial cells have a capacitiy for movement independent of the underlying stromal tissue. The movement of the uterine epithelial cells rotates the blastocyst lying upon them into the correct orientation.

4) Before implantation the blastocyst is subject to rotary movements by myometrial contractions. There is a morphogenetic field across the vertical axis of the uterus. When the inner cell mass, as a result of random movements, arrives in the correct orientation attachment is initiated in all, or in a previously determined part of the trophoblast wall.

5) The initial attachment between the blastocyst an the uterine epithelium is random. Thereafter the trophoblast cells which differentiate from the wall of the blastocyst anchor it to the underlying epithelium by cytoplasmic processes. As a result of selective shortening of these prosesses the blastocyst is rotated into its correct orientation.

6) The surface of the blastocyst has a uniform potential for attaching to the uterine epithelium and attachment occurs at random immediately after the loss of the zona pellucida. The inner cell mass is free to move round the inside of the trophoblast shell. The final position of the inner cell mass is determined either by a morphogenetic gradient across the vertical axis of the uterus or by changes in the trophoblast associated with its attachment to the underlying tissues, or by both.

The specific involvement of the uterus as indicated in hypotheses 1, 2, 3 and 4 can be investigated by determining the orientation of the blastocyst implanting in an extra-uterine site. Sixteen blastocysts were transplanted into the anterior chamber of the eye. In this location the blastocysts can rotate freely before implanting. All 10 of the blastocysts recovered. 24 hrs. later had implanted with the inner-cell-mass directed away from the underlying host tissue (Fig. 11 b).
This result proves that the uterus is not essential in the control of orientation of the implanting blastocyst.

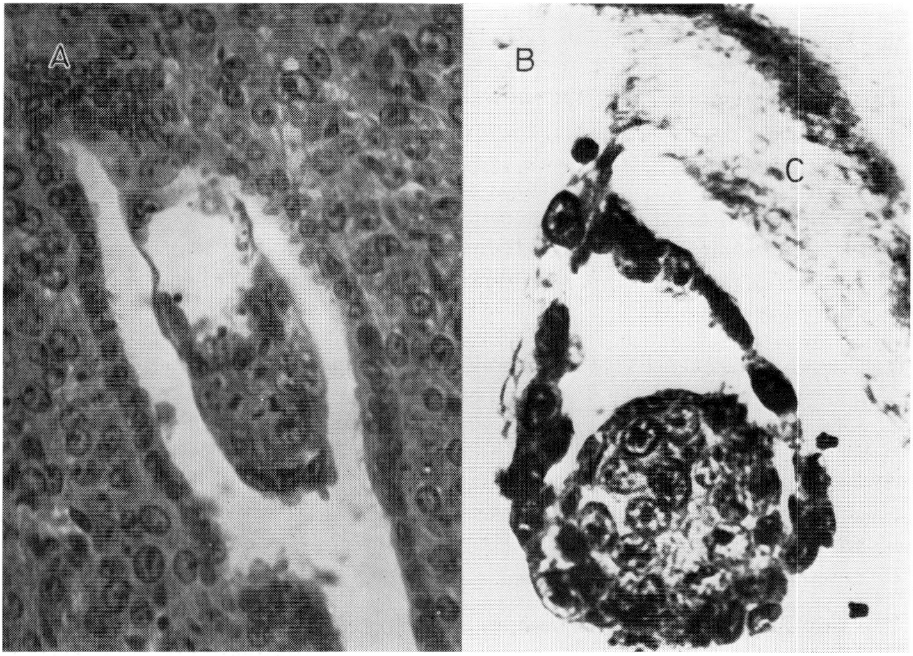

Fig. 11. Mouse blastocysts implanting in
A) the uterus, and
B) the anterior chamber of the eye.
Note the orientation of the inner cell mass is the same in both situations. C cornea.

Light and electron microscope studies showed that between the blastocyst an the uterine epithelium interlocking microvilli would prevent blastocyst rotation and argue against hypothesis 5.

It is concluded that the only mechanism fulfilling the available observations is one (hypothesis 6) in which the inner-cell-mass is free to travel round the inside of the trophoblast shell once the latter has become fixed to the uterine epithelium [23].

The immunological role of the uterus during pregnancy

It is generally accepted that the uterus is not an immunologically privileged site. This assumption is based mainly on the investigations of *Schlesinger* [31] and *Poppa* et al. [30]. However, in none of these studies was the experimental model analogous to the situation which exists during pregnancy. *Schlesinger* transplanted tumour homografts to the uterine lumen, but these outgrew the confines of the "physiological" uterus and penetrated the myometrium. Rejection in this site is not pertinent to pregnancy. *Poppa* et al. used transplants of parathyroid gland, which remained in the uterine lumen but did not elicit a decidual reaction. At implantation, if the correct hormonal status exists, the egg, or various foreigen objects, will bring about a dramatic metamorphosis of the endometrium, such that the egg (or foreign object) is surrounded by a mass of tightly packed decidual cells. (Fig. 12 b) As *Poppa* et al. say, their experiments "do not rule out the rich decidual layer of normal pregnancy as possible contributor to the postulated barrier between mother and fetus".

We [22] reinvestigated the problem in the following way. Two strains of mice were selected which were known to show major histocompatibility differences to one another. Mice of one strain (C3H) were immunized by skin grafts and spleen cell injections from the other strain (C57BL). Blastocysts from the immunizing strain (C57BL) were then transplanted either to the lumen of the pseudopregnant uterus or to the kidney of the immunized (C3H) mice (Fig. 12 a + b).

It was found that all the blastocysts transplanted to the kidney failed to develop, whilst those grafted to the uterus implanted and developed normally. Failure of the ectopic allogeneic blastocysts to develop was shown to be due to a specific immunity against them because syngeneic (C3H) blastocysts developed normally on the kidney of these immunized C3H recipients.

The most likely explanation to account for these results and those of *Schlesinger* [31] and *Poppa* et al. [30] is that the decidual tissue of the uterus (of course absent on the kidneys) protects the implanting blastocysts from immunological destruction. Later in pregnancy it seems the conceptus develops its own immunological protective mechanism in the form of a sialo-mucin or fibrinoid layer that surrounds the trophoblast cells [21].

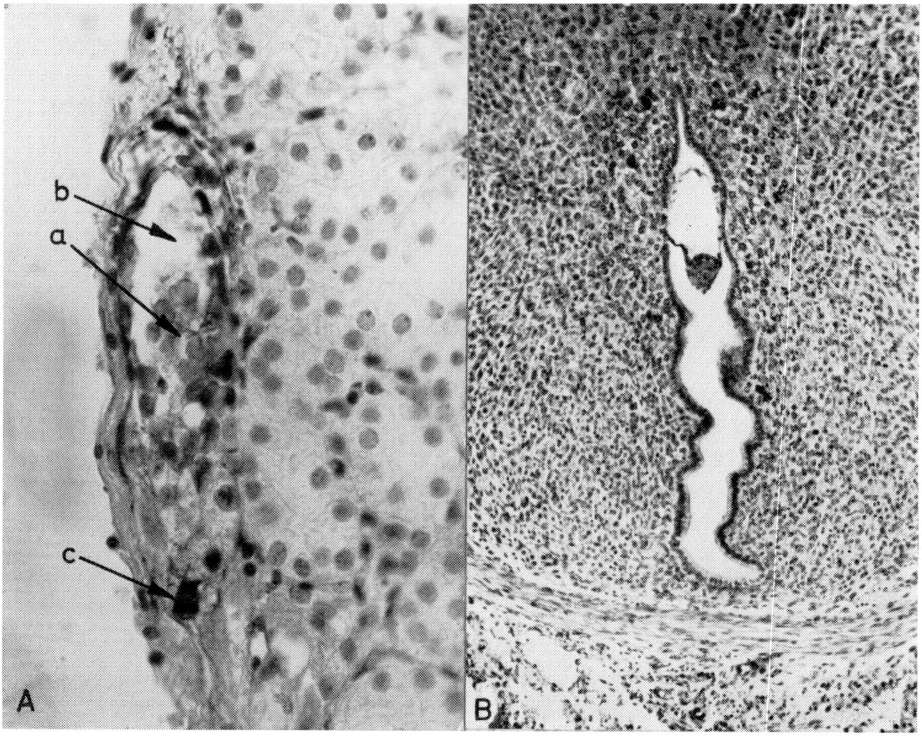

Fig. 12. Blastocyst developing in
A) the kidney, and
B) the uterus.
Note the complete lack of decidual tissue in the ectopic site.

Trophoblast invasion and its control

The mammalian egg, endowed as it is with small food reserves, must rapidly make contact with maternal blood supply, its main source of nourishment. Many species, including man and mouse, have highly invasive trophoblast which burrows down through the decidua until the maternal blood vessels are reached and breached. Once the maternal blood has been tapped, the trophoblast in all but pathological conditions ceases to invade. Is this abrupt change in activity, which in the mouse occurs on about the 9th day of pregnancy, caused by an active suppression by maternal tissue or does it simply indicate an intrinsic "running down" of the invasive activity of trophoblast? To answer this question it is necessary to investigate the invasiveness of trophoblast when transplanted beyond the confines of the uterus.

The blastocyst or the ectoplacental cone from an early conceptus, transplanted to an ectopic site give rise to trophoblast. The invasive abilities of this tissue have been studied in the kidney [8, 11, 12] (Fig. 13), spleen [13] (Fig. 7), testis [3, 14] and brain [17]. Space does not permit a full description of the pattern of trophoblast invasion of these organs. However, it is true to say that in all situations the amount of destruction wrought in the ectopic host organs is almost always in excess of that inflicted on the uterus during normal pregnancy. In some situations, for example the cryptorchid testis, trophoblast invasion could be so servere as to leave very little undamaged testcular tissue.

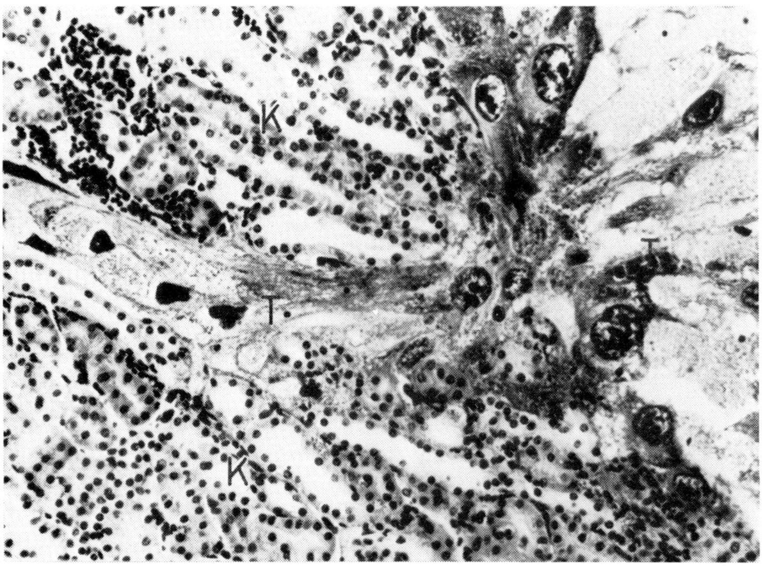

Fig. 13. Trophoblast invading deeply into the kidney tissue.
T trophoblast
K kidney tissue

It is clear that the full invasive proclivities of trophoblast are not realised during normal pregnancy. Further work suggested that the decidual tissue was probably involved in the suppression of trophoblast activity. Decidual tissue is invariably absent from ectopic situations but always surrounds the conceptus *in utero*.

Experiments in which the ectoplacental-cone was introduced into the uterus of a cycles mouse (in which the decidua does not differentiate) showed that the trophoblast that develops invaded deeply into the muscular wall of the uterus. In servere

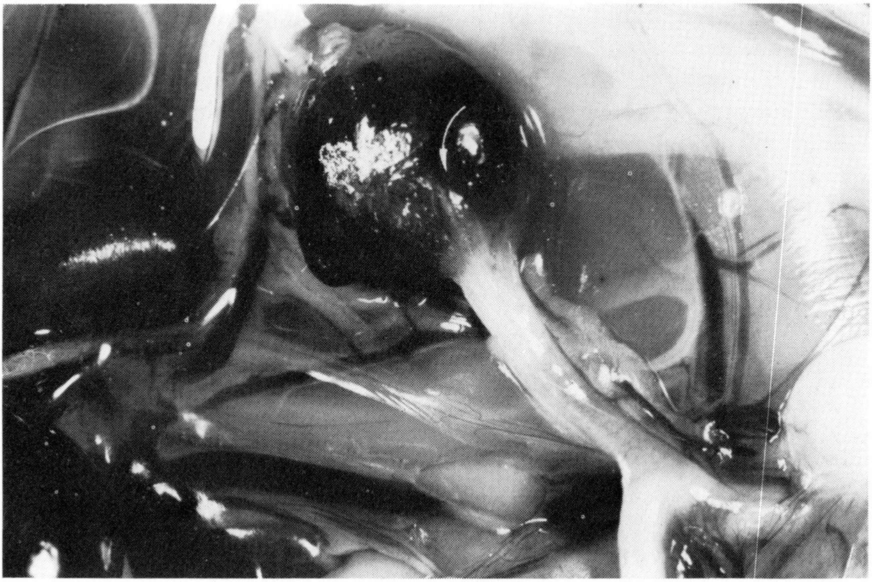

Fig. 14. Destruction of mouse uterus wall by uninhibited trophoblast (arrow). The uterus contained no decidual tissue.

cases the uterus in the area of the graft was virtually destroyed (Fig. 14). It seems, then, that the presence of decidual tissue curtails the invasiveness of the trophoblast during normal pregnancy. Even at term decidual tissue, although reduced to a thin layer, always envelopes the trophoblast. These results may shed some light on the etiology of the *placenta accreta,* a grave complication of human pregnancy. In this condition the trophoblast invades beyond its physiological boundary and erodes the deeper tissues of the uterus. It is of interest that the condition is always associated with the scant development, or absence of, decidual tissue.

The way the decidua contains trophoblast invasion is unknown, but we are in a position at least to speculate a little further The decidual swelling induced in a pseudopregnant mouse uterus by some physical agent such as the injection of oil, or by gently pinching the uterus, is as far as can be judged similar to that produced by the normal implanting blastocyst. The decidual swelling enlarges and then, commencing more or less at the point where the egg would implant, the central region becomes necrotic and breaks down (Fig. 15). This region of necrosis enlarges progressively toward the mesometrial pole of the uterus, leaving a blood-and cellular debris-filled cavity. Now if we assume that trophoblast can only invade moribund and degenrating decidua, invasion would be stopped short at the healty tissue, a small layer of which (the decidua basalis) persits throughout pregnancy. It seems reasonable to suggest that this spontaneous and intrinsic necrosis occurs during normal pregnancy but that it is not seen simply because of its continual ingestion by the expanding trophoblast [20].

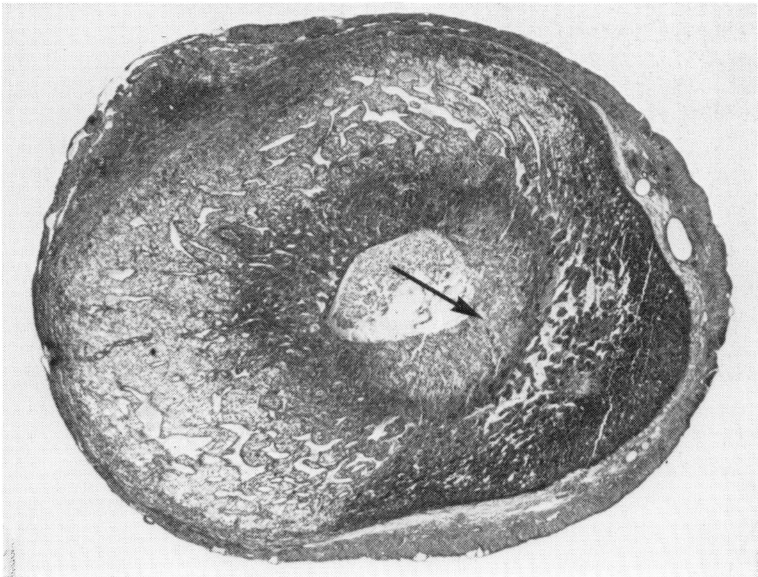

Fig. 15. Section through an artificially induced decidual swelling. Note the 'implantation cavity' and the area of necrotic decidua (arrow). The intact decidual tissue will degenerate progressively and lead to an enlargement of the cavity.

Conclusion

The various aspects of conception reviewed here are joined by a common feature; that they have all been investigated using the extra-uterine development of the eggs as an experimental tool.

The list is not complete, for example I have not included those investigations in which we have studied the immunological properties of trophoblast [5, 20] nor of the fate of mouse eggs tranplanted to the rat [12]. However, I hope it is sufficient to demonstrate the catholic value of the technique.

References

[1] *Astwood, E. B. & Greep, R. O.* (1938). A corpus luteum stimulating substance in the rat placenta. Proc. Soc. exp. Biol. Med **38,** 713.

[2] *Averill, S. C., Ray, E. W. & Lyons, W. R.* (1950). Maintenance of pregnancy in hyphophysectomised rats with placental implants. Proc. Soc. Exp. Biol. Med. **75,** 3.

[3] *Billington, W. D.* (1965). The invasiveness of transplanted mouse trophoblast and the influence of immunological factors. J. Reprod. Fertil. **10,** 343-52.

[4] *Billington, W. D., Graham, C. F. & McLaren, A.* (1968). Extra-uterine development of mouse blastocysts cultured *in vitro* from early cleavage stages. J. Embryol. exp. Morph. **20,** 391-400.

[5] *Bradbury, S., Billington, W. D. & Kirby, D. R. S.* (1965). A histochemical and electron microscopical study of the fibrinoid of the mouse placenta. J. Roy. Micr. Soc. **84,** 199-211.

[6] *Canivenc, R.* (1956). Le placenta de Rat et son activite endocrine. Archs. Anat. Histiol. Embryol. **39,** 1.

[7] *Cerruti, R. A. & Lyons, W. R.* (1960). Mammogenic activities of the mid-gestational mouse placenta. Endocrinology, **67,** 884.

[8] *Fawcett, D. W.* (1950). The development of mouse ova under the capsule of the kidney. Anat. Rec. **108,** 71-91.

[9] *Fawcett, D. W., Wislocki, G. B. & Waldo, C. M.* (1947). The development of mouse ova in the anterior chamber of the eye and in the abdominal cavity. Amer. J. Anat. **81,** 413-443.

[10] *Kirby, D. R. S.* (1960). The development of mouse eggs beneath the kidney capulse. Nature, Lond. **187,** 707-708.

[11] *Kirby, D. R. S.* (1962). The role of the uterine environment on the development of mouse eggs. J. Embryol. exp. Morph. **10,** 496-506.

[12] *Kirby, D. R. S.* (1962). Reciprocal transplantation of blastocysts between rats and mice. Nature, Lond. **194,** 786.

[13] *Kirby, D. R. S.* (1963). The development of mouse blastocysts transplanted to the spleen. J. Reprod. Fertil **5,** 1-12.

[14] *Kirby, D. R. S.* (1963). The development of the mouse blastocyst transplanted to the scrotal and cryptorchid testis. J. Anat. **97,** 119-130.

[15] *Kirby, D. R. S.* (1965). The role of the uterus in the early stages of mouse development In Preimplantation Stages of Pregnancy. Ed. *G. E. W. Wolstenholme & M. O'Connor.* Ciba Found. Symp.

[16] *Kirby, D. R. S.* (1965). Endocrinological effects of experimentally induced extra-uterine pregnancies in virgin mice. J. Reprod. Fertil. **10,** 403-412.

[17] *Kirby, D. R. S.* (1965). The invasiveness of the trophoblast. In The Early Conceptus, Normal and Abnormal. Ed. *W. W. Park,* Livingstone, Edinburgh and London.

[18] *Kirby, D. R. S.* (1966). In Egg Implantation (p. 19). Ed. *G. E. W. Wolstenholme & M. O ' Connor.* Ciba Found. Study Group 23.

[19] *Kirby, D. R. S.* (1967). Ectopic autografts of blastocysts in mice maintained in delayed implantation. J. Reprod. Fertil. **14,** 515-517.

[20] *Kirby, D. R. S.* (1968). The immunological consequence of extra-uterine development of allogeneic mouse blastocysts. Transplantation **6,** 1005-1009.

[21] *Kirby, D. R. S., Billington, W. D., Bradbury, S. & Goldstein, D.* (1964). Antigen barrier of the mouse placenta. Nature. Lond. **204,** 548-9.

[22] *Kirby, D. R. S., Billington, W. D. & James, D. A.* (1966). Transplantation of eggs to the kidney and uterus of immunised mice. Transplantation **4,** 713–718.

[23] *Kirby, D. R. S.. Potts, D. M. & Wilson, I. B.* (1967). On the orientation of the implanting blastocyst. J. Embryol. exp. Morph. **17,** 527-532.

[24] *Kirby, D. R. S. & Cowell, T.* (1968). Trophoblast-host interactions. In Epithelial-Mesenchymal intreactions. Ed. *R. Fleichsmajer* and *R. E. Billingham.* Williams & Wilkins, Baltimore.

[25] *Lyons, W. R., Li, C. H., Ahmad, W. & Rice-Wray, E.* (1967). Mammotrophic effects of human hypophyseal growth hormone preparations in animals and man Proceedings of the 2nd International Proceedings on Growth Hormone, Milan.

[26] *Mayer, G. & Canivenc, R.* (1950). Autografts de placenta chez la ratte. C. R. Soc. Biol. **144,** 410.

[27] *Newton, W. H. & Beck, N.* (1939). Placental activity in the mouse in the absence of the pituitary gland. J. Endocr. **1,** 65.

[28] *McLaren, A.* (1968). A study of blastocysts during delay and subsequent implantation in lactating mice. J. Endocr. **42,** 453-463.

[29] *Nicholas, J. S.* (1931). Experiments on early stages of rat embryos. Science **74,** 573.

[30] *Poppa, G., Simmons, R. L., David, S. D. & Russell, P. S.* (1964). The uterus as a recipient site for parathyroid homotransplantation. Transplantation **4,** 496-502.

[31] *Schlesinger, M.* (1962). Uterus of rodents as site for manifestation of transplantation immunity against transplantable tumours. J. Nat. Cancer. Inst. **28,** 927-45.

[32] *Wallace, C. J.* (1917). Transplantation of ectopic pregnancy from fallopian tube to cavity of uterus. Surg. Gynae. Obstet. **24,** 578.

[33] *Zeilmaker, G.* (1968). Luteotrophic activity of the ectopic mouse trophoblast. J. Endocr. **41,** 455-456.

Hormonal Requirements for Egg-Implantation

A. Psychoyos

I. N. S. E. R. M., Hôpital Saint-Antoine, Paris, France

Summary: Induction of the egg-implantation process in the Rat was found to depend on a sequential interplay between the two ovarian hormones. A secretion of progesterone for 48 hours and the offer of a minimal amount of estrogen at the end of this period constitues the basic hormonal sequence inducing the irreversible occurence of receptivity and subsequent non-receptivity in the endometrium.

Data obtained in ovariectomized Rats treated with the ovarian hormones at different sequences show that changes of the progesterone-estrogen time relationships influence the uptake of estradiol by the endometrium and the ability of uterine nuclei for RNA synthesis.

Studies on the role of the non-estrogenised uterus on the growth-arrest of blastocysts during delayed implantation offer evidence favoring the hypothesis that during the period of delay an active inhibition is exerced by the endometrium upon the blastocyst.

Recent data suggests that the supernatant of extracts prepared from non-estrogenised uteri contain a factor able to inhibit the *in vitro* uptake of RNA precursors by the rat and rabbit blastocysts.

During the last years, considerable progress has been made in understanding the hormonal interplay which controls egg-implantation. The fact that, in the Rat and the Mouse, the most common laboratory animals, the ovary happens to be the only source of hormones with estrogenic activity, has helped to determine with a maximum precision the hormonal requirements of the nidation process, in these species. At that, not only in terms of quantity, but especially in terms of timing. A model of hormonal regulation of this process was then revealed, which may be valuable, in its major lines, in a large scale of other species.

Let us first have a view of the various events occuring in the Rat on the 4th and 5th days of pregnancy.

In those animals kept under standard light conditions, that is, light on from 5 in the morning to 7 in the evening, ovulation takes place around 2 in the morning of the estrus day. We consider this day as the first day of pregnancy, or, according to *Shelesnyak's* terminology, day L0. The eggs enter the cornua late on the evening of the 4th day. During the night they reach the stage of blastocyst. They loose their zona pellucida around noon of the 5th day. Macromollecular dyes injected intraveinously early in the afternoon of this day, show spread of the dye at the sites at which the blastocysts are present. Decidual metamorphosis of the first subepithelial cells follows some hours later and we consider that the initiation of the implantation is achieved around noon of the 6th day.

Histological examination of specimens taken from uteri of animals sacrificed at noon of the 5th day, show the blastocysts laying their embryonic pole on the antimensometrial epithelium, while a generalised edema of the stroma develops (Fig. 1). A little later in this day the increase of the capillary permeability of the

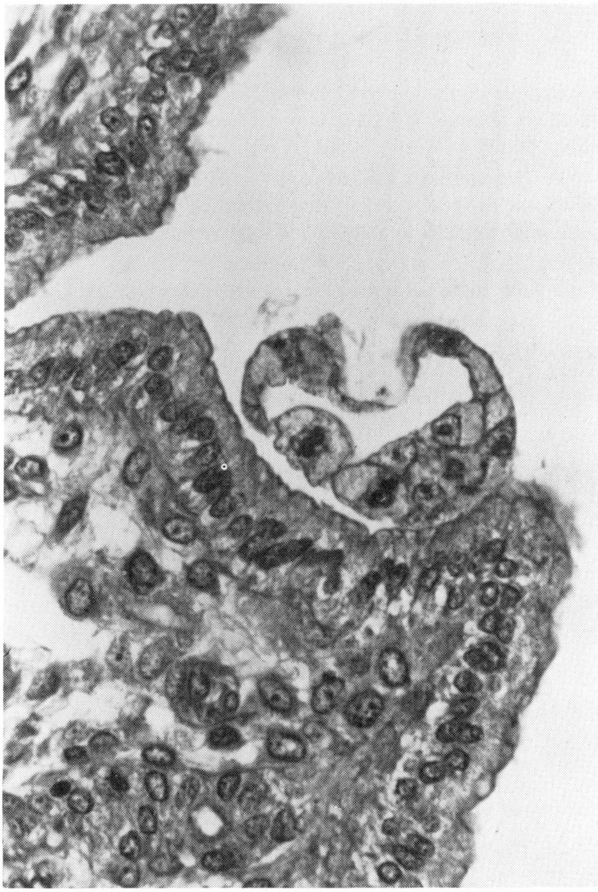

Fig. 1

Longitudinal section of the uterus of rat on the 5th day of pregnancy; the blastocyst is laying its embryonic pole on the antimesometrial epithelium.

surrounding the egg stroma, starts. Sections of uteri at this moment show the blastocysts compressed between the now contiguous walls of the swelling endometrium (Fig. 2). An hypothesis would be that increased permeability is initiated

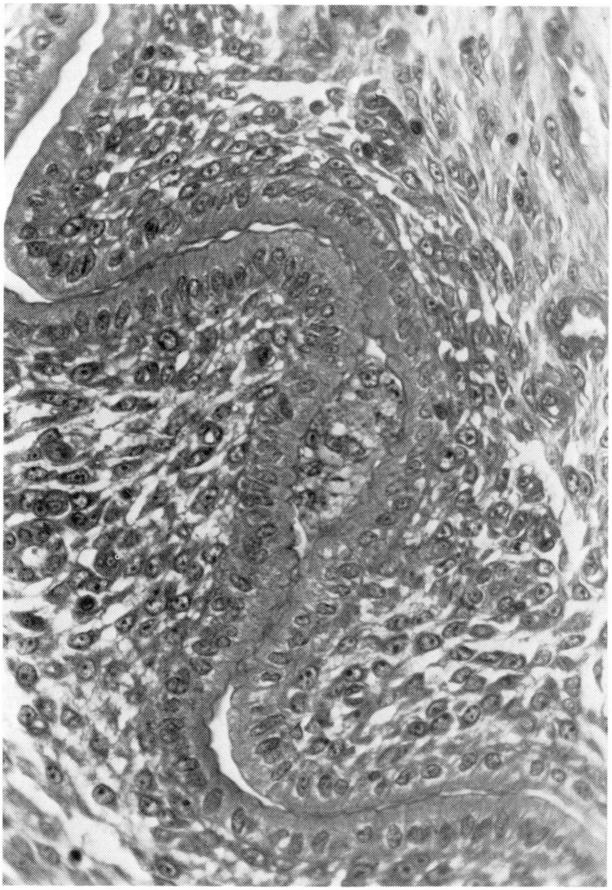

Fig. 2
Longitudinal section through a uterine region showing increased capillary permeability (5 p.m. of the 5th day of pregnancy). The blastocyst is compressed between the wells of the swelling epithelium.

by chemical stimuli offered by blastocelic fluid released from the ovum under the compression. The permeability change progresses rapidly and in about 2 hours covers the region which will be decidualised.

The decidual response to other than trauma stimuli can be obtained only during a short period of the 5th day. The brievity of this period of high endometrial sensitivity was demonstrated by *Kraicer* and *Shelesnyak* using Pyrathiazine intraperitoneally [1]. *De Feo,* injecting intraluminally various substances, placed the maximal endometrial response at noon [2, 3]. The time limits during which the permeability of the endometrial capillaries can be increased by trauma, were found to parallel closely those limits of the endometrial sensitivity for deciduoma formation [4].

Increased sensitivity of the endometrium is accompanied by changes of the biochemical profile of the uterus. Transient elevation of the uterine content in nucleic acids and proteins were found to occur this day [5]. Quantitative evaluation of the activity of some lysosomal enzymes in the endometrium, revealed a high activity of rubonuclease and acid cathepsine [6].

Parallel to these uterine modifications is the uterine receptivity to blastocysts. Successful transfer can be performed only until the evening of the 5th day [7]. Blastocysts transfered later are expelled from the vagina within a few hours. A state of non receptivity succeeds the short period of receptivity [8].

The hormonal basis of these uterine modifications has been patterned from the results of experiments involving ovariectomy at various moments in early pregnancy. When ovariectomy is performed late in the evening of the 4th day, implantation occurs normally [9, 10], whereas, when performed earlier [9–13], progesterone alone is unable to assure implantation. The blastocysts remain in this case unimplanted until one single injection of a minute dose of estrogen [14] is added to the progesteronic treatment. The blue test, indicating the local increase of the permeability appears some 24 hours after the administration of estrogen. Since, during normal pregnancy, the blue spots are obtained in the afternoon of the 5th day [9], it has been presumed that the endogenous estrogen inducing the implantation process intervenes on the afternoon of the 4th day.

In fact, any procedure interfering with the presence of estrogen at this moment, postpones or prevents implantation.
This is the case not only for early ovariectomy, but also for hypothalamo-pituitary disconnection [15–17], hypophysectomy [10, 12, 18] or use of antiestrogens [19].

The precision in the timing of the estrogenic intervention has been interpreted to reflect a particular activity of the ovary on the 4th day. However, from the available data, it has been concluded [8] that this precision is rather related to the uterine responsiveness to the presence of estrogen. In fact, it is only on the afternoon of the 4th day of normal pregnancy, or, in ovariectomized, pregnant animals, at least 48 hours after preparation of the endometrium by progesterone, that the estrogen acts. We consider therefore that it is a precise hormonal sequence which is required for the induction of the specific modifications leading to implantation namely: Progesterone for at least 48 hours and the presence of estrogen at the end of this period. If this sequence is not accomplished, either because of insufficient progesteronic preparation or by postponement of the presence of estrogen, the uterus remains in a state which we defined in the past, as neutral. During this state the eggs do not implant and are retained in the uterus in diapause. The decidual reaction can be induced with no time limits by traumatic stimuli.

By intervening at the end of a previous progesteronic preparation, the same minute dose of estrogen which induces egg-implantation, becomes responsible for subsequent uterine hostility, as illustrated by the following experiment: ovariectomy and ligation of one oviducte are applied in Rats on the 2nd day post-coitum, and 5 mg of progesterone are given daily. On the 5th day post-coitum, the animals receive sub-cutaneously, various doses of estradiol, and on the 8th day, the occurence of implantations in the normal horn is verified. Blastocysts, taken from animals in delayed implantation, are then transfered into the sterile horn and estradiol (0.1 μg) is injected on the next day. A new laparatomy is performed on the 12th day to check occurence of implantations in the sterile horn.

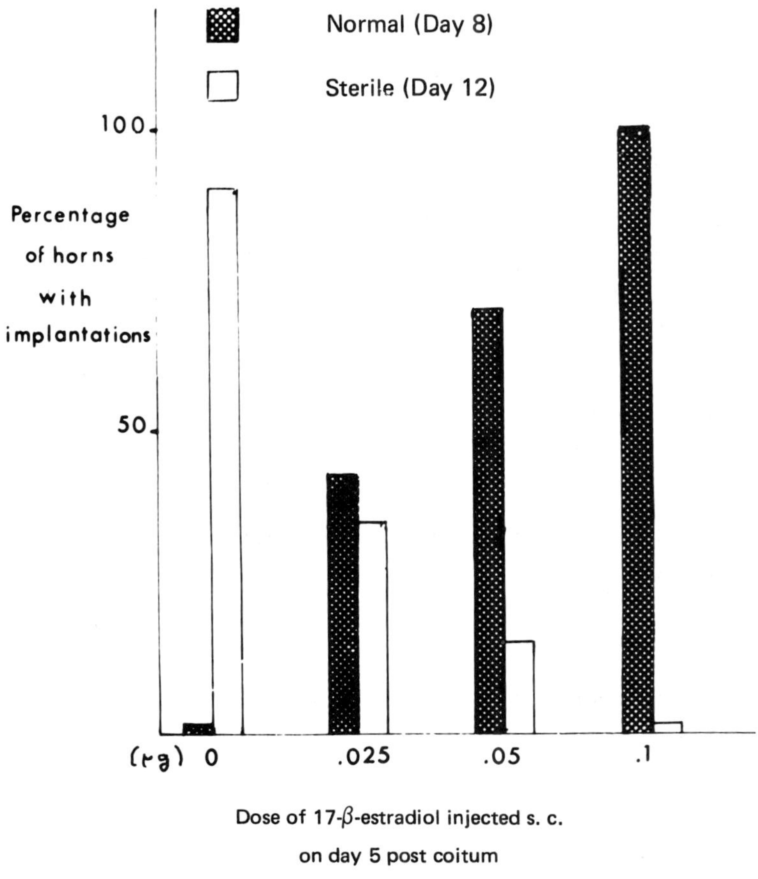

Fig. 3. Uterine receptivity to blastocysts after a single injection of various doses of estradiol in the progesterone treated ovariectomized rat (see text).

As can be seen in Fig. 3, the number of implantations induced in the non-ligated horn by the first estrogen, increases according to the dose injected. The chances of implantation of the transfered blastocysts in the sterile horn are reduced correspondingly. Thus, in animals in which no estrogen is injected on day 5, all sterile horns show implantations. With the intermediate doses of 0.025 to 0.05 μg two sets of asynchronously developping embryons are obtained.

The implantation of the first set, induced in the normal horn by the first injection of estrogen, very often concerns a part of the blastocysts of this horn while the rest remain in delay. They implant later, when the blastocysts of the sterile horn implant, after the second injection of estrogen [20]. Thus, asynchrony in implantation occurs not only between the two horns but also at the same horn (Fig. 4).

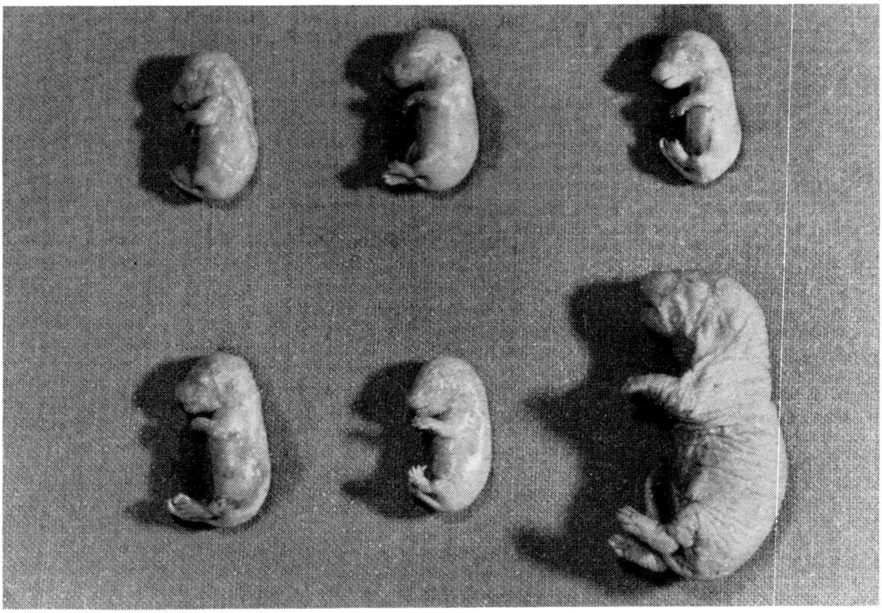

Fig. 4. Asynchronously implanted embryos in an early ovariectomized and progesterone treated pregnant rat, having received 0,03 μg of estradiol s.c. on day 5 post-coitum (see text). At the top the embryos of the *sterile* horn; at the bottom the embryos of the normal horn.

The threshold of endometrial sensitivity to estrogen seems therefore to vary along the length of the uterine horns, so that small doses of this hormone given systemically, may induce a receptive state locally, while the rest of the uterus remains in the neutral state until more estrogen is offered.

Based on the results we shall report below, we can interprete these local differences in the uterine responsiveness to estrogen, by postulating local changes in the number of available receptor molecules for estradiol.

Data obtained in ovariectomized Rats, treated with the ovarian hormones at different sequences, show that changes of the progesterone estrogen time relationships influence the uptake of estradiol by the endometrium [21].

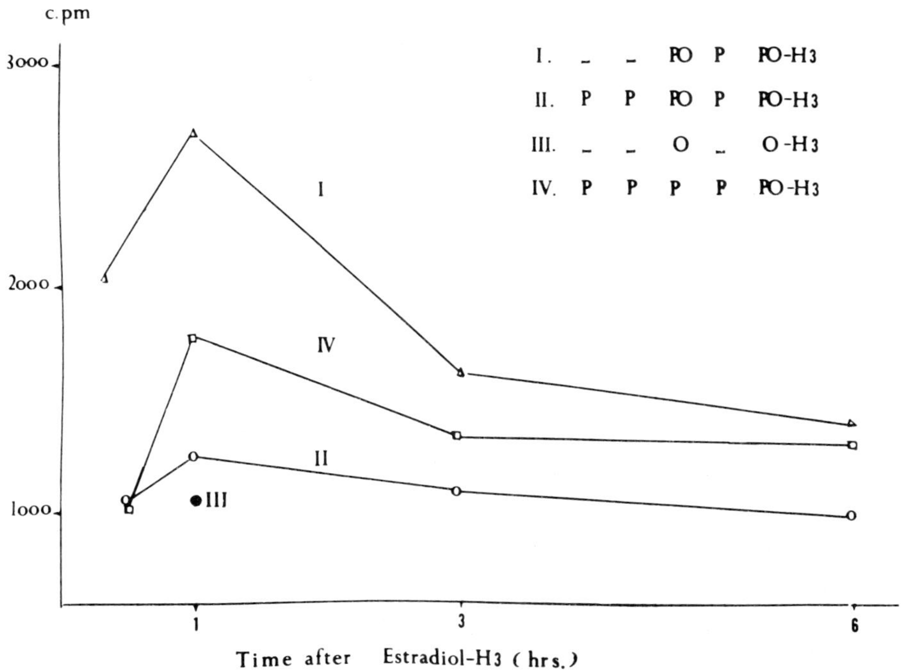

Fig. 5. Incorporation of estradiol-^{3}H into the endometrium of ovariectomized rats treated with various hormonal sequences (*Psychoyos* et al., 1968).

Fig. 5 shows the radioactivity levels 30′, 1,3 and 6 hours after the injection of 0.1 μg of tritiated estradiol. Major changes in the uptake of tritiated hormone by the endometrium were readily seen in uteri of animals sacrificed one hour after the injection. In the group III of this figure, the tritiated hormone was given 48 hours after one injection of non-labelled estradiol. Taking this group as a base line, we see that the daily injection of progesterone during this 48 hours period (group I), increases significantly the uptake of the labelled hormone. There is no such effect when the progesteronic treatment was started 48 hours before the first injection of estradiol (group II). A high uptake was also obtained in group IV, in which the labelled hormone was given at the end of a five day treatment with progesterone alone. In this later group, as in group I, the radioactive hormone complets the basic progesterone-estrogen sequence which induces the uterine state of receptivity. In group II this sequence has been completed by the first estrogen injection so that when the radioactive hormone was given the uterus was already in the non-receptive state. A parallelism is therefore revealed between the afinity of endometrium for estradiol and its fonctional state. Fig. 6 illustrates the hormonal treatments applied

Fig. 6. Implantation rate of blastocysts transfered to ovariectomised Rats, pretreated with the ovarian hormones at different sequences. The basic progesterone-estrogen sequence (see text) is underlined.

	Treatment prior to transfer (Day 0)						Percentage of implanted blastocysts
on days	-5	-4	-3	-2	-1	0	
	-	-	-	P	P	PO	44
	-	P	P	P	P	PO	71
	-	P	P	P	PO	P	66
	-	P	P	PO	P	P	0
	-	P	P	PO	P	PO	0
	-	-	-	PO	P	PO	63

P Progesterone 5mg s.c.
0 17-β-Estradiol 0.1 μg s.c.

to ovariectomized Rats before the transfer of delayed blastocysts and the obtained results. These blastocysts implant if transfered within 24 hours of the completion of the basic hormonal sequence, but fail to implant if transfered later. In other terms, implantation takes place only within 24 hours from what seems to be a saturation of the receptor-sites by estradiol.

An idea of what occurs immediately after such saturation, can be given by following the ability of the uterine nuclei to synthetize RNA. Fig. 7 shows the mean values of the radioactivity found in the nuclear fraction of homogenates prepared from uteri of ovariectomized Rats, treated with progesterone for various periods. These animals had received an intraperitoneal injection of estradiol 30 minutes and tritiated Uridine 10 minutes before the sacrifice. The radioactivity reflects the degree of estrogenic stimulation of nuclear RNA synthesis and is expressed as per cent of the mean value obtained in animals treated with estrogen only. We can see that

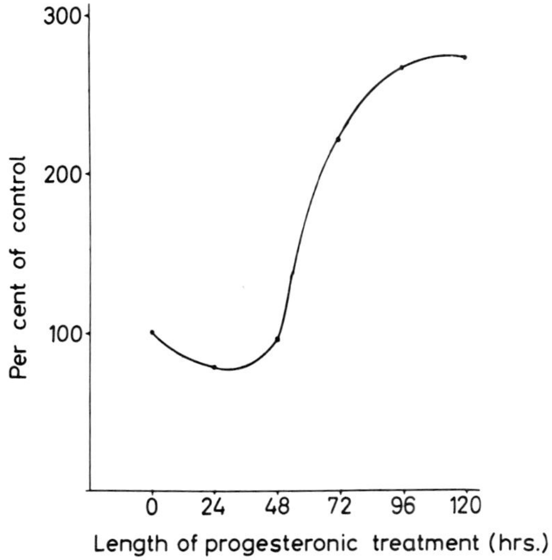

Fig. 7
Estrogen induced incorporation of uterine nuclear RNA, in ovariectomized rats treated with progesterone for various periods.

until 48 hours after the beginning of the progesterone treatment, there are no changes in the degree of the stimulatory effect of estradiol, wherease after 48 hours, the incorporation of Uridine is suddenly increased. Thus, an increased nuclear RNA synthesis caracterizes the receptive uterus resulting from a serie of sequential interactions, in which progesterone appears to act as regulator and estrogen as inducer.

Recent histochemical [22], ultrastructural [23] and autoradiographic [24] studies of animals in delayed implantation, have shown that an important stimulation of various synthetic activities takes place in the blastocyst after the administration of estradiol. An uptake of estradiol was reported for mouse blastocysts *in vitro* [25], suggesting a direct effect of this hormone. However, it must be emphasized, in this respect, that estrogen is not necessary for blastocyst growth, except when they are in contact with the endometrium. Blastocysts in diapause recover their proliferating activity, either when transfered into ectopic sites [22, 26, 27], or, when cultivated *in vitro* [28, 29]. Moreover, *Segal* and his collaborators with RNA extracted from estrogenised uteri replace this hormone in inducing the whole implantation process [30]. That also suggests that the stimulation of the growth of the blastocyst which follows the administration of estrogen is mediated by the dependant of this RNA uterine change.

Fig. 8 shows the results obtained [31] when exposing for 2 hours blastocysts in delay and normal blastocysts of the 5th day to labelled Uridine *in vitro*. Tested 10 minutes after their recovery from the uterus, the 2 hours uptake of Uridine by

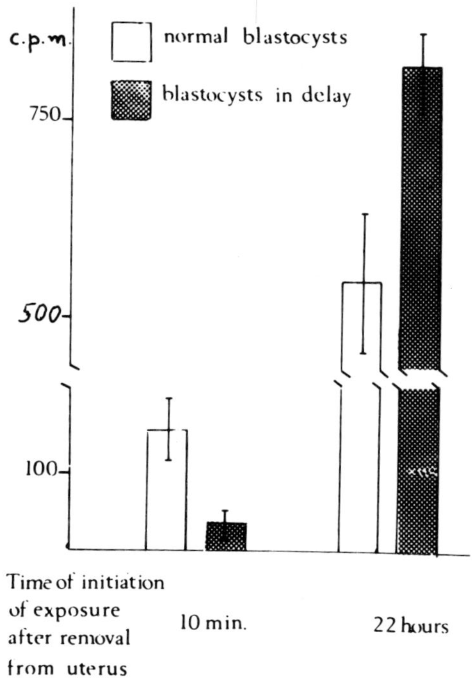

Fig. 8
In vitro, two hours uptake of Uridine-5-^{3}H (IO μC/ml) by normal or in delay blastocysts. The blastocysts were exposed to the tritiated uridine 10 minutes or 22 hours after their recovery from the uterus.

blastocysts in diapause was found about 75 % inferior to the one shown by normal blastocysts. When tested 22 hours after their recovery, this difference in the uptake was found inversed. Blastocysts in diapause have shown a radioactivity 30 % superior to the one obtained for normal blastocysts and 20 times superior to that of blastocysts in diapause tested 10 minutes from their recovery.

These results suggest that during the delay, the blastocysts, in terms of their activity for RNA synthesis are under a repression which ceases once they are withdrawn from the uterine environment.

The following results suggest that the supernatant obtained by centrifugation at 40 000 xg for 2 hours of homogenates prepared from uteri of ovariectomized, progesterone treated Rats, contains a factor, which could be responsible for such a repression. When this supernatant is added to the medium in which are incubated rat or rabbit blastocysts, their uptake of tritiated Uridine is highly decreased. Fig. 9 shows the results obtained with rabbit blastocysts. It can be seen that this inhibition is abolished by heating the supernatant at 80 °C for 10 minutes. It is significantly reduced when the donnor Rats had received in addition to the progesteronic treatment, estradiol 4 hours before the sacrifice.

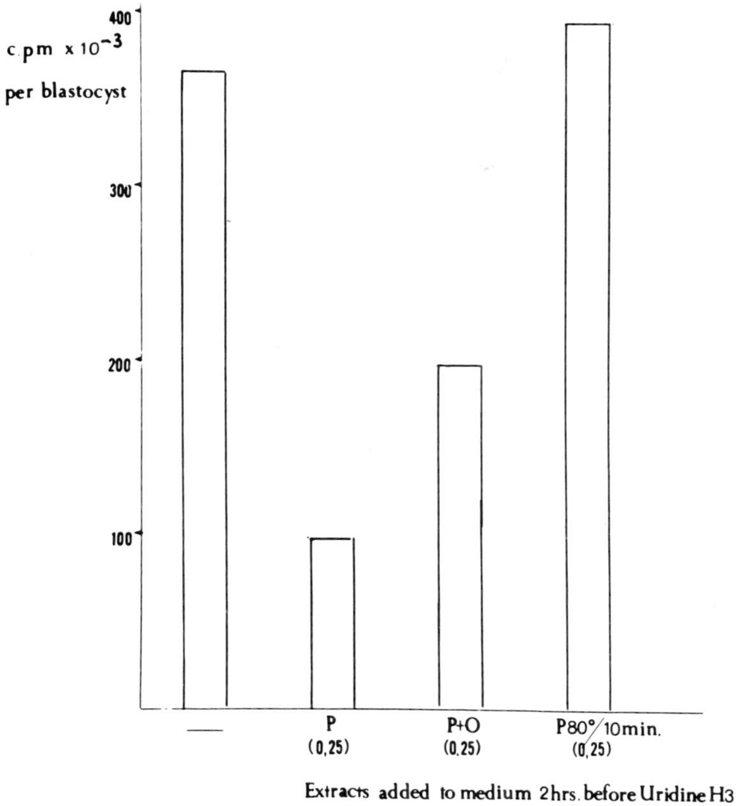

Fig. 9. Effect of extracts prepared from uteri of ovariectomized Rats treated with ovarian hormones, on the *in vitro* two-hours uptake of Uridine-^{3}H by rabbit blastocysts:

P (uterine extract from ovariectomized, progesterone treated rats).

P + 0 (In addition to the progesteronic treatment the donnors received estradiol 4 hours before the sacrifice).

P 80 °C / 10 min. (P extract treated at 80 °C).

Our working hypothesis is that the active factor exerces its effect at the transcriptional level, depriving the blastocyst of information necessary for the synthesis of some proteins indispensable for its further development. Estrogen should promote indirectly the synthesis of these proteins, by annulling the repression exerted by the non-estrogenised endometrium.

It has happened before to look at egg-implantation mechanisms through the prism of fashionable concepts, and I feel necessary to apologize for my recent insistence on giving explanations of these mechanisms inspired by principles of the Molecular

Biology. I have to close, however, this paper by our last results [32], suggesting that the information for decidual metamorphosis of the endometrium must be also coded on the genes and transmitted by DNA-dependant RNA.

In order to investigate a such possibility, we examined the uterine response to local instillation of graduates doses of total extracts of uterine RNA. This RNA was prepared from uteri of ovariectomized Rats, treated only with progesterone. The endometria of these animals were either intact or traumatized 72 hours before the sacrifice. The recipients were ovariectomized, then treated during five days with progesterone alone. The tested material was introduced into the lumen of both uterine horns in 0.1 ml of saline.

No effect was observed with 85 or 170 μg of RNA extracted from non-traumatized uteri or from liver. But, as shown in Fig. 10, a linear dose response effect was observed with these doses when RNA was extracted from decidualized uteri which had been traumatized 72 hours before the sacrifice.

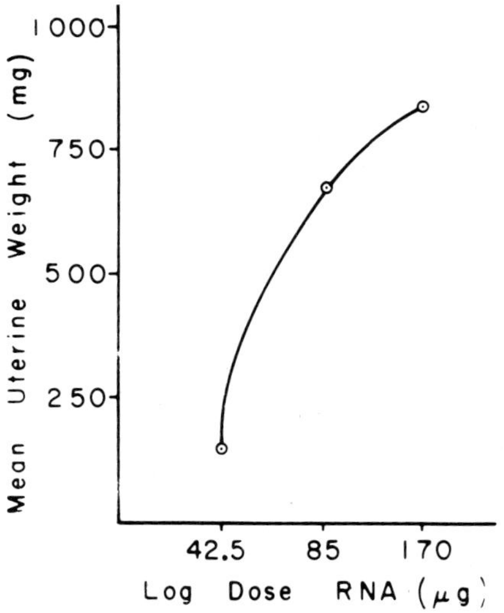

Fig. 10
Uterine response to local instillation of graduated doses of RNA extracted from decidualized uteri; traumatized 72 hours before sacrifice. Donors and recipients were ovariectomized rats, treated with progesterone alone.

Our understanding of the way by which the two ovarian hormones control the egg-implantation process is only at the beginning. We are just at the point of realising that this control is exerted through a molecular language, the whole of which remains to be decoded.

Acknowledgement: Supported by research grants from the Population Council, New York, N. Y.

References

[1] *Kraicer, P. F. & Shelesnyak, M. C.,* C. R. Acad. Sci. 1959, 348, 3213.

[2] *De Feo, V. J.:* Endocrinology 1963, **73**, 4, 488.

[3] *De Feo, V. J.:* In: Cellular Biology of the Uterus 1967, 252. Ed. *R. M. Wynn* Appleton Century Crofts, New York.

[4] *Bitton, V., Vassent, G. & Psychoyos, A.:* C. R. Acad. Sci. 1965, **261**, 3474.

[5] *Shelesnyak, M. C. & Tic, L.:* Acta Endocrin. 1963, **42**, 465.

[6] *Wood, C. & Psychoyos, A.:* C. R. Acad. Sci. 1967, **265**, 141.

[7] *Noyes, R. W. & Dickmann, Z.:* J. Reprod. and Fert. 1960, **1**, 186.

[8] *Psychoyos, A.:* Ciba Foundation Study Group n° 23, 1966, 4, J. and A. CHURCHILL Ltd. London.

[9] *Psychoyos, A.:* C. R. Acad. Sci. 1960, **251**, 3073.

[10] *Zeilmaker, G. H.:* Acta Endocrinol. 1963, **44**, 355.

[11] *Canivenc, R. & Laffargue, M.:* C. R. Acad. Sci. 1956, **242**, 2857.

[12] *Cochrane, L. R. & Meyer, R. K.:* Proc. Soc. Exp. Biol. and Med. 1957, **96**, 155.

[13] *Psychoyos, A. & Alloiteau, J. J.:* C. R. Soc. Biol. 1962, **156**, 46.

[14] *Psychoyos, A.:* C. R. Acad. Sci. 1961, **253**, 1616.

[15] *Everett, J. W.:* Anat. Rec. 1956, **124**, 287.

[16] *Gale, C. C. & McCann, S. M.:* J. Endocrinol. 1961, **22**, 107.

[17] *Psychoyos, A.:* J. Endocrinol. 1963, **27**, 337.

[18] *Alloiteau, J. J.:* C. R. Acad. Sci. 1961, **253**, 1348.

[19] *Shelesnyak, M. C., Kraicer, P. F. & Zeilmaker, G. H.:* Acta Endocrinol. 1963, 42, 225.

[20] *Psychoyos, A.:* C. R. Acad. Sci. 1962, **254**, 1504.

[21] *Psychoyos, A., Alberga, A. & Baulieu, E. E.:* C. R. Acad. Sci. 1968, **266**, 1407.

[22] *Psychoyos, A.:* Arch. Anat. Microsc. Morph. Exp. 1967, **56**, Supple. 3–4, 616.

[23] *Potts, M. & Psychoyos, A.:* C. R. Acad. Sci. 1967, **264**, 370.

[24] *Prasad, M. R. N., Dass, C. M. S. & Mohla, S.:* J. Reprod. Fert. 1968, **16**, 97.

[25] *Smith, D. M.:* J. Endocrinol. 1968, **41**, 17.

[26] *Kirby, D. R. S.:* Transplant. 1968, **6**, 9, 1005–1010.

[27] *Mayer, G. & Duluc, A. J.:* C. R. Acad. Sci. Paris 1968, **267**, 5, 509–510.

[28] *Gwatkin, R. B. L.:* J. Cell Physiol. 1966, **68**, 335.

[29] *Bitton-Casimiri, V. & Psychoyos, A.:* C. R. Acad. Sci. 1968, **267**, 762.

[30] *Segal, S. J., Wada, K. & Schuchner, E.:* Cited by *Segal, S. J.* In: Cellular Biology of the uterus 1967. Ed. *R. M. Wynn* Appleton Century Crofts, New York.

[31] *Psychoyos, A. & Bitton-Casimiri, V.:* C. R. Acad. Sci. 1969, **268**, 188.

[32] *Psychoyos, A.:* Proc. Third Intern. Congr. of Endocrinology, Mexico 1968. Exc. Med. Found.

Discussion

D. R. S. Kirby: This suppression of metabolic activity of the blastocyst during delayed implantation is a very exciting discovery and elegantly demonstrated. In your experiments, Dr. *Psychoyos,* you give progesterone to the ovariectomized rats. Do you think this is essential to demonstrate the inhibitory effect? In the mouse, ovariectomized shortly after mating and given no further treatment of any kind, viable blastocysts can be recovered 53 days *post coitum (Kirby,* unpublished).

A. Psychoyos: I have the impression that this is essential in preserving the maximum possible number of blastocysts into the uterus. It may be that progesterone increases the adhesiveness of the epithelium. I agree that you can also find a few blastocysts living for a long time in the uterus without progesterone but we find many more in animals treated with this hormone. This is the reason why we inject progesterone. I donot think that progesterone makes any difference to the suppression of metabolic activity of the blastocysts during the delay.

K. Fotherby: Could I ask Dr. *Psychoyos* about his estrogen-incorporation experiment? Were you measuring the incorporation into the nuclear-binding protein or were you measuring the incorporation into the endometrial cells? Was it the specific binding or was it . . .

A. Psychoyos: In the experiments I have been refering to here we measured the uptake of estradiol by the endometrium.

K. Fotherby: So this was not a measure of the amount bound to the specific receptor protein?

A. Psychoyos: No, but I suppose that variations of this uptake reflect the variations of the bound amount.

K. Fotherby: And you are suggesting that progesterone treatment increased the uptake of estradiol?

A. Psychoyos: Yes, according to the hormonal sequence applied before the injection of the tritiated estradiol. Once we had to do with ovariectomized animals not treated with progesterone, no uptake was observed after the second injection of estradiol. I suppose because the receptor sites were already saturated by the first injection. The daily injection of progesterone during this 48 hours period changes this picture.

K. Fotherby: I disagree with you, because I donot think with your experimental design you ought to be talking about receptor sites. I am interested in the progesterone effect because I think there is evidence which shows that, in fact, progesterone does not effect estradiol binding to the nuclear receptor protein. I am not sure how you were recording your data. If your progesterone experiment is done on the same basis, of measuring the uptake into the endometrial cells, you may be getting more estrogen taken up because under progesterone-estrogen treatment you are increasing the amount of endometrium. So that you should be expressing your uptake in terms of, say, DNA concentrations in your endometrial cells.

A. Psychoyos: This may be true and for this reason I consider conclusive only the comparison between animals in which both ovarian hormones were administred. As for the calculations of the DNA concentration in these endometria we found about the same values.

R. L. Brinster: I think Dr. *Psychoyos* has brought up a very interesting problem. I question somewhat the last point having to do with what is the molecule that is bringing the message to the egg in this case, rather than to the endometrium. It has to do with binding. One of the ways you can get at this, I believe, is to measure the estrogen-uptake of the blastocysts, perhaps *in vitro,* and if estrogen does bind to the blastocysts, as has been shown in the case of the mouse by *Smith* and *Biggers,* (1968) to determing the increase in uridine uptake in the blastocyst? I was wondering if you had any information about this in the case of the rat, and if there is anything new you can tell us about it?

A. Psychoyos: Personally, I donot have any experience in the uptake of estrogens by the blastocyst of the Rat. From preliminary data of ours on the *in vitro* effect of estradiol on the blastocyst uptake of Uridine, it seems that estradiol decreases this uptake in the dose of 0.025 μg/ml.

But we have to repeat these experiments with various doses. I would be very glad to discuss the results you were refering to on the mouse blastocysts. I think that even if it was found that there is an estrogen-uptake this should not reflect a direct effect of this hormone on the blastocyst's growth. In the Rat, normal blastocysts of the 5th day, that is blastocysts stimulated by the endogenous estrogen, when transfered into the uterus of ovariectomized animals treated with progesterone stop their development and remain in diapause until the injection of estrogen to the recipient. This same arrest of development happens even to blastocysts recovered from the uterus on the sixth day, when giant cells are present. (*Psychoyos*, C. R. Acad. Sc. 1961, 252, 2306). This phenomenon offers a basic argument to the concept that during delayed implantation, an active inhibition is excered by the endometrium on to the blastocyst. I believe that estrogen rather than acting directly on the blastocyst in annulling this inhibition.

H. R. Lindner: I would like to mention a few observations which might support the idea that the enhanced estrogen-binding activity induced by progesterone pre-treatment in Dr. *Psychoyos'* experiments may not be due to the same estrogen-receptor demonstrable in the untreated ovariectomized rat and referred to by Dr. *Fotherby* in this Discussion. We have recently been interested in non-steroidal estrogens of plant origin, such as the isoflavone genistein and the coumarin derivative coumestrol. With these compounds we can mimic many of the known effects of estrogens in the uterus, like the 6-hour-water-inhibition test, the 4-day uterine growth test, or the bluing reaction on injecting Evans blue that seems to be characteristic of the decidualizing uterus. Yet when we used the same system which you were using to delay implantation, that is ovariectomy two days after mating and maintenance with a long-acting progestagen, we completely failed to induce implantation with these non-steroidal estrogens at dose-levels that would give all the other estrogen responses (*Perel* and *Lindner,* unpublished). Estradiol-17 β (0.06 μg) and diethylstilbestrol (1 μg) were fully effective in inducing implantation in these rats. This discrepancy suggests that the estrogen receptors responsible for the implantation-inducing effect of estrogen might be dissimilar from those which are involved in the induction of uterine growth.

A. Psychoyos: Yes, that is exactly my point of view. This morning, I spoke about estriol. Something like 40 μg of estriol is needed and this is about eight hundred times more than the dose of estradiol required to induce implantation.

H. R. Lindner: How much do you need for the estradiol effect? Our rats require less than 0.08 μg injected systemically.

A. Psychoyos: About 0.05 μg subcutaneously and the tenth of this dose locally.

H. Lindner: Yes, but in this case it is also a weaker estrogen with respects to other end points of estrogen assay. Where serial, with the non-steroid estrogens you seem to have a dissociation between the uterine growth stimulation and implantation effects.

A. Psychoyos: In inducing implantation it seems to me that estradiol is much more specific than any other estrogen.

K. W. Humphrey: In the mouse in a similar situation you need very much more estriol than you do estradiol. With estradiol the minimum dose I have used is 0.00125 μg which is as effective as the optimum dose, of 0.025 μg. With estriol you need in the region of 2.5 or 10 γ.

With non-steroidal estrogens – anti-estrogens-I have obtained implantation with DMS, VII, 100 A, V 11, 55 a, also V I0993 and a great variety of nov-steroidal estrogens. So these results would differ from Dr. *Lindner's.*

R. G. Edwards: Could you explain very briefly the relationship between the estrogen surge and the secretion of these uterus-specific proteins that *Beier* and *Daniel* have described? Do these follow the estrogen surge?

A. Psychoyos: By the term estrogen surge I understand the saturation of the uterine receptors for estradiol. In the Rabbit where these experiments have been done, we dc not have any precise information even as to the necessity of estrogen for implantation.

I do not know if these proteins exist in the Rat. It has to be found out.

D. W. Schomberg: If I remember correctly, you suggested that there may be a direct effect of estrogen on the blastocyst if it is in contact with the endometrium and not obviously *in vitro* or in ectopic sites. Would you enlarge upon that?

A. Psychoyos: That is as I have said a working hypothesis. What I believe is that in the absence of estrogen a uterine factor represses the synthesis by the blastocyst of RNA necessary for its further development.

Estrogen should promote indirectly this synthesis by annulling the repression. It may be that we are dealing here with a model of regulation similar to the one suggested by *Talwar* and his collaborators (Proc. Nat. Acad. Sci. 1963, **52**, 1059) for estrogen action. I am rather thinking of a cytoplasmic component of the uterine epithelium arriving to the trophoblast by an intracellular passage. In our study with *Malcolm Potts* on the ultrastructure of the embryomaternal contact during delayed implantation, we observed that the microvilli of the epithelium were in a very close contact with those of the trophoblast and a communication in some parts is probable. It may be that maternal cytoplasmic components are thus passing into the blastocyst. We may have there a system in which one group of cells work in regulating directly the synthetic activities of another.

D. R. S. Kirby: The cytoplasmic continuity, though, has not been confirmed by other workers has it – it is only *Potts,* I think, that sees cytoplasmic continuity in the breakdown of the intercellular bridges. The Scandinavian people have not shown any . . . I do not say it invalidates your hypothesis though.

A. Psychoyos: You mean during delayed implantation or at the beginning of implantation after estrogen?

D. R. S. Kirby: This is, I think, both during delayed, and at the beginning of implantation.

A. Psychoyos: I have not a big experience in electron microscopy but considering the number of microvilli even in only one cell I really cannot see how it can be affirmed that there is no communication.

J. F. Sommerville: I am certain that the steroid biochemists will help you in this interesting work in the future by measuring the absolute tissue concentrations of individual estrogens even in sub-cellular fractions. The moment that one administers estradiol one has, in effect, administered a galaxy of metabolic products some of which await identification and it should be of interest to define to what extent these products are involved in the effects which you describe.

A. Psychoyos: That is one of the things I would like to see realised. For the moment, what we try to prove with Professor *Baulieu* is that the number of available receptor – molecules for estradiol is involved in the uterine receptivity for nidation.

D. R. S. Kirby: I think this concept of the blastocyst during delay being actively suppressed rather than simply deprived of the necessities of life is an exciting concept, and am sure we will hear very much more about it.

The Histochemistry of the Human Endometrium and the Problems of Nidation

H. Schmidt-Matthiesen

Universitäts-Frauenklinik, Göttingen, Germany

Concerning the nidation of the ovum we have to respect

1. the *surface secretion,*
2. the *subepithelial vessels,* and
3. *other factors* which could be important for the penetration of the trophoblast into the endometrial stroma.

1. Surface secretion

In the early secretory phase the *surface epithelium is covered with secretion* on which the ovum moves to its implantation site. Qualitatively the findings are fairly constant. Of all substances which are present in the surface layer glycogen shows the most marked variations. These and other variations of the surface may occur in an otherwise completely normal endometrium; the contrast between the surface and the glandular epithelium is often obvious, for example concerning the *activity of acid phosphatases* and the *distribution of phosphatides.* These differences are surprising, because autoradiography studies on nucleic acid and protein metabolism indicated that both, – surface and glands –, show equal potentials *in vitro.* Perhaps, special features of vascular supply are responsible for the obvious differences *in vivo.*

In Fig. 1 we see the *histochemical analysis* of the surface secretion, -or better-, of the *surface layer.* The proteins consist partly of enzymes and are partly combined with aminosugars and neuraminic acid which form predominantly acid mucoids. There are also glycogen, nucleic acids and lipids, mainly phosphatides.

One can only speculate on the *importance of the surface epithelium* and of different *components of the secretion.* Very viscous mucoids may possibly help adhesion of the blastocyst. They contain many electrolytes and may therefore act as ion-exchangers. Direct metabolic exchange between the secretion and the ovum is possible:

Manuscript received: 15 March 1969

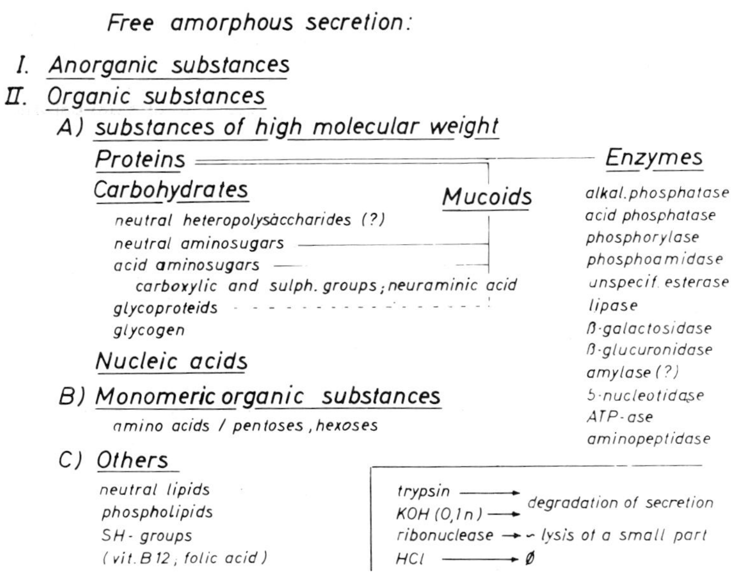

Fig. 1

A direct, partly selective transfer of numerous ions, hexoses and other compounds of the secretion into the blastocyst was already demonstrated in experiments an animals. Furthermore, we have to think of chemotactic influences on the localisation of the nidation, influences upon cleavage and evolution of ovum.

2. Subepithelial vessels

A rich supply of oxygen and the neighbourhood of blood vessels are necessary for the metabolism of the travelling ovum and the functional capacities of the surface epithelium. The *subepithelial vascular network* is shown in Fig. 2. Its obvious super-

Fig. 2

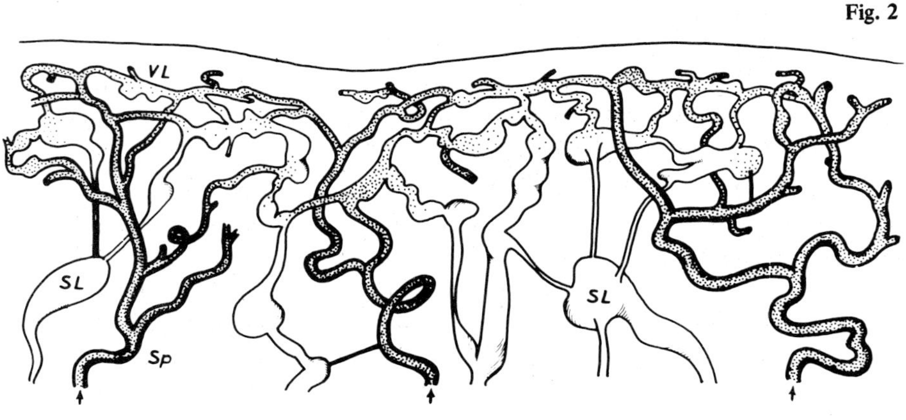

ficial localisation, lacunar distension of smaller blood vessels and thinness of their wall contribute to an increased permeability, especially under the influence of estrogens. The lacunar parts in their histochemical staining resemble arterial capillaries and not veins.

3. The environment of the implanting ovum

In assessing the nutritional requirements of the implanting ovum the endometrial tissue components must be considered. We have to distinguish between primary histopoetic substances like secretion and the ground substance- and the secondary products of the histiolysis. Both are histiotrophic.

In the beginning of implantation acid mucoids and glycoporteids have gradually accumulated in the epithelium. The *secretion caps* are completely separated from the resting cell (Fig. 3). The caps form secretion globules, which accumulate in the lumen.

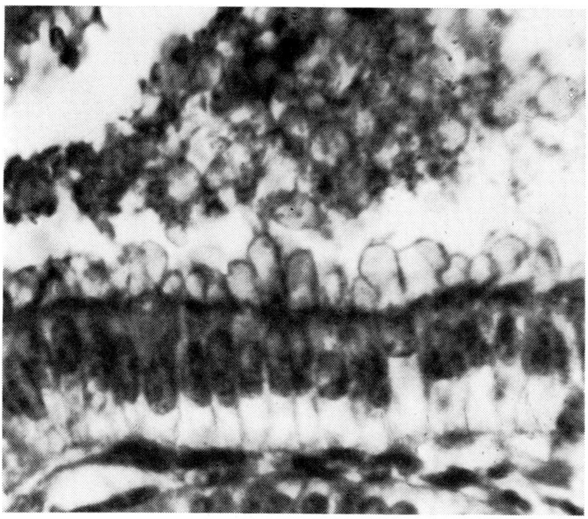

Fig. 3

Later, the globules fuse together. Finally, they desintegrate: Glycogen and other incorporated material has been liberated.

In Fig. 4 we see the *chronological course* of the changing of secretion.

The most impressing change in the endometrium is the *transformation of the intercellular ground-substance.* Until the 12th day of the cycle the ground-substance is very viscous and extremely rich in substances which give clear histochemical reaction for acid mucopolysaccharides. From the 12th day until the beginning of implantation the preparatory transformation of ground-substance has been completed, represented by a steady breakdown of high molecular components. The breakdown

Changing of secretion as function of time

Secretion of corpuscular character (globes)
→ decay of secretion globes
→ liberation of glycogen, till 22th day completed
→ degradation or transformation of glycogen
→ loose of association between proteins and aminosugars
→ degradation of the associated partners
→ decrease of enzyme activity
→ secretion has been homogenous

Fig. 4

affects firstly the acid mucopolysaccharides, than the neutral ones and finally proteins. On the 21st day the ground-substance is optically almost empty and contains only traces of high molecular compounds. The results of this transformation are very important (Fig. 5):

Histochemical tests show a marked *loss of viscosity,* for the final stage almost similar to that of water. Consequently, the connective tissue – id est the structural framework and ground-substance –, is unusually *loose, distored and vulnerable.*

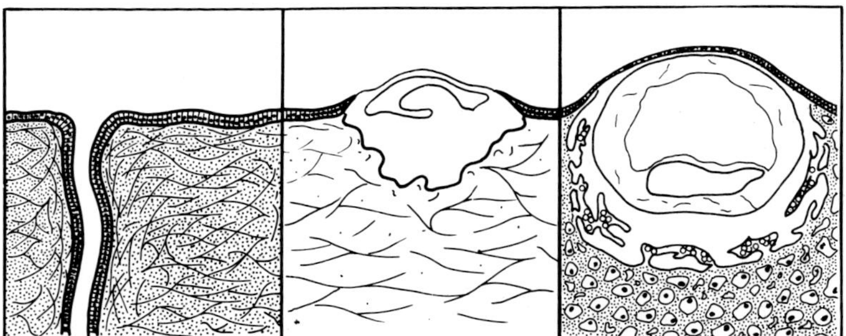

Proliferationsphase	21. Tag; Implant. ca. 36 h	26. Tag; Implant. ca. 144 h
Fasern: dicht	aufgelockert	perizell., wabige Neubildg.
Grundsubst.: hochmolekul. →	verflüssigt, Sol →	hochmolekular, Gel
hochviskös →	niedrig viskös →	hochviskös
saure MPS, neutr. MPS, Glyco-Pr., Proteine — Abbau →	niedermolekul. Abbauprodukte (Aminozuck. u.a.)	→ Polymerisation (u.a. Sulfo-MPS)
Diffusionsvorg. erschwert →	extrem begünstigt →	erschwert
fibrinolyt. Aktiv. niedrig →	sehr hoch →	Abfall, → später Ø
a	b	c

Fig. 5

As a result of the loss of viscosity there are marked *changes in diffusibility* which must affect reactive kinetics in the intercellular space. As result of breakdown processes large *amounts of low molecular compounds* accumulate in the interstitial space. A metabolic utilisation of this material by the implanting ovum must be considered.

The *breakdown* of the ground-substance coincides in time with the *increase of the content of glycogen.*

At the same time, the *stroma cells* show a considerable contents of *neutral fats. Glycogen* begins to accumulate in the deeper but not the superficial layers. This stroma material can be utilised by the trophoblast after incorporation.

4. The penetration of trophoblast

The penetration of the ovum into the endometrium often has been understood as a carcinoma-like invasion. I think, this is not correct. Firstly, we look to the conditions in the *intercellular space* (Fig. 5 b). Because of the breakdown of the macromolecular ground-substance of the stroma, the latter has lost its originally high viscosity and resistance. Now, here are optimal conditions for the growth of the trophoblast. Only the *fibrils* of the stroma should be of greater resistance. In connection with the penetration of the ovum there must exist a proteolystic potency on the border between the trophoblast and the stroma fibrils. Now we have to ask for the origin of these forces favouring the implantation process. I think, they do not belong exclusively to the trophoblast but also are represented by the endometrium itself (Fig. 6).

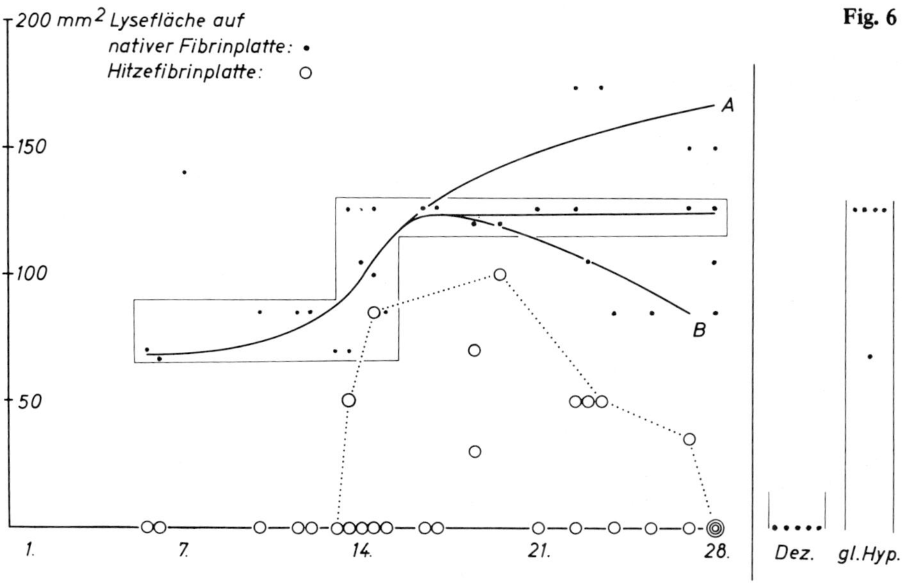

Fig. 6

We remember that during the implantation time the normal endometrium not only possesses a high concentration of *activators of fibrinolysis* (the upper diagram-) but also of *proteases* (the lower pointed diagram-) which are localised in the lysosomes of stromal and glandular cells. We know something about the facts which can induce the liberation of lysosomal enzymes. These and other findings inaugurate a *new hypothesis of early nidation* (Fig. 7).

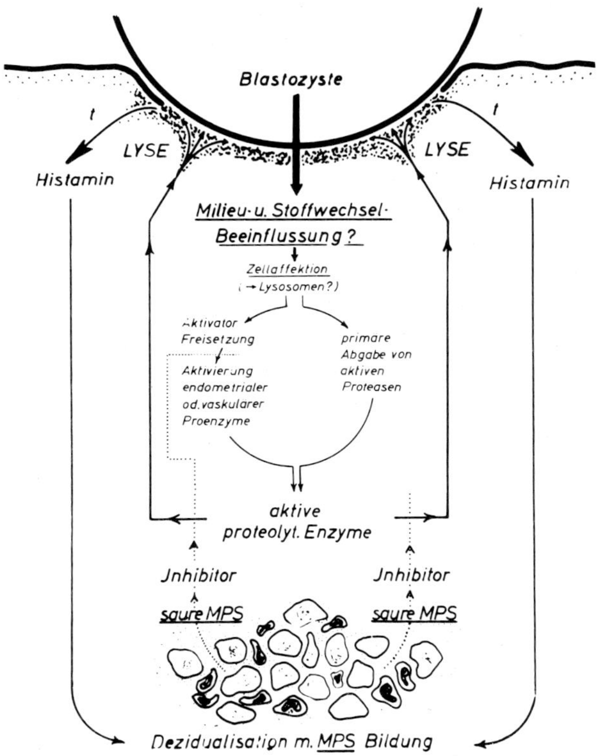

Fig. 7

It is possible that the blastocyst only influence the medium and the metabolism, followed by a liberation of lysosomes and activation of fibrinolysis and unspecific proteolysis in the contact area.

In connection with the proteolytic processes histamin will be produced. It induces the decidual stroma reaction. The latter one is correlated with the building up of sulpho-mucopolysaccharides. In respect to our investigations we may assume that these substances react as inhibitors of fibrinolysis and proteolysis. Definitely, –

tested against fibrin-, the homogenate of the young decidua is proteolytic inactive, in extreme contrast to the normal secretory endometrium. The decidualised tissue components – especially the sulpho-mucopolysaccharides-stop the deep growth of the trophoblast like a barrier.

In humans, decidualisation is not the supposition but the consequence of nidation and the correlated proteolytic processes.

References

1) Monographs:

Das normale menschliche Endometrium. Thieme, Stuttgart 1963.

Die gewebsgebundene Fibrinolyse und Proteolyse in Endometrium und Myometrium, Dezidua und Plazenta, Kollum- und Korpuskarzinom, Karger, Basel N. Y. 1967.

The normal human endometrium. USA-Ausgabe, McGraw-Hill Verlag, N. Y. 1968.

2) Special Papers:

Histochemische Untersuchungen der Endometrium-Grundsubstanz. Acta histochem. **13**, 129–164 (1962).

Die Vaskularisierung des menschlichen Endometrium. Arch. Gynäk. **196**, 575–597 (1962).

Histochemische Studien am Sekret der Endometriumdrüsen. Acta histochem. **16**, 28–45 (1963).

Aufbau und funktionelle Ausreifung des Endometriums. Med. Bild-Dienst ROCHE 1965, H. 2.

Histochemie des menschlichen Endometrium vor der Nidation. Vth World Congress on Fertility and Sterility, Stockholm Internat. Congr. Series, Excerpta Medica 1966.

Endometrium und Nidation beim Menschen. Z. Geburtsh. u. Gynäk. **168**, 113 (1968).

Cytochemical Studies on Oviducts of the Rat During the Estrous Cycle and Progestation

B. L. Lobel, E. Levy* and H. R. Lindner

Department of Biodynamics, The Weizmann Institute of Science, Rehovot and
*Department of Reproduction, Veterinary Institute, Beth-Dagan, Israel

Summary: Autoradiographic and histochemical studies were carried out on oviducts and uteri of pregnant rats during progestation and nidation (day of mating to day 9 p.c.). Pulse labelling with tritiated thymidine (radioactive DNA precursor) during progestation showed no uptake of tritiated thymidine by the cellular elements of the tubal wall (epithelium, stroma, blood vessels and smooth muscle), while in the uterus, it revealed a sequence of cell renewal in the luminal and glandular epithelium and stroma; this synthetic activity may be associated with ovarian hormonal secretion during progestation.

On the day of mating the uterus was heavily infiltrated with inflammatory cells and the luminal epithelium showed signs of degeneration. The oviducts showed no such reaction in spite of the fact that they were undoubtedly exposed to the same hormonal influences as the uterus and to contact with sperm; in fact, spermatozoa were seen adherent to the tubal epithelial cells as late as day 5 p.c. Both hydrolytic (e.g. acid and alkaline phosphatase and ATP-ase) and steroid dehydrogenase enzymic activities increased in the oviducts with the passage of the fertilised ova towards the uterus. Succinic and 3β-hydroxy-Δ5-steroid dehydrogenase activities were demonstrable in the granulosal cells surrounding the fertilized ova within the oviducts. Following regeneration of the uterine epithelial cells, these showed a steady increase in enzymic activities.

Uptake of tritiated thymidine and enzymic activities were demonstrable in blastocysts on day 5 p.c., while they were still "free" in the uterine lumen and encased in the zona. During the next 24 hours the blastocysts became abherent to the uterine epithelium and focal endometrial decidualization took place resulting in the formation of niduses. The cells comprising the niduses revealed morphological and enzymic characteristics not previously demonstrable in the endometrial stromal cells. Prior to this differentiation, the stromal cells incorporated tritiated thymidine, while the definitive decidual cells did not. When the niduses had developed into implantation chambers, they became extensively vascularised. At the same time, blastocysts rapidly expanded and primary giant cells developed from the trophoblastic layer. The giant cells penetrated between and beneath the uterine epithelial cells, which then degenerated and were phagocytosed by these giant cells and by leucocytes. Inner cell mass, trophoblastic layer and giant cells incorporated thymidine. Nevertheless, mitotic figures were seen only in trophoblast and inner cell mass and not in giant cells, though the latter increased in size and number. Concurrently, these giant cells exhibited activities of 3β-hydroxy-Δ^5-steroid dehydrogenase and $NADPH_2$ diaphorase enzyme systems necessary for the biosynthesis of steroids.

Manuscript received: 15 March 1969

Introduction

In the rat, capacitation of spermatozoa, fertilization of ova and cleavage occur in the ampullary loops of the oviduct. During this time the spermatozoa, the ovulated eggs and subsequently the fertilized ova are bathed in tubal fluids, which contain substances necessary for the successful completion of these processes [19, 21, 22]. It has been shown that tubal fluid is the product of the oviduct itself, without significant contributions from the peritoneal cavity or the uterus. *Marcus & Saravis* [19] have studied the production of tubal fluid in the rhesus monkey and observed that secretion by epithelial cells, transudation from the vascular system and passage of protein-rich lymph contribute to its formation; and that the volume and pressure are under the control of ovarian hormones. This investigation of the oviducts of the rat was undertaken to elucidate some cytochemical properties of the tubal epithelial cells and, if possible, to relate such properties to secretory activities.

Materials and methods

Oviducts from cyclic, pregnant and ovariectomized rats were studied. The rats used were from the colony of the Biodynamics Institute, originally Wistar stock. The rats were housed in air conditioned quarters and a normal (14/10) light cycle obtained. Purina laboratory chow and fresh water were available *ad lib.* Vaginal smears were taken daily and only rats exhibiting 4-day cycles were used.

Estrous Cycle

20 rats were used. 5 rats were killed on each of the following stages of the estrous cycle; proestrus (0_1), estrus (Co) and diestrus (L_1 & L_2).

Progestation

20 rats were caged with males of known fertility and insemination verified by the finding of a vaginal plug or spermatozoa in the vaginal smear. This day was designated as day 1 post-coitum and the days following as day 2, day 3 p.c. etc. Of these pregnant rats 5 were killed on each day of progestation from day 1 to day 4 p.c.

Ovariectomized rats

35 rats were ovariectomized on diestrus of the cycle. Four weeks later 10 rats received progesterone 5 mg in 0.1 ml sesame oil i.m.; 10 rats received estradiol benzoate 5 γ in 0.1 ml peanut oil i.m.; and 10 rats both hormones in the doses described above. Five rats of each group were killed at 24 and at 36 hours after injection. Five rats received only the vehicle (peanut oil) and were killed 36 hours later.

Autoradiography

Eight cyclic and 8 progestational rats were used for the autoradiographic studies, 2 rats for each day of the cycle (0_1, Co, L_1, L_2) and progestation (day 1 to day 4 p.c.). The rats received one i.p. injection of 200 μc tritiated thymidine in 0.1 ml distilled water and were killed one hour later.

The rats were killed by cervical dislocation, the oviducts removed and rapidly frozen or fixed. Frozen oviducts were cut at 10–15 μ in a cryostat maintained at -20 °C. The sections were incubated for demonstration of activities of phosphorylase; succinic dehydrogenase; NADPH & $NADPH_2$ diaphorases, and 3β-17β-20β-hydroxysteroid dehydrogenases. For 3β-OHSD both dehydro*epi*androsterone (DHA) and pregnenolone were used as substrates. For 17β-OHSD estradiol 17β was used, and for 20β-OHSD 20β-hydroxypregn-4-en-3-one was used as substrate. Control sections were incubated in media lacking the substrates and in media lacking the co-factor. In addition, oviducts from rats on day 1 p.c. were incubated for demonstration of the steroid dehydrogenase activities in media prepared at pH 7, 7.6, 8, 8.5 and 9. Polyvinylpyrrolidone (PVP) was added to the incubation media. Controls were incubated at each pH level with addition of the vehicle for the steroid (dimethylformamide) only. Details of the methods, and a critical evaluation of the validity of the histochemical techniques have been published previously [17]. The fixatives employed were neutral buffered formalin (NBF), alcohol-formalin-acetic acid (AFA) and Bouin's.

The tubes in NBF were fixed overnight, rinsed, cut in a freezing microtome and incubated for demonstration of alkaline and acid phosphatase and adenosine triphosphatase ATP-ase activities. Oviducts fixed in AFA were processed to paraffin and the sections stained with H. & E.,trichrome, and by the PAS technique for demonstration of glycogen. Oviducts from rats which had received tritiated thymidine were fixed in Bouin's and autoradiographs prepared as described previously [16].

Results

Autoradiography

Few labelled cells or mitotic figures were observed in the oviducts at any time.

Estrous cycle

Careful scrutiny of autoradiographs of oviducts from rats during the stages of the estrous cycle revealed only minor differences in labelling (Fig. 1–4). At proestrus some labelled epithelial and stromal cells were observed mainly in the ampullary region, while in estrus and diestrus labelled cells and mitotic figures were found in

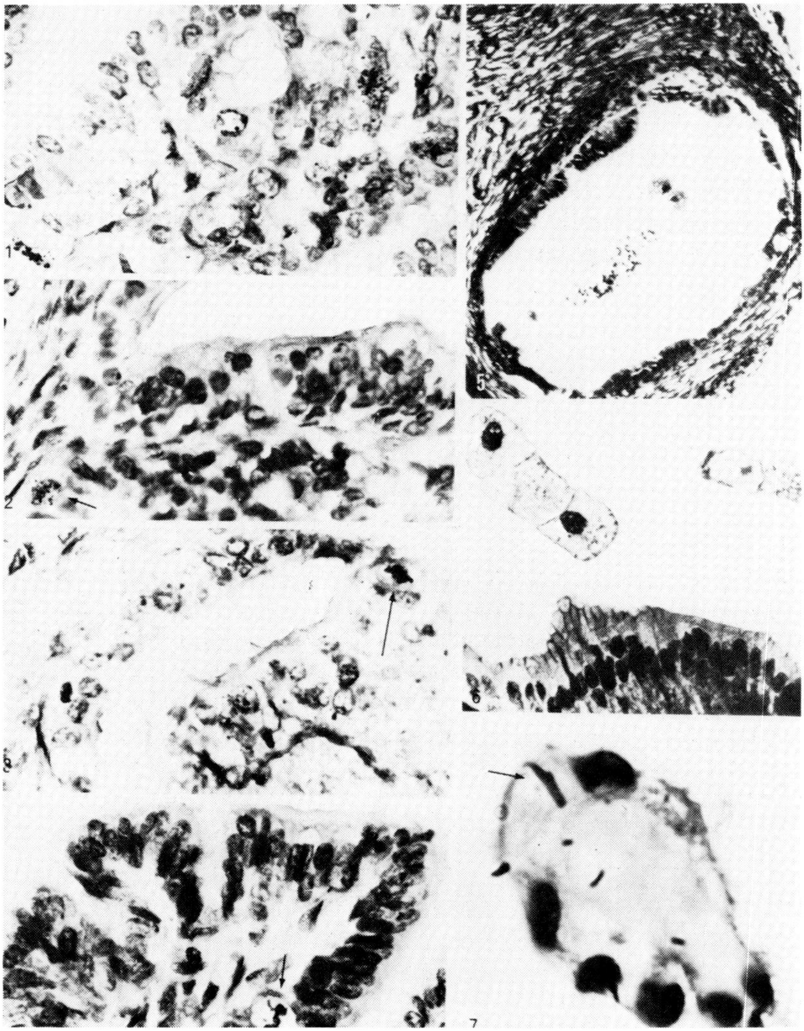

Fig. I. *Autoradiographs of oviducts*
I, 1–I.4 x circa 400

I, 1. Oviduct on proestrous, note labelled epithelial cells. The dark mark at the left edge of the picture is an artefact.

I, 2. Oviduct on estrus. Note labelled stromal cell (at arrow).

I, 3. Oviduct on diestrus (L_1); note mitotic figure (arrow) and labelled epithelial cells.

I, 4. Oviduct on diestrus (L_2); note mitotic figure.

I, 5. x circa 100. Oviduct on d_2 p. c. Note moderately dilated loop and few remaining granulosal cells in the lumen near the ovum.

I, 6. x circa 400. Fertilized ova in isthmus on d_4 p. c. Note no labelling of fertilized ova or tubal epithelium and no infiltration.

I, 7. x circa 1000. Oviduct on d_4 p. c. Note head of spermatozoan adherent to tubal epithelium (arrow) and no infiltration.

the middle loops. Infiltration with inflammatory or wandering cells was not observed during the cycle, neither in the epithelium nor the stroma of the oviducts (Fig. I, 1–4). Morphological differences between the stages of the estrous cycle (which might be used to identify these stages with any degree of certainty) were not seen in either autoradiographs or histological sections of oviducts.

Progestation

Few labelled nuclei were found in the oviducts at any time of progestation. Likewise, infiltration with inflammatory or wandering cells was not observed in the tubal walls during progestation (Fig. I, 5–7). Spermatozoa were found in the oviducts from day 1 to 4 p.c. within the lumen, or apparently adherent to epithelial cells (Fig. I, 7).

Ovariectomized rats

Four weeks after ovariectomy, the oviducts had undergone shrinkage, which affected the whole tubal wall. The epithelial cells were low, the stromal layer was reduced in width, and appeared less cellular with relatively more collagen visible as a band beneath the epithelial cells. Villi were low or absent. The smooth muscle fibers were also shrunken. After administration of estrogen the oviducts appeared stimulated and the height of the epithelial cells had increased (Fig. VIII, 9). After treatment with progesterone, there was marked dilation of the tubal loops in the ampullary region (Fig. VIII, 1). Following administration of both hormones, tubal loops between the ampullar and isthmus were dilated (Fig. VIII, 2), the epithelium appeared high and secretory (Fig. VIII, 10) and many tubal folds were again observed. Hormone effects at 24 and 36 hours were similar.

Distribution of enzymic activities in oviducts

Alkaline phosphatase

(see Fig. II). Activity occurred in blood vessels in the tubal walls and villous processes, in apices of some tubal epithelial cells and in some cilia.

Acid phosphatase

(see Fig. III). Activity in the oviducts was seen as a fine granular precipitate in some epithelial, stromal and endothelial cells.

Adenosine triphosphatase

(see Fig. IV, 3). Activity occurred in blood vessels, in the smooth muscle fibers and the apices of cellular membranes of some epithelial cells.

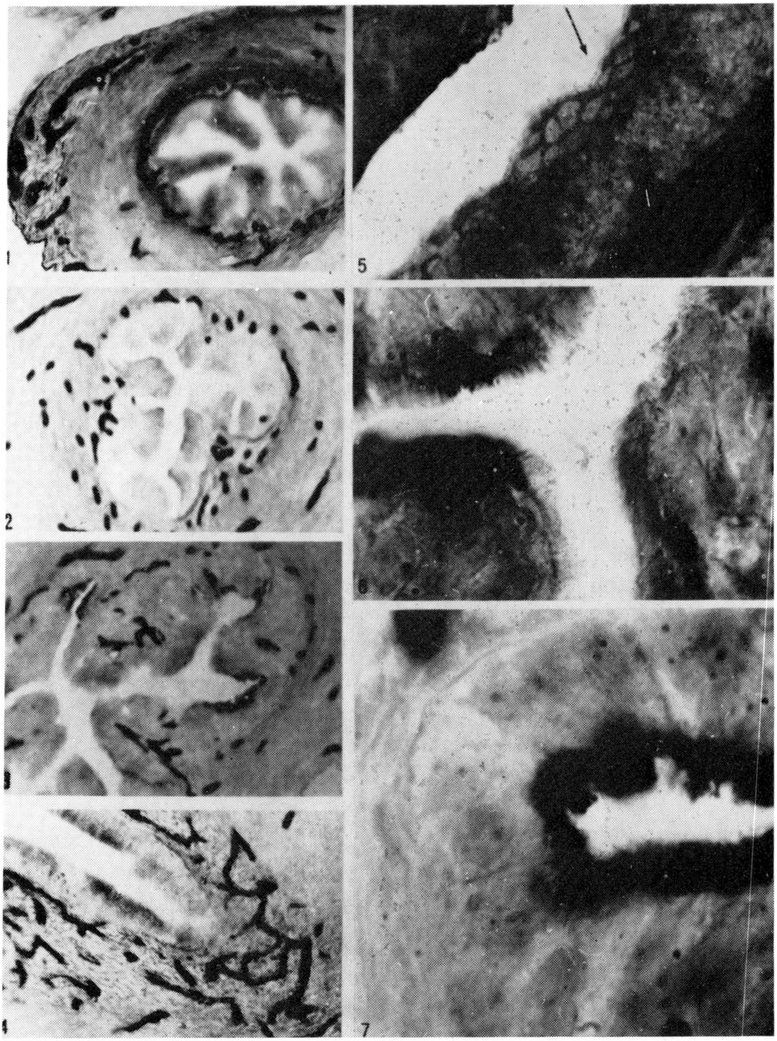

Fig. II. *Sections of oviducts incubated for demonstration of alkaline phosphatase activity*

II, 1–4. x circa 100, II, 5–7 x circa 1000.

II, 1. Oviduct on proestrus, loop near isthmus. Note strong activity in blood vessels only.

II, 2. Oviduct on estrus, similar to Fig. I, 1.

II, 3. Oviduct on diestrus (L_1) activity appears also in apices of epithelial cells.

II, 4. Oviduct on diestrus (L_2) section of tube near isthmus. Moderate but more widespread activity in epithelial cells.

II, 5–7. Oviducts on diestrus (L2) higher magnifications of epithelial cells from tubal loop mid-portion of oviducts showing variations in enzymic activity.

II, 5. Tangential section through epithelium showing enzymic activity in borders of epithelial cells (arrow) and blood vessels (lower right).

II, 6. Ciliated epithelial cells with moderate or no enzymic activity.

II, 7. Strong enzymic activity in apices of epithelial cells.

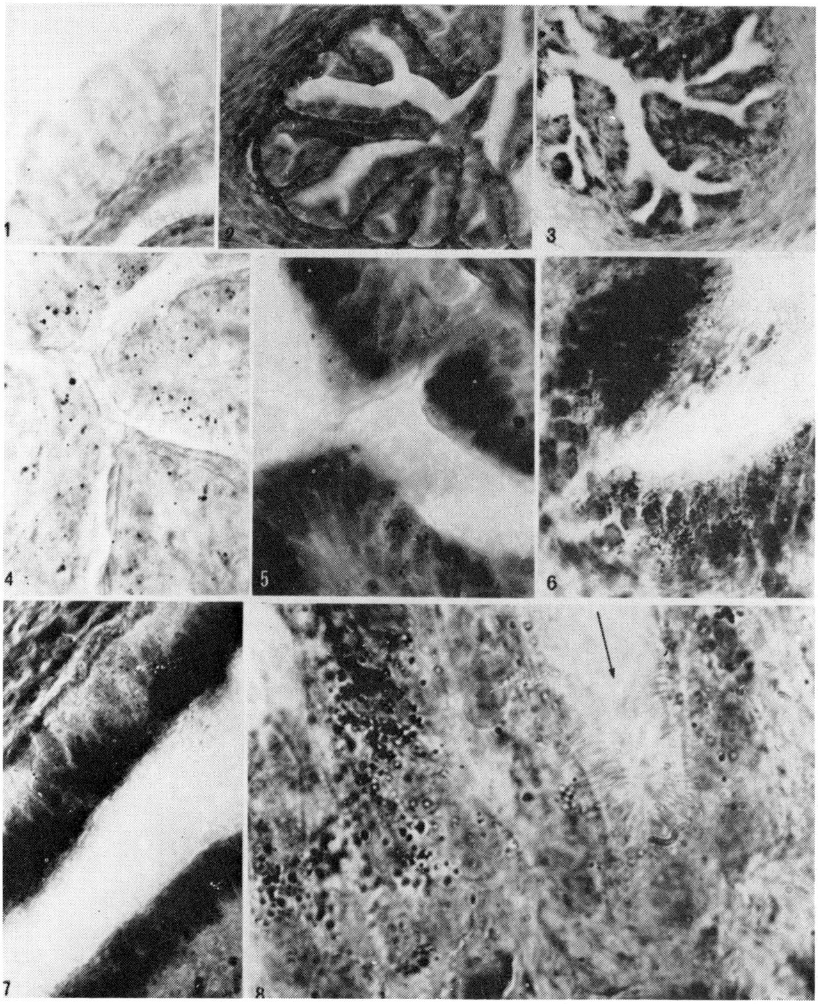

Fig. III. *Sections of oviducts incubated for demonstration of acid phosphatase activity.*

III, 1–3. x circa 100. Oviducts on proestrus.

III, 1. Note lack of reaction in the fimbria.

III, 2. Shows minimal activity in epithelium of ampullary loops, but some reaction product in subepithelial stromal cells.

III, 3. Loops from middle of oviduct. Note strong activity in apices of some epithelial cells as well as stromal cells.

III, 4. x circa 400. Ampulla on estrus; note isolated reactive granules in epithelial cells.

III, 5. x circa 400; loops from middle of tube on diestrus (L_1) showing moderate activity, which has increased on L_2.

III, 6 and 7. x circa 400 tube near isthmus on L_2.

III, 8. Loop from mid-portion of tube on L_2 x circa 1000. Note that ciliated cells (arrow) have almost no enzymic activity, while the non-ciliated cells show many reactive foci.

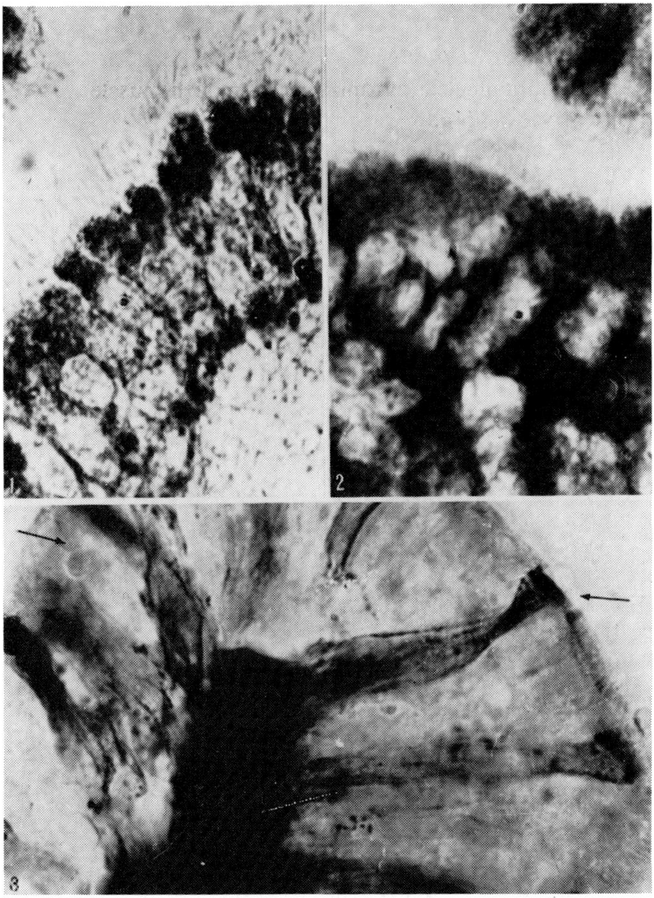

Fig. IV. *Sections of oviducts on proestrus x circa 1000*

IV, 1. Section incubated for demonstration of succinic dehydrogenase activity. Note strong activity in supranuclear cytoplasm of ciliated epithelial cells.

IV, 2. Section incubated for demonstration of $NADPH_2$ diaphorase activity, which is strong in non-ciliated cells; tangential section.

IV, 3. Section incubated for demonstration of ATP-ase activity. Note strong activity in villous core. Note reaction product in cellular membranes of some epithelial cells only (arrows).

Succinic dehydrogenase

(see Fig. IV, 1). Activity was demonstrable in tubal epithelial cells, where it was strongest and concentrated in the supranuclear cytoplasm; in smooth muscle layers and stromal cells, where it was moderate.

Phosphorylase

Activity was seen in the blood vessels and muscular coats of the oviducts and in apices of some epithelial cells of villous processes (Fig. V, 5).

NADPH and $NADPH_2$ diaphorases

(see Fig. IV, 2). The activities appeared mainly in the tubal epithelial cells.

3β, 17β and 20β-hydroxysteroid dehydrogenases

(see Fig. VI and VII). Activity was confined to tubal epithelial cells.

Enzymic activity in oviducts during the estrous cycle, progestation and after experimental procedures

Alkaline phosphatase

In proestrous and estrous (Fig. I, 1 and 2), activity occurred in the blood vessels. In diestrus activity was also noted in the apices of many epithelial cells, both ciliated and secretory, particularly in the loops near the ampulla (Fig. I, 3–7). In early progestation (Fig. V, 6) activity was similar to that of proestrus and estrus, but on subsequent days (2, 3 & 4 p.c.) activity was strong in epithelial cells in the middle loops of the tube. In ovariectomized rats, activity was demonstrable in the blood vessels and only a faint, spotty reaction occurred in epithelial cells in tubal loops near the uterus. After administration of estrogen little change in distribution or intensity of reaction was observed. However, after progesterone or both hormones, enzymic activity was stronger and more widespread.

Acid phosphatase

Activity was consistently weaker in the fimbria and ampullary epithelium (Fig. III, 1, 2 and 4), than in that of loops nearer to the uterus and activity increased from proestrus to diestrus in these loops (Fig. III, 3, 5–8). A similar increase was observed during progestation from day 1 to day 4 p.c. Following ovariectomy, no activity was detectable in fimbria and ampullary loops. However, enzymic activity remained in the loops near the uterus. After administration of estrogen, the distribution in

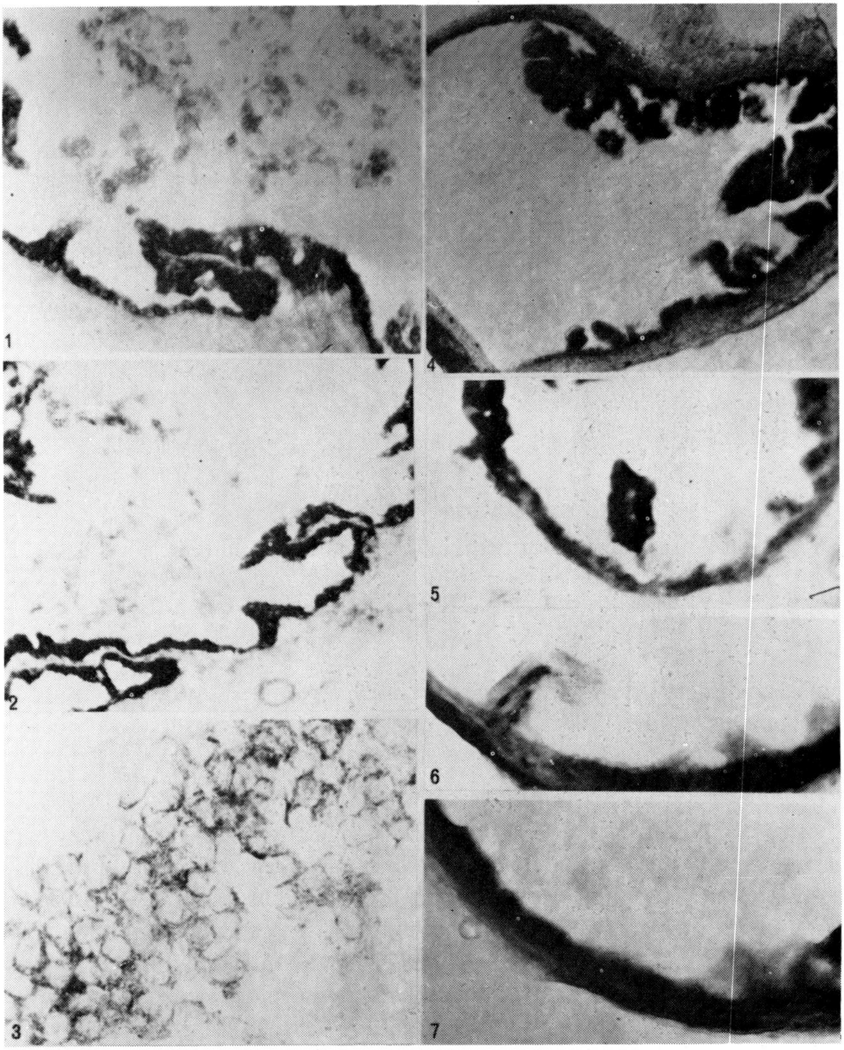

Fig. V. *Sections of oviducts on day 1 p. c.*

V, 1 and 2. x circa 100 demonstrate activities of NADPH and $NADPH_2$ diaphorases respectively. Note activity in the epithelium and moderate activity in the granulosa cells in the lumen.

V, 3. x circa 400 demonstrates 3β-hydroxysteroid dehydrogenase activity in granulosal cells.

V, 4. x circa 40 succinic dehydrogenase activity is strong in the epithelial cells of the dilated loop.

V, 5. x circa 100. Phosphorylase activity in epithelial cells and villi.

V, 6. x circa 100. Section incubated for ATP-ase activity which is strong in the smooth muscle of the wall of the distended loop, and in the blood vessels in the villis.

V, 7. x circa 100. Alkaline phosphatase activity is almost absent from the epithelium (lower right) of the distended loop.

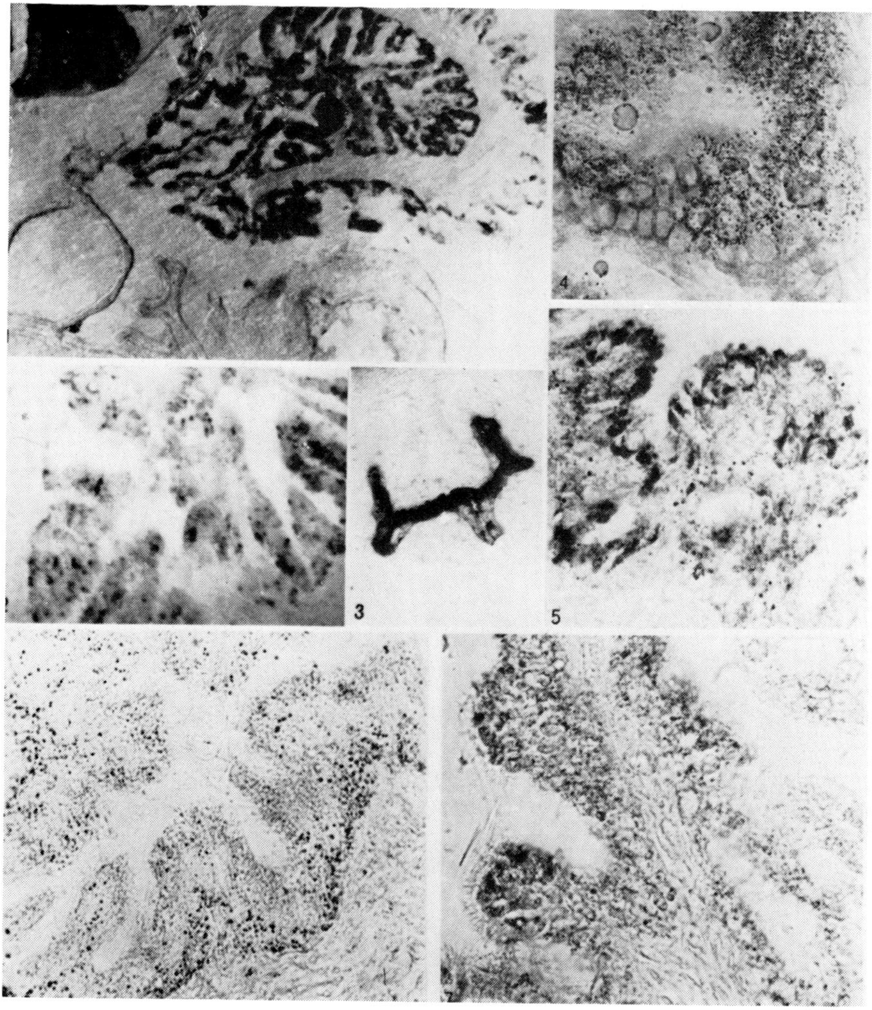

Fig. VI. *Sections of oviducts incubated for demonstration of 3β-hydroxysteroid dehydrogenase activity.*

VI, 1. x circa 40. Section incubated with DHA as substrate, diestrus. Note part of reactive corpus luteum in upper left hand corner and strong reaction in ampulla of oviduct.

VI, 2. x circa 100, incubation with pregnenolone as substrate, reaction present but beaker.

VI, 3. x circa 100. Middle loop in diestrus, incubation after rinsing of the section with acetone, DHA as substrate. Reaction appears stronger due to folding of the section across the middle.

VI, 4–7. x circa 400. Sections incubated with DHA as substrate. VI 4 pH 7.6; VI, 5 pH8; VI, 6, pH8.5 and VI, 7, pH9.

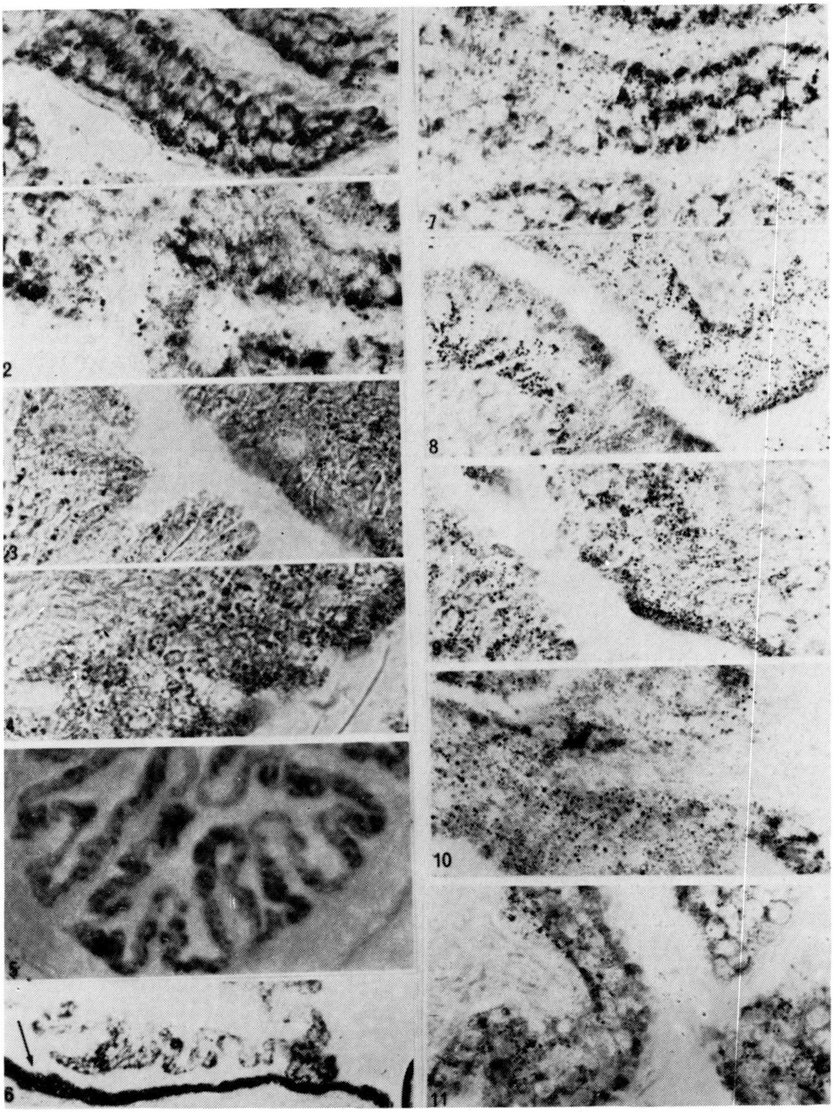

Fig. VII. *Sections of oviducts incubated for demonstration of 17β- and 20β hydroxysteroid dehydrogenase activities.*

VII, 1–4 and **7–11.** x circa 400. **VII, 5** and **6.** x circa 100. All from oviducts on day 1 p. c.

VII, 1–4. Incubations at different pH levels for demonstration of 17β-OHSD activity. VII, 1 pH7.3. VII, 2 pH7.6. VII, 3, pH8. VII, 4, pH9.

VII, 5 and **6.** Incubation for demonstration of 17β-OHSD activity after rinsing the section with acetone. Note strong activity in germinal epithelium of the ovary in VII, 6 (arrow).

VII, 7–11. Sections incubated for demonstration of 20β-OHSD activity at different pH levels. VII, 7, pH7.3. VII, 8, pH7.6. VII, 9, pH8. VII, 10, pH8.5; VII, 11, pH9.

enzymic activity was similar, but the reaction appeared weaker (Fig. VIII, 5), probably due to the increase in cell size. After administration of progesterone or estrogen and progesterone the distribution of the reaction was comparable to that seen in diestrus of the cycle, and many reactive foci appeared in those cells exhibiting acid phosphatase activity (Fig. VIII, 3 and 4).

Adenosine triphosphatase

Enzymic activity was similar during the cycle and progestation. Variations in activity in the smooth muscle were not noted. Epithelial cells of fimbira and ampulla exhibited weaker activity than those of the rest of the oviduct (Fig. V, 7). An interesting finding was that of activity in the cellular membranes of epithelial cells tentatively identified as "secretory cells" (Fig. IV, 3). After ovariectomy activity in the shrunken blood vessels, smooth muscle coats and epithelial cells of the atrophic tubes remained, but was weak. Administration of estrogen stimulated activity particularly in cells of the ampullary region, while progesterone and progesterone plus estrogen stimulated enzymic activity in cells of loops nearer the uterus.

Succinic dehydrogenase

Variations in enzymic activity during the cycle and progestation were not observed. Activity was consistently strong in the apices of epithelial cells, particularly those of ciliated cells in the ampullary region (Fig. IV, 1). Moderate activity was shown by stromal and muscle cells throughout the cycle and progestation (Fig. V, 4). After ovariectomy the shrunken epithelial cells still retained strong enzymic activity. However, that of the stromal layer and muscular coat appeared less, perhaps because of the decrease in cellularity and shrinkage of the muscle fibers. Administration of progesterone only did not appear to effect enzymic activity. The greater staining reaction observed after estrogen or estrogen plus progesterone probably parallelled the increase in cell size.

Phosphorylase

In the cycle, variations in intensity of enzymic activity were observed in the tubal epithelial cells. The uneven and spotty reaction seen in the tubal epithelial cells in proestrus and estrus became stronger and more widespread, i.e. more epithelial cells throughout the length of the tube showed reactivity in diestrus. Similarly, during progestation (Fig. V, 5) from day 1 p.c. onwards an increasing number of tubal epithelial cells exhibited enzymic activity till day 4 p.c. when the fertilized ova entered the uterus and thereafter reactivity declined. In ovariectomized rats the reaction remained strong only in the smooth muscle cells in the blood vessels and muscle coat. After administration of estrogen the smooth muscle fibers enlarged to their original size and enzymic activity became more widespread. Following administration of progesterone or estrogen and progesterone enzymic activity became stronger and more widely distributed.

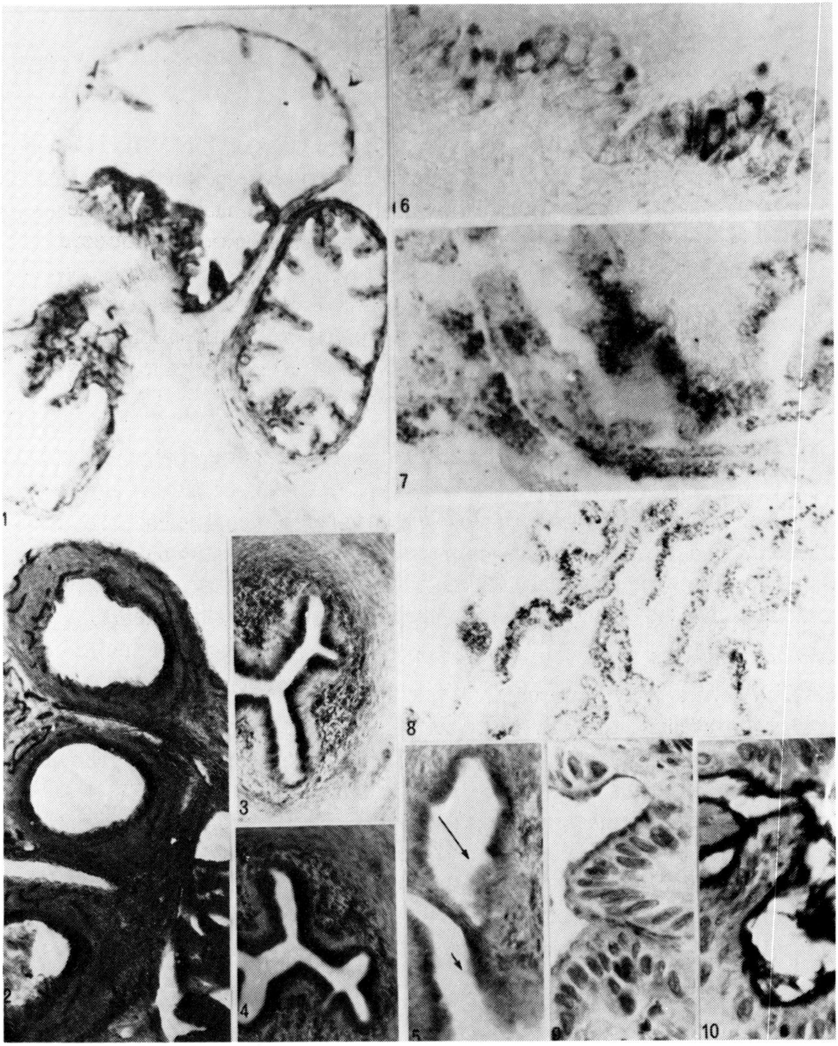

Fig. VIII. *Sections of oviducts of ovariectomized rats after administration of hormones.*

VIII, 1. x circa 40; progesterone only; alkaline phosphatase activity occurs in the blood vessels in the distended ampullary loops.

VIII, 2. x circa 400. Estrogen plus progesterone, alkaline phosphatase activity; note moderate dilation of middle loops.

VIII, 3. x circa 100 progesterone only. Acid phosphatase activity in peripheral cells of middle loop.

VIII, 4. x circa 100; estrogen plus progesterone. Strong acid phosphatase activity in epithelial cells.

VIII, 5. x circa 100, estrogen only; weak acid phosphatase activity in the epithelial cells (arrows).

VIII, 6, 7 and 8. Progesterone only. VIII, 6 x circa 400, 3β-OHSD activity. VIII, 7 x circa 100, 20β-OHSD activity. VIII, 8 x circa 100 17β-OHSD activity.

VIII, 9 and 10. Paraffin sections x circa 400 stained by PAS for demonstration of glycoproteins VIII, 9 estrogen only. VIII, 10, estrogen plus progesterone. Note secretory product (PAS positive) in apices of epithelial cells and lumen of tube.

NADPH and $NADPH_2$ diaphorases

In proestrus and estrus (Fig. IV, 2) as well as in early progestation (Fig. V, 1 and 2) the reaction was strongest in the ampullary region. In the second half of the cycle and later in progestation reactivity in the middle loops appeared more pronounced, while that in the ampullary region showed no change. Following ovariectomy, the activity of the diaphorases decreased greatly, only a few reactive grains were seen in the apices of some epithelial cells. An increase in activity of the diaphorases was noted after administration of hormones and no significant differences in intensity or distribution were detected between the three treatment groups.

3β-hydroxysteroid dehydrogenase

A reaction product was obtained with both DHA and pregnenolone as substrates with and without pretreatment of the sections with acetone (Fig. VI, 1–3). However, activity with DHA as substrate was consistently stronger (Fig. VI, 1 and 2). During the cycle activity was demonstrable mainly in fimbria and ampulla with very weak or nor reactivity in middle loops and isthmus. In early progestation the distribution was the same as in the cycle, while later in progestation, apparently more reactive cells appeared in the middle loops of the oviducts. Granulosal cells shed with the ova at ovulation and found in the ampullary loops surrounding the ova exhibited activity of 3β-hydroxysteroid dehydrogenase, (Fig. V, 3). Such activity may represent an abortive attempt at luteinization similar to that seen in atretic follides. The granulosa of a healthy follicle does not normally exhibit 3β-OHSD activity until after rupture and corpus luteum formation.

17β-hydroxysteroid dehydrogenase

The results were similar to those described for 3β-hydroxysteroid dehydrogenase (Fig. VII, 1, 5 and 6).

20β-hydroxysteroid dehydrogenase

Though the distribution was similar to that of 3β- and 17β-hydroxysteroid dehydrogenases, the reaction was much weaker at all times (Fig. VII, 7).

Following ovariectomy, the reaction of the steroid dehydrogenases (3β, 17β and 20β) decreased greatly – only few reactive grains remained in the apices of some of the tubal epithelial cells. After administration of hormones an increase in activities of the steroid dehydrogenases was observed in the three treatment groups (Fig. VIII, 6–8). However, after progesterone only a more widespread reaction was seen throughout the oviduct.

Incubations at different pH levels

Oviducts were incubated in media prepared at pH 7.3, 7.6, 8, 8.5 and 9 for demonstration of 3β, 17β and 20β hydroxysteroid dehydrogenase activities. A reaction was observed at all the pH levels investigated, both as a granular precipitate and a cytoplasmic staining reaction (Fig. VI, 1 and 4–7; and Fig. VII, 1–4 and 7–11). In all incubations, the strongest reaction was obtained in the ampullary region of the oviduct. At pH 7.3 and 7.6 the reaction appeared mainly as a cytoplasmic staining; at pH 8, 8.5 and 9 mainly a granular precipitate was observed.

Discussion

The oviduct is undoubtedly a target organ under the direct control of ovarian hormones since it regresses on withdrawal of these hormones (ovariectomy) and responds to the administration of estrogen and progesterone. This response is characterized by growth of the tube as a whole, by secretory activity of its epithelial cells and by motility of the cilia, the villous processes and the muscular coats. The origin of the oviduct is similar to that of the uterus – the Mullerian duct provides the common primordium for both structures. A remarkable feature of the oviduct is the fact that there are no synchronized variations in overall cellular morphology or enzymic activity occurring along the whole length of the oviduct during the estrous cycle (for review, see [24] and progestation. However, a certain degree of uniformity in enzymic activity was observed in the epithelial cells in three segments of the oviducts: the fimbria and ampulla, the middle loops and the isthmus. Physiologically, each segment provides the environment for a different step in the reproductive process: the fimbria, which pick up the ovulated eggs, are lined by tall ciliated epithelial cells which set up a fluid current in the direction of the ampulla. The ampullary loops have a thinner wall than the rest of the tube, permitting rapid distention by the fluid in which capacitation, denudation and fertilization are accomplished. During their passage through the middle loops (2 days) the ova undergo cleavage and absorb substances provided by the tubal secretion, for example, amino acids [3]. The isthmus has the thickest muscular coat, exerts a sphincter-like action and thus regulates the entry of the morulae into the uterus.

The oviducts are lined by the ciliated and non-ciliated or secretory cells. At any time during the cycle or progestation secretory cells exhibiting low, moderate or intense enzymic activities could be found in the middle loops and isthmus. In the ampullary region, the epithelial cells showed more or less the same levels of enzymic activity. Degeneration and replacement in the tubal epithelium are minimal as shown by the failure of the epithelial cells to incorporate the tritiated thymidine (labelled DNA precursor) and the paucity of mitotic figures.

Enzymic activity

(a) Hydrolytic enzymes

Activity of these enzymes has been associated with secretory processes and the varying levels detectable in the epithelial cells may reflect stages in the secretory cycle of the cell; the type of secretion being apocrine with accumulation of secretory products and their extrusion from the cellular apices. Electron microscope studies on the human oviduct have shown an increase in RNA particles and dilation of the endoplasmic reticulum under the influence of estrogen [7]. *Leonard* [15] has reported a doubling of uterine phosphorylase activity in spayed rats after administration of estrogen. In this study it was observed that the secretory cells contained more acid phosphatase granules, presumably lysosomes, than the ciliated cells. Lysosomes are readily demonstrable in other secreting cells, including those of endocrine tissues, where their function is related to secretory and not to autolytic processes. Progesterone has a labilising effect on the lysosomal membrane [25]. In mice, it has been shown that the rate of uterine secretion is increased by estrogen partly by the stimulation of synthesis of enzyme proteins, while the administration of progesterone caused an increased flow a of highly diluted secretion [10]. It is possible that these hormones exercise similar effects in the oviducts of rats, as shown by the results of the experiments with ovariectomized animals: the administration of estrogen and progesterone resulted in increased enzymic activity in the epithelial cells and moderate dilation of the tubes. After progesterone only the ampullary loops were markedly dilated due to distention with fluid. It would appear that in the oviduct of the rat estrogen stimulates the formation of secretory products and progesterone influences their extrusion from the cell. The tubal secretion provides the current in which the ova are carried towards the uterus. The flow is aided by the beating of the cilia (accelerated by estrogen), movements of the villi and peristaltic contractions of the tubal smooth muscle. As shown by *Harrington* [9] and *Chang* [5, 6], estrogen accelerates the rate of transport of ova through the oviducts while progesterone counteracts the estrogenic affects and slows egg transport, possibly by an inhibiting effect on the smooth muscle comparable to its action on the uterine musculature.

(b) Dehydrogenases

The presence of a variety of steroid dehydrogenases in other than steroid producing cells has been demonstrated by biochemical and histochemical methods. *Levy* et al. [15] demonstrated the presence of 3β-hydroxysteroid dehydrogenase in the oviducts of rats and suggested that the reaction may have been due to the presence of a broad spectrum alcohol dehydrogenase. However, in this investigation only dimethylformamide was used as solvent for the steroid substrate, hence it is likely that the reaction observed did rest on activity of a steroid dehydrogenase. A number of speculations has been made as to the significance of steroid dehydrogenases in non-steroid-hormone producing cells. Such enzymic activity has been associated with the peripheral metabolism of steroid hormones (see [1]), and with the target organ control. As an example of the latter, *Ferguson* and *Christie* [8] cite the increase in 17β-hydroxysteroid dehydrogenase activity in the yolk sac epithelium in the cat which paralleled the accumulation of secretion in the lumen. *Kim* and *Hermann* [12] have demonstrated the presence of steroid sulfatase and 3β- and 17β-hydroxysteroid dehydrogenases in skin and have suggested that *in situ* synthesis of a biologically active steroid from an inactive precursor may occur, thus proposing a link between enzymic activity and hormonal requirements in endocrine-sensitive tissues. *Bryson* & *Sweat* [4] have shown that human proliferative endometrium has the enzymic mechanisms to convert progesterone to eight different products. *Kinson* and *Kochakian* [13] using labelled testosterone as substrate have demonstrated low levels of NADPH and $NADPH_2$-17β-hydroxysteroid dehydrogenase activity in skeletal muscle. *Sweat, Bryson* and *Young* [23] performed incubations of human proliferative endometrium and myometrium and demonstrated the ability of both tissues to transform 17β-estradiol to estrone, though the capacity of endometrium was approximately 40 x that of myometrium. The authors suggested that the influence of estrogen in metabolic processes in, for example, muscle, hair, bone and fat may be due to transformation of minute quantities of hormone for which only low levels of enzymic activities would be required. *Jensen* [11] has reported that whole rat uteri convert small quantities of estradiol to estrone and postulated that this conversion may parallel the hormonal activity of the estrogen in the endometrium. These studies do not indicate the particular cell or cells involved in the transformation, but it should be borne in mind that uterine epithelium and glands form a significant part of endometrial tissue and are continuous with tubal epithelium.

Both uterine [20] and tubal [2] fluids have a pH more alkaline than that of blood peritoneal and vaginal secretions. The pH of uterine fluids of rats treated with estrogen and with estrogen and progesterone did not differ significantly pH 7.08 and pH 7.780 respectively. The pH of oviduct fluid in the rat was found to be 8.04. Spermatozoa survive best in an alkaline medium with a pH of 8–9.5, [18]. It is of interest to note that the activity of the dehydrogenases was histochemically demonstrable at these pH levels. (Only the oviducts are treated here, as studies on uterus and blastocysts are already published: *Lobel, B. L., E. Levy & M. C. Shelesnyak:* Acta Endocrin. (Kbh) **56,** (1967), Supp. 123).

References

[1] *Baillie, A. H., M. M. Ferguson & D. MckHart:* Developments in steroid histochemistry. Academic Press N. Y. and London 1966.

[2] *Blandau, R., L. Jensen & R. Rumery:* Fertil. & Steril. 9 (1958), 207.

[3] *Brinster, R. L.,* J. Exp. Zool. **158** (1965), 69.

[4] *Bryson, M. J. & M. L. Sweat:* Endocrinology **81** (1967), 729.

[5] *Chang, M. C.:* Endocrinology **79** (1966), 939.

[6] *Chang, M. C.:* Endocrinology **81** (1967), 1251.

[7] *Clyman, M. J.:* Fertil. & Steril. **17** (1966), 281.

[8] *Ferguson, M. M. & G. A. Christie:* J. Endocrinol. **38** (1967), 291.

[9] *Harrington, F. E.:* Endocrinology **77** (1965), 635.

[10] *Homberger, F., P. Bernfeld, A. Tregier, M. S. Grossman & P. Harper:* Ann. N. Y. Acad. Sci. **106** (Article 2) (1963), 683.

[11] *Jensen, E. V.:* Proc. Canad. Cancer Res. Conf. **6** (1965), 143.

[12] *Kim, M. H. & W. L. Hermann:* J. Clin. Endocrin. and Metab. **28** (1968), 187.

[13] *Kinson, J. A. & C. D. Kochakian:* Endocrinology **74** (1964), 233.

[14] *Leonard, S. L.:* Endocrinology **63** (1958), 853.

[15] *Levy, H., H. W. Deane & B. L. Rubin:* Endocrinology **65** (1959), 932.

[16] *Lobel, B. L., E. Levy, E. S. Kisch & M. C. Shelesnyak:* Acta Endocrin. (Kbh) **55** (1967), 451.

[17] *Lobel, B. L. & E. Levy:* Acta Endocrin. (Kbh) **59** (1968), Supp. 123, 8.

[18] *Mann, T.:* The Biochemistry of Semen, London, 1954, Methuen Co., Limited.

[19] *Marcus, S. L. & C. A. Saravis:* Fertility and Sterility **16** (1965), 785.

[20] *Marcus, L. M.:* Am. J. Obstet. Gynec. **19** (1965), 114.

[21] *Mastroianni, L. & J. Ehteshamzade:* J. Reprod. Fertil. **8** (1964), 145.

[22] *Suzuki, S. & L. Mastroianni:* Am. J. Obstet. Gynec. **93** (1965), 465.

[23] *Sweat, M. L., M. J. Bryson & R. E. Young:* Endocrinology **81** (1967), 167.

[24] *Velardo, J. T. & C. G. Rosa:* In: Handbuch der Histochemie Band VII/3 Gustav Fischer Verlag Stuttgart (1963), 45.

[25] *Weissmann, G.:* Fed. Proc. (1964), 1009.

Discussion

J. M. Bedford: Could I just ask very briefly – do you imply then that there are pH levels of 9 in the ovum?

B. L. Lobel: I do not imply that at all, I only say that enzymic activity was demonstrable at these pH levels. We have made no measurements of pH levels in the oviducts of the rat.

D. R. S. Kirby: The leucocyte that was attached to the blastocyst. Did it look as though the blastocyst was suffering as a result of this contact?

B. L. Lobel: No, it did not. The blastocyst itself was indistinguishable from other blastocysts which had no leucocytes attached to them.

K. Fotherby: We have been interested in the biochemical aspects of uterine sensitisation and have measured the activity of a number of enzymes in the rat uterus. I cannot discuss all of these because there is not the time, but if I could show one figure (Fig. IX). This shows the data on the phosphatases during the early stages of pregnancy in the rat; the most striking change is the ten-fold increase in alkaline phosphatase activity between days 6 and 8. There are other changes which we see occasionally, but this is by far the most marked finding.

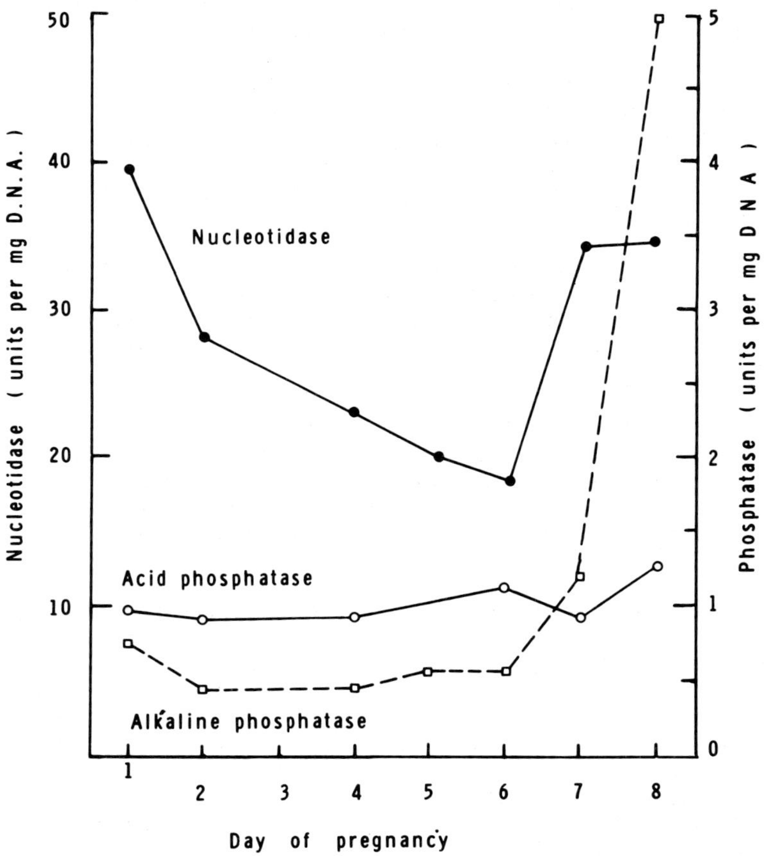

Fig. IX

I would like to make one comment about a number of the papers we ve heard today. People have been talking about the progestogenic phases and estrogenic phases; I think we now have data on the levels of progesterone and 20 α-di-hydroprogesterone in the rat and rabbit during early pregnancy. I think until we have this information we ought to be careful about using the words progestogenic phase and estrogenic phase.

R. L. Brinster: Dr. *Bedford,* did you ask about the pH in the rabbit oviduct?

J. M. Bedford: Yes – I thought there was an inference that was pertinent.

R. L. Brinster: I mentioned this in my presentation briefly. There have been a number of reports in the literature about pH in the Fallopian tube of various species. Yet, I feel that over the years there has been very little evidence, actually, to suggest that the pH is anything other than that in the blood plasma. The report you quoted by *Blandau, Jensen* and *Rummery* (1958) was a good piece of work, but still there is a lot of difficulty in doing pH measurements on small volumes of fluid. Some workers have tried to avoid the small volume problem by using fluid from ligated tubes. This fluid can hardly be expected to accurately represent the fluid in the normal tube. The work of *Hamner* and *Williams* (1965) in the rabbit and *Restall* (1966) in the sheep suggest a bicarbonate concentration in the tube of about 1.76 mg/ml which is not greatly different from that found in blood serum. If the pCO_2 is the same as venous blood, which we would expect it to be, then you would have a pH of 7.4. Work in my laboratory by Dr. *Brunton* now shows, contrary to some of the literature which suggests bicarbonate is secreted by the tube, that bicarbonate is really not the major ion secreted into the lumen of the rabbit Fallopian tube. This paper should be appearing in a period of about o or 8 months. Therefore I think that there is very little evidence right now that the pH is anything other than that in the bloodstream.

H. R. Lindner: I meant only to comment on Dr. *Fotherby's* figure (Fig. IX). It is in agreement with experience in our laboratory. – the most striking change demonstrable histochemically in the uterus during the pre-implantation phase of pregnancy is this rise in alkaline phosphatase activity, and the interesting feature is that this is strictly confined to focal areas where decidualization is to take place.

K. Fotherby: To the placental sites.

H. R. Lindner: Yes. Now, with the kind permission of the Chairman, I intended to provide some light entertainment by showing a series of slides illustrating the morphological features of blastocyst attachment and trophoblastic invasion by electron microscopy, almost as a film, with minimal comment (see *Sumie Tachi, C. Tachi* and *H. R. Lindner,* J. Reprod. Fertil., in press, for reproductions).

One point shown in these slides might be relevant to Dr. *Kirby's* interesting presentation. We paid particular attention to the mode of contact of the trophoblast with endometrial cells, to find out whether the electron microscope provided any indication, at these early phases of pregnancy, of a separation of fetal and maternal elements by a barrier akin to the periplacental fibrinoid demonstrated by Dr. *Kirby's* group at later stages of pregnancy in the mouse. No such evidence was obtained. During the early invasion phase, trophoblast cells were seen in direct contact, and in fact forming tight unctions, with maternal epithelial cells (see Fig. X a, taken at midnight on the sixth day postcoitum). Following disintegration of the basement membrane of the uterine epithelium, trophablast cells were seen in intimate contact with stromal elements (see Fig. X b, taken on the afternoon of day 8 p.c), with their cell membranes in close apposition (within 100–200 A). Thus there appears to be at least a transient phase when no physical barrier between trophoblast and maternal tissues is evident that might prevent access of fetal antigens to maternal lymphoid cells. Of course, such a barrier may develop later. It is conceivable that an initial immunological interaction between blastocyst and maternal tissues may have a subsidiary role in decidual induction. At the same time Reichert's membrane, which lines the inner aspect of the trophblast, increases markedly in thickness on day 8 p.c. (Fig. XI, a–b), but this structure does not surround the whole of the embryo.

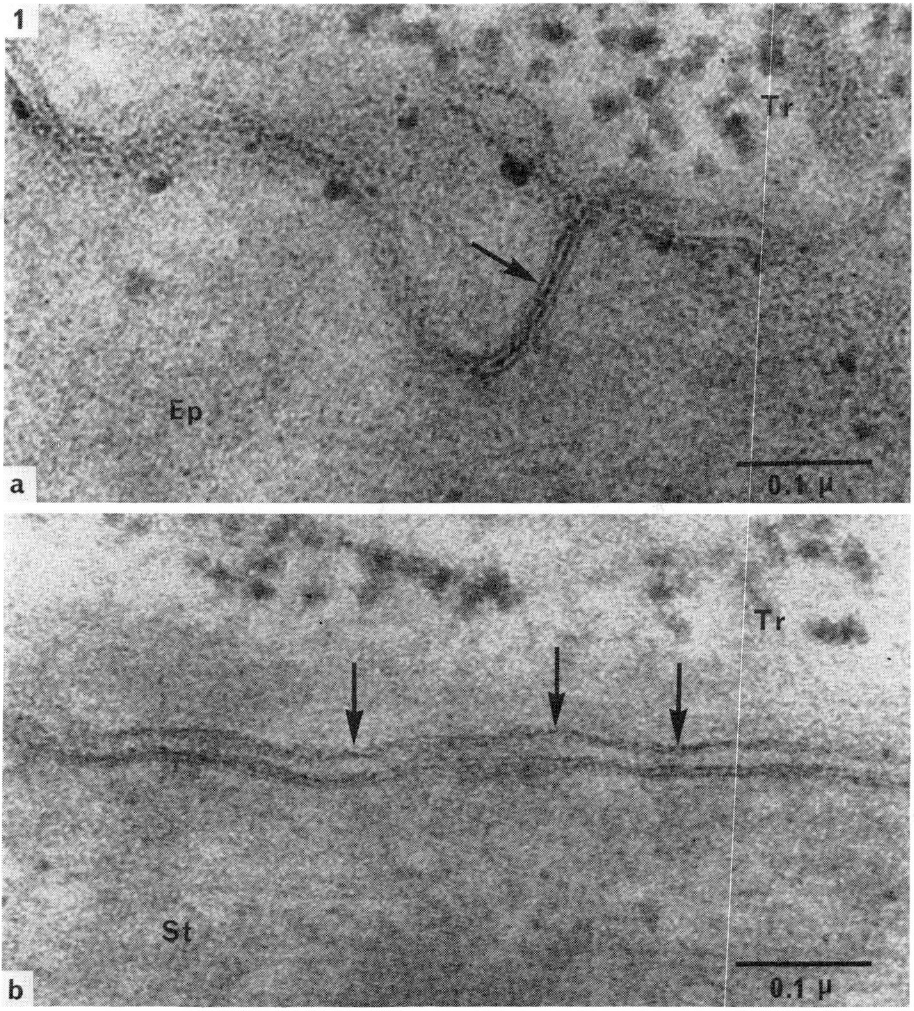

Fig. X

D. R. Kirby: I darenot return of Oxford if they thought I would sold out on fibrinoid. I am utterly convinced that this is an important barrier and your findings that it is not present early in pregnancy do, of course, fit in perfectly with our theory of its development and function.

R. G. Edwards: We have heard a lot about leukocytes in the uterus, especially those that phagocytose spermatozoa. A recent paper claims that the origin of these cells is actually in the uterine wall, and not the same as those circulating in the serum. Would this explain why you seemed to find no phagocytosis in certain systems, e.g. where you get sperms in the uterus over the winter in bats?

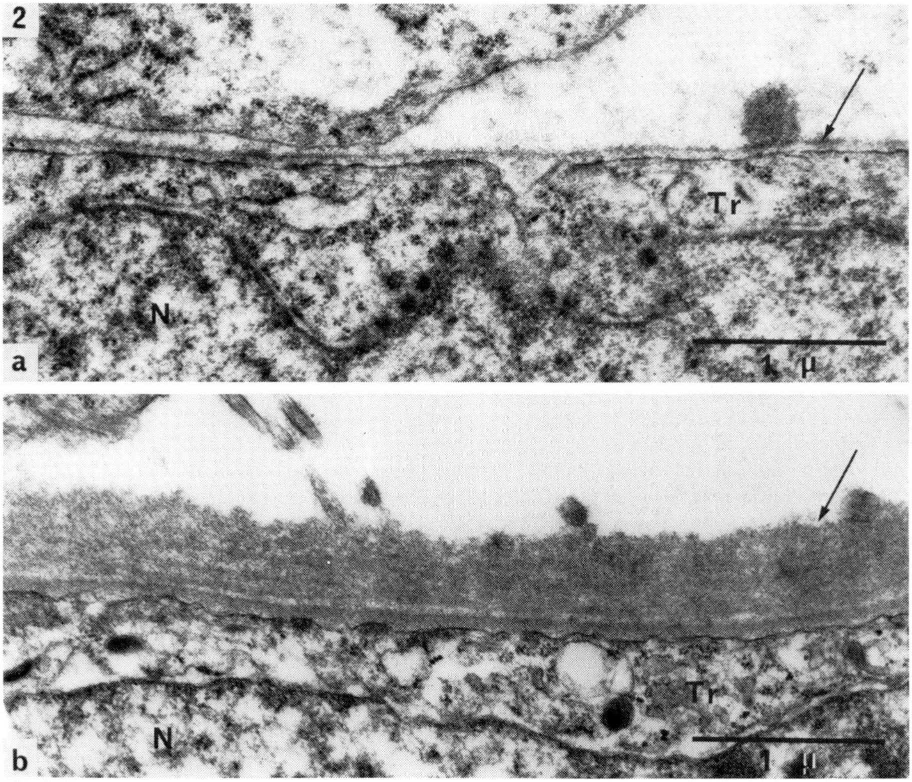

Fig. XI

B. L. Lobel: I do not know in what species these results were obtained but I would certainly not think that these were true for the rat. On the contrary, I think we have shown with autoradiographic studies that the leukocytes infiltrating the wall of the uterus definitely come from the peripheral blood stream and, of course, from the bone marrow, and are not produced within the uterus itself. I do not know in what species the results you refer to were obtained, but certainly this is true for the rat.

R. J. Ericsson: To comment on leukocytes – I think the paper that Dr. *Edwards* referred to was done in the rat. The rat has a major fluctuation of the eosinophilic cells from around 150 000, in diestrus to perhaps around 8 million in estrus. The question is, did you type the leukocytes?

B. L. Lobel: During the cycle or during pregnancy?

R. J. Ericsson: During your study did you make attempts to distinguish whether they were neutrophils or eosinophils?

B. L. Lobel: Oh! Yes, of course, you see a progression during progestation. On the first day, on day 0 or day 1, whichever terminology you use, you find a mixture of leucocytic cells, but mainly neutrophilic leukocytes; subsequently you get the eosinophils, and then towards the end of the progestational period and at the time of implantation you get a lot of lymphocytes. And you can see this very well just by studying plain sections. You do not even need any sophisticated techniques for it. But in the rat this is very obvious. There is a changeover in the type of population of migratory cells during the progestational period. But the eosinophils – and this has been shown by many workers, especially Swedish workers, do invade the uterus of the rat in response to estrogen.

A. Psychoyos: You demonstrated so beautifully that there is a spreading of the thymidine uptake in the stroma cells during decidualisation. I want to ask you if you think this spreading proves that the message for decidualisation is passing from one cell to the other, and if you have any indication, especially from your studies concerning enzymes, to prove that there is a gradient in the transformation. A diffusion of the information between cells could probably explain the metamorphosis of the whole endometrium we obtain by the intraluminal injection of RNA.

B. L. Lobel: Proof, I do not know. But observations, yes. I would say from our observations, and we have looked at a lot of uteri, that it appears that the reaction definately spreads outwards – or inwards, whichever way you are looking. If you are looking from the lumen, it is from the lumen outwards into the endometrium. If you have enough rats, and use enough thymidine and you kill them at specific intervals – you cannot of course do it in the same rat, because you cannot kill the same rat twice – but if you use enough animals you can see that the reaction does spread outwards. This was also shown in the figures. First the uptake of the tritiated thymidine, in other words, first the cells proliferate, and then only do you get enzymic activity in these cells. The activity definitely spreads outwards, and this is true for all the enzymes that we looked at so far.

The Effect of Estrogens and Related Substances on Embryonic Viability

R. P. Blye

Ortho Research Foundation, Raritan, New Jersey, USA

Summary: The mamalian embryo is exquisitely sensitive to alterations in the normal balance of estrogen and progesterone particularly prior to and during the process of nidation. Evidence is lacking for a direct embryotoxic action for estrogen or progesterone. The presence of zygotes in the reproductive tract at a location abnormal for their age and development, a condition found following estrogen administration during early pregnancy in a number of species, rapidly reduces the chance for embryonic survival.

Interference with the estrogen/progesterone ratio required for the initiation of nidation precludes the continuance of normal pregnancy although blastocyst survival *in utero* may continue for extended periods.

Most of the non-steroidal estrogens, estrogen-antagonists or their analogs appear to alter the rate of zygote transport or interfere with the process of implantation. A notable exception may be Fenestrel, a member of a series of cyclohexenecarboxylic acids, which affects lysis of the zygote in the oviduct or interferes with chorioallantoic placentation in the rat depending upon the time of medication.

A number of factors control pre- and post-nidatory embryonic viability and development in mammalian species. Intrinsically a delicate balance of estrogen and progesterone is responsible for the production of the proper *milieu interieur* within the reproductive tract and also for regulation of the passage of the zygote from the ampulla of the oviduct to the site of implantation in the uterus. Even after nidation occurs the continuance of pregnancy is dependent upon these same sex hormones whether produced by the gonads or the placenta. Although the ordinary sequence of events which contribute to successful gestation may be altered by a variety of factors, I should like to direct our attention to the influence of estrogens and related substances on embryonic viability.

The ability of estrogens to interrupt pregnancy was known as early as 1926 [46, 58]. It remained for *Burdick* and his coworkers [4, 47, 63, 64] to demonstrate that the disruptive effect of estrogens on early pregnancy was associated with either prolonged retention of zygotes in the oviduct (tube-locking) or accelerated transport whereby they were rapidly expelled from the reproductive tract, both of which were dose

Manuscript received: 2 May 1969

and species dependent. The findings of these investigators remained unchallenged for the next 20 years until *Greenwald* [26, 27, 29] showed that small doses of estrogen in the rabbit expelled zygotes while large doses caused tubal retention. No dose level of estrogen could be found which blocked zygote transport in the rat: effective doses always caused rapid flushing of zygotes from the reproductive tract. [28, 31]. *Harper* [34] recently reported that interruption of pregnancy in the rat following oral administration of dienestrol on any of the first 3 days of gestation was due to the expulsion of zygotes from the reproductive tract by day 5. Treatment on day 4 caused the flushing of some embryos from the tract by day 5, but also affected the position of the blastocysts within the uterine lumen and prevented the onset of the decidual reaction.

The effect of a single injection of estradiol cyclopentyl propionate on zygote transport in several species is summarized in Fig. 1 (date of *Greenwald* [31]). It is quite clear that early pregnancy is particularly vulnerable to exogenous estrogen. Man may be an exception. Notwithstanding the report of *Morris* and *Van Wagenen* [43], where

	DOSE OF ESTROGEN TO *		
SPECIES	ACCELERATE	RETAIN	INTERRUPT PREGNANCY (ED_{80})
GUINEA PIG	50-100	250	10
HAMSTER	100	250	25
MOUSE	≥1	1	1
RABBIT	25	100	50
RAT	≥10	NO DOSE	10

Fig. 1. Effect of estrogen on zygote transport in several species.
μg of estradiol cyclopentylpropionate (Data of *Greenwald, G. S.* 1967 Anat. Rec. **157,** 163–172).

no pregnancies followed the administration of high doses of stilbestrol or ethinyl estradiol 4–6 days after coitus and vague claims, largely from the college community, of the ability of estrogen-containing oral contraceptives to thwart pregnancy when ingested in large quantities after exposure, it has been shown that human pregnancy is remarkably tolerant to massive doses of estrogen (see [2] for review). I do not wish to dwell on the subject here as it has been adequately covered by previous speakers at this symposium. However, interference with zygote transport does appear to be a major, if not the sole, means by which most non-steroidal estrogens and related substances interrupt pregnancy. Loss of embryonic competence by zygotes placed prematurely in the uterus has been clearly demonstrated [45].

Let us turn now to some of the substances which are more or less related to estrogens. Chlorotrianisene or TACE was synthesized as a nonsteroidal estrogen at the laboratories of *Wm. S. Merrell.* Subsequently, a number of analoges were prepared and studied for antifertility properties. The first of these was called ethamoxytriphetol or simply MER-25 (Fig. 2). *Segal* and *Nelson* [51] reported that the compound interrupted early pregnancy in the rat and *Chang* [5] confirmed these observations and found similar activity in the rabbit. Whereas these workers concluded

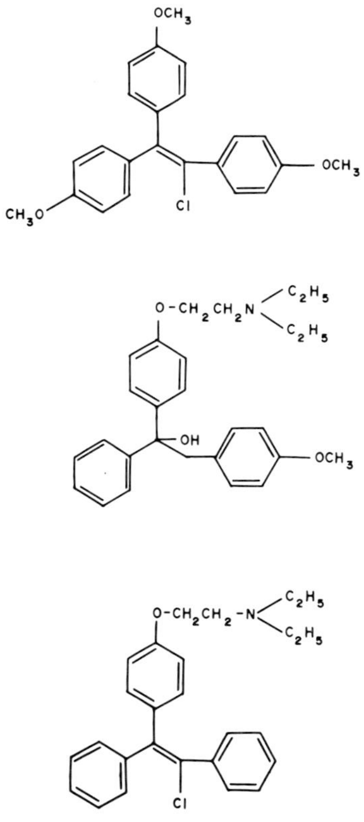

CHLOROTRIANISENE (TACE)

ETHAMOXYTRIPHETOL (MER-25)

CHLOMIPHENE (MRL-41)

Fig. 2

that MER-25 effected degeneration of the fertilized ova, it would appear that this was the result of altered zygote transport rather than due to any specific antizygotic action since MER-25 is clearly an estrogen antagonist and also exhibits limited estrogenic activity [40]. The next analog prepared at the Merrell laboratories was clomiphene or MRL-41. This compound was anti-gonadotropic, anti-estrogenic and estrogenic [35] and has been extensively investigated as an antifertility agent [1, 8, 9, 48,

52, 54]. While MRL-41 was most effective in interrupting pregnancy when administered during the period of zygote transport, it became clear that there was little evidence for a direct cytotoxic effect on the embryo [6, 49, 55, 60]. Most of the degenerating zygotes observed in these studies could be attributed to their presence in the reproductive tract at a location abnormal for their age and development.

A number of compounds closely related to clomiphene have been prepared by I.C.I. Perhaps the most interesting was a pair of cis-trans isomers, denoted by the numbers I.C.I. 46474 (trans) and I.C.I. 47699 (cis), whose structures are shown in Fig. 3. The cis form exhibited conventional estrogenic activity which probably accounted for

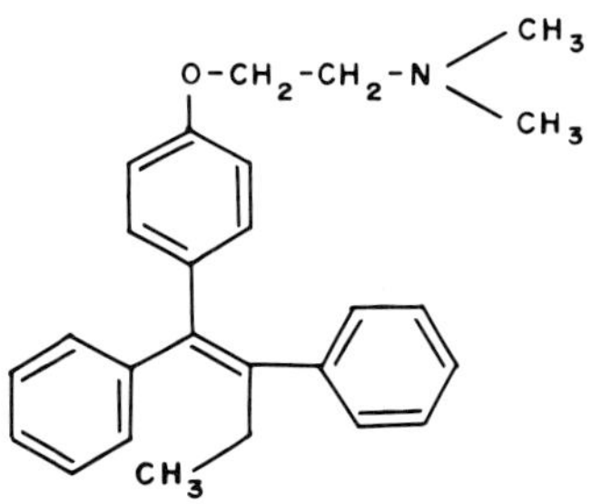

I.C.I. 46474 (trans)
I.C.I. 47699 (cis)

Fig. 3

its ability to terminate early pregnancy in rats [32]. On the other hand the trans form was only weakly estrogenic but also anti-estrogenic, and far more potent in interfering with gestation than the cis isomer. Nonetheless, a dose level sufficient to prevent implantation was found to accelerate passage of the zygotes through the oviduct when administered on the 1st or 2nd day of gestation but not when given on day 3 or 4 [32]. A direct effect on the blastocysts appeared to be ruled out by the successful implantation of embryos from rats treated on day 4 following administration of sufficient estrogen within seven days or after transfer to pseudopregnant hosts on day 5 [33, 61].

A series of dihydronaphthalenes and 2,3-diphenylindenes have been synthesized at the Upjohn Laboratories and examined for antifertility activity [36–39]. Fig. 4 shows the chemical structure of two of these compounds which have been studied most extensively. Their similarity to clomiphene is obvious. *Duncan* and his associates [10] found the indente derivative, U-11555A, effective in inhibiting pregnancy in rats when administered on any of the first four days following mating and suggested that it had a direct cytotoxic effect on the zygote. This drug was also found to possess atypical estrogenic activity and rather substantial anti-estrogenic potency.

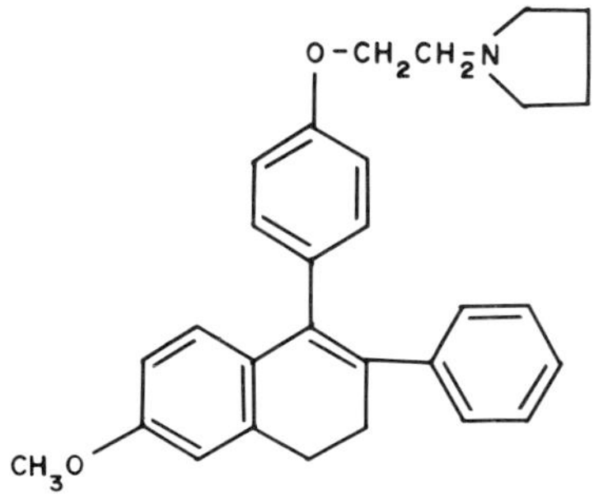

NAFOXIDINE (U-11100 A)

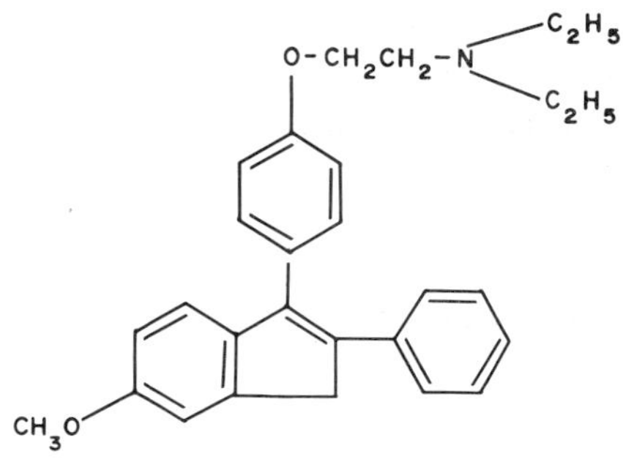

(U-11555 A)

Fig. 4

The latter probably accounts for the finding that U-11555A could prevent estrone-induced nidation in rats in which implantation was experimentally delayed although no deleterious effect on the blastocyst could be demonstrated [12, 50]. Additional studies in the rat [65], mouse [60], and rabbit [6] indicated that the primary anti-fertility activity of U-11555A was due to altered tubal transport of zygotes and not to a cytotoxic effect on the embryo. The dihydronaphthalene, U-11100A, exhibited biological properties similar but not identical to those of U-11555A [11, 14, 23, 24]. *Chang* and *Yanagimachi* [7] found that U-11100A altered the normal rate of tubal transport of zygotes in the rabbit and *Greenwald* [30] concluded that the drug was not embryocidal in the rat but caused acceleration of fertilized ova transport and interfered with decidualization.

Certain members of a series of cyclohexylidenemethanes synthesized by the Ferrosan laboratories [42] are also structurally related to TACE and have received considerable attention as potential abortifacients. Two of the more active compounds, F-6066 and F-6103 (Fig. 5) have been extensively reviewed by *Einer-Jensen* [19]. Both compounds were found to be weakly estrogenic and to antagonize the effect of progesterone in the Clauberg rabbit test with F-6103 appearing some ten times more active than F-6066. Pregnancy could be interrupted at almost any time in mice, rats and rabbits following subcutaneous injection for at least two days regardless of when treatment was initiated. The localization of F-6066 in active corpora lutea and the antiprogestational activity of these drugs would seem to rule out a direct embryotoxic action in the species studied.

Departing now from substances related to TACE or clomiphene, a series of oxazolidinethiones was synthesized at the Upjohn laboratories several of which terminated early pregnancy in rats [66]. The most active compound of the series, U-11634, (Fig. 6) was found to be a unique inhibitor of decidual growth in that it did so in the absence of demonstrable sex hormone or anti-sex hormone activity [15–18].

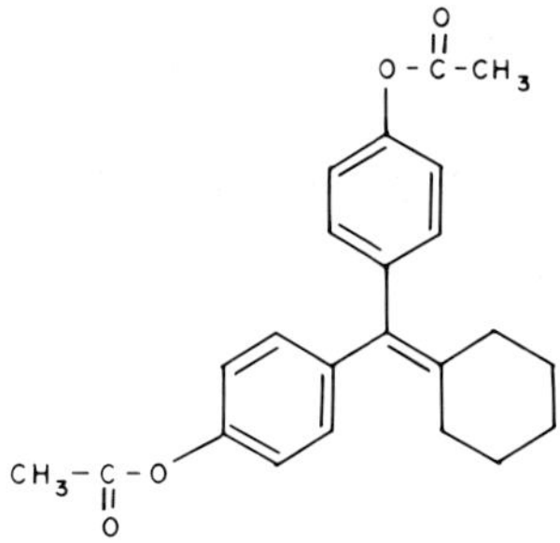

BIS-(p-ACETOXYPHENYL)
CYCLOHEXYLIDENEMETHANE
(F-6066)

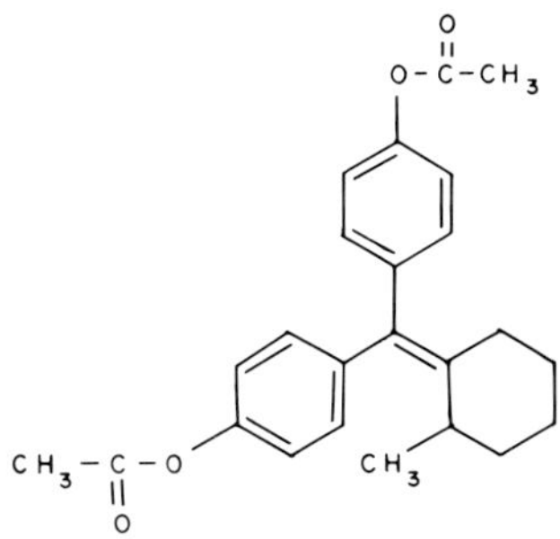

BIS-(p-ACETOXYPHENYL)-2-METHYL-
CYCLOHEXYLIDENEMETHANE
(F-6103)

Fig. 5

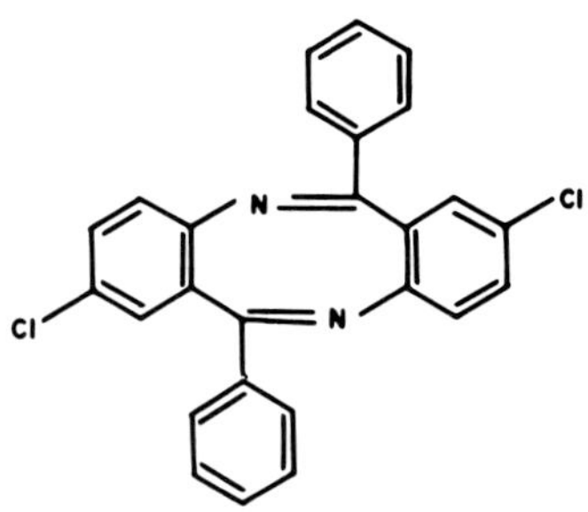

(U-10293)

Fig. 6

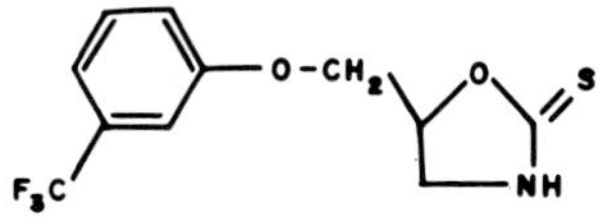

(U-11634)

Although the drug was found to exhibit antithyroid properties in female rats [62], no blastolytic or embryocidal activity has been reported. Upjohn workers have also prepared a diazocine compound known as U-10293 which terminated pregnancy when administered on any day prior to nidation [13]. U-10293 was found to exhibit many of the biological properties of classical estrogens and thus its ability to prevent nidation was not unexpected.

Modification of the basic stilbene molecule has led to a number of interesting compounds which exhibit varying degrees of estrogenic, anti-estrogenic and antifertility activity [20–22, 41, 55, 57]. Saturation of the olefinic bond of a stilbene derivative yields the corresponding dihydroxybibenzyl compounds for which a number of isomeric forms are possible depending upon the stilbene precursor (Fig. 7). *Emmens*

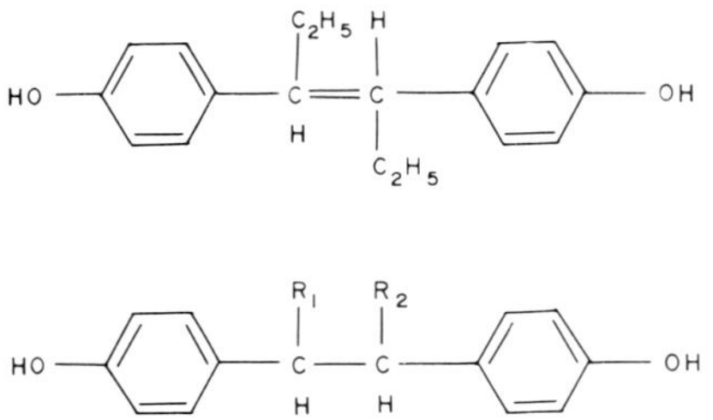

trans-DIETHYL STILBESTROL

SUBSTITUTED 4,4' DIHYDROXYBIBENZYLS

Fig. 7

and his associates [25] prepared a number of 4-4′-dihydroxybibenzyls and reported that the erythro form of the methyl-ethyl (erythro-MEA) exhibited the greatest separation of estrogenic and antifertility activity. These authors suggested that the high anti-estrogenic potency of erythro-MEA may account for its ability to interfere with gestation when administered prior to implantation which, if true, would militate against a direct blastocidal action for these substances.

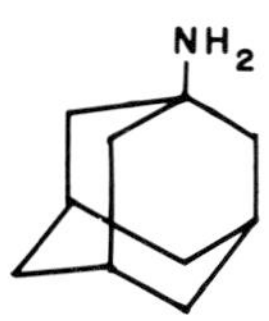

AMANTADINE

Fig. 8

More recently, *Soupart* [59] found that amantadine hydrochloride (Symmetrel, E. I. DuPont de Nemours & Co., Fig. 8) had a direct effect on developing embryos of the sea urchin *Arbacia punctualata* where cleavage and DNA synthesis were inhibited.

Several years ago the synthesis of a series of cyclohexenecarboxylic acids was initiated at the Ortho Research laboratories in the search for new non-steroidal estrogens. One member of this series, Fenestrel (Fig. 9), was found to exhibit potent antizygotic activity in the rat [3] and interrupted pregnancy in the rabbit and monkey [43, 44]. The compound was estrogenic in the rat, mouse and rabbit and exhibited respectively 2.4, 0.68 and about 7 times the oral estrogenic potency of ethynylestradiol in these three species.

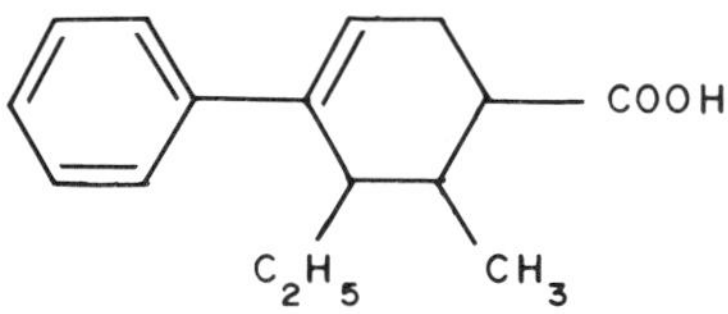

FENESTREL (ORF – 3858)

Fig. 9

Adult male rats of a Wistar-derived strain were caged overnight with proestrus females. The appearance of sperm in vaginal washings the following morning was considered evidence of positive mating and constituted day 0 of pregnancy. This numbering system was used in view of the fact that pregnancy was less than 12 hours old at any time prior to noon on the day when vaginal sperm were found. Fenestrel or ethynyl estradiol was dissolved in sesame oil and administered by gavage at various periods during pregnancy. Rats were examined on days 9–15 for the appearance of implantation sites or were sacrificed earlier when the reproductive tract could be flushed with saline to recover zygotes.

DOSE (μg / kg)	TREATMENT ROUTE	TREATMENT DURATION[a] (DAYS)	NO. RATS PREG./TOTAL	NUMBER IMPLANTATION SITES NORMAL	RESORBING	TOTAL
VEHICLE	I.G.	0,1	22/26	247	0	247
10	I.G.	-1,0	3/5	10	1	11
10	I.G.	0	3/3	26	0	26
10	I.G.	0,1	0/15	0	0	0
10	I.G.	1,2	1/5	0	1	1
10	I.G.	2,3	3/5	0	25	25
10	I.G.	3	3/5	5	13	18
10	I.G.	3,4	5/5	12	33	45
10	I.G.	4	4/5	50	0	50
10	I.G.	5-9	4/5	28	0	28
25	I.G.	-1	4/6	21	0	21
25	I.G.	0	2/12	3	2	5
25	I.G.	0-3	0/4	0	0	0
25	I.G.	1	0/5	0	0	0
25	I.G.	2	1/5	9	0	9
25	I.G.	3	5/5	27	33	60
25	I.G.	4	4/5	57	0	57

Fig. 10. Effect of ORF–3858 on early pregnancy in the rat.

a) Day 0 = Day sperm observed in vaginal washings.
 Day 1 = Day prior to day 0.

The data summarized in Fig. 10 indicate that Fenestrel was most effective in terminating pregnancy when administered prior to day 3 regardless of whether a single or 2-day administration period was used. The drug appeared to become less effective as medication was instituted progressively later in gestation. The most efficacious treatment was 10 μg/kg administered on each of the first two days of pregnancy.

DOSE (μg /kg)	TREATMENT[a] ROUTE	DURATION DAYS	NO. RATS PREG./TOTAL	NUMBER NORMAL	IMPLANTATION RESORBING	SITES TOTAL
CONTROL	I.G.	0,1	22/26	247	0	247
2	I.G.	0,1	5/5	64	0	64
4	I.G.	0,1	5/5	55	0	55
5	I.G.	0,1	3/5	21	0	21
8	I.G.	0,1	1/5	5	0	5
10	I.G.	0,1	0/15	0	0	0
25	I.G.	0,1	0/6	0	0	0
2.5	S.C.	0,1	4/5	39	0	39
5.0	S.C.	0,1	3/5	16	12	28
10.0	S.C.	0,1	1/5	0	1	1
CONTROL	I.G.	0	10/10	128	10	138
8	I.G.	0	10/10	72	8	80
16	I.G.	0	4/10	21	0	21
32	I.G.	0	1/10	2	0	2

Fig. 11. Effect of ORF–3858 on early pregnancy in the rat.
a) Day 0 = Day sperm observed in vaginal washings.

Fig. 11 shows the relationship between dose and pregnancy interruption following oral and subcutaneous administration on these two days and after oral medication on day 0 only. Some of this data is shown graphically in Fig. 12 which illustrates a very steep dose-response curve.

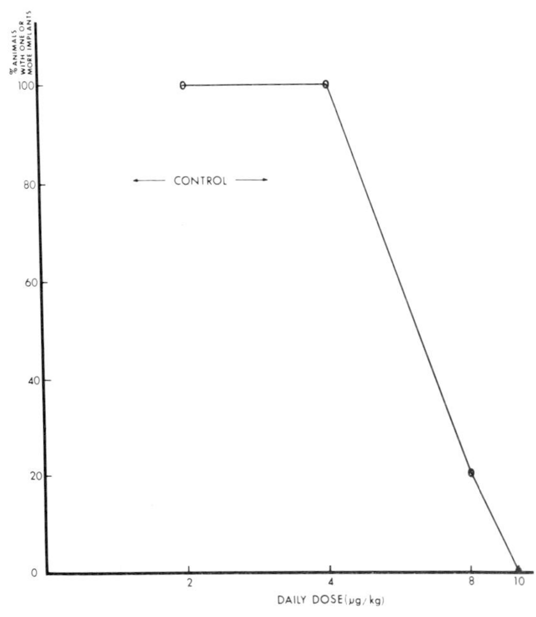

Fig. 12
Antizygotic Activity of ORF 3858 in the rat after oral administration on days 0 + 1 of gestation.

Ethynyl estradiol failed to alter the course of pregnancy when administered on each of the first two days of gestation except at dose levels above 100 μg/kg (Fig. 13). Even at a dose level of 500 μg/kg, some 20 times the equivalent estrogenic dose for

DOSE (ug/kg)	ROUTE	DURATION[a] (DAYS)	NO. RATS PREG./TOTAL	NUMBER NORMAL	IMPLANTATION RESORBING	SITES TOTAL
CONTROL	I.G.	0,1	22/26	247	0	247
5	I.G.	0,1	5/5	68	0	68
25	I.G.	0,1	4/4	40	0	40
50	I.G.	0,1	4/5	35	0	35
100	I.G.	0,1	5/5	66	0	66
200	I.G.	0,1	1/5	4	0	4
250	I.G.	0,1	1/5	0	3	3
300	I.G.	0,1	1/4	0	3	3
500	I.G.	0,1	5/10	17	3	20

Fig. 13. Effect of ethynyl estradiol on early gestation in the rat.
a) Day 0 = Day sperm observed in vaginal washings.

complete interruption of pregnancy with Fenestrel, normal nidated conceptuses were found. This would tend to indicate that the absence of implantation following administration of Fenestrel early in pregnancy was not due to the estrogenic potential of the drug. This was substantiated by experiments in which the reproductive tract was flushed with saline following medication with 10 μg/kg on days 0 and 1 of pregnancy (Figs. 14–21). The earliest abnormalities in zygotes from treated rats were seen at 60 hours (post-fertilization) and consisted of small indentations in the surface of the blastomeres (Fig. 17). The drug may interfere with cleavage mechanisms at this time for the abstriction of small bodies, such as that appearing in Fig. 17, was observed in several embryos. Defects in blastomere surfaces always heralded lytic destruction of the zygote. This first appeared as small areas of liquification in the cytoplasm and was frequently observed in one blastomere before the other(s) (Fig. 18). Indentation of the blastomere surface of a 4-cell zygote is shown in Fig. 19. Fig. 20 shows a 4-cell zygote, the cytoplasm and cell walls of which have undergone extensive lysis. Curiously, the nucleus of each blastomere appears morphologically intact. Apparently the cytolyzed contents of the zona pellucida cause its rupture (Fig. 21) and numerous empty zona were encountered at 84 hours.

Fenestrel slightly hastened the transport of zygotes through the oviduct at the dose level used to induce lytic destruction (Fig. 22). However, normal embryos were never recovered in flushings beyond 60 hours. Considering the ultimate demise of the zygote, it is not surprising that the reproductive tract contained little evidence of conception by the fourth day of gestation.

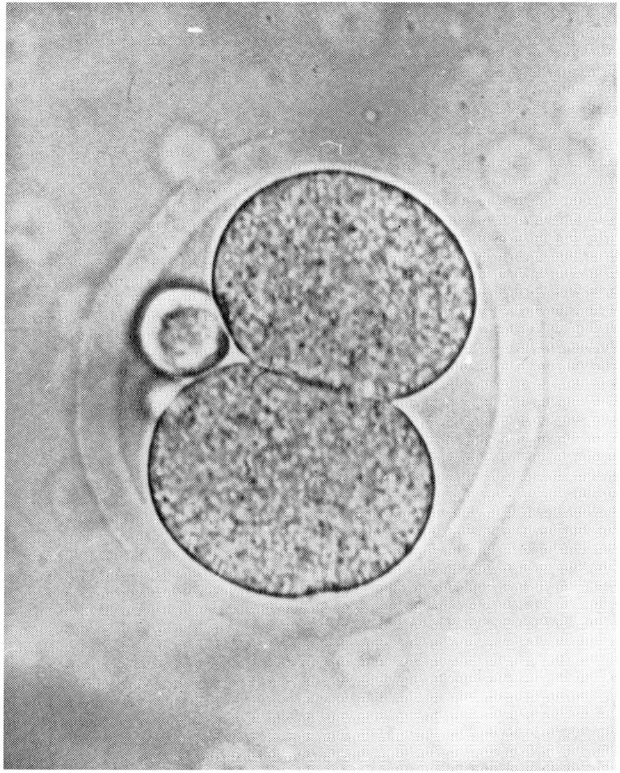

Fig. 14
Control, 2-cell rat zygote photographed directly from tubal washings obtained approximately 60 hours after ovulation. Approx. 695 X

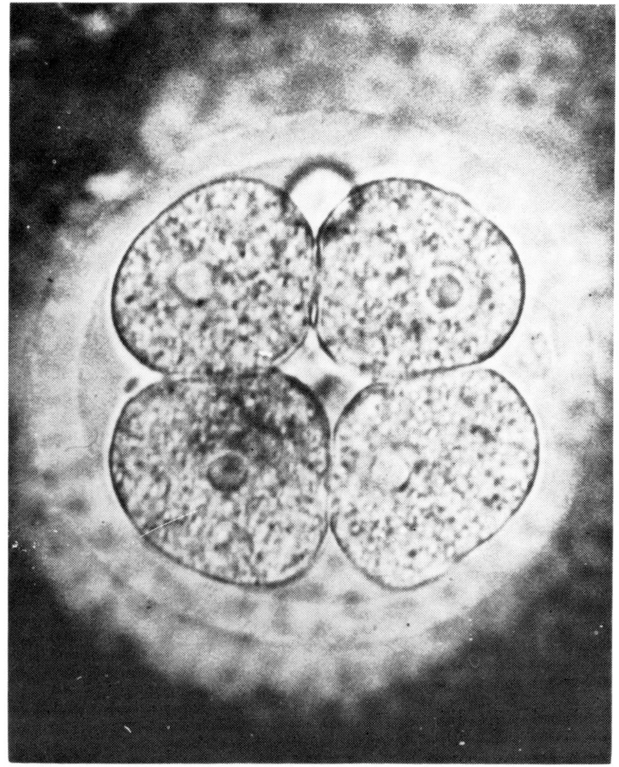

Fig. 15
Control, 4-cell rat zygote photographed as described for Fig. 14 Two-, three- and four-cell zygotes were encountered 60 hours after ovulation. Approx. 695 X

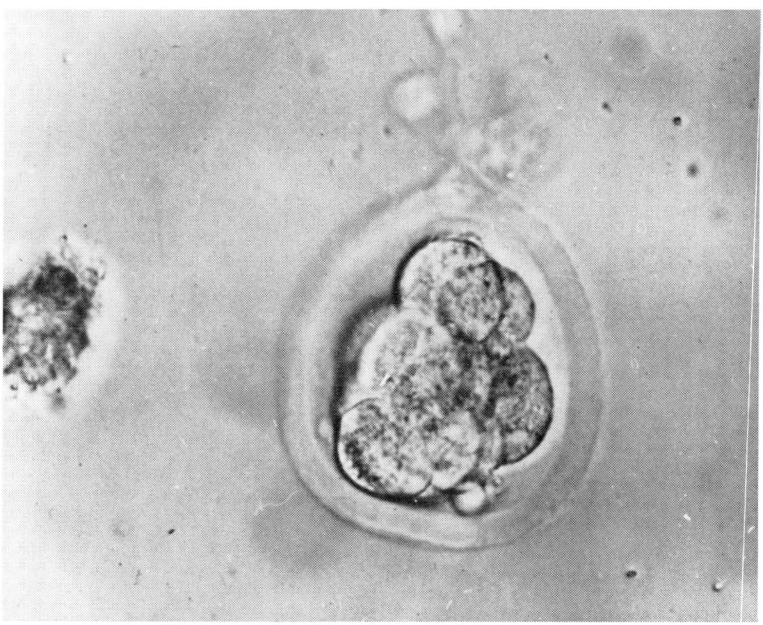

Fig. 16
Control, 8-cell rat zygote photographed directly from tubal washings obtained approximately 84 hours after ovulation. Approx.695 ×

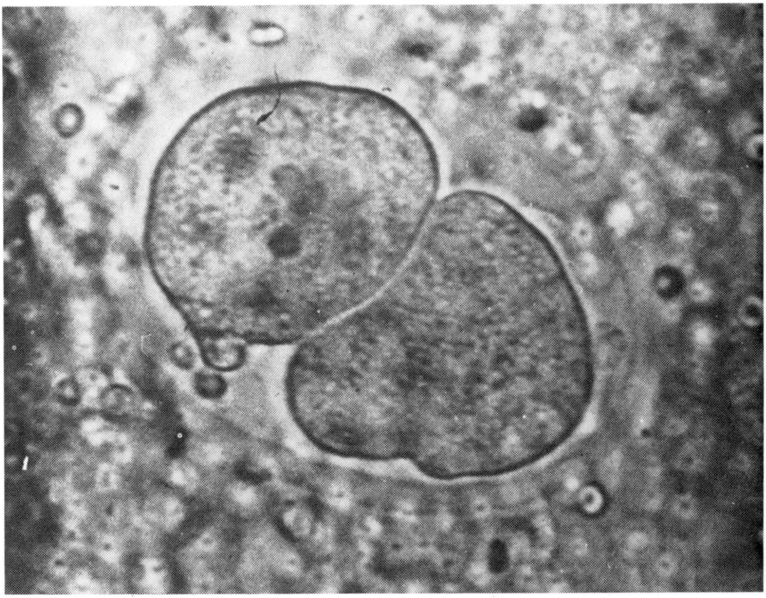

Fig. 17
Two-cell zygote obtained directly from tubal washings approximately 60 hours after ovulation from a rat which had received 10 μg/kg. Fenestrel on the first two days of pregnancy. Note irregular outline of blastomeres and abstriction of small body. Approx.695 ×

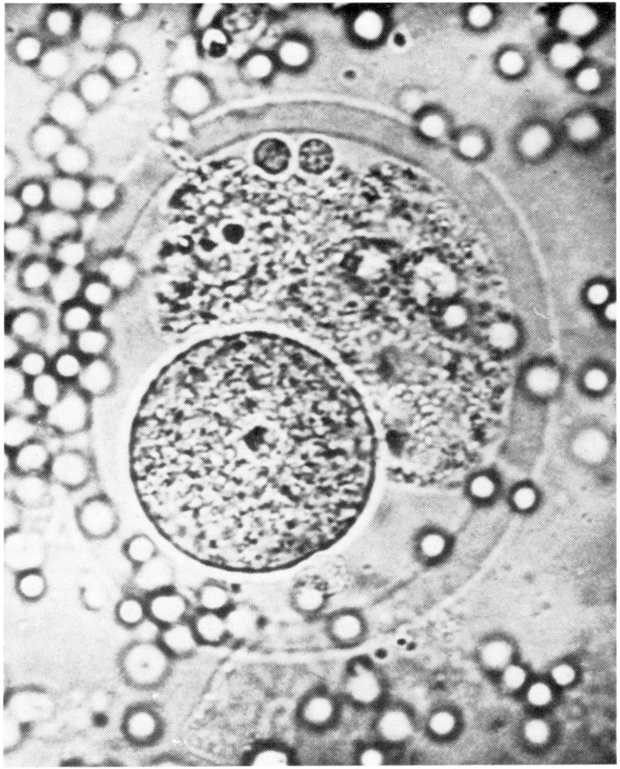

Fig. 18

Two-cell rat zygote obtained as described for Fig. 17. Note the lysis of one blastomere before the other. Approx. 695 ×

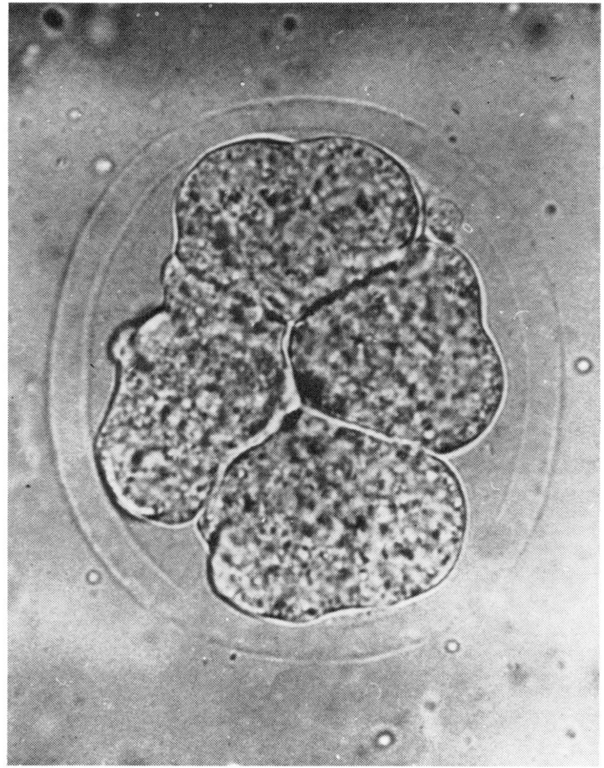

Fig. 19

Four-cell rat zygote obtained as described for Fig. 17. The irregular outline of the blastomeres so prominent in this photograph always preceded lytic degeneration. Approx. 695 ×

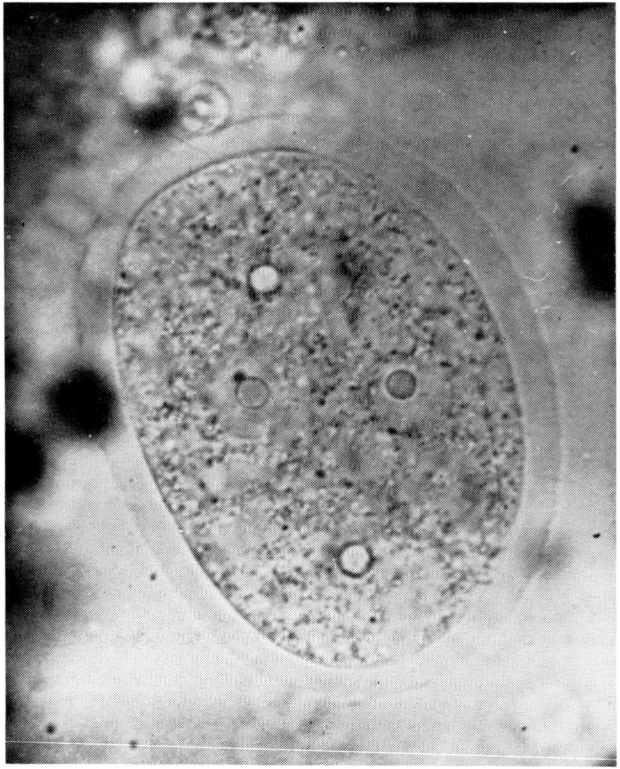

Fig. 20

Four-cell rat zygote obtained as described for Fig. 17. showing extensive areas of cytolysis and complete dissolution of the blastomere walls. Nuclei appear intact. Approx. 695 ×

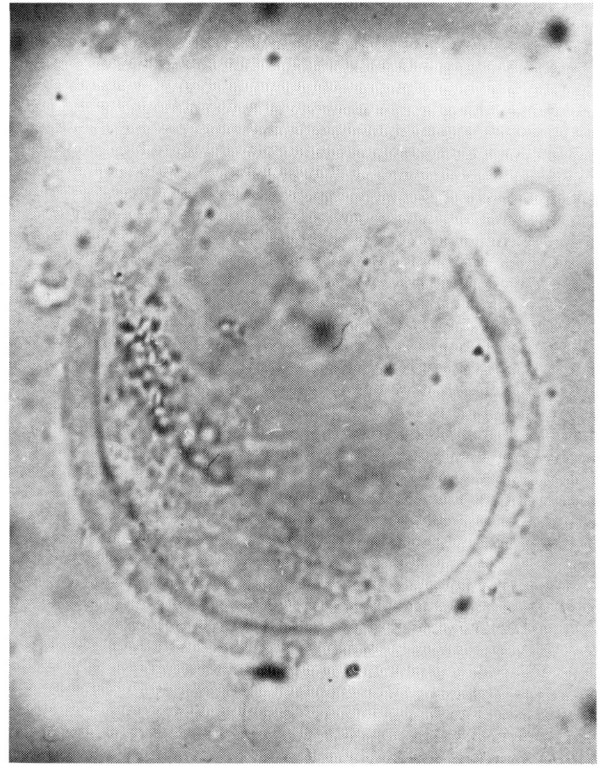

Fig. 21

Ruptured zona pellucida obtained from tubal washings 84 hours after ovulation from a rat which had received 10 μg/kg. Fenestrel on the first two days of pregnancy. Empty zona were frequently seen at this time and presumably represent the ultimate demise of the zygote following maternal medication. Approx. 695 ×

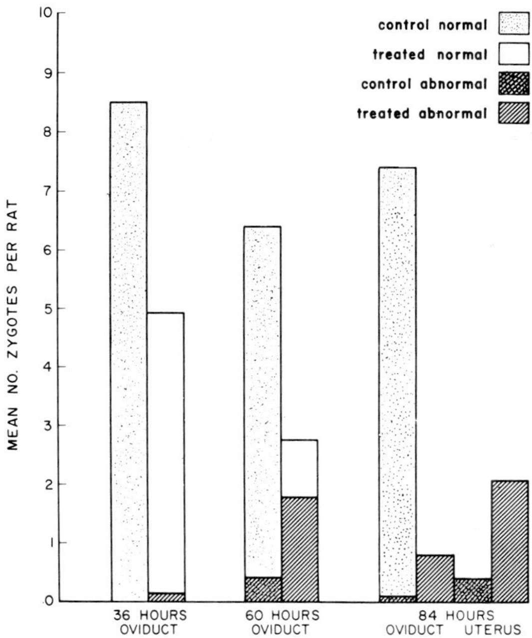

Fig. 22. Number and position of rat zygotes following oral administration of ORF−3858 on the 1st two days of pregnancy.

DOSE (µg/kg)	TREATMENT DURATION (DAY)	ROUTE	DAY[a] OF AUTOPSY	NO. RATS PREG./TOTAL	NUMBER OF IMPLANTATIONS NORMAL	RESORBING[b]	TOTAL
CONTROL	0	I.G.	9-10	10/10	128	10	138
25	0	I.G.	9	0/5	0	0	0
25	1	I.G.	8-9	0/5	0	0	0
25	2	I.G.	9	1/5	9	0	9
25	3	I.G.	8-11	5/5	27	33	60
25	4	I.G.	9-10	4/5	57	0	57
25	3	I.G.	15	3/5	11	17	28
25	4	I.G.	14	4/5	2	46	48
25	5	I.G.	15	2/4	16	9	25
25	7	I.G.	15	3/5	21	7	28
25	9	I.G.	15	4/5	46	0	46
100	3	I.G.	14-15	2/10	0	8	8
100	4	I.G.	13-15	3/5	0	30	30
100	5	I.G.	15	4/4	4	19	23
100	7	I.G.	15	5/5	31	27	58
100	9	I.G.	15	5/5	14	42	56

Fig. 23. Effect of ORF−3858 on Placental Development in the rat.

a) Day 0 = Day sperm observed in vaginal washings.

b) Resorbing = Resorbing and / or evidence of placental hemorrhage.

The sensitivity of the rat zygote to lytic destruction by Fenestrel appeared to end abruptly at about day 2. Administration of the drug on day 3 or later failed to interfere with implantation and normal numbers of implantations could ususally be found (Fig. 23). However, when rats medicated with a single oral dose of 25 μg/kg on day 3 or 4 were autopsied about 10 days later most of the conceptuses were undergoing resorption. A dose of 100 μg/kg was even more effective (Fig. 24). Examination of

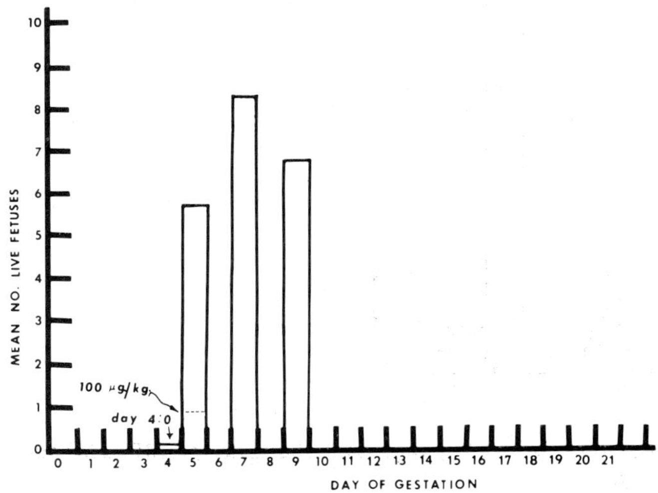

Fig. 24. Effect of a single dose of ORF–3858 on the course of pregnancy in the rat. (25 μg/kg, except where noted).

the uteri of these animals revealed marked placental hemorrhage which, when sufficiently extensive, caused embryonic death and subsequent resorption. Placental hemorrhage always preceded embryonic death. Thus Fenestrel may affect the anlage of the trophoblastic elements of the blastocyst destined to participate in chorioallantoic placentation rather than produce a maternal environment inimical to embryonic survival.

References

[1] *Barnes, L. E. & Meyer, R. K.* (1962): Effects of ethamoxytriphetol, MRL-37, and clomiphene on reproduction in rats. Fertil. & Steril. **13**, 471–480.

[2] *Blye, R. P.* (1967): Toxicity and steroid hormones in the dog. 22nd Annual Conference of the Bureau of Biological Research, Rutgers University (in press).

[3] *Blye, R. P. & Homm, R.* (1967): Antizygotic activity of 2-methyl-3-ethyl-4-phenyl-4-cyclohexene carboxylic acid. Fed. Proc. **26**, 486 (Abs).

[4] *Burdick, H. O. & Pincus, G.* (1935): The effect of estrin injections upon the developing ova of mice and rabbits. Am. J. Physiol. **111**, 201–208.

[5] *Chang, M. C.* (1959): Degeneration of ova in the rat and rabbit following oral administration of 1-(p-2-diethyl-aminoethoxyphenyl)-1-phenyl-2-p-anisylethanol. Endocrinology **65**, 339–342.

[6] *Chang, M. C.* (1964): Effects of certain antifertility agents on the development of rabbit ova. Fertil. & Steril. **15**, 97–106.

[7] *Chang, M. C. & Yanagimachi, R.* (1965): Effect of estrogens and other compounds as oral antifertility agents on the development of rabbit ova and hamster embryos. Fertil. & Steril. **16**, 281–291.

[8] *Davidson, O. W., Wada, K. & Segal, S. J.* (1965): Effects of clomiphene at different stages of pregnancy in the rat. Fertil. & Steril. **16**, 195–201.

[9] *Davidson, O. W., Schuchner, E. B. & Wada, K.* (1965): Effect of clomiphene on rat zygotes. Fertil. & Steril. **16**, 495–501.

[10] *Duncan, G. W., Stucki, J. C., Lyster, S. C. & Lednicer, D.* (1962): An orally effective mammalian antifertility agent. Proc. Soc. Exp. Biol. Med. **109**, 163–166.

[11] *Duncan, G. W., Lyster, S. C., Clark, J. J. & Lednicer, D.* (1963): Antifertility activities of two diphenyl-dihydronaphthalene derivatives. Proc. Soc. Exp. Biol. Med. **112**, 439–442.

[12] *Duncan, G. W. & Lyster, S. C.* (1963): Effect of a diphenylindene derivative (U-11555A) on blastocyst survival *in utero*. Fertil. & Steril. **14**, 565–571.

[13] *Duncan, G. W., Lyster, S. C. & Wright, J. B.* (1965): Reproductive mechanisms influenced by a diazocine. Proc. Soc. Exp. Biol. Med. **120**, 725–728.

[14] *Duncan, G. W. & Forbes, A. D.* (1965): Blastocyst survival and nidation in rats treated with estrogen antagonists. J. Reprod. Fertil. **10**, 161–167.

[15] *Duncan, G. W., Johnston, R. L. & Lyster, S. C.* (1966): The inhibition of pregnancy with an oxazolidinethione (U-11634). J. Reprod. Fert. **11**, 85–95.

[16] *Duncan, G. W. & Forbes, A. D.* (1966): The effect of U-11634 on the vaginal cytology of rats. J. Reprod. Fert. **11**, 459–461.

[17] *Duncan, G. W., Cornette, J. C., Lyster, S. C., Northam, J. I. & Wyngarden, L. J.* (1966): Biochemical aspects of rat deciduomata as affected by an oxazolidinethione. Am. J. Physiol. **211**, 184–192.

[18] *Duncan, G. W., Wyngarden, L. J. & Cornette, J. C.* (1968): Interaction of estradiol and nafoxidine hydrochloride with progesterone in deciduomata formation. J. Reprod. Fert. **15**, 53–63.

[19] *Einer-Jensen, N.* (1968): Antifertility properties of two diphenylethenes. Acta Pharmacol. Toxicol. **25**, Supp. 1.

[20] *Emmens, C. W., Cox, R. I. & Martin, L.* (1959): Estrogen inhibitors of the stilboestrol series. J. Endocr. **18,** 372–380.

[21] *Emmens, C. W., Cox, R. I. & Martin, L.* (1964): The estrogenic and anti-estrogenic activity of compounds related to diethylstilboestrol. Acta endocr. Copenh. Suppl. **90,** 61–69.

[22] *Emmens, C. W.* (1965): The estrogenic and antifertility activities of diethylstilboestrol and dimethylstilboestrol. Acta endocr. Copenh. **49**, 83–89.

[23] *Emmens, C. W. & Martin, L.* (1965): Biological activities of U-11100A. J. Reprod. Fertil. **9**, 269–275.

[24] *Emmens, C. W., Miller, B. G. & Owen, W. H.* (1968): Biological activities of U-11555A. J. Reprod. Fert. **15**, 33–37.

[25] *Emmens, C. W., Collins, D. J., Hobbs, J. J., Miller, B. G. & Owen, W. H.* (1968): Anti-estrogenic and antifertility properties of some 4-4′-dihydroxybibenzyls. J. Reprod. Fertil. **16**, 1–6.

[26] *Greenwald, G. S.* (1957): Interruption f pregnancy in the rabbit by the administration of estrogen. J. Exp. Zool. **135**, 461–481.

[27] *Greenwald, G. S.* (1959): The comparative effectiveness of estrogens in interrupting pregnancy in the rat. Fertil. & Steril. **10**, 155–161.

[28] *Greenwald, G. S.* (1961): The anti-fertility effects in pregnant rats of a single injection of estradiol cyclopenthylpropionate. Endocrinology **69**, 1068–1073.

[29] *Greenwald, G. S.* (1963): Interruption of early pregnancy in the rabbit by a single injection of estradiol cyclopentylpropionate. J. Endocr. **26**, 133–138.

[30] *Greenwald, G. S.* (1965): Effects of a dihydronaphthalene derivative (U-11100A) on pregnancy in the rat. Fertil. & Steril. **16**, 185–194.

[31] *Greenwald, G. S.* (1967): Species differences in egg transport in response to exogenous estrogen. Anat. Rec. **157**, 163–172.

[32] *Harper, M. J. K., & Walpole, A. L.* (1967): A new derivative of triphenylethylene: effect on implantation and mode of action in rats. J. Reprod. Fertil. **13**, 101–119.

[33] *Harper, M. J. K. & Walpole, A. L.* (1967): Mode of action of I.C.I. 46,474 in preventing implantation in rats. J. Endocr. **37**, 83–92.

[34] *Harper, M. J. K.* (1968): Interference with early pregnancy in rats by estrogen: mechanism of action of dienestrol. Anat. Rec. **162**, 433–452.

[35] *Holtkamp, D. E., Greslin, J. G., Root & Lerner, L. J.* (1960): Gonadotrophin inhibiting and anti-fecundity effects of chloramiphene. Proc. Soc. Exp. Biol. Med. **105**, 197–201.

[36] *Lednicer, D., Babcock, J. C., Lyster, S. C., Stucki, J. C. & Duncan, G. W.* (1961): Derivatives of 2,3-diphenylindenes as antifertility agents. Chem. Ind. 2098–2099.

[37] *Lednicer, D., Babcock, J. C., Lyster, S. C. & Duncan, G. W.* (1963): Derivatives of 1,2-diphenyl-3,4-dihydronaphthalene as antifertility agents. Chem. Ind. 408.

[38] *Lednicer, D., Lyster, S. C., Aspergren, B. D. & Duncan, G. W.* (1966): Mammalian antifertility agents. III. 1-aryl-2-phenyl-1,2,3,4-tetrahydro-1-naphthols, 1-aryl-2-phenyl-3,3-dihydronaphthalenes, and their derivatives. J. med. Chem. **9**, 172–178.

[39] *Lednicer, D., Lyster, S. C. & Duncan, G. W.* (1967): Mammalian antifertility agents. IV. Basic 3,4-dihydronaphthalenes and 1,2,3,4-tetrahydro-1-naphthols. J. med. Chem. **10**, 78–84.

[40] *Lerner, L. J., Holthaus, F. J. & Thompson, C. R.* (1958): A non-steroidal estrogen antagonist 1-(ρ-diethyl aminoethoxyphenyl)1-phenyl-2-ρ-methoxyphenol ethanol. Endocrinology **63**, 295–317.

[41] *Martin, L., Cox, R. I. & Emmens, C. W.* (1963): Further studies on the effects of estrogens and anti-estrogens on early pregnancy in mice. J. Reprod. Fert. **5**, 239–247.

[42] *Miquel, J. F., Wahlstam, H., Olsson, K. & Sundbeck, B.* (1963): Synthesis of unsymmetrical diphenylalkenes. J. med. Chem. **6**, 774–780.

[43] *Morris, J. Mcl. & Van Wagenen, G.* (1966): Compounds interfering with ovum implantation and development. III. The role of estrogens. Am. J. Obst. & Gynec. **96**, 804–813.

[44] *Morris, J. McL., Van Wagenen, G., McCann, T. & Jacob, D.* (1967): Compounds interfering with ovum implantation and development. II. Synthetic estrogens and anti-estrogens. Fertil. & Steril. **18**, 18–34.

[45] *Noyes, R. W., Dickmann, Z., Doyle, L. L. & Gates, A. H.* (1963): Ovum transfer synchronous and asynchronous, in the study of implantation. In Delayed Implantation edited by *A. C. Enders*, Univ. Chicago Press, Chicago, 197.

[46] *Parkes, A. S. & Bellerby, C. W.* (1926): Studies on the internal secretions of the ovary. 2. The effects of injection of the estrus producing hormone during pregnancy. J. Physiol. **62**, 145–155.

[47] *Pincus, G. & Kirsch, R.* (1936): The sterility in rabbits produced by injections of estrone and related compounds. Am. J. Physiol. **115**, 219–228.

[48] *Prasad, M. R. N., Kalra, S. P. & Segal, S. J.* (1965): Effect of Clomiphene on blastocysts during delayed implantation in the rat. Fertil. & Steril. **16**, 101–105.

[49] *Prasad, M. R. N. & Kalra, S. P.* (1967): Mechanism of anti-implantation action of Clomiphene. J. Reprod. Fert. **13**, 59–66.

[50] *Schlough, J. S. & Meyer, R. K.* (1965): Effect of anti-estrogens on estrogen-induced ova implantation in the ovariectomized rat. Fertil. & Steril. **16**, 106–112.

[51] *Segal, S. J. & Nelson, W. O.* (1958): An orally active compound with anti-fertility effects in rats. Proc. Soc. Exp. Biol. Med. **98**, 431–436.

[52] *Segal, S. J. & Nelson, W. O.* (1961): Antifertility action of chloramiphene. Anat. Rec. **139**, 273 (Abst.).

[53] *Segal, S. J. & Nelson, W. O.* (1961): An orally active compound with antifertility effects in rats. Proc. Soc. Exp. Biol. Med. **98**, 431–436.

[54] *Segal, S. J. & Davidson, O. W.* (1962): Prolonged antifertility action of clomaphene in delayed implantation. Anat. Rec. **142**, 278 (Abst.).

[55] *Staples, R. E.* (1966): Effect of clomiphene on blastocyst nidation in the rat. Endocrinology **78**, 82–86.

[56] *Stone, G. M. & Emmens, C. W.* (1964): The action f estradiol and dimethyl-stilboestrol on early pregnancy and deciduoma formation in the mouse. J. Endocr. **29**, 137–145.

[57] *Stone, G. M. & Emmens, C. W.* (1964): The effect of estrogens and anti-estrogens on deciduoma formation in the rat. J. Endocr. **29**, 147–157.

[58] *Smith, M. G.* (1926): On the interruption of pregnancy in the rat by the injection of ovarian follicular extract. Bull. Johns Hopkins Hosp. **39**, 203–214.

[59] *Soupart, P.:* (Personal communication).

[60] *Thomson, J. L.* (1968): Effect of two non-steroidal antifertility agents on pregnancy in mice. J. Reprod. Fert. **16**, 363–369.

[61] *Walpole, A. L.* (1968): Non-steroidal drugs in relation to ovualtion and implantation. J. Reprod. Fert. Suppl. **4**, 3–14.

[62] *Webster, H. D., Johnston, R. L. & Duncan, G. W.* (1967): Toxicologic and interrelated studies with an oxazolidinethione contraceptive. Toxic. Appl. Pharmac. **10**, 322–333.

[63] *Whitney, R. & Burdick, H. O.* (1936): Tube-locking of ova by estrogenic substances. Endocrinology **20**, 643–647.

[64] *Whitney, R. & Burdick, H. O.* (1938): Acceleration of the rate of passage of fertilized ova through the fallopian tubes of rabbits by massive injections of Progynon B. Endocrinology **22**, 639–642.

[65] *Yoshida, K. & Craig, M.* (1968): The effect of an estrogen antagonist (U 11555A) on early pregnancy in rats. Fertil. & Steril. **19**, 637–643.

[66] *Youngdale, G. A., Duncan, G. W., Emmert, D. E. & Lednicer, D.* (1966): Synthesis and antifertility activity of 5-(phenoxymethyl)-2-oxazolidine thiones. J. med. Chem. **9**, 155–159.

Discussion

D. R. S. Kirby: I am a little confused about your reference to yolk-sac placentation, and chorio-allantoic placentation in the rat. What do you mean by yolk-sac placentation?

R. P. Blye: It is well established that in the rat, two types of placenta develop. There is a primary placenta, which is classically referred to as the 'yolk-sac placenta' and a secondary or definitive placenta, which is called the 'chorio-allantoic placenta'.

D. R. S. Kirby: But the initial attachment is by the trophoblast or, in other words, the chorionic epithelium. The yolk-sac does not differentiate until much later, and then it functions, not as an organ of attachment but as a enveloping membrane. If the drug is going to affect the chorio-allantoic placenta, you would have determined the response sooner than you were in fact suggesting you did.

R. P. Blye: We would expect this and find it very strange inasmuch as these early ...

D. R. S. Kirby: It is this time lag that I find very intriguing.

R. P. Blye: Right. I have no explanation for it at all. One would expect to see interference with nidation and then with early or primary placentation.

H. R. Lindner: There is a crucial point in the abstract of your paper which I missed in your presentation – you hint there that you have done transfers of these Fenestrel – treated blastocysts to untreated hosts and that they still die under those circumstances.

R. P. Blye: We have done some transfer studies; quite frankly I expected them to be further along than they are now. In the rat we have succeeded in doing transfers from uterus to uterus. In other words, we have transferred embryos from the uterus of treated mothers on day 4 or 5, to the uterus of pseudopregnant non-treated recipients. We have indeed found that these embryos implant but subsequently undergo degeneration, just as they would have done if they had been left in the treated mother. Unfortunately, our controls also exhibited some post transfer resorption and this is why I did not present this data.

H. R. Lindner: If this point is not defnitively settled then I would be tempted to consider an alternative interesting possibility, namely that your drug treatment somehow interferes with the development of the immunological barrier, that is with the establishment of a privileged environment for the blastocyst, and thus gives rise to later rejection of the embryo. Perhaps a bit far-fetched, but ...

R. P. Blye: One would expect that the drug is affecting the maternal environment, but knowing what we do about the metabolism of the drug (it is metabolized rather rapidly in the rat, and about 80 % is excreted within 48 hours) one cannot conceive of an effect (on the mother) which persists for a week or ten days and during which two very important events occur, apparently normally, implantation and very early placentation.

D. R. S. Kirby: You must give it, Dr. *Blye,* to a rat with an extra-uterine pregnancy. This would tell you whether the drug's primarily affecting the embryo or the uterine environment.

R. P. Blye: I would like to do that.

C. E. Adams: Some years ago we experimented with 6 mercaptopurine and 8-azaguanine to determine their effect, if any, on 2–8 cell rabbit eggs *in vitro.* No immediate effect was observed; in fact, the eggs continued cleaving. Subsequently, we transferred similar eggs to the tubes of recipients, which had been treated with either drug the previous evening. Six hours after transfer, the eggs were recovered and re-transferred to an untreated animal, where a high proportion failed to develop beyond implantation. We inferred that even short exposure to such drugs during the tubal phase of development was sufficient to interfere with the embryos' later development (*Adams, Hay & Lutwak-Mann,* 1961, J. Embryol. exp. Morph. **9**, 468–491. The action of various agents upon the rabbit embryo).

R. P. Blye: That is very interesting. I wish I had some information on whether or not these zygotes were exposed to the drug that was administered. We do not know the concentration of the drug (Fenestrel) or its metabolites in the oviduct at the low doses that we have used.

E. Steinberger: Could I ask a question and make a comment? In animals where you observed resorptions, did you get any full term pregnancies; in other words, did you get any offspring at all? And were the offspring normal?

R. P. Blye: Dr. *Steinberger,* we undertook, of course, many experiments which you have not seen here. Rarely do we get any offspring at all when we give a dose of the drug which is sufficient to show some effect. In other words. we see almost an all-or-none effect. We have found some reduction in viable fetuses at borderline doses. Quite frankly, it is astounding to see, for example, a perfectly normal embryo surrounded by maybe ten resorbing conceptuses, and one wonders, of course, what is so unique about the one that seems to be completely unaffected.

E. Steinberger: This is a most interesting finding. Almost 20 years ago we demonstrated similar effect with nitrofuranes. Administration of these compounds on the third day of pregnancy produced exactly the same picture with placental hemorrhage and resorption. However, occasionally one of the embryos continued to develop and a perfectly normal offspring resulted. Furthermore, when we used for mating with normal females males that were treated with an alkylating agent, we also found degeneration of the embryos with the placental hemorrhage.

Preliminary Observations on in vitro Steroid Biogenesis in the Metrial Gland of the Rat

E. Steinberger and M. Ficher

Division of Endocrinology and Reproduction, Research Laboratories, Albert Einstein Medical Center, Philadelphia, USA

Summary: A brief review of the cytology and development of the metrial gland during pregnancy will be given. The possibility that the metrial gland may be concerned or involved with steroid metabolism will be introduced. Experiments to obtain evidence for possible participation of the metrial gland in steroid biosynthesis will be presented. Specifically, two sets of experiments will be discussed.

1) Metrial glands from rats on the 18th day of pregnancy were used in these experiments. The capacity of the metrial gland tissue to metabolize radiolabelled progesterone was determined. The experiments were performed by incubating *in vitro* for 3 hours minced metrial glands with tritiated progesterone and isolating and characterizing the metabolites. In these experiments evidence for formation of 20α-dihydroprogesterone, androstenedione, testosterone and estrogens was obtained. In most instances the identification of these metabolites was obtained by means of repeated chromatography, derivatives formation and isotopic dilution and recrystallization to constant ratio or specific activity.
2) In this set of experiments tritiated acetate was used as the precursor and the steroids formed were isolated and identified by the techniques mentioned above. Evidence for formation of progesterone, testosterone, androstenedione and estrone and estradiol were obtained.

The various hypotheses concerning physiologic activity and formation of biologically active substances (e.g., relaxin, steroids) by the metrial glands at various stages of pregnancy will be discussed.

Introduction

The metrial gland is located in the subplacental site of the uterus in the region of attachment of the mesometrium to the uterus. It is in this areas that the granulated cells of the metrial gland were first observed [5, 10]. *Weill* [10] and later *Gérard* [6] suggested endocrine function for these cells. *Selye* and *McKeown* [8], while accepting possible endocrine function of the metrial gland, suggested that the high glycogen content of the granulated cells may indicate also a nutritive role of the metrial gland for the developing embryo. *Velardo* et al. [9] suggested that the granules of the

Manuscript received: 15 March 1969

metrial gland cells may contain relaxin. This suggestion was supported by an immunohistologic study [4]. Also, a number of investigators demonstrated hydrolytic enzymes in the granules, generally associated with lysosomes (see, for example [3, 4]). *Bulmer* [3] stressed the importance of the hydrolytic enzymes in the metrial gland cells, but did not conclude concerning their possible role. He agreed with *Dallenbach-Hellweg* et al. [4] that relaxin may be present in metrial gland cells and that the hydrolytic enzymes in the granulated cells may be related to the relaxin content, but could not accept the suggestion that the metrial gland cells produce this hormone.

Baker [2] studied the histochemical changes occurring in metrial gland cells from the time of nidation through lactation. He described three periods of development, each marked by specific histochemical characteristics. The first period, commencing with the developing of decidual reaction in the uterine mucosa, is characterized by appearance of rounded connective tissue cells with strongly basophilic cytoplasm; the basophilia being due to presence of ribonucleic acid. About the 9th or 10th day of pregnancy eosinophilic granules and glycogen appear in the basophilic cells, which become arranged in "cuffs" around the blood vessels. The amount of eosinophilia and glycogen reaches a peak between the 13th and 15th day of pregnancy. By the 17th day of pregnancy, lipid granules can be detected in cells at the periphery of the "cuffs". As the quantity of the eosinophilic granules and glycogen decreases with progression of the pregnancy, sudanophilia increases and persists up to the 22nd day of lactation. The appearance of lipids was interpreted by *Baker* as an indication of involvement of the metrial gland with steroid metabolism. *Bulmer* [3], however, interpreted the increase of sudanophilia in the metrial gland of late pregnancy as due to "breakdown of bound lipids in the granules".

The metrial gland cells in late pregnancy possess some characteristics of cells involved in steroid metabolism. Their morphologic appearance is not incompatible with such function; they contain lipid granules which show cytochemical reaction for cholesterol [2], and the limited electron microscopic studies of these cells suggest the presence of smooth endoplasmic reticulum [7, 11] usually associated with steroid producing cells. These considerations led us to study *in vitro* the metabolism of progesterone in incubates of metrial gland tissue and the possibility of *de novo* synthesis of steroids from acetate in such incubates.

Materials and methods

Female Long-Evans rats obtained from a stock bred in our laboratory were sacrificed on the 18th day of pregnancy. The uteri containing the conceptus were excised. The metrial glands, yellow bodies about 3 to 4 mm in diameter located at the mesometrial region of the uterus beneath the placenta, were dissected from the surrounding uterine tissue and the placenta, and were immediately placed in chilled beakers. About

60 mg (wet weight) of metrial gland tissue was transferred into incubation flasks containing 1.5 ml of Hank's culture medium [1]) and the appropriate radiolabelled precursor. The tissue was thoroughly minced with a fine scissors and the flasks were incubated in a Dubnoff shaker at 37 °C in atmosphere of 95 % air plus 5 % CO_2 for three hours. At the completion of the incubation period, ^{14}C labelled tracers - testosterone, androstenedione, progesterone, estrone and estradiol (about 40,000 dpm of each) and 10 ml of ethyl acetate were added to each incubation flask and the material was frozen.

Chemicals

Analytical grade solvents were redistilled before use. Steroids (Steraloids, Inc.) were purified on silica gel column and recrystallized to correct melting point. Where indicated, the radiolabelled steroids, 7α-^{3}H-Progesterone, ^{3}H-Na acetate, 4-^{14}C-Testosterone, 4-^{14}C-Androstenedione, 4-^{14}C-Estrone and 4-^{14}C-17β-Estradiol (New England Nuclear Corp.) were purified by means of paper and thin layer chromatography.

Extraction procedure

The incubation mixture was thawed and extracted with ethyl acetate, partitioned with hexane - 70 % methanol, and finally extracted with chloroform. The appropriate cold carriers (50 μg each of androstenedione, progesterone, 17α-OH-progesterone, 20α-dihydroprogesterone and testosterone) were added to the dry residue obtained from the chloroform extract.

Paper chromatography and identification of metabolites

The dry residue obtained from the chloroform extraction was chromatographed in Bush A system (heptane : methanol : water, 100 : 80 : 20) against a number of appropriate standards. The Δ^4-3-ketosteroid (carriers and standards) were located on the paper chromatography strips by means of shortwave ultraviolet lamp. On the chromatography strips containing the standard steroids, the androgens were located by spraying with Zimmermann's reagent and the estrogens by spraying with a mixture of ferric chloride-potassium ferricyanide. The chromatograms from the chloroform extracts were scanned in a Packard model 7200 radiochromatogram scanner. The highly polar radioactive peaks obtained after the first chromatography in Bush A system were eluted and rechromatographed in Bush 6 system (toluene : methanol : water,

[1]) Fortified with NaNADP and glucose-6-phosphate.

100:75:25); the less polar peaks were eluted and rechromatographed in Bush 3 system (heptane:benzene:methanol:water, 66:34:80:20). Identification of the different steroids was accomplished by formation of derivatives and by the technique of isotopic dilution and recrystallization to constant ratio $^3H/^{14}C$ or constant specific activity, after addition of 50 to 100 mg of appropriate carrier steroid. Crystallization was carried out in several solvent mixtures. The labelled compound was considered radiochemically pure and identical with the added carrier steroid if the ratio $^3H/^{14}C$ or the specific activity (dpm/μM) in 2 to 3 consecutive samples of the crystals did not deviate by more than 5–8 % from the mean value.

Acetylation

The dry material was dissolved in0.1 ml pyridine, to which 0.1 ml acetic anhydride was added, leaving the reaction mixture overnight at room temperature; 0.2 ml of methanol was added to stop the reaction and the solution was evaporated to dryness.

Measurement of radioactivity

Radioactivity in samples of eluted steroids was measured by counting in Packard model 3003 liquid scintillator spectrometer, set for simultaneous counts of ^3H and ^{14}C, with efficiencies of about 25 % and 0 % for ^3H, and about 54 % and 7 % for ^{14}C in the respective channels. The samples were prepared by dissolving 1.0 ml of the dried eluate in 10 ml scintillation solution (0.4 % w/v) of 2.5–bis-[2′(5′-tert-buthyl-benzoxsazolyl)]-thiophene in toluene. The radioactivity of the substrates and the tracers was measured in the same instrument but setting it for single counts, with efficiencies of about 37 % for ^3H and about 86 % for ^{14}C. Quenching was usually absent in the samples counted by the procedure mentioned above, but in certain cases it was necessary to correct for quenching by using an internal standard and the automatic external standardization provided in the instrument; both methods gave the same correction factor.

Calculations

The percent conversion of progesterone to estradiol, testosterone and androstenedione was calculated by the following formula:

$$\frac{^{3}\text{H dpm in last crystals} \times \text{dpm } ^{14}\text{C steroid added}}{^{14}\text{C dpm in last crystals} \times \text{dpm } ^{3}\text{H-P (in incubation medium)}} \times 100 = \%\ \text{conversion}$$

Results

A. Incubation with ^{3}H-progesterone

The metrial gland tissue was incubated with 25 x 10^6 dpm of ^{3}H-progesterone. The chloroform extracts were chromatographed in Bush A system. Scans of these chromatograms revealed radioactive peaks with mobility similar to that of the following standards: Estradiol, testosterone and 17α-hydroxyprogesterone, androstenedione, progesterone and 20α-dihydroprogesterone. Distribution of radioactivity among these peaks is summarized in Fig. 1.

Fig. 1. Distribution of radioactivity on the first chromatogram (Bush A) of an extract prepared from an incubation of metrial gland tissue with ^{3}H-progesterone.

Zones eluted	Corresponding standards	Recovered radioactivity dpm x 10^{-4}	%
Origin	Estradiol	47.7	8.0
I	Testosterone + 17α-OH-P	19.2	3.2
II	20α-dihydroprogesterone	27.0	4.5
III	Androstenedione	14.8	2.5
IV*	–	63.4	10.6
V	Progesterone	382.0	63.7
VI*	–	26.1	4.4
			96.9

*) These peaks did not correspond in mobility to any of the standards used in this experiment.

a) Area corresponding in chromatographic mobility to testosterone and 17α-hydroxyprogesterone

The radioactivity in this area was eluted and the material rechromatographed in Bush 3 system. Scan of this chromatogram revealed a single radioactive peak. This peak was eluted, acetylated and rechromatographed in Bush A system. A small amount of radioactivity remained in the area corresponding in mobility to 17α-hydroxyprogesterone, but most of the radioactivity moved into an area corresponding in mobility to testosterone acetate. This material was eluted for isotopic dilution and recrystallization to constant ratio (^{3}H/^{14}C). Results of the recrystallization study suggest strongly that this material is testosterone acetate (see Fig. 2).

b) Area corresponding in chromatographic mobility to 20α-dihydroprogesterone

The radioactivity in this area was eluted and rechromatographed in Bush 3 system. A single peak with mobility corresponding to that of 20α-dihydroprogesterone was

Fig. 2. Recrystallization to constant ratio $^{3}H/^{14}C$ or specific activity of material with chromatographic behavior of estradiol, testosterone, androstenedione and 20α-dihydroprogesterone obtained from incubates of ^{3}H-progesterone with metrial gland tissue.

	Estradiol			Testosterone			Androstenedione			20α-dihydroprogesterone (c) Specific activity		
Cryst.	R $^{3}H/^{14}C$	Δ(a)	Δ% (b)	R $^{3}H/^{14}C$	Δ	Δ%	R $^{3}H/^{14}C$	Δ	Δ%	dpm/μM	Δ	Δ%
1 xls	9.09			1.99			6.80			718		
2 xls	5.17*	.005	0.1	1.77*	+.07	4.1	5.73*	+.43	8.1	606*	+11	1.8
3 xls	5.18*	.005	0.1	1.63*	−.07	4.1	5.02*	−.28	5.3	584*	−11	1.8
4 xls							5.14*	−.16	3.0			
Mean: (d)	5.175	± .005		1.70	± .07		5.30	± .22		595	± 11	
% Conversion (e)	.83			.26			.82			--		

xls crystals
(a) deviation from mean
(b) % deviation from mean
(c) to constant specific activity
(d) mean ± S.E. derived from the figures marked *
(e) Percent conversion from progesterone to estradiol. testosterone and androstenedione.

obtained. This material was eluted and isotopic dilution and recrystallization to constant specific activity was performed. The recrystallization data suggest that this compound is 20α-dihydroprogesterone (see Fig. 2).

c) Area corresponding in chromatographic mobility to androstenedione

This material was handled identically to the above. The recrystallization data are summarized in Fig. 2.

d) The radioactivity remaining at the origin of the chromatogram

This material was eluted and rechromatographed in Bush 6 system. Several minor peaks and one major peak were resolved. The major peak corresponded in mobility to estradiol. This material was rechromatographed and recrystallized to constant ratio ($^3H/^{14}C$). The crystallization data provided evidence that this material is estradiol (Fig. 2).

B. Incubation with ^{3}H-Na acetate

About 50 mg of metrial gland tissue was incubated with 7.5 x 10^6 dpm of ^{3}H-Na acetate. The chloroform extract was chromatographed in Bush A system. Radioscan of this chromatogram revealed a picture similar to that found in incubates with ^{3}H-progesterone, except for relatively smaller amounts of radioactivity associated with the area, corresponding in mobility to progesterone, and no radioactivity in the area corresponding in mobility to 20α-dihydroprogesterone. The material in the areas of the various radioactive peaks was handled in similar fashion as in the case of incubates with ^{3}H-progesterone, i.e., repeated chromatography, derivative formation and isotopic dilution and recrystallization to constant specific activity or constant ratio ($^3H/^{14}C$). The data indicating incorporation of acetate into the various steroids are summarized in Fig. 3. The recrystallization data are summarized in Fig. 4. The data provide evidence of *de novo* synthesis of progesterone, androstenedione, estradiol and testosterone.

Fig. 3. Incorporation of ^{3}H-Na Acetate into steroids following incubation with metrial gland tissue.

Zones eluted	corresponding standards	^{3}H-Na acetate incorporation (dpm/g tissue)*
Origin	Estradiol	184,000
I	Testosterone	372,000
II	Androstenedione	372,000
III	Progesterone	274,000

*) Based on the radioactivity eluted directly from the first chromatogram (Bush A) and corrected for recovery with the respective ^{14}C tracers.

Fig. 4. Recrystallization to constant ratio $^3H/^{14}C$ of material with chromatographic behavior of estradiol, testosterone, androstenedione and progesterone obtained from incubates of ^3H-Na acetate with metrial gland tissue.

	Estradiol			Testosterone			Androstenedione			Progesterone		
Cryst.	R $^3H/^{14}C$	Δ (a)	Δ% (b)	R $^3H/^{14}C$	Δ	Δ%	R $^3H/^{14}C$	Δ	Δ%	R $^3H/^{14}C$	Δ	Δ%
1 xls	4.06			.220			.521			.510		
2 xls	4.28*	–.29	6.3	.364			.470	+.033	7.6	.459*	+.020	4.6
3 xls	4.92*	+.35	7.6	.226			.435*	–.002	0.4	.401*	–.038	8.6
4 xls	4.51*	–.06	1.3	.184*	–.007	3.7	.408*	–.029	6.6	.459*	+.020	4.6
5 xls				.198*	+.007	3.7						
Mean: (c)	4.57	± .187		.191	± .007		.437	± .018		.439	± .019	

xls crystals
(a) deviation from mean
(b) percent deviation from mean
(c) mean ± S.E. derived from the figures marked *

Comment

The role of the metrial gland in physiology of pregnancy has been debated for the past fifty years. A nutritive function for the granulated cells of the metrial gland, as well as an endocrine function specifically related to relaxin, has been suggested. The gland has been shown to be an important site of synthesis of hydrolytic enzymes, whose possible role in the physiology of the pregnancy remains obscure. In the work reported in this paper, experimental evidence was sought for the suggestion of *Baker* [2] that during late stages of pregnancy the metrial gland may be involved in steroid metabolism.

The data presented here do not provide definite proof for major steroid biogenetic activity in the metrial gland. However, conversion of progesterone to biologically active metabolites (estradiol, androstenedione and testosterone) has been demonstrated. Since only a relatively small amount of progesterone was metabolized by the metrial gland tissue, the physiologic importance of this phenomenon remains to be determined. On the other hand, the possibility exists that the biogenesis of steroids in the metrial gland does not proceed via the Δ_4 pathway or that progesterone is the major end product. The latter possibility is not attractive in light of the data obtained in experiments where ^{3}H-Na acetate was used as the precursor, since incorporation of acetate into progesterone was actually less than into androstenedione or testosterone. However, the experiments with ^{3}H-Na acetate – illustrating incorporation of acetate into a number of steroids (estradiol, progesterone, androstenedione and testosterone) – provide considerable evidence for presence of steroid biosynthetic pathways in metrial gland tissue. The formation of Δ_5 metabolites was not investigated in this study and thus the possibility of the presence of a major Δ_5 pathway cannot be ruled out.

The over-all levels of steroid metabolic activity (conversion of progesterone) and *de novo* biosynthetic activity (incorporation of acetate) in the metrial gland incubates were low. The possibility exists that the stage of development of the metrial gland selected in this study (18th day of pregnancy) was not that of the highest steroid metabolic activity. Further experiments with different steroidal precursors, and possibly more importantly with metrial gland tissue at different stages of pregnancy will be required to provide proof of the participation of the metrial gland in steroid metabolism and the possible physiologic role of such biosynthetic activity.

Acknowledgement: The expert technical assistance of Mrs. *Mae Diamond* and Mr. *William Lafferty* is gratefully acknowledged. This work was supported by a research grant from The Ford Foundation.

References

[1] The following trivial names and abbreviations have been used in this paper:

Androstenedione (A)	Δ^4-androstene-3,17-dione
Estradiol (E_2)	$\Delta^{1,3,5(10)}$-estratriene-3,17β-diol
Estrone (E_1)	$\Delta^{1,3,5(10)}$-estratriene-3-ol-17-one
17α-hydroxyprogesterone (17-OH-P)	Δ^4-pregnen-17α-ol-3,20-dione
20α-dihydroprogesterone (20α-DP)	Δ^4-pregnen-20α-ol-3-one
Progesterone (P)	Δ^4-pregnen-3,20-dione
Testosterone (T)	Δ^4-androstene-17β-ol-3-one

[2] *Baker, B. L.* (1948): Histochemical variations in the metrial gland of the rat during pregnancy and lactation. Proc. Soc. Exp. Biol. Med., **68**, 492–496.

[3] *Bulmer, D.* (1968): Further studies on the granulated metrial gland cells of the pregnant rat. J. Anat. **103**, 479–489.

[4] *Dallenbach-Hellweg, G., Battista, J. V. & Dallenbach, F. D.* (1965): Immunohistological and histochem ical localization of relaxin in the metrial gland of the pregnant rat. Am. J. Anat. **117**, 433–450.

[5] *Duval, M.* (1891): Le placenta des rongeurs. Le placenta de la souris et du rat. J. de l'Anat. et de la Physiologie, Paris, **27**, 24–73, 344–395, 512–612.

[6] *Gérard, P.* (1925): Sur la glande myométriale de la souris et du rat. C.R. Soc. de Biol., **93**, 457–459.

[7] *Peach, R. & Bulmer, D.* (1965): Cytochemistry and electron microscopy of the granulated metrial gland cells of the rat placenta. J. Anat. **99**, 415–416.

[8] *Selye, H. & McKeown, T.* (1935): Studies on the physiology of the maternal placenta of the rat. Proc. Roy. Soc. of London, **119**, 1–31.

[9] *Velardo, J. T., Dawson, A. B., Olsen, A. G. & Hisaw, F. L.* (1935): Sequence of histological changes in the uterus and vagina of the rat during prolongation of pseudopregnancy associated with the presence of deciduomata. Am. J. Anat. **93**, 273–305.

[10] *Weill, P.* (1919): Glande myométriale endocrine dans l'uterus de la rat gestante. C.R. Soc. Biol., **82**, 1433–1435.

[11] *Wislocki, G. B., Weiss, L. P. Burgos, M. H. & Ellis, R. A.* (1957): The cytology, histochemistry and electron microscopy of the granular cells of the metrial gland of the gravid rat. J. Anat. **91**, 130–140.

Discussion

G. A. Overbeek: May I ask if this gland is only present during pregnancy? You have only talked about pregnant animals.

E. Steinberger: This "gland" (actually when you look at the pregnant uterus it looks like a gland) appears as a structure detectable by the naked eye around the 10–12th day of pregnancy and persists during lactation. Because of the late hour I did not want to hypothesise, particularly about its presence during lactation.

H. G. Goslar: Have you analysed the lipids correlating to sudan black staining and have you found anisotropic substances?

E. Steinberger: Let me repeat the question. Have I analysed the sudan ...

H. G. Goslar: The lipids.

E. Steinberger: The lipids were not extracted and analyzed. What was the other question.

H. G. Goslar: Have you found anisotropic substances? For instance, cholesterol?

E. Steinberger: As you noticed, I mentioned the formation of several steroids, but there were a number of other steroids, or their metabolites. These were not analyzed.

H. G. Goslar: Do these glands contain sialic acid?

E. Steinberger: I do not know.

R. I. Dorfman: I think of course this is a most dramatic piece of work. I would just like to ask: if you incubate the same substrates, for example of indifferent tissue like liver ...

E. Steinberger: Well I doubt that I would pick on liver. It is not very indifferent, I am afraid we find all kinds of steroid conversions there.

R. I. Dorfman: Like hydroxylations ...

E. Steinberger: That is right. No, as a matter of fact, one of the more important aspects to make our findings more meaningful would be first of all to follow through the possible biosynthetic activity during the different stages of the development of this gland. This would partly answer your question. Secondly, one should study uterine tissue adjacent to the metrial gland. I tried to be careful and cautious during my presentation not to indicate that we have proven steroid biosynthetic activity of physiologic significance in the metrial gland, but only to point out that this could be the case.

B. L. Lobel: It is well known that the giant cell layer of the fetal placenta in the rat is right adjacent to the uterine decidue.
Now you seem quite sure that the two may be separated completely. In the rat, the giant cells produce steroid hermones, therefore some contamination of the material gland tissue used might be enough to give a low conversion.
The other question concerns the lipids. In the metrial gland cells they are quite different in aspect from those seen within steroid-hormone producing cells. In fact the size and distribution of lipid droplets is more like that in macrophages.
An other point is that right after delivery in rats, whether they lactate or not, there is a rapid degneration of the metrial gland.

E. Steinberger: Correct. I really do not wish to speculate. This needs to be determined. The metrial gland does not degenerate after delivery. As a matter of fact it remains quite large and "active looking" for a period of time, as far as I can remember, looking at the metrial glands of lactating rats. I did not look at metrial glands of rats that did not lactate. I agree with you that the distribution of lipids is not the classic one for steroid producing cells. Some of the cells may be macrophages, except they do not look like it.

J. F. Sommerville: Have you studied the capacity for steroid transformation ***in vitro*** of adjoining tissues notably the mesometrial decidua and the myometrium ?

E. Steinberger: No, it is being examined, but I do not have any information at the moment.

J. F. Sommerville: I was thinking of the work by Dr. *Sweat* and his colleagues in Utah upon the transformation of progesterone by uterine fibroblasts (*Sweat* et al., 1958) and human proliferative endometrium (*Bryson & Sweat,* 1967; *Bryson & Young,* 1967) and of our own studies of steroid transformations in human endometrium using computerised techniques and numerous labelled substrates (*Collins* et al., Biochem J., **112**, 1969). I am glad that you agree about the need to study these tissues since the capacity to perform these transformations ***in vitro*** is widely distributed – even in human skin.

E. Steinberger: I agree with every word you said. However, there are very few tissues capable of incorporating acetate, one of the classic "sex" steroids.

The Biological Activity of Chlormadione Acetate (CAP) and Related Steroids

R. I. Dorfman

Syntex Research, Palo Alto, California, USA

Summary: Chlormadinone acetate (CAP) is a highly active progestational agent 35 times more active than norethindrone by the oral route and 69 times more active than progesterone by subcutaneous injection.

The antiovulatory activity of CAP by the subcutaneous route was 40 times progesterone. Compounds of the 17-acetoxyprogesterone type showed a high correlation between progestational and antiovulatory activity, both tests being performed in the rabbit. Similar high antiovulatory activity was found for CAP by the oral route. The compound was 35 times more active than norethindrone by both the progestational and antiovulatory tests.

By means of the parabiotic assay method for inhibition of gonadotropins in the rat, chlormadinone acetate was found to be particularly inactive.

A recent study by *Chang* [47] demonstrated that progestational agents including CAP administered to isolated immature rabbits treated with HCG (human chorionic gonadotropin) disturbed sperm transport and capacitation. Additional effects of CAP as an antifertility agent and perhaps a major factor relates to the abnormal transportation of eggs including expulsion of the eggs from the uterus.

It appears, therefore, that continuous low dosage of chlormadinone acetate without estrogens may approach a rather ideal hormonal contraceptive, being minimally effective as a restrainer of the hypothalamic-anterior pituitary-ovarian axis while having important antifertility activity on sperm deposited in the female tract and a profound influence on egg transport.

Chlormadinone acetate (6-chloro-17α-hydroxypregna-4, 6-diene-3, 20-dione-17-acetate), originally reported from the Syntex Laboratories by *Ringold, Batres, Bowers, Edwards* and *Zderic* [16], is a highly active progestational agent when administered both by the subcutaneous and oral routes. In the rabbit McPhail assay, a progestational assay, by the oral route, this steroid was some 35 times more active than norethindrone (17α-ethynyl-17β-hydroxyestr-4-en-3-one). By the subcutaneous route, CAP was 69 times more potent than progesterone. In this original publication it was demonstrated that the progestational activity of 6α-chloro-17-acetoxy-progesterone was increased by the introduction of both the Δ^1 and Δ^6 double bonds.

Manuscript received: 10 April 1969

The present paper reviews the biological activity of CAP, a steroid which has become one of the more important progestational agents used for contraceptive purposes, especially more recently on a low dose continuous basis.

Antiovulation in the rabbit

Ovulation in experimental animals may be inhibited by an active corpus luteum [12] and by treatment with progesterone or other progestational agents. *Pincus* and *Chang* [15] developed a relatively standardized technique using the mated doe and reported that a single injection of progesterone may prevent ovulation for 24 hours. Oral doses of 2 to 10 mg of ethynyl testosterone were effective. These investigators reported that 17α-hydroxyprogesterone, usually considered to be inactive as a progestational agent, was an effective antiovulatory substance at the one and 4 mg dose levels when administered intravaginally. 17α-Methylprogesterone was effective subcutaneously (5 mg) and intravaginally (5 mg) and inactive by gavage.

Barnes, Schmidt and *Dulin* [1] used a rabbit ovulation inhibition test based on the work of *Makepeace, Weinstein* and *Friedman* [12] using an ED_{50} end-point. Progesterone by injection at a dose of 1 mg produced 100% inhibition of ovulation. 17α-Acetoxyprogesterone was 7 times as active as progesterone, and 6α-methyl-17-acetoxyprogesterone was assessed as 20 times as active as the same standard.

Kincl and *Dorfman* [8, 9] made an extensive study of antiovulatory steroids in the rabbit and demonstrated that CAP was particularly active by both the oral and injectable routes by the following bioassay procedure. Adult estrus female rabbits were injected subcutaneously or administered by gavage once with 1 ml of the test compound suspended in the aqueous medium composed of sodium chloride (0.9%), polysorbate 80 (0.4%), carboxymethyl cellulose (0.5%) and benzyl alcohol (0.9%). Eighteen to 24 hours after treatment, the female rabbits were bred. The ovaries were examined for ovulation points 18–24 hours after mating. The results of each treatment group are recorded as the mean number of ovulations per rabbit.

The relative potencies of the highly active antiovulatory steroids are listed in Fig. 1 and arranged in the order of decreasing potencies. Specifically, the relative potencies varied from about 8 to 0.2 times that of norethindrone (17α-ethynyl-19-nortestosterone), the standard. Four 6-chloro derivatives of 17α-acetoxyprogesterone had significantly higher antiovulatory activity than norethindrone. 6-Chloro-Δ^6-dehydro-17-acetoxyprogesterone was 8 times more active than the standard norethindrone.

Fig. 2 indicates a rather close correlation between progestational activity by injection and antiovulatory activity. CAP was most active by both parameters, being some 3 to 4 times more active than progesterone.

CAP was the most active antiovulatory compound when the compounds were administered by gavage (Fig. 3). The dose-response data on the norethindrone (standard) is presented in Fig. 4.

Fig. 1. Relative potencies of anti-ovulatory steroids administered by subcutaneous injections [8].

Steroid	Total Number of Rabbits	Relative Potencies (Norethindrone = 1.0)
6-Chloro-Δ^6-dehydro-17α-acetoxy-progesterone (CAP)	20	8
6α-Chloro-$\Delta^{1,6}$-bisdehydro-17α-acetoxyprogesterone	11	5
6-Chloro-16α-methyl-Δ^6-dehydro-17α-acetoxyprogesterone	9	2.8
6α-Chloro-Δ^1-dehydro-17α-acetoxy-progesterone	18	$\geqslant$2
Δ^6-Dehydro-17-acetoxyprogesterone	5	2
Progesterone	30	0.2

Fig. 2. Correlation of anti-ovulatory and progestational activity [8] (Subcutaneous Injection).

Steroid	Anti-Ovulatory	Progestational (Progesterone = 1)
6-Chloro-Δ^6-dehydro-17α-acetoxy-progesterone	40	69
6α-Chloro-Δ^1-dehydro-17α-acetoxy-progesterone	$\geqslant$10	15
Δ^6-Dehydro-17α-acetoxyprogesterone	10	ca. 10

Fig. 3. Relative potency of anti-ovulatory steroids administered by gavage [8].

Steroid	Number of Rabbits	Relative Potency (Norethindrone = 1)
6-Chloro-Δ^6-dehydro-17α-acetoxy-progesterone (CAP)	33	35
6α-Methyl-Δ^1-dehydro-17α-acetoxy-progesterone	17	$\geqslant$10
6-Fluoro-Δ^6-dehydro-17α-acetoxy-progesterone	23	6–12
6-Methyl-Δ^6-dehydro-17α-acetoxy-progesterone	9	5
Δ^6-Dehydro-17α-acetoxyprogesterone	23	$\geqslant$3
6α-Methyl-17α-acetoxyprogesterone	14	2.6
6-Chloro-$\Delta^{1,6}$-bisdehydro-17α-acetoxy-progesterone	2	$\geqslant$2
6α-Fluoro-16α-methyl-17α-acetoxy-progesterone	7	$\geqslant$1.25

Fig. 4. The anti-ovulation activity of norethindrone administered by gavage.

Norethindrone Dose, mg	Number of Rabbits	Mean Ovulation Points ± S.E. (Range)
0	26	5.2 ± 0.3 (0–10)
0.312	11	6.2 ± 0.4 (4–8)
0.625	11	1.6 ± 0.9 (0–7)
1.25	17	2.1 ± 0.6 (0–6)
2.5	3	0

Among the progestational agents there is no necessary correlation between progestational activity and antiovulatory activity. This is true for 17-acetoxyprogesterone (where the ratio of antiovulatory activity to progestational activity is < 0.1 -(Fig. 5). Similar lack of correlations are reported for Δ^1-dehydro-17*a*-acetoxyprogesterone (ratio = 0.04), 6*a*-chloro-Δ^1-dehydro-17*a*-acetoxyprogesterone (ratio = 0.005) and 3-deoxy-17*a*-acetoxyprogesterone (ratio = 0.15). It is of interest that the relative antiovulatory potency was less than the progestational activity.

Although instances are presented wherein lack of correlation between antiovulatory and progestational activity is indicated from the data in Fig. 5, it should be emphasized that among some steroids correlation of the two parameters has been found. The most potent progestational agent, 6-chloro-Δ^6-dehydro-17*a*-acetoxyprogesterone, had identical relative potencies of 35 by both tests. If one considers the error of estimate of the relative potencies, the following compounds may be considered to show correlations rather than noncorrelations: Δ^6-dehydro-17*a*-acetoxyprogesterone (ratio $\geqslant$ 2), 6*a*-fluoro-17*a*-acetoxyprogesterone (ratio = 1), 6*a*-chloro-17-acetoxyprogesterone (ratio = 0.3), 6*a*-fluoro-Δ^6-dehydro-17*a*-acetoxyprogesterone (ratio = 0.5), 6*a*-methyl-Δ^1-dehydro-17*a*-acetoxyprogesterone (ratio $\geqslant$ 1.3) and others listed in Fig. 5.

The dose required to produce a 50% inhibition in ovulation has been calculated for four compounds by both the gavage and subcutaneous routes (Fig. 6). Included in this summary table is a ratio of the subcutaneous 50% effective dose to the gavage dose. The highly active 6-chloro-Δ^6-dehydro-17*a*-acetoxyprogesterone was effective by both routes at the level of 10 micrograms giving the ratio for this parameter of one. All the other compounds appear to have a ratio of less than one, indicating that the compounds are less active by gavage than by injection.

Unsaturation at position 6 resulted in a compound at least 300 times more potent than the reference standard, 17*a*-acetoxyprogesterone (by gavage), while the same change at carbons 1 produced no significant alteration in relative potency. The introduction of the 6*a*-methyl group without unsaturation gave a value of 260. The 6-methyl substituted compound of either the Δ^1- or Δ^6-unsaturated derivative resulted

Fig. 5. The relationship between anti-ovulatory and progestational activity. (Administered by gavage) [8].

Steroid	Anti-Ovulatory Activity	Progestational Activity
	Norethindrone = 1	
17α-Acetoxyprogesterone	< 0.01	0.1
Δ^1-Dehydro-17α-acetoxyprogesterone	0.02	0.5
Δ^6-Dehydro-17α-acetoxyprogesterone	$\geqslant 3$	1.5
6α-Fluoro-17α-acetoxyprogesterone	0.5	0.5
6α-Fluoro-Δ^6-dehydro-17α-acetoxy-progesterone	9	18
6α-Chloro-17α-acetoxyprogesterone	0.6	2
6α-Chloro-Δ^1-dehydro-17α-acetoxy-progesterone	0.3	6
6-Chloro-Δ^6-dehydro-17α-acetoxy-progesterone	35	35
6αMethyl-Δ^1-dehydro-17α-acetoxy-progesterone	10	8
6-Methyl-Δ^6-dehydro-17α-acetoxy-progesterone	5	15
6α-Fluoro-16α-methyl-17α-acetoxy-progesterone	$\geqslant 1.25$	ca. 1
3-Deoxy-17α-acetoxyprogesterone	0.03	0.2
6α-Bromo-17α-acetoxyprogesterone	0.15	0.7
6-Chloro-$\Delta^{1,6}$-bisdehydro-17α-acetoxyprogesterone	$\geqslant 2$	30
6α-Methyl-17α-acetoxyprogesterone	2.6	2
16α-Methyl-17α-acetoxyprogesterone	0.03	0.1

in highly active compounds. The Δ^1-6α-methyl compound was equal to or more than 10 times more active than norethindrone and 1,000 times more active than 17α-acetoxyprogesterone, and the Δ^6-methyl derivative had a relative potency of 500 times the latter compound (Fig. 7).

The effect of the double modification of 17α-acetoxyprogesterone is summarized in Fig. 8, where both a 6-halogen group and unsaturation was effected. The potency (by gavage) of 6-chloro-$\Delta^{1,6}$was not properly evaluated but judged at this time to be

Fig. 6. Comparative gavage and subcutaneous activity of steroids in the anti-ovulatory test [8].

Steroid	A Gavage Dose (mg)	B Subcutaneous Injection (mg)	Ratio = A/B
	To Produce 50% Inhibition		
Norethindrone	0.5	0.09	0.18
6α-Chloro-Δ^1-dehydro-17α-acetoxy-progesterone	1.65	$\leqslant$ 0.05	$>$ 0.09
6-Chloro-Δ^6-dehydro-17α-acetoxy-progesterone (CAP)	0.01	0.01	1
Δ^6-Dehydro-17α-acetoxyprogesterone	$\leqslant$ 0.17	0.05	$\geqslant$ 0.30

Fig. 7. Influence of unsaturation, the 6-methyl group or a combination of both on the anti-ovulatory activity of 17-acetoxyprogesterone (gavage) [8].

Changes in the Molecule	Change in Relative Potency from 17α-Acetoxyprogesterone
Δ^1	1
Δ^6	$\geqslant$ 300
6α-Methyl	260
Δ^1-6α-Methyl	$\geqslant$ 1000
Δ^6-6-Methyl	500

Fig. 8. Influence of unsaturation and 6-halogen substitution on the anti-ovulatory activity of 17-acetoxyprogesterone (gavage) [8].

Changes in the Molecule	Change in Relative Potency from 17α-Acetocyprogesterone
Δ^6-6-Chloro	3500
Δ^6-6-Fluoro	900
$\Delta^{1,6}$-6-Chloro	$\geqslant$ 200

at least equal to 200 times the parent compound; 6-fluoro-Δ^6-dehydro-17α-acetoxy-progesterone had a potency of 900 and the corresponding 6-chloro-Δ^6-steroid was 3,500 times more potent than 17α-acetoxyprogesterone.

The addition of a 16α-methyl to the two compounds that were studied resulted in no great change in activity. The potency of 6α-fluoro-17-acetoxyprogesterone was 0.5 norethisterone and the 16α-methyl derivative was judged to be = 1.25. In view of the relatively high error of estimation, this difference may not be significant. The potency of 16α-methyl-17-acetoxyprogesterone was rated as 0.03 and is only somewhat more active than 17α-acetoxyprogesterone recorded as 0.01 or lower.

Effect of CAP on the ovary

Chlormadinone acetate in a single oral dose inhibits ovulation in the adult mated rabbit, whereas no inhibition was achieved with a ten-fold greater doese when ovulation was induced by HCG [17] (Fig. 9). Similarly, ovulation could not be inhibited by CAP in the hypophysectomized rat in which ovulation was stimulated by the administration of PMS followed by HCG. The animals were hypophysectomized by the pharyngeal-eosophagal route at the age of 30 days and PMS (30 I.U. in saline) was injected subcutaneously on the same day. The steroids dissolved in sesame oil were injected on the first and second post-priming days. Ten I.U. of HCG (in saline) were given by intravenous injection on the second post-priming day. At autopsy 24 hours later, the oviducts were removed, flushed, and the number of ova present was counted.

Fig. 9. Failure of chlormadinone acetate to block ovulation in HCG-stimulated adult rabbits [17] (administered by gavage).

dose, μg	No. of Rabbits Ovulating/Total	%	Average No. of Ovulations (Range)
0	28/28	100	5.1 (2–9)
30	7/8	88	4.8 (0–9)
100	12/14	86	3.9 (0–8)
300	14/14	100	6.2 (3–11)
30 x 3 days	8/8	100	5.5 (4–7)
100 x 3 days	8/8	100	4.9 (2–9)
300 x 3 days	7/8	88	5.0 (0–8)

Antifertility assay in the rat

Kincl and *Dorfman* [10] studied the antifertility activity of various steroids in the adult mated rat. The test compounds were administered once daily for seven days, beginning on the day of proestrus. Autopsy was on day nine of the test, at which time the number of implantation sites was counted. Estrogens were very active whereas various progestational agents and possible progesterone metabolites were found to produce no effect. These include progesterone (15 mg) administered subcutaneously and norethindrone (17α-ethynyl-17β-hydroxyestr-4-en-3-one) (1.2 mg) and chlormadinone acetate (5 mg) given by gavage.

Inhibition of gonadotropin in the parabiotic rat

The method is that described by *Shipley* [18]. Comparative results are summarized in Fig. 10 and show that CAP is particularly inactive by this assay.

Fig. 10. Inhibition of gonadotropin by various steroids measured in the parabiotic rat [7a]

Steroid	Effective daily Dose, μg Oral	Subcutaneous
Norethindrone (estrogen free)	40	30
Progesterone		3000 [18]
Chlormadinone acetate	>4000	
Medroxyprogesterone	1000	100
Ethyndiol diacetate	15	
Mestranol	0.1	0.05 [7a]
Ethynyl estradiol	0.1	0.01
Estradiol	1.0	0.01 [7a]

Comparative antiovulatory activity of CAP in the rat and rabbit

The relative antiovulatory activities in the rat and in the rabbit have been reported [7, 8] (Fig. 11). The antiovulatory activity in the rat was measured by a modification of the *McCormack* and *Meyer* [14] method using 30 day old rats instead of 22 day old rats [7]. Ovulation was stimulated with 30 I.U. of PMS. The steroids were given

Fig. 11. Comparison of anti-ovulatory activity of subcutaneously injected steroids in the rat and rabbit. (Subcutaneous injection);

Steroid	Relative Potency Rat [7] Anti-Ovulatory	Rabbit Anti-Ovulatory [8]
Progesterone	1	1
19-Norprogesterone	15	15–30
Norethindrone	1.7	5
Chlormadinone acetate	< 1	40

for two consecutive days and on the day following the last administration, the oviducts were removed, the ova flushed and counted. Progesterone has been arbitrarily assigned the value of 1 and the activity of the three synthetic steroids was calculated by graphic estimate. The activity of 19-norprogesterone was approximately the same in both tests, whereas norethindrone and more notably chlormadinone acetate were found to be much more active in the rabbit antiovulatory assay than in the rat test. It has been suggested that in the rat the steroids are acting on the hypothalamus whereas in the rabbit the central nervous system is involved [17].

Influence of chlormadinone acetate on fertility in the rabbit

Chlormadinone acetate (Fig. 12) (after the data of *Bennet* et al. [2] prevented fertilization completely at the 1 mg daily dosage. It is not certain whether the fertility of the eggs (20 %) at the high dose of 10 mg is significant. In the same study, CAP significantly decreased the number of spermatozoa recovered from both the uterus and Fallopian tubes (Fig. 13) suggesting a cervical barrier.

Fig. 12. The effect of chlormadinone acetate (CAP) on the percentage of eggs fertilized in the female tract of the rabbit following artificial insemination and gonadotropin stimualtion . (Data of *Bennett* et al. [2].

CAP mg/kg/day 3 days	(No. of Rabbits)	Eggs fertilized %	Mean No. of Eggs/Rabbit	Mean No. of Corpora Lutea/Rabbit
0	(6)	85	8.5	8.8
0.001	(6)	80	5.5	6.3
0.01	(5)	45	6.4	8.2
0.1	(6)	40	6.4	6.4
1.0	(4)	0	6.2	6.2
10	(4)	20	9.3	10

Fig. 13. The effect of chlormadinone acetate (CAP) in the number of spermatozoa received from the female reproductive tract of rabbits after artificial insemination [2].

CAP mg/kg/day	(No. of Rabbits)	Mean No. of Spermatozoa in Millions	Mean No. of Spermatozoa recovered	
			Uterus	Fallopian Tubes
0	(12)	199	383,000	3,100
0.1	(1)	456	143,000	250
1.0	(3)	338	63,000	420
10	(3)	236	18,000	420

Previous studies of the action of chlormadinone acetate on the inhibition of pregnancy in the rabbit [2] have now been extended with the aim of determining something of the mechanism of action of this low dose contraceptive [3]. Chlormadinone acetate has been investigated over different dosage schedules using two different indices of pregnancy inhibition. The antifertility effect of preovulatory dosage with CAP, ceasing at the time of ovulation induction, was compared to that continued up to the eighth day of pregnancy (Fig. 14). When chlormadinone acetate was investigated in this way, it was found that no increased antifertility potency was achieved (Fig. 15) by extending dosage beyond ovulation. This would suggest that if the compound possessed postovulatory dosing antifertility effects, they were masked by a lower preovulatory dose threshold. Further investigation (Fig. 16) then showed that with preovulatory treatment, although fertilization was inhibited by chlormadinone acetate (0.2 mg/kg/day), lower doses (0.09 mg/kg/day) still completely inhibited pregnancy. More detailed studies (Fig. 17) revealed the antifertility effect to be upon inhibition of implantation, mediated through inhibition of blastocyst development.

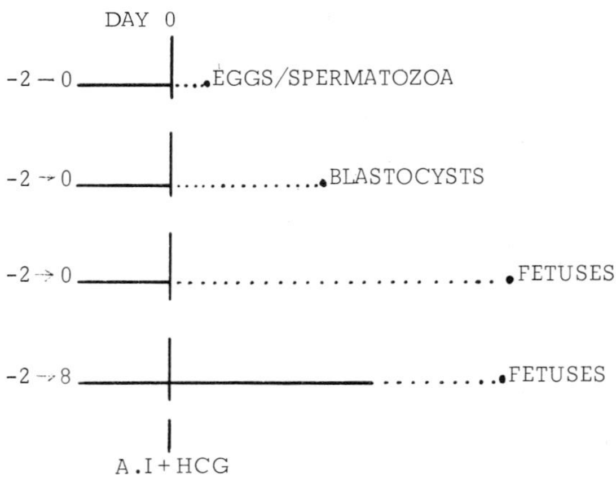

Fig. 14. Summary of dosing and recovery schedules.

Fig. 15. Comparison of effects on implantation of preovulatory dosing with chlormadinone acetate ceasing at ovulation (–2→8).

Dose mg/kg/day	% Inhibition of Implantation –2→0	–2→8
–	10	43
0.01	19	22
0.1	100	100
1.0	100	100

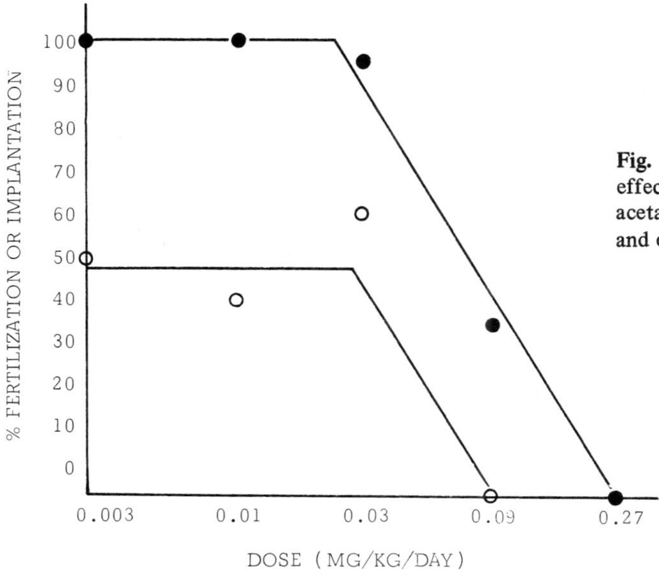

Fig. 16. Comparison between effects of chlormadinone acetate on fertilization (•—•) and on implantation (0-0).

Fig. 17. Effect on blastocyst volume of maternal dosage with chlormadinone acetate (−2 → 0).

Dose mg/kg/day	Age of Blastocysts (Days)	% Recovery	Mean Volume (mm^3)
0.09	6.5	17	12
–	6.5	98	62
0.09	7.0	10	41
–	7.0	98	262

Thus chlormadinone acetate effects inhibition of pregnancy in the rabbit by different mechanisms at high and at low dose levels. It is also apparent that the steroid produces both activities by preovulatory dosage alone.

These findings illustrate a remarkable correspondence between the effects of chlormadinone acetate in the rabbit and women. At dose levels in excess of those necessary for ovulation inhibition, 17-acetoxyprogestin inhibits sperm migration through the cervix in the rabbit and in women. At lower dose levels (i.e., on the borderline of ovulation suppression) the inhibition of sperm migration is in doubt in the human and is not the effective agent in the rabbit. It is interesting in this context that the dose level necessary for complete suppression of ovulation in the rabbit by chlormadinone acetate is greater than 0.09 mg/kg/day.

Low dose continuous treatment with CAP in women

Chlormadinone acetate is effective in conception control in women at low dose levels (0.5 mg/day) by continuous dosage throughout the cycle. Ovulation is not suppressed in the majority of cases. The uterine endometrial pattern is considered to be changed in women because of the increased incidence of breakthrough bleeding in the first few months of treatment. This must at least imply a decrease in maternal receptivity for implantation at the endometrial level if not changes in the secretory pattern of the uterine glands which could affect the blastocyst development.

Highly acceptable antifertility control can be obtained with 0.5 mg continuous treatment with chlormadinone acetate (Fig. 18). In three different studies in various centers of the US and Mexico the Pearl Index for the method varied from 1.2 to 2.4. One short-coming of this method is the occurrence of shortened cycles. Bleeding intervals are illustrated in Fig. 19 illustrating that some 65.3 percent had cycles of 25 to 35 days.

Martinez-Manautou and *Giner-Velazques* [13] did culdoscopic studies in the post-ovulatory phase in a group of fifty women. In 37 of the cases, a corpus luteum could be demonstrated through direct observation; in 6 cases the morphology of the

Fig. 18. The effectiveness of chlormadinone acetate as a continuous minipill at a daily dose of 0.5 mg.

	No. of Treatment Months (Lunar)	Pregnancies		Index: all Pregnancies	Index: Method Failures
		Total No.	No. of Method Failures		
Pilot Study USA	5,611	19	11	4.0	2.4
Pilot Study Mexico	33,064	114	36	4.1	1.3
Multi-Center USA	11,880	26	12	2.6	1.2

Fig. 19. The bleeding intervals of women on daily dosage of 0.5 mg of chlormadinone acetate [13].

Days	Cycles	%
21 to 24	3597	12.8
25 to 35	18374	65.3
36 to 59	5559	19.7
60 – –	628	2.2
Total	28158	100.0

ovaries indicated inactivity of the gonad; in 3 cases follicular activity without formation of a corpus luteum was observed; in 3 cases ovarian cysts and in one case there was a corpus luteum not of recent origin.

In 18 of the 37 cases in which the culdoscopy showed a probable corpus luteum, a biopsy of the ovary was performed and the presence of an active corpus luteum confirmed in 12 of them. Three of the biopsies corresponded to normal ovarian tissue, two of them showed thecaluteal cysts, and one an involuted corpus luteum (Fig. 20).

Fig. 20. The culdoscopic studies of women receiving a daily dose of 0.5 mg chlormadinone acetate [13].

Diagnosis	Number of Cases	%
Corpus luteum present	37	74.0
Inactive ovaries	6	12.0
Follicular activity	3	6.0
Ovarian cysts	3	6.0
Old corpus luteum	1	2.0
Total	50	100.0
	Ovarian Biopsies	
Active corpus luteum	12	66.8
Ovarian tissue without corpus luteum	3	16.6
Thecalutein cyst	2	11.1
Involutive corpus luteum	1	5.5
Total	18	100.0

One thousand and seventy endometrial biopsies were performed in women receiving continuous daily doses of 0.5 mg chlormadinone acetate. The histological endometrial pictures show the following percentages: during the first twelve cycles of therapy, 78.4% were diagnosed as normal or irregular secretory; 9.6% irregular; 7.6% proliferative; and 4.4% inactive. This pattern remained practically unchanged after 52 cycles. This study was performed during days 23 to 28 of the menstrual cycle. The large percentage of normal secretory and irregular secretory endometria supports the fact that ovulation occurred. Endometrial inactivity was present only in a low percentage of the cases.

With the purpose of determining the effect of the compound upon cervical mucus, studies were undertaken in a group of 115 women who had been receiving the compound for periods ranging between one to seven months. Postcoital tests and additional studies, such as ferning, stretching and soldium chloride content, were done.

In general, the cervical mucus obtained from chlormadinone acetate-treated women represented a physicochemical barrier to sperm penetration. The postcoital tests showed normal results in 5.2% of the cases. Ferning is below ++ in 88.7% of the cases. Viscosity is increased and the concentration of sodium chloride does not show the ovulatory peak normally present in women without therapy. In all samples the amount of leukocytes and epithelial cells increased in women being treated.

It is very likely that all the alterations occurring at the level of cervical mucus may play an important role in inhibiting sperm migration and survival. Therefore, interference to reach the Fallopian tubes is increased and fertilization becomes extremely difficult. These findings may be supported by the studies reported by *Gutierrez-Najar, Giner-Velazquez,* and *Martinez-Manautou* [5] who were unable to recover spermatozoa by tubal flushing in 90% of a group of 24 women receiving continuous chlormadinone acetate, while in 15 controls sperm was recovered from the tubes in 93% of the cases (Fig. 21). However, the participation of other parameters which could also play a role in the contraceptive efficacy of this compound deserve further exploration.

Fig. 21. Postcoital spermatozoa recovery from fallopian tubes on patients receiving daily 0.5 mg chlormadinone acetate for contraception [13].

		Spermatozoa (1)	
	Number of Cases	Recovered Number of Cases (%)	Absent Number of Cases (%)
Control Group	15	14 (93.3)	1 (6.7)
Chlormadinone acetate	24	2 (8.3)	22 (91.7)
Total	39	16	23

(1) $P < 0.001$

Summary and conclusion

Chlormadinone acetate (CAP) is a highly active progestational agent 35 times more active than norethindrone by the oral route and 69 times more active than progesterone by subcutaneous injection.

The antiovulatory activity of CAP by the subcutaneous route was 40 times progesterone. Compounds of the 17-acetoxyprogesterone type showed a high correlation between progestational and antiovulatory activity, both tests being performed in the rabbit. Similar high antiovulatory activity was found for CAP by the oral route. The compound was 35 times more active than norethindrone by both the progestational and antiovulatory tests.

By means of the parabiotic assay method for inhibition of gonadotropins in the rat, chlormadinone acetate was found to be particularly inactive.

On the basis of studies in our laboratory as well as elegant experiments of *Chang* [4], CAP has been demonstrated to produce antifertility effects in the rabbit by disturbing sperm transport, capacitation, ova transport, and implantation.

Continuous low dosage (0.5 mg per day) of chlormadinone acetate without estrogens is a highly effective contraceptive agent in women and is only minimally effective as a restrainer of the hypothalamic-anterior pituitary-ovarian axis. On the basis of our present knowledge, the contraceptive effect is due to some of the same factors demonstrated in the rabbit. Outstanding effects include the formation of a sperm barrier due to a viscous cervical mucous, a hostile endometrium for implantation. and abnormal transport rates for both sperm and ova.

References

[1] *Barnes, L. E., Schmidt, F. L. & Dulin, W. E.:* Proc. Soc. Exptl. Biol. **100,** 820 (1959).

[2] *Bennett, J. P., Vallance, D. K. & Vickery, B. H.:* Postgraduate Med. J., Dec., 1967 Supplement. Fertility Control and Ovarian Hormone Replacement, pp. 39–42.

[3] *Bennett, J. P., Vickery, B. H. & Dorfman, R. I.:* (1969). (Unpublished).

[4] *Chang, M. C.:* Endocrinology **81**, 1251 (1967).

[5] *Gutierrez-Najar, A., Giner-Velazques, J. & Martinez-Manautou, J.:* Presented before the Sixth Annual Meeting of the A.A.P.P.P., San Antonio, Texas, April, 1968.

[6] *Kincl, F. A.:* Endokrinologie **44**, 67 (1963).

[7] *Kincl, F. A.:* Vth Congr. Int. per la Reprod. Animale, Trento, Vol. **III,** 294 (1964), (Abst.).

[7a] *Kincl, F. A., Birch, A. J. & Dorfman, R. I.:* Soc. Exptl. Biol. Med. **117,** 549–552 (1964).

[8] *Kincl, F. A. & Dorfman, R. I.:* Acta Endocrinol. **42**, Suppl. 73 (1963).

[9] *Kincl, F. A. & Dorfman, R. I.:* Steroids **2**, 521 (1963 b).

[10] *Kincl, F. A. & Dorfman, R. I.:* J. Reprod. Fertility **10**, 105 (1965);

[11] *Lipschutz, A. & Fuguerose, S.:* Acta Endocrinol. **26**, 371 (1957).

[12] *Makepeace, A. W., Weinstein, G. L. & Friedman, M. H.:* Am. J. Physiol. **119,** 512 (1937).

[13] *Martinez-Manautou, J. & Giner-Velazques, J.:* 4th Asian Cong. Obst. Gynecol., Singapore, November, 1968.

[14] *McCormack, C. E. & Meyer, R. I.:* Gen. Comp. Endocrinol. **3**, 300 (1963).

[15] *Pincus, G. & Chang, M. C.:* Acta Physiol. Latino-Americano **3**, 177 (1953).

[16] *Ringold, H. J., Batres, E., Bowers, A., Edwards, J. & Zderic, J.:* J. Am. Chem. Soc. **81,** 3485 (1959).

[17] *Rudel, H. W. & Kincl, F. A.,* Acta Endocrinol. Suppl. **105**, 12 (1966).

[18] *Shipley, E. G.:* In: Methods in Hormone Research, edited by *R. I. Dorfman,* Academic Press, New York, 1962, Chapt. 5, 179.

Discussion

E. Johansson: Together with Drs. ***Larsson-Cohn, Wide*** and ***Gemzell*** have I studied the anti-ovulatory effect of norethindrone and chlormadinone, 0.5 mg a day continuously in healthy women. On Fig. 22 with 0.5 mg norethindrone, you have first a normal control cycle with progesterone, pregnanediol, total estrogens and LH, thereafter three cycles during norethindrone treatment with low progesterone levels during the "luteal phase", but increased excretion of estrogens. On Fig. 23 a small luteal complex is seen in the first treated cycle but in the next cycle it is gone. And if I could show you Fig. 24 you see a complete abolishment of the luteal complex during treatment, but there is an increased excretion of total estrogens.

When we now treated women with the same dose of chlormadinone we got normal progesterone levels during the luteal phase in most of the cases. Thus, women seem to react more strongly to norethindrone than to chlormadinone as judged from the suppression of the luteal function, quite opposed to the findings in the rabbit (Fig. 25).

K. Fotherby: I would like to make a comment about Dr. ***Dorfman's*** paper, The big advantage, of course, of the mini-pill is that it is not supposed to inhibit ovulation. As Dr. ***Johansson*** has shown this is probably true; it does however have effects upon the ovary – this is also apparent from Dr. ***Johansson's*** figure. We have done a fairly large series with a related progestogen-norgestrel-which has an antifertility effect in a dose of only 50 μg per day as opposed to Dr. ***Dorfman's*** 500 μg per day of chlor-madinone. We were struck in our early investigations by the fact that pregnanediol excretion was low in women taking this small dose of norgestrel. We just completed a series where the corpus luteum was removed at operation in women who had been treated with norgestrel, and in every case a normal corpus luteum was present, with normal histological structure. But these women were not producing pregnanediol, so there is a functional disturbance of the corpus luteum in these women taking the mini-pill. This is very marked with norgestrel; it may be less marked with chlormadinone, but there are some cases which show that chlormadinone has this effect, also.

F. G. Sulman: I wonder whether the effect of "waning" which we have...

R. I. Dorfman: The effect of what?

F. G. Sulman: ... of waning, fading ... if this effect is only due to the explanation you gave: readjustment of the hypothalamus – pituitry axis. We did some experimental work on rats and rabbits with regard to inactivation of prednisone and stilbestrol.

We found that we can induce liver enzymes to "learn" to decompose these substances quicker, so we called them "trained" enzymes; today we use the word "induced" enzymes. There is perhaps little difference between these terms but I prefer the word "trained" to "induced". You have for instance no "trained" enzymes in the woman in the second month of pregnancy, when she starts vomiting; but you have them in the fourth month, when she stops vomiting, because she is trained her enzymes to decompose pregnancy steroids. Here, she has "trained" existing enzymes, but she has not "induced" ***new*** enzymes. The same seems to pertain to the inactivation mechanism of prednisone and stilbestrol which we studied in animals, by giving them together with SKF-525-A, a biotransformation blocker. This set-up showed that by adding a biotransformation blocker you would not get the result of "training" liver enzymes.

H. R. Lindner: Chlormadinone acetate has some structural resemblance to Provera (6α-methyl-17α-acetoxyprogesterone, Upjohn). Provera produces a very marked reduction in adrenal size in the rat. Does chlormadinone acetate have a similar side-effect?

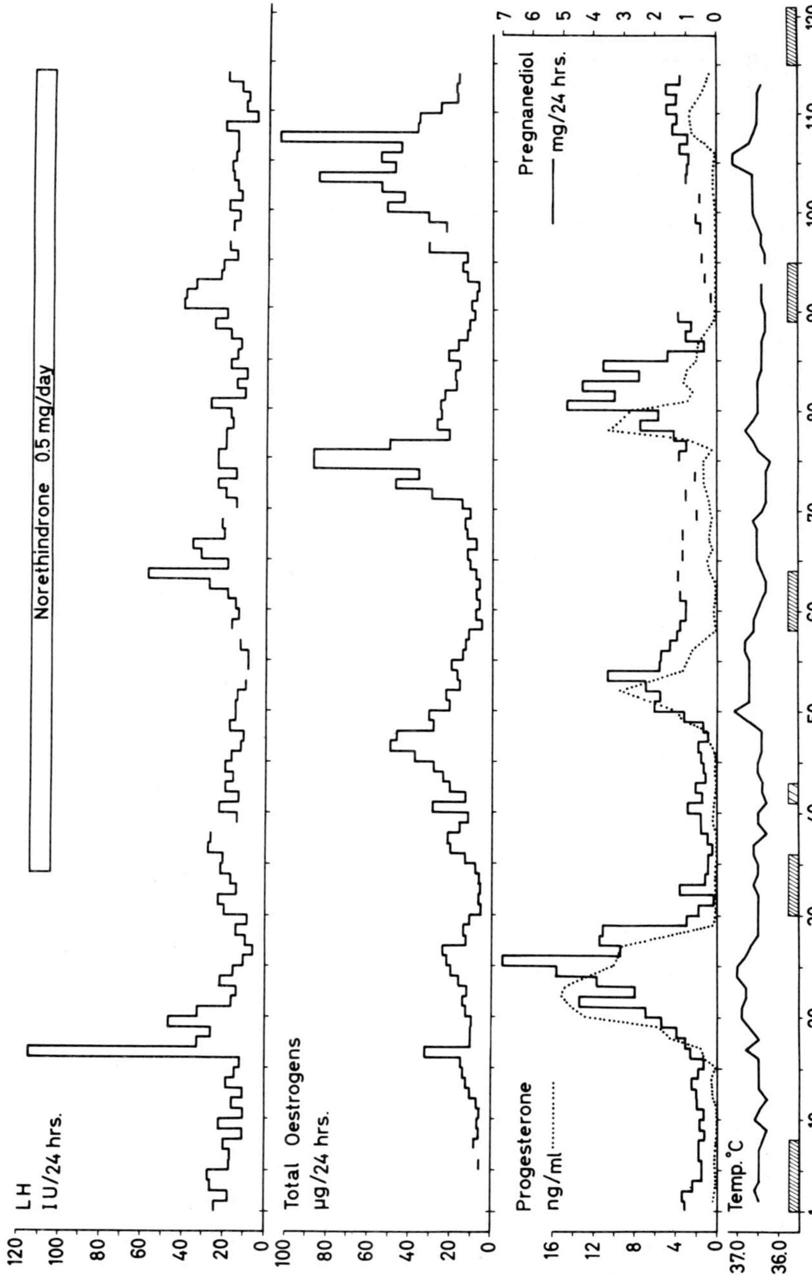
Norethindrone 0.5 mg/day
LH
I U/24 hrs.
Total Oestrogens
µg/24 hrs.
Progesterone
ng/ml
Pregnanediol
mg/24 hrs.
Temp. °C
Day of investigation
120
100
80
60
40
20
0
37.0
36.0
1
10
20
30
40
50
60
70
80
90
100
110
120

Fig. 22

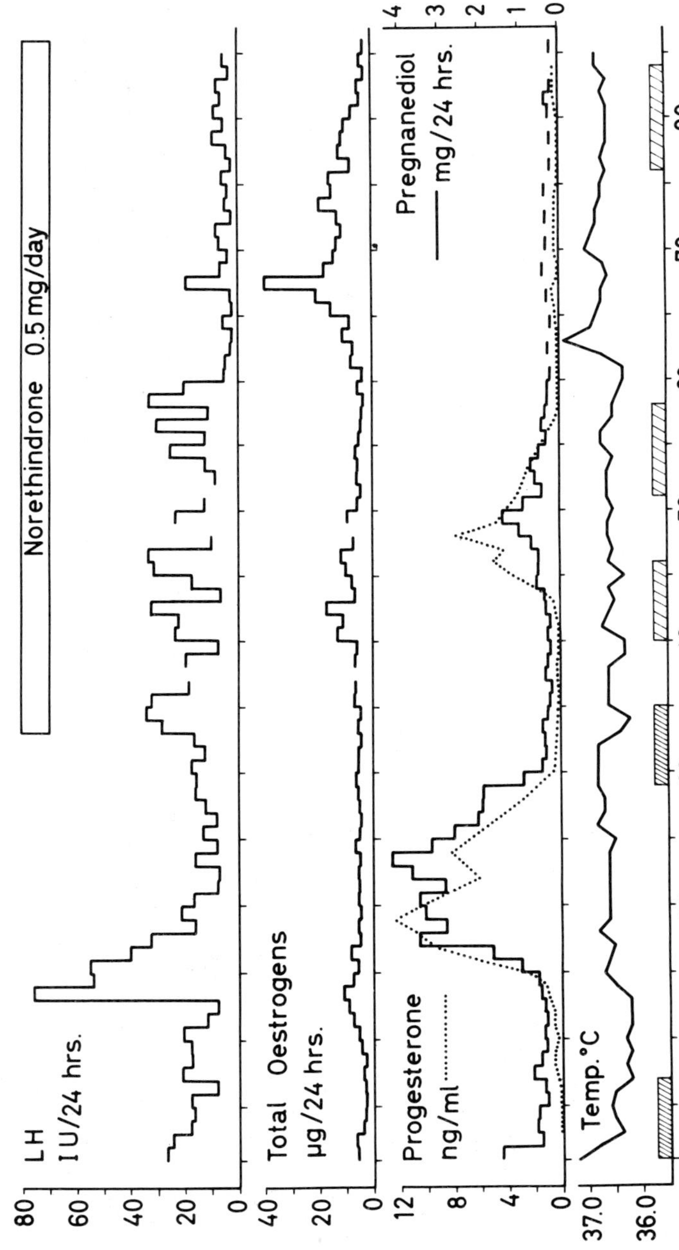
Norethindrone 0.5 mg/day
LH
I U/24 hrs.
Total Oestrogens
µg/24 hrs.
Progesterone
ng/ml
Pregnanediol
mg/24 hrs.
Temp. °C
Day of investigation

Fig. 23

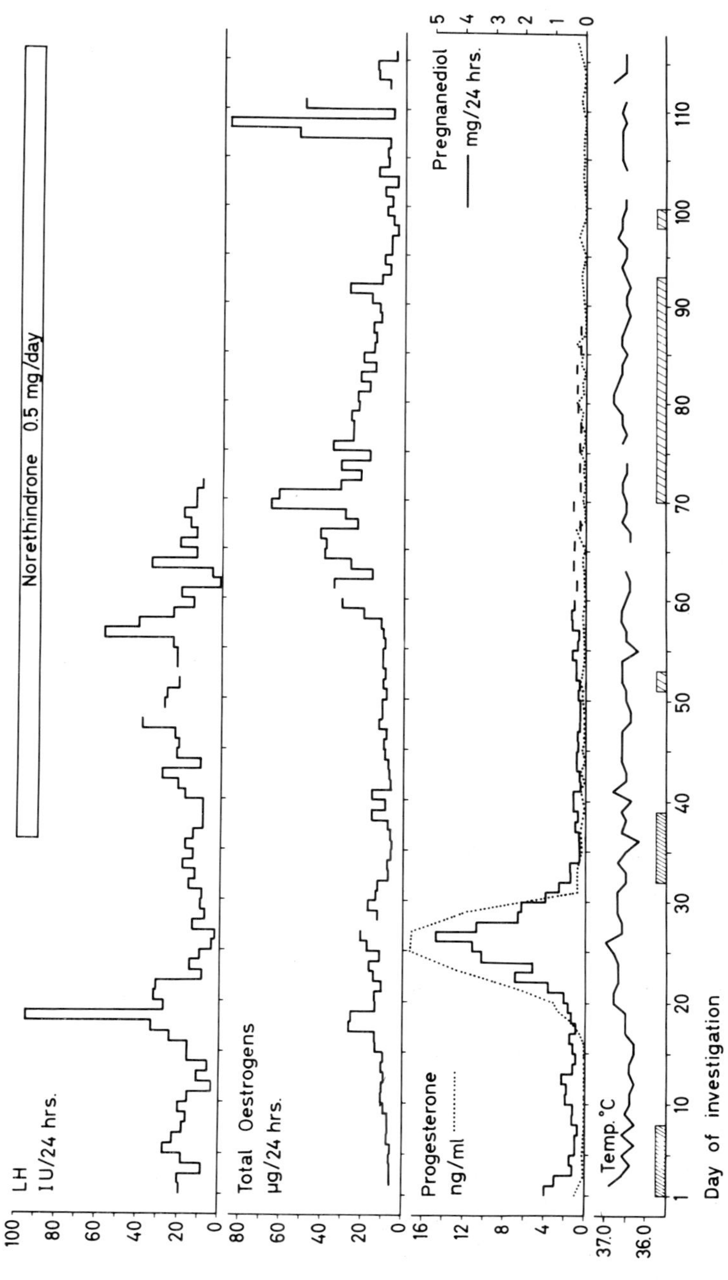
Norethindrone 0.5 mg/day
LH
I U/24 hrs.
Total Oestrogens
µg/24 hrs.
Progesterone
ng/ml
Pregnanediol
mg/24 hrs.
Temp. °C
Day of investigation

Fig. 24

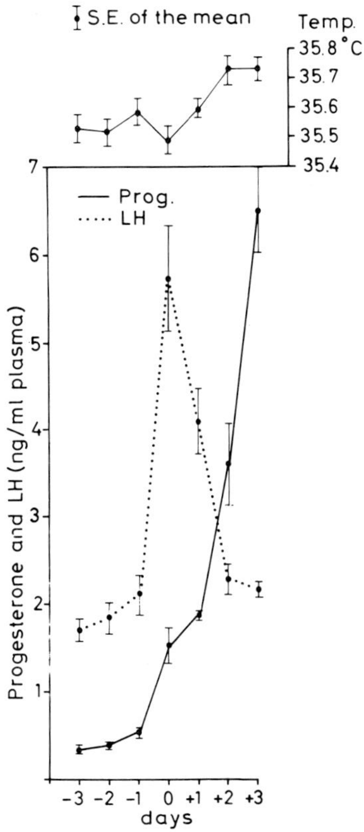

Fig. 25. A composite graph of plasma LH and progesterone and the basal body temperature during the mid-cycle portion of 16 ovulatory menstrual cycles. The points are mean± S.E.

R. I. Dorfman: In experimental animals such as the rat CAP is less active than provera in the adrenal involution assay. In man both compounds are essentially inactive on the adrenal.

M. C. Chang: Would like to ask you what the half-life of your CAP is.

R. I. Dorfman: The half-life of CAP is relatively great especially as compared to the naturally occuring progesterone.

M. C. Chang: Is there a difference between the half-life of CAP given by injection and orally administered ?

R. I. Dorfman: We do not have any direct data to evaluate this point at this time.

M. C. Chang: How does it compare with Provera ?

R. I. Dorfman: It would be of the same order.

K. W. Humphrey: Prof. *Emmens* and his group has been looking at anti-estrogens and contraception and we are very interested to hear your "gut feeling" that anti-estrogens are necessary, but *Emmens* has found that assays of compounds as anti-estrogens using the mouse uterine weight gain test are not very reliable. Have you looked at CAP and estrogens in the rodent vagina ?

R. I. Dorfman: No, we have not done such studies.

Regulation of Corpus Luteum Function by Gonadotropins and Related Compounds

J. A. McCracken, M. E. Glew and L. K. Levy

Worcester Foundation for Experimental Biology, Shrewsbury, Massachusetts, USA

Summary: The direct local effects of pituitary gonadotropins and related compounds on the secretion of ovarian steroids and on ovarian blood flow will be described using sheep with the left ovary autotransplanted to a jugulo-carotid skin loop in the neck. This preparation permits long term access to both the arterial and venous sides of the ovarian circulation in the unstressed, unanaesthetized animal [11, 29].

It has been established that, based on steroid secretion rates, the sheep with transplanted ovaries show abnormally long cycles with retention of the corpus luteum [3]. Some animals however have shown fluctuations in progesterone and estradiol secretion indicating some cyclicity of ovarian function. However only a few animals have shown behavioral estrous. Because of this failure to show estrus regularly the ewes with ovarian transplants were treated with progesterone then given PMS subcutaneously in order to induce ovulation at a given time. After PMS there was a rapid rise in estrogen and androgen secretion representing follicular growth over several days. On the other hand progesterone secretion did not begin to rise until the sixth day indicating that ovulation had occurred. The rise in estrogen and androgen secretion following PMS was very consistent and occurred whether or not the corpus luteum had been surpressed with progesterone injections. i.e. the response occurred whether or not there was a high endogenous progesterone secretion.

Treatment with PMS it possible to carry out further experiments with gonadotrophic hormones at a known time after ovulation. Controlled infusions directly into the ovary via the carotid artery were made with each of the three purified ovine pituitary hormones [28].

With ovine LH a dramatic increase occurred on both steroid secretion and blood flow through the ovary. The increase in steroid secretion after LH is in agreement with several *in vitro* studies by various workers. The dramatic increase in blood flow through the transplanted ovary is similar to the findings for the rat ovary after systemic LH [59].

In contrast the infusion of ovine prolactin at doeses imilar to or greater then LH had no effect on either blood flow or steroid secretion. The infusion of FSH also had no effect on blood flow and on some occasions appeared to diminish progesterone secretion by the ovary.
The effects of ACTH, Prostaglandin, and Histamine on the ovary will also be described.

Manuscript received: 15 March 1969

In order to further investigate the effect of the uterus on corpus luteum function, the left ovary, fallopian tube, uterine horn, cervix, and anterior vagina were autotransplanted together into the jugulo-carotid loop and the vagina was exteriorized at the base of the neck. With this preparation the sheep showed cycles of normal duration with behavioral estrus and it was possible to observe changes at estrus in the color and mucus secretion of the cervix. In addition progesterone secretion showed a pattern identical to that reported by other workers for the sheep ovary *in situ* [29a]. Additional evidence for a auterine luteolysin will be presented in a sheep with the congenital absence of one uterine horn. This animal gives additional support to the concept that the uterus is involved in the regulation of the life span of the corpus luteum since surgical interfèrence in this case cannot be invoked to explain the prolonged luteal phase exhibited by this animal [27].

Introduction

Many of the steroid biosynthetic pathways have been elucidated by use of *in vitro* preparations of slices, homogenized or minced ovarian tissue [42, 43, 50]. Most of these workers found that gonadotrophic stimulation of the ovary *in vivo,* prior to its removal for *in vitro* studies, was highly effective in augmenting the biosynthetic capacity of the preparation.

Mason, Marsh and *Savard* [33] and *Savard* and *Mason* [45] have shown that slices of bovine luteal tissue *in vitro* will produce progesterone and that this can be stimulated by the addition to the incubation medium of LH, HCG, and FSH but not by ACTH, prolactin or serum albumin. More recently, *Marsh* and *Savard* [31] have demonstrated that the addition of cyclic adenosine 3′,5′-monophosphate to *in vitro* preparations of bovine corpora lutea significantly stimulates steroid synthesis. They suggested that this mechanism is involved in the mode of action of LH. *Savard* and coworkers have further shown that the conversion of acetate-1-^{14}C to progesterone-^{14}C by luteal slices is increased fivefold by the addition of LH [44, 46].

In recent years a number of *in vivo* studies have been carried out in different species of animal to determine the effects of pituitary gonadotropins on the secretion of steroids from the ovary (Fig. 1). In most cases the ovarian vein was cannulated *in situ* in the anaesthetized animal and ovarian steroid secretion was determined after various gonadotropins were given, either systemically [17] or via the ovarian artery [7]. Using these techniques there have been a number of conflicting reports as to the effects of gonadotropins on the secretion of steroids from the ovary. For example *Short, McDonald* and *Rowson* [49] were unable to demonstrate any marked effects on the ovarian secretion of progesterone in the ewe following large doses of LH, FSH or prolactin given systemically. On the other hand *Domanski, Skrzeczkowski, Stupnicka, Fitko,* and *Dobrowolski* [8] by infusing the ovary of the ewe via its arterial supply have demonstrated an increased secretion of progesterone following large

Fig. 1. The effect of gonadotropins on steroid secretion by the ovary *in vivo.*

Species	Route of Administration	Gonadotropin	Effect on Steroid Secretion	References
Rabbit	Systemic	LH	Stimulation	[17]
Bitch	Systemic	HCG	Stimulation	[41]
Bitch	Systemic	HCG	Stimulation	[56]
Bitch	Ovarian artery *in situ*	HCG	Stimulation	[36]
Monkey	Systemic	HCG	Stimulation	[16]
Rat	Systemic	LH Prolactin	Stimulation No effect	[32]
Rat	Systemic	LH Prolactin	Slight Depression No effect	[60]
Ewe	Systemic	LH FSH Prolactin	No effect No effect No effect	[49]
Ewe	Ovarian Artery *In situ*	LH Prolactin FSH	Stimulation Stimulation Depression	[8]
Ewe	Systemic (Chronic)	LH (crude) Prolactin	C.L. maintenance No effect	[35]
Ewe	Ovarian artery (transplanted ovary)	LH Prolactin FSH	Stimulation No effect Slight depression	[28]

doses of LH and prolactin but not by FSH. Two other *in vivo* studies have demonstrated that LH but *not* prolactin will cause an increase in steroid secretion, in the rabbit [17] and in the rat [60].

These above studies can only be carried out in the anaesthetized animal over relatively short periods of time. In addition the interpretation of these results is complicated by the possible adverse effects of cannulating the ovarian vessels on blood flow and steroid secretion. In addition anaesthetics particularly the barbiturates are known to affect the release of trophic hormones from the Pituitary [10a].

Lindner, Sass, and *Morris* [25] described a technique whereby a cannula was inserted into the ovarian vein of the ewe then led out through a stab incision in the flank. It was thus possible to obtain ovarian venous blood in the conscious animal over a period of several days. However cannulae frequently became blocked and it was not possible to reach the arterial circulation of the ovary.

It would therefore appear that in order to gain access to both the arterial and the venous sides of the ovarian circulation, the ovary would have to be moved to a more accessible site. In 1958, *McDonald, Goding* and *Wright* [30] described a technique for transplanting the adrenal gland of the sheep to a jugulo-carotid skin loop in the neck with vascular anastomoses. A similar technique for the ovary was recently developed at the Worcester Foundation [11, 29]. In this procedure the right ovary is removed and the left ovary with its vascular pedicle is transplanted with vascular anastomoses to a skin loop in the neck (Fig. 2). In this preparation occlusion of the vessels in the cranial section of the loop caused the ovary to be supplied exclusively by the carotid artery and caused the ovarian venous blood to drain into the jugular vein where it was easily collected by means of a wide bore catheter. It was thus possible to obtain ovarian venous blood on a long-term basis in the unstressed conscious animal.

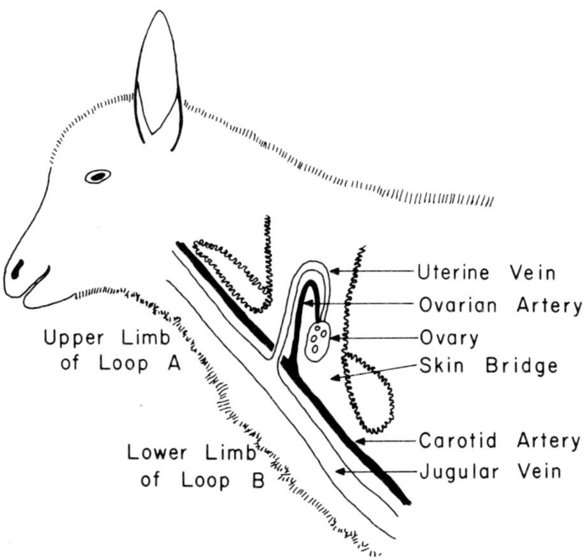

Fig. 2. Diagram showing ovary in jugulo-carotid loop. (From [29]).

Although these animals exhibited some degree of ovarian function the diestrous interval was very long with infrequent estrous behavior, probably as a result of removing the ovary from the luteolytic influence of the uterus. The spontaneous secretion of steroids was investigated to determine the basic pattern of steroid secretion [3]. Progesterone and 20α-dihydroprogesterone were determined by GLC [26] and the androgens and estrogens by double isotope derivative methods [2, 20, 38]. Two basic patterns of spontaneous steroid secretion emerged. As shown in Fig. 3

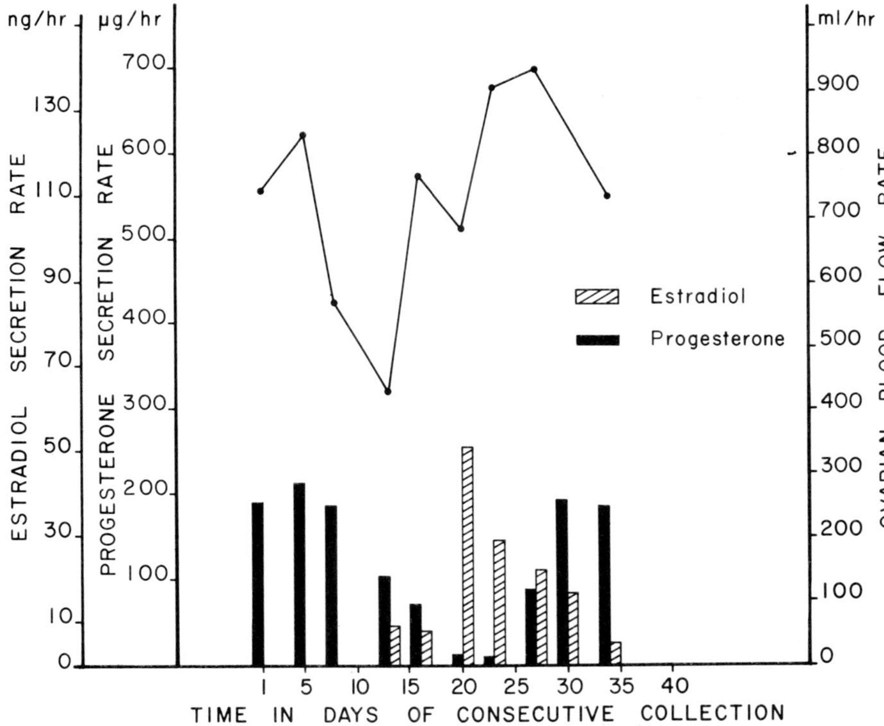

Fig. 3. Spontaneous secretion of steroids over a period of thirty days. (From [3]). Ewe No. 013

this animal showed a drop in progesterone secretion over a 30-day collection period. During this minimum period of progesterone secretion a peak of estradiol occurred which strongly suggested follicular growth, ovulation, and subsequent luteal development. Fig. 4 illustrates another type of animal in which there was an abnormally prolonged secretion of progesterone indicating retention of the corpus luteum. Estradiol showed intermittent peaks of secretion suggesting periodic discharge of pituitary gonadotropin.

The most likely reason for this alteration in ovarian function is the physical separation of the ovary from the local effects of a uterine luteolysin. A number of studies by, e.g., *Wiltbank* and *Casida* [58] and by *Moor* and *Rowson* [34] have shown that when a uterine horn of the ewe is removed the adjacent ovary shows retention of the corpus luteum for a prolonged period. The reason for the absence of behavioral estrus in ewes with transplanted ovaries is not clear but it may be that the estrogen secretion seldom rose high enough or the progesterone secretion never fell low enough to induce a behavioral response at estrus.

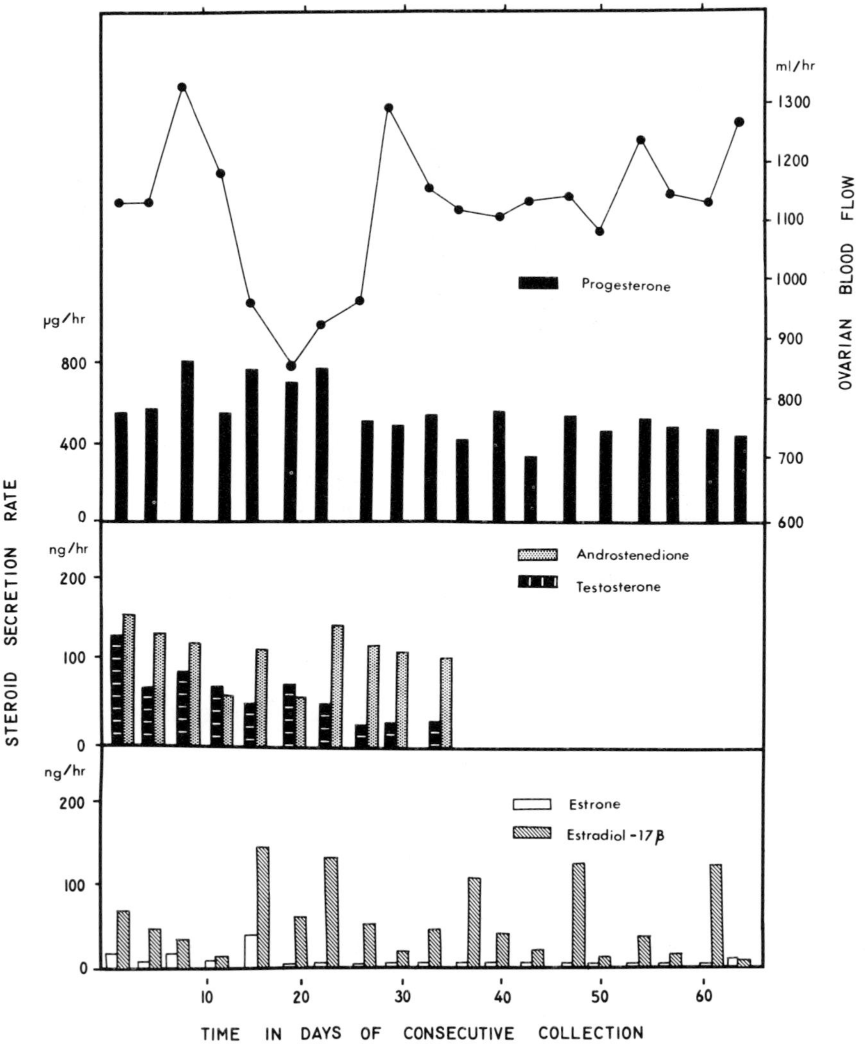

Fig. 4. Spontaneous secretion of steroids over a period of sixty days. (From [3]). Ewe No. 025

Because of the irregular occurrence of estrous behavior it was decided to induce ovulation artificially by giving PMS systemically. This treatment was performed in six ewes and an example of the pattern of steroid secretion following PMS is shown in Fig. 5. This animal was in anestrus at the time and showed a sustained rise in both androgens and estrogens following PMS. On the other hand progesterone did not

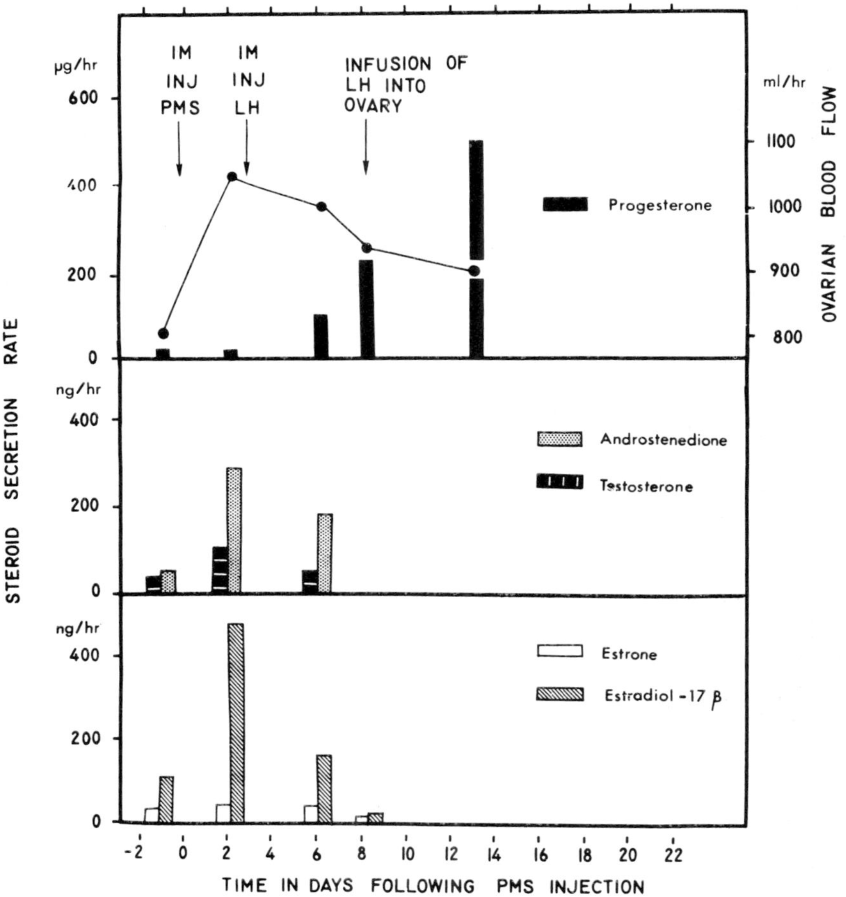

Fig. 5. The effect of PMS in a ewe during anestrus. (From [3]). Ewe No. 006.

begin to increase until the 6th day after PMS strongly suggesting that ovulation had occurred. This rise in estrogen secretion following PMS occurred even in the presence of a high endogenous secretion of progesterone as shown in Fig. 6. It is unlikely that this animal ovulated after PMS as the high progesterone levels would probably inhibit ovulation by preventing the release of pituitary LH. Certainly there was no marked increase in progesterone secretion up to 11 days later.

This treatment with PMS allowed experiments to be performed at a known time after ovulation. Infusions of the three pituitary gonadotropins were made directly into the ovary via its arterial supply during both the follicular and the luteal phase [28].

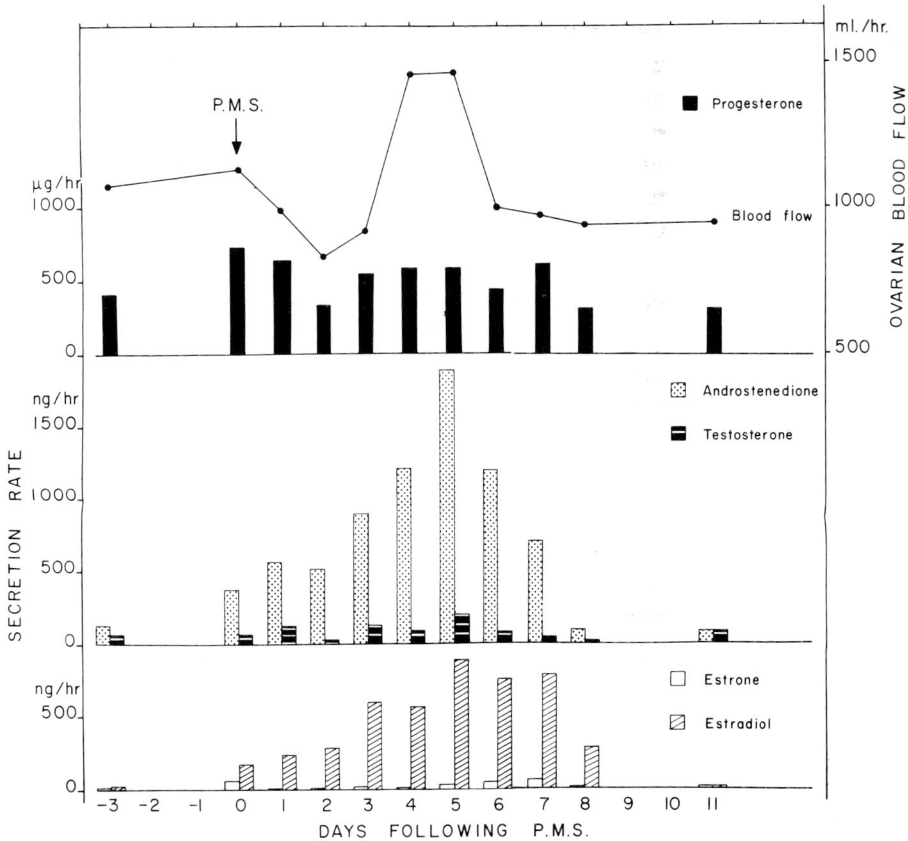

Fig. 6. The effect of systemic PMS in a ewe during the luteal phase. (From [3]). Ewe No. 025

On some occasions the follicular phase was superimposed on the luteal phase so that both follicular and luteal elements were present at the same time. The technique for carrying out the intra-arterial infusion of gonadotropins is shown in Fig. 7. The carotid artery was cannulated by means of an intra-arterial needle as previously described [28]. A pneumatic cuff was placed on the upper part of the loop and under these conditions the carotid arterial blood, containing the infusate, supplied the ovary exclusively. The ovarian venous blood was collected via the jugular cannula at various time intervals throughout the infusion.

Four infusions of each of the three purified ovine pituitary hormones (supplied by the NIH. Endocrinology Study Section, Bethesda, Maryland) were carried out at different times after PMS. The effect of infusing LH at various dose levels during

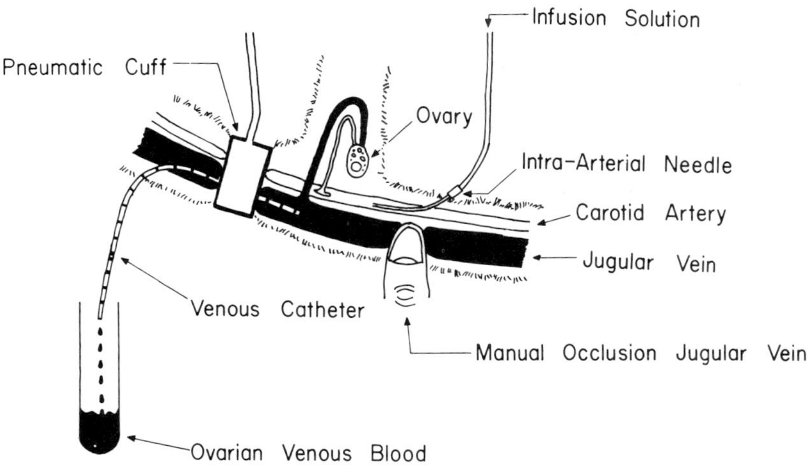

Fig. 7. Diagram of technique for continuous intra-arterial infusion of transplanted ovary and periodic collection of ovarian venous blood. (From [29]).

the early luteal phase of the cycle is shown in Fig. 8. It can be seen that both progesterone secretion and ovarian blood flow increased with LH at 0.1 μg/hour and also at subsequent dose levels. Estradiol secretion did not appear to be increased until the 10 μg/h infusion. In another experiment LH was infused over a longer period of time, again in the early luteal phase of the cycle (eight days after systemic PMS). The results are shown in Fig. 9. There was an increase in the secretion of all steroids except estrone and blood flow increased throughout the infusion. The greatest increase in steroid secretion occurred with androstenedione (600 %) followed by testosterone (400 %), estradiol (50 %) and progesterone (50 %). However the greatest increase in mass of secreted material occurred with progesterone which was present in 1000-fold greater concentration than the other steroids.

In contrast to the dramatic effects of LH, the infusion of FSH at dose levels varying from 0.1 μg to 100 μg/h had essentially no effect on steroid secretion or blood flow. In one out of four infusions there appeared to be a transient fall in progesterone secretion. During infusions of prolactin in doses similar to or greater than the LH infusions, no response was elicited in terms of steroid secretion or blood flow [28].

The lack of response of the ovary to short-term infusions of FSH is in keeping with the classical concepts formulated by *Greep* and his coworkers [13] that FSH alone will induce follicle growth but not the secretion of estrogens. The stimulation of ovarian steroidogenesis following the acute administration of FSH in some studies *in vitro* [12, 61] and *in vivo*, *Aakvaag* and *Eik-Nes* [1] could be explained by the LH content of the FSH preparations used. Even highly purified preparations of

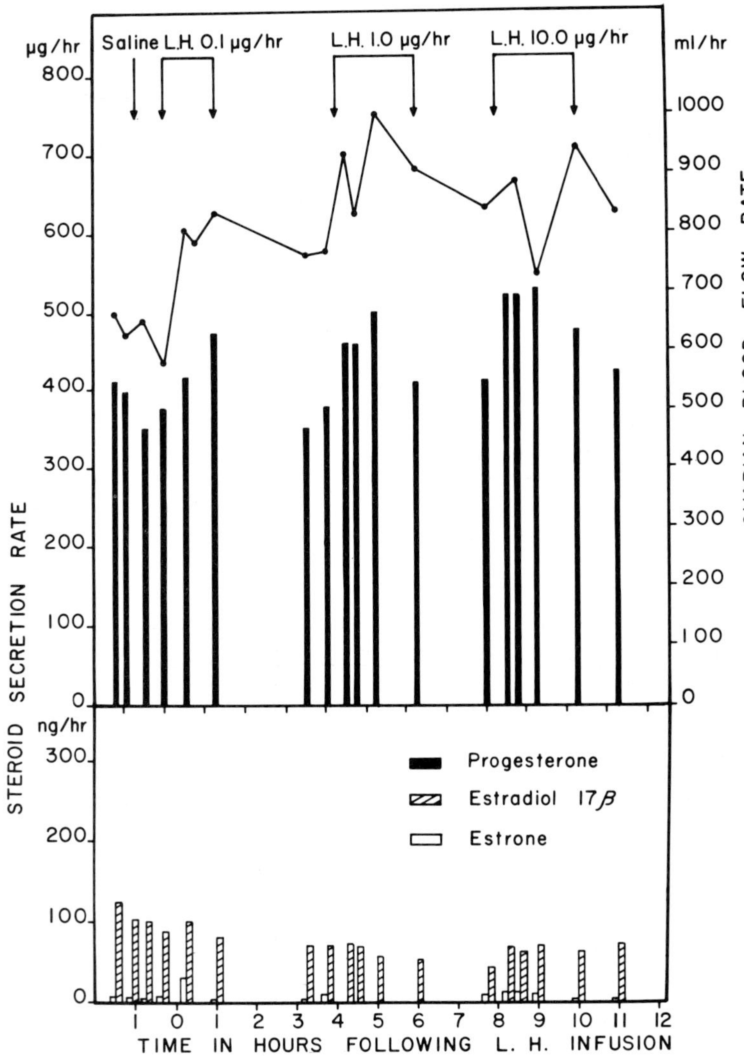

Fig. 8. The effect of the intra-arterial infusion of L. H. on blood flow and steroid secretion in ovarian transplant. (From [28]). Ewe No. 039

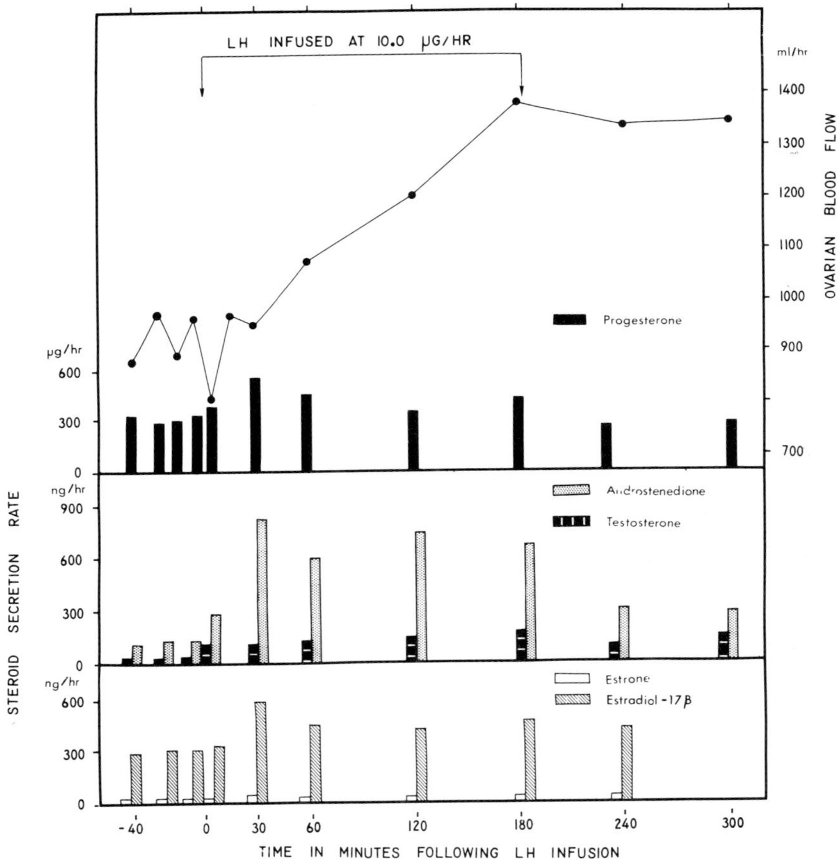

Fig. 9. The effect of the intra-arterial infusion of L.H. on blood flow and steroid secretion in ovarian transplant. (From [29]). Ewe No. 025.

FSH contain LH like activity as tested by the ovarian ascorbic acid depletion test [36a]. However, when human FSH was incubated with gamma globulin against human chorionic gonadotropin, the incubate produced growth and enlargement of the ovaries but no effect on estrogen secretion as judged by uterine weight [10]. In a similar fashion the stimulating effect of an FSH preparation *in vitro* on the conversion of androstenedione to estrogens was blocked by adding anti-serum prepared against HCG [22].

In the present study an effect following FSH might have been expected from the LH content alone (0.4%). It may be that the inhibitory effect of FSH which we observed in one infusion and which was also observed by *Domanski* et al. [8] may have been negated by active stimulation due to its LH content.

The lack of response of the transplanted ovary to infusions of prolactin is in agreement with numerous other studies *in vitro* [6, 9, 23, 24] and *in vivo* [17, 35, 60]. However *Domanski* et al. [8] reported that prolactin stimulated progesterone secretion in the sheep ovary *in vivo* and *Romanoff* [40] reported that prolactin increased progesterone secretion from the isolated perfused bovine ovary. The difference between these results and ours may be explained by the relatively large doses employed (1.0 to 27.0 mgm per hour) which may have had enough LH contamination to cause stimulation.

It may be that, as with FSH [19], prolactin has a synergistic action with LH on ovarian steroid secretion especially when given over longer periods of time. *Hilliard, Spies, Lucas,* and *Sawyer* [18] have shown that in the rabbit, prolactin treatment can restore the ovary to a state receptive to LH quite rapidly after depletion of precursors caused by LH stimulation. It may be necessary to give prolactin in conjunction with LH in order to demonstrate a long-term effect. In this connection *Nalbandov* and *Cook* [35] were able to demonstrate that crude LH but *not* purified LH was able to mainfain the functional integrity of the corpus luteum in the hypophysectomized ewe.

The stimulating effects of LH which were observed in this study with the transplanted ovary are in keeping with a number of reports using *in vitro* studies (see Introduction). The results reported in this paper agree particularly well with the findings of *Hammerstein, Rice* and *Savard* [14] who demonstrated that an increased incorporation of ^{14}C-acetate occurred in estrogen as well as progesterone when human corpora lutea slices were incubated with HCG or human pituitary LH. In addition *Kaltenbach* et al. [23] using sheep luteal slices *in vitro* showed that LH, but not FSH or prolatin, caused a significant increase in progesterone concentration.

Although *in vivo* there are conflicting reports for FSH and prolactin almost all studies with LH have consistently shown a increase in steroid secretion by the ovary (see Fig. 1). In some studies where prolactin has also been demonstrated to have an effect *in vivo* [8] and during *in vitro* perfusion, *Bartosik, Romanoff, Watson* and *Scricco* [4], the positive effects could be explained by the LH content in the relatively high doses of prolactin used in these studies.

The dramatic increase in blood flow which was observed with as little as 0.1 μg LH infused into the transplanted ovary is in agreement with the findings of *Wurtman* [59] who reported a doubling of the blood flow through the immature rat ovary when LH was given systemically as a single injection. He was able to mimic this effect when histamine was given instead. No effect was obtained following the administration of FSH, prolactin, serotonin or epinephrine. The work of *Szego* and *Gitin* [55] suggests that this action of LH may depend on the release of histamine or other vaso-active amines. Since they were able to dem onstrate that the histamine contentof the rat ovary dropped markedly after the administration of LH.

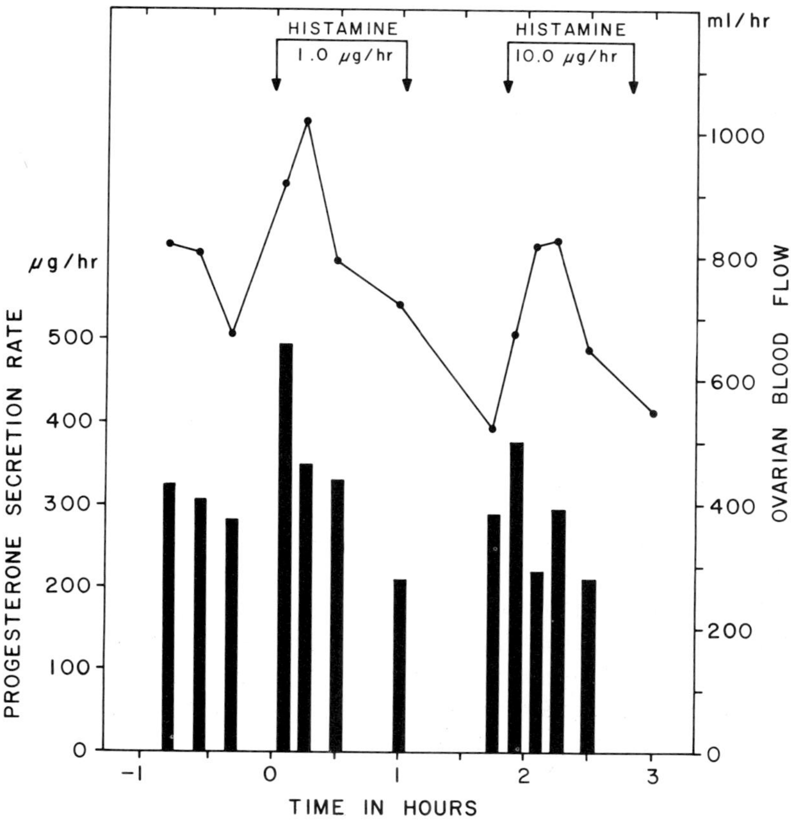

Fig. 10. The effect of histamine on ovarian blood flow and progester one secretion during the luteal phase. Ewe No. 056.

We have infused a number of vaso-active compounds through the transplanted ovary and the effect on blood flow is shown in Fig. 10, 11 and 12. It can be seen that both histamine (Fig. 10) and prostaglandin (Fig. 11) temporarily increased blood flow through the ovary even when infused in very small doses. Prostaglandins may very

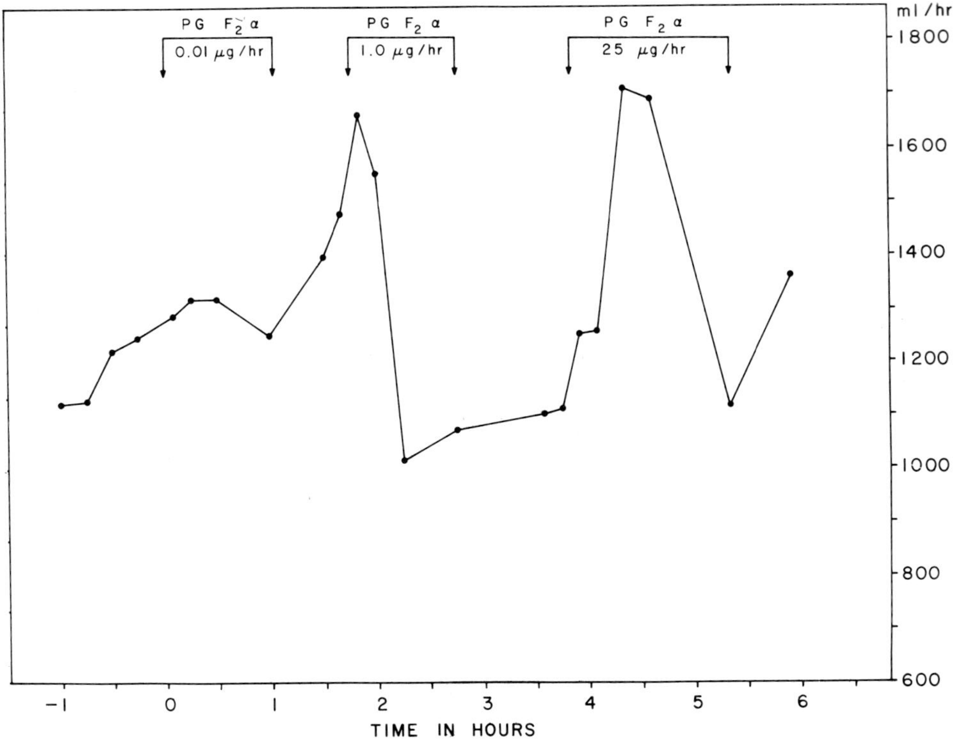

Fig. 11. The effect of P.G.$F_2\alpha$ on ovarian blood flow during luteal phase. Ewe No. 060.

well be involved in ovarian steroidogenesis as *Bedwani* and *Horton* [5] have shown that prostaglandin-E_2 will stimulate ovarian tissue incubated *in vitro. Pharriss, Wyngarden* and *Gutknecht* [37] have shown that prostaglandin F_{2a} has a luteolytic effect in rats when given in high doses. ACTH (Fig. 12) also caused an increase in ovarian blood thus emphasizing the fact that experiments of this nature must be performed on unstressed animals. This phenomenon has also been observed in the ovary of the bitch following the systemic administration of ACTH [52].

The hyperaemia induced by trophic hormones on their target organs appears to be a widespread phenomenon [53], e.g., estrogens and the uterus [54]. TSH and the thyroid [51] and ACTH and the adrenal gland [57]. Hyperemia by increasing critical materials via local transport mechanisms is probably an intrinsic part of many tropic hormones.

Short [48] has suggested that a uterine luteolysin may be more important for the control of corpus luteum function in the sheep than a luteotropin. Hysterectomy performed in the luteal phase of the cycle of ewes results in a marked extension of

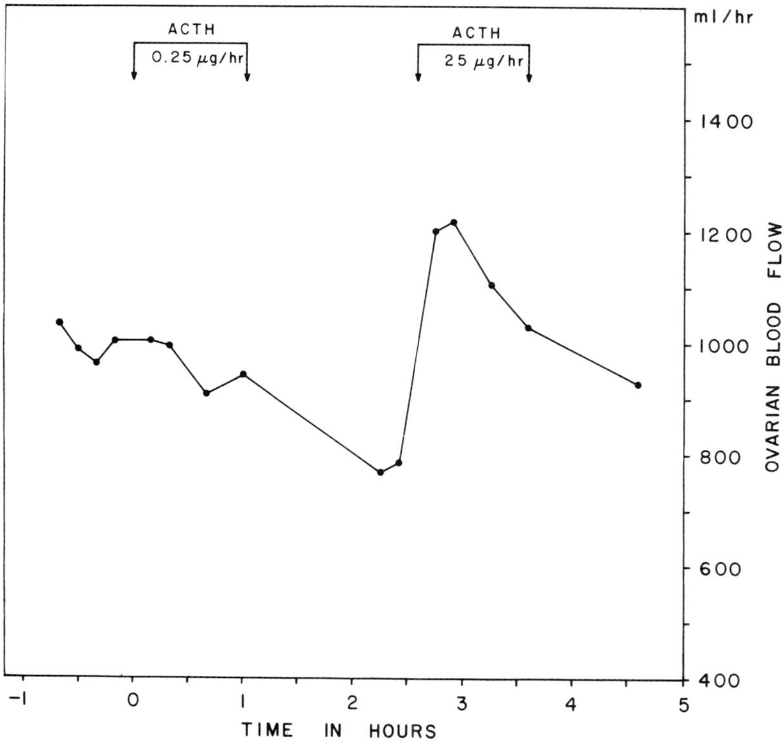

Fig. 12. The effect of ACTH on ovarian blood flow during luteal phase. Ewe No. 060.

the life span of the corpus luteum [58]. The luteolysin secreted by the uterus *in situ* would then only have a systemic effect on the corpus luteum on the transplanted ovary in the sheep. When the ovary and its pedicle are simply divided from the uterus *in situ* in the sheep a prolongation of the life span of the corpus luteum is apparent [21]. In this study a similar finding has been made in controls ewes (*McCracken* and *Baird,* unpublished observation). During the course of transplantation studies in the ewe a case of congenital absence of the left uterine horn was discovered [27]. This animal gives additional support to the concept of a uterine luteolysin since surgical interference in this instance cannot be invoked to explain the prolonged luteal phase exhibited by this animal.

In order to further investigate the local effects of the uterus on the ovary of the ewe the entire reproductive tract was transplanted to the jugulo-carotid skin loop in the neck with vascular anastomoses [29a]. This included the ovary, fallopian tube, left uterine horn, body of the uterus, cervix and anterior vagina. The procedure was essentially the same as for transplantation of the ovary alone. In this case an additional arterial supply for the uterus and vagina was necessary. This consisted of the uterine artery which arises from a branch of the internal illiac artery. The continuity of the uter-ovarian vein was retained and served as a common venous drainage for the uterus and ovary. The ovarian artery (by the patch method) and the uterine artery (by an end-to-side anastomosis) were anastomosed to the left carotid artery. The utero-ovarian vein was joined to the jugular vein by a routine end-to-side anastomosis. The body of the uterus was loosely attached to subcutaneous muscle by two sutures and the vagina was exteriorized at the base of the neck and sutured to the skin. At the time of anastomosis a good blood supply was observed from the cut end of the uterp-ovarian vein when arterial blood from the carotid was allowed to flow alternately through the uterine artery or through the ovarian artery. In addition blood was observed flowing from the veins round the sectioned end of the vagina. A similar type of preparation has been described briefly by *Harrison, Heap,* and *Linzell* [15].

With this preparation the estrous cycle was normal in duration (16 days) and the animal showed behavioral estrus, i.e., it allowed the ram to mount and mate. In addition it was very easy to observe changes in the color and mucus secreting capacity of the cervix which was quite superficial. During pro-estrus the cervix became swollen and began to discharge thin mucus. With the onset of true estrus (in 12–24 hrs.) the cervix deepened in color and became swollen and edematous. The mucus secretion was more copious and viscous in nature. During a period of close observation in which the subject was teased with a vasectomized ram every 4 hours, the duration of pro-estrus was 17 hours ± 2 hours and the duration of estrus was 22 ± 2 hours which is normal for the intact ewe [39].

Blood samples of utero-ovarian venous blood were collected by percutaneous puncture of the jugular vein at regular intervals over a period of 35 days and analyzed for steroid hormones as described previously and the results are shown in Fig. 13. The collection period covered the course of two ovarian cycles each of approximately 16 day duration. The values and pattern of the progesterone secretion rates were very comparable to those found during *in situ* studies of the cycling ewe [25, 47]. Estradiol secretion showed a very distinct peak between the two cycles and coincided exactly with pro-estrus and estrus. The blood flow (in this case utero-ovarian) showed a spectacular increase during the time of maximum estradiol secretion. Since this increase did not occur when the ovary transplanted by itself (see Fig. 3) secreted a similar or higher amount of estradiol, then this rise must represent an increase in uterine blood flow. This is compatable with the concept of estrogen causing hyperaemia of its target organ, the uterus [53]. Presumably in this case it was largely mediated via the systemic circulation although a local diffusion-mechanism cannot be completely ruled out. Compared to the transplanted ovary preparation the utero-ovarian preparation showed a much more precipitous drop in progesterone secretion

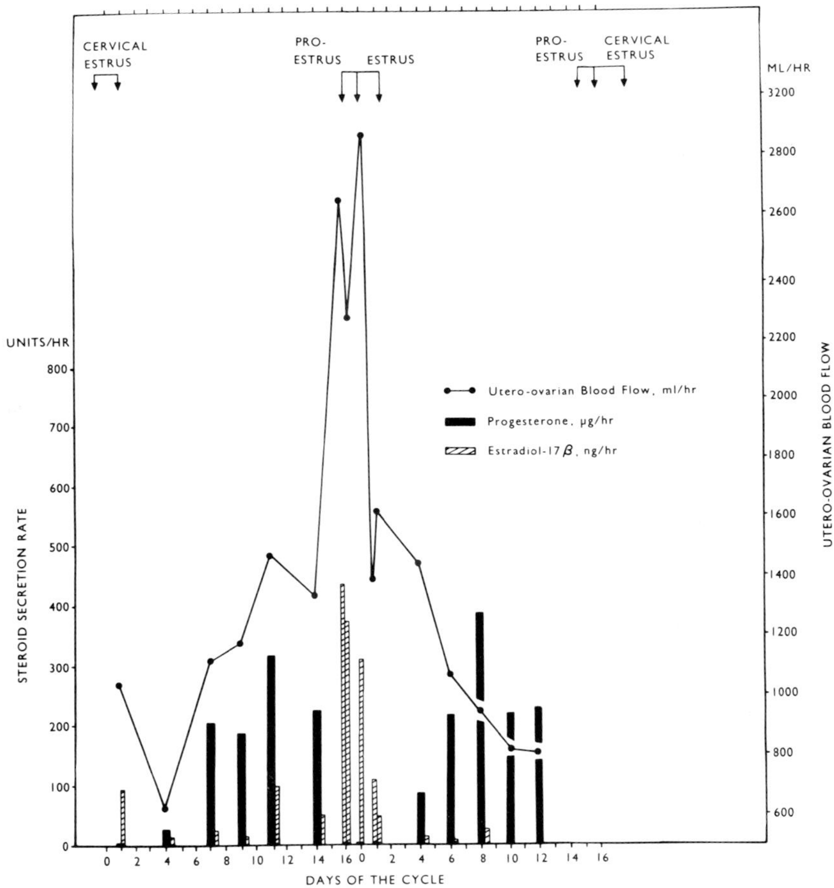

Fig. 13. Spontaneous steroid secretion and utero-ovarian blood flow during two cycles in animal with transplanted uterus and ovary. Ewe No. 077.

to virtually undetectable levels and estradiol secretion showed a much sharper peak. It was observed in the animal shown in Fig. 13 that the concentration of LH in utero-ovarian vein plasma rose from < 1.0 ng/ml to 50 ng/ml 12 hours after the peak of estradiol secretion [29a]. It may be that because of the absence of a local luteolysin in the ovarian transplant preparation the correct ratio of estradiol to progesterone was rarely present, thus accounting for the low incidence of behavioral estrus. It may well be that the utero-ovarian transplant will prove to be a more physiological preparation for the further investigation of gonadotropins and related compounds.

Acknowledgement : This study was supported by The National Science Foundation, Biological and Medical Science Division, Grant No. GB-6231 X and the Atomic Energy Commission, Contract No. AT (30-1). The financial support of G. D. Searle and Co. is gratefully acknowledeged.

I wish to thank several collaborators who have helped me during the various stages of this study: Dr. *J. R. Goding,* Dr. *D. T. Baird,* Dr. *Y. Ichikawa,* Dr. *A. Uno,* Dr. *B. V. Caldwell,* Dr. *S. A. Tillson,* Dr. *I. H. Thorneycroft,* and Dr. *R. J. Scaramuzzi.* I am grateful to Mr. *Lawrence Underwood* for valuable surgical assistance; to Mr. *C. Barber* and Mrs. *Grace Osborne* for drawings and photography; and to Miss *K. Carter* for typing the manuscript.

References

[1] *Aakvaag, A. & Eik-Nes, K. B.* (1969): Formation of steroids in the dog ovary. In "The Gonads". Ed. *K. McKerns.* Pub. Appleton-Century-Crofts, New York. ch. 4 pp. 93-118.

[2] *Baird, D. T.* (1968): A method for the measurement of estrone and estradiol-17β in peripheral human blood and other biological fluids using ^{35}S pipsyl chloride. J. Clin. Endo. & Metab. **28,** 244–258.

[3] *Baird, D. T., Goding, J. R., Ichikawa, Y. & McCracken, J. A.* (1968): The secretion of steroids from the autotransplanted ovary in the ewe; spontaneously and in response to systemic gonadotropin. J. Endocr. **42,** 283–299.

[4] *Bartosik, D., Romanoff, E. B., Watson, D. J. & Scricco, E.* (1967): Luteotropic effects of prolactin in the bovine ovary. Endocrinology **81,** 186–194.

[5] *Bedwani, J. R. & Horton, E. W.* (1968): The effect of prostaglandin E_1 and E_2 on ovarian steroidogenesis. Life Sciences **7,** 389–393.

[6] *Catchpole, H. R.* (1963): In Gonadotropins – their chemical and biological properties and secretory control Ed. *H. H. Cole.* Pub. W. H. Freeman and Co., San Francisco. Physiology of the Gonadotropic Hormones p. 40–70.

[7] *De Paoli, J. & Eik-Nes, K. B.* (1963): Metabolism *in vivo* of [7α-^{3}H] pregnenolone by the dog ovary. Biochim. & Biophys. Acta **78,** 457–465.

[8] *Domanski, E., Skrzeczkowski, L., Stupnicka, E., Fitko, R. & Dobrowolski, W.* (1967): The effect of gonadotropins on the secretion of progesterone and estrogens by the sheep ovary perfused *in situ.* J. Reprod & Fertil., **14,** 365–372.

[9] *Duncan, G. W., Bowerman, A. M., Anderson, L. L., Hearn, W. R. & Melampy, R. M.* (1961): Factors influencing the in vitro synthesis of progesterone. Endocrinology **68,** 199–207.

[10] *Eshkol, A. & Lunenfeld, B.* (1967): Purification and separation of follicle stimulating hormone (FSH) and luteinizing hormone (LH) from human menopausal gonadotropin (HMG). Acta Endocrinol. **54,** 91–95.

[10a] *Everett, J. W. & Sawyer, C. H.* (1950). A 24 hour periodicity in the LH release apparatus of female rats disclosed by barbiturate sedation. Endocrinology. **47,** 198–218.

[11] *Goding, J. R., McCracken, J. A. & Baird, D. T.* (1967): The study of ovarian function in the ewe by means of a vascular autotransplantation technique. J. Endocrinol. **39,** 37–52.

[12] *Gospodarowicz, D.* (1964): The action of follicle stimulating hormone and of human chorionic gonadotropin upon steroid synthesis by rabbit ovarian tissues *in vitro*. Acta Endocrinol. **47**, 293–305.

[13] *Greep, R. O.* (1961): Physiology of the anterior hypophysis in relation to reproduction. In "Sex and Internal Secretions". Ed. *W. C. Young.* Pub. Williams and Wilkins, Co., Baltimore. Vol 1, pg. 240–301.

[14] *Hammerstein, J., Rice, B. F. & Savard, K.* (1964): Steroid hormone formation in the human ovary. I. Identification of steroids formed *in vitro* from acetate-1-^{14}C in the corpus luteum. J. Clin. Endocrin. **24**, 597–605.

[15] *Harrison, F. A., Heap, R. B. & Linzell, J. L.* (1968): "Ovarian function in the ewe after auto-transplantation of the ovary and uterus to the neck". J. Endocrinology **40,** Proceedings, xiii.

[16] *Hayward, J. N., Hilliard, J. & Sawyer, C. H.* (1963): Preovulatory and postovulatory progestins in monkey ovary and ovarian vein blood. Proc. Soc. exp. Biol. Med., **113,** 256–259.

[17] *Hilliard, J., Endröczi, E. & Sawyer, C. H.* (1961): Stimualtion of progestin release from rabbit ovary *in vivo.* Proc. Soc. exp. Biol. Med., **108,** 154–156.

[18] *Hilliard, J., Spies, H. G., Lucas, L. & Sawyer, C. H.* (1968): Effect of prolactin on progestin release and cholesterol storage by rabbit ovarian interstitium. Endocrinology, **82,** 122–131.

[19] *Hisaw, F. L.* (1947): Development of the graafian follicle and ovulation. Physiol. Reviews **27,** 95–119.

[20] *Horton, R.* (1965): Estimation of androstenedione in human peripheral blood with ^{35}S-thiosemicarbizide. J. Clin. Endo. & Metab. **25,** 1237–1245.

[21] *Inskeep, E. K. & Butcher, R. L.* (1966): Local component of utero-ovarian relationships in the ewe. J. Animal Science **25,** 1164–1168.

[22] *Kaiser, J.* (1964): *In vitro* stimulation of steroidosynthesis in human ovary by human FSH and LH. Acta Endocrinol. **47,** 676–688.

[23] *Kaltenbach, C. C., Cook, B., Nisewander, G. D. & Nalbandov, A. V.* (1967): Effects of pituitary hormones on progesterone synthesis by ovine luteal tissue *in vitro.* Endocrinol. **81,** 1407–1409.

[24] *Kilpatrick, R., Armstrong, D. T. & Greep, R. O.* (1964): Maintenance of the corpus luteum by gonadotropins in hypophysectomized rabbit. Endocrinol. **74,** 453–461.

[25] *Lindner, H. R., Sass, M. B. & Morris, B.* (1964): Steroids in the ovarian lymph and blood of conscious ewes. J. Endocrin. **30,** 361–376.

[26] *McCracken, J. A. & Goding, J. R.* (1966): Progesterone secretion by the jugulo-carotid ovarian autotransplant in the sheep. Proc. 2nd Int. Congr. Hormonal Steroids Abstract **473,** 259–260.

[27] *McCracken, J. A. & Caldwell, B. V.* (1969): Corpus luteum maintenance in a ewe with one congenitally absent uterine horn. In press. J. Reprod & Fertil.

[28] *McCracken, J. A., Uno, A., Goding, J. R., Ichikawa, Y. & Baird, D. T.* (1969): The *in vivo* effects of gonadotropins on the secretion of steroids from the transplanted ovary of the ewe. In press. J. Endocrinol.

[29] *McCracken, J. A. & Baird, D. T.* (1969): The study of ovarian function by means of transplantation of the ovary in the ewe. In "The Gonads". Ed. *K. McKearns.* Pub. Appleton-Century-Crofts, New York. Ch. 7 pp. 175–209.

[29a] *McCracken, J. A., Caldwell, B. V., Tillson, S. A., Thorneycroft, I. H. & Scaramuzzi, R. J.* (1969). Ovarian steroid secretion and LH levels in sheep after autotransplantation of the ovary, uterus, cervix, and vagina to the neck. Proceedings of 2nd Meeting of Society for the Study of Reproduction. Davis, California.

[30] *McDonald, I. R., Goding, J. R. & Wright, R. D.* (1958): Transplantation of the adrenal gland of the sheep to provide access to its blood supply. Aust. J. Expt. Biol. Med. Sci. **36**, 83–96.

[31] *Marsh, J. M. & Savard, K.* (1966): The stimulation of progesterone synthesis in bovine corpora lutea by adenosine 3′5′-monophosphate. Steroids **8**, 133–148.

[32] *Marsh, J. M., Telegdy, G. & Savard, K.* (1966): Effect of gonadotropin on steroidogenesis in rat ovaries at diestrus. Nature (London) **212**, 950–952.

[33] *Mason, N. R., Marsh, J. M. & Savard, K.* (1962): An action of gonadotropin ***in vitro.*** J. Biol. Chem. **237**, 1801–1806.

[34] *Moor, R. M. & Rowson, L. E. A.* (1966): Local uterine mechanisms affecting luteal function in sheep. J. Reprod. & Fertil. **11**, 307–310.

[35] *Nalbandov, A. V. & Cook, B.* (1968): Annual Review of Physiology **30**, 253–254.

[36] *Nishizawa, E. E. & Eik-Nes, K. B.* (1964): On the secretion of progesterone and androstenedione by the canine ovary in animals stimulated with human chorionic gonadotropin. Biochim. Biophys. Acta, **86**, 610–621.

[36a] *Parlow, A. F.* (1961). Bioassay of pituitary luteinizing hormone by depletion of ovarian ascorbic acid. In "Human Pituitary Gonadotrophins" pp. 300–312. Ed. *A. Albert.* Pub. Charles C. Thomas, Springfield, Illinois.

[37] *Pharriss, B. B., Wyngarden, L. J. & Gutknecht, G. D.* (1968): A concept for uterine control of luteal activity in the rat. Proc. 1st Annual Meeting Soc. Study Reprod., Vanderbilt.

[38] *Riondel, A., Tait, J. F., Gut, M., Tait, S. A. S., Joachim, E. & Little, B.* (1963): Estimation of testosterone in human peripheral blood using ^{35}S-thiosemicarbazide. J. Clin. Endocrin. **23**, 620–628.

[39] *Robinson, T. J.* (1959): The estrous cycle of the ewe and the doe. In Reproduction in Domestic Animals, pp. 291–333, Ed., *Cole, H. H.* and *Cupps, P. T.,* Pub. Academic Press, New York.

[40] *Romanoff, E. B.* (1966): Steroidogenesis in the perfused bovine ovary. J. Reprod. Fert., Suppl. **1**, 89–99.

[41] *Romanoff, E. B., Deshpande, N. & Pincus, G.* (1962): Rate of ovarian progesterone secretion in the dog. Endocrinology, **70**, 532–539.

[42] *Ryan, K. J. & Smith, O. W.* (1961): Biogenesis of estrogens by the human ovary. I – Conversion of acetate-1-C^{14} to estrone and estradiol. J. Biol. Chem. **236**, 705–709.

[43] *Ryan, K. J. & Smith, O. W.* (1961): Biogenesis of estrogen by the human ovary. II – Conversion of progesterone-4-C^{14} to estrone and estradiol. J. Biol. Chem. **236**, 710–714.

[44] *Savard, K. & Casey, P. J.* (1963): Differing actions of NADPH and tropic hormone (LH and ACTH) on steroidogenesis. Federation Proc. **22**, 530.

[45] *Savard, K. & Mason, N. R.* (1963): Studies of gonadotropin action on steroidogenesis. 45th Meeting Endocrine Soc., Abst. 66, p. 32.

[46] *Savard, K. & Casey, P. J.* (1964): Effects of pituitary hormones and NADPH on acetate utilization in ovarian and adrenocortical tissues. Endocrinology, **74**, 599–610.

[47] *Short, R. V.* (1964): Ovarian steroid synthesis and secretion *in vivo*. Recent progress in Hormone Research, **20,** 303–340.

[48] *Short, R. V.* (1967): Annual Review of Physiology **29,** 384–389.

[49] *Short, R. V., McDonald, M. F. & Rowson, L. E. A.* (1963): Steroids in the ovarian venous blood of ewes before and after gonadotropic stimulation. J. Endocrin. **26,** 155–169.

[50] *Solomon, S., Vande Wiele, R. & Lieberman, S.* (1956): The *in vitro* synthesis of 17α-hydroxy-progesterone and androstenedione from progesterone by bovine ovarian tissue. J. Am. Chem. Soc. **78,** 5453–5454.

[51] *Solomon, D. H., Prujan, R. L. & Triplett, H. W.* (1963): Effect of thyrotropin on fractional blood flow of chick thyroid gland. Amer. J. Physiol. **205,** 549–554.

[52] *Stark, E., Varga, B. & Ács, Z. S.* (1967): An extra-adrenal effect of corticotropin. J. Endocrin. **37,** 245–252.

[53] *Szego, C. M.* (1965): Role of histamine in mediation of hormone action. Federation Proceedings, **24,** 1343–1352.

[54] *Szego, C. M. & Roberts, S.* (1953): Steroid action and interaction in uterine metabolism. Recent Progr Hormone Res. **8,** 419–469.

[55] *Szego, C. M. & Gitin, E. S.* (1964): Ovarian histamine depletion during acute hyperaenmic response to luteinizing hormone. Nature (London) **201,** 682–684.

[56] *Telegdy, C. & Huszár, L.* (1962): The effect of FSH and HCG on the dog's ovarian progesterone and androstenedione secretion *in vivo*. Acta Physiol. Acad. Sci. Hungarice, **21,** 339–345.

[57] *Urquhart, J.* (1965): Influence of adrenal blood flow rate on adrenocortical response to ACTH. Fed. Proc. **24,** 448.

[58] *Wiltbank, J. N. & Casida, L. E.* (1956): Alteration of ovarian activity by hysterectomy. J. Anim. Sci. **15,** 134–140.

[59] *Wurtman, R. J.* (1964): An effect of LH on the fractional perfusion of the rat ovary. Endocrinology **75,** 927–933.

[60] *Yoshinaga, K., Grieves, S. A. & Short, R. V.* (1967): Steroidogenic effects of luteinizing hormone and prolactin on the rat ovary *in vivo*. J. Endocr. **38,** 423–430.

[61] *Yuhara, M., Cohen, A. I & Frieden, E. H.* (1963): Stimulation by pituitary gonadotropin of the synthesis of estrogen by rat ovary slices *in vitro*. Proc. Soc. Exp. Biol. & Med. **113,** 907–909.

Uterine Factors Influencing Corpus Luteum Function

B. V. Caldwell

Worcester Foundation for Experimental Biology, Inc., Shrewsbury, Mass., USA

Summary: In many mammalian species the functional life-span of the corpus luteum appears to be under the direct control of the uterus since uterine removal results in a marked prolongation of luteal function. In sheep and hamsters the transplantation of uterine or endometrial tissue to ectopic sites significantly reduces this prolonged period although these procedures are unable to completely re-establish normal ovarian periodicity. Extracts prepared from sheep endometrium or hamster uterus are also effective in shortening the life span of corpora lutea when injected into hysterectomized pseudopregnant hamsters, further suggesting that the uterus produces a hormonal substance involved in the regulation of luteal function.

Surgical procedures such as sub-total hysterectomy, unilateral hysterectomy, or the vascular autotransplantation into the neck of sheep of an ovary; a uterine horn; or an ovary with a uterine horn, all support the concept that the uterus produces a "luteolytic" substance whose primary effect is local in nature, *i.e.* acting on the adjacent ovary. Some experiments indicate that the "luteolysin" may have a systemic component to its activity. The pathway(s) by which this effect may be mediated to the ovary remains to be clucidated as does the mechanism (s) elucidated by which it may cause the regression of the corpus luteum at a specific time during the reproductive cycle.

The isolation and further characterization of the "uterine luteolysin" is hampered by the lack of a specific and sensitive bioassay system. However, the sheep in which ovaries have been transplanted to the neck may serve as a suitable preparation for assaying the activity of uterine extracts.

It is necessary to consider the possible role of the embryo in the overall process of luteal life-span and function since the normal regression of the corpus luteum must be pre-empted in most species if pregnancy is to be maintained. Experiments in sheep suggest that the initial effect of the developing conceptus may be to inhibit the production of the "uterine luteolytic hormone".

In comparison, a great deal more has been learned about the processes which control the growth, maturation and ovulation of follicles with the subsequent formation of a secretory corpus luteum, than is known about the mechanism(s) which control the functional life-span of this luteal tissue. It has been over fifty years since *Loeb* [1] first reported that the presence of the uterus was essential to assure normal ovarian periodicity in guinea pigs, noting that corpus luteum function was markedly elongated following hysterectomy. It was postulated at that time that the uterus may be the site for the formation of a substance (hormone) that was necessary in the control of luteal function. Despite the many advances in methodology and great increases in our knowledge about most other aspects of reproductive physiology, we are today able to add little to the original hypothesis of Dr. *Loeb.* Even after the enormous concentration of research efforts directed toward this subject, the "uterine luteolytic hormone" remains a concept, as yet unproven to the satisfaction of most investigators.

There have been four reviews in the past year [2, 3, 4, 5] which have considered in detail the subject of utero-ovarian interrelationships and the reader is referred to these manuscripts for a more extensive treatment of the literature, in particular the recent paper by *Anderson, Bland* and *Melampy* [3] which is most noteworthy for its completeness. The present manuscript will deal primarily with the "uterine luteolytic hormone" concept and will describe the numerous investigations which have been aimed at clarifying the role of the uterus in the regulation of ovarian periodicity. An attempt will be made to view the data in relation to the procedures normally used to establish the existence of a new hormone.

Historically, the experimental procedures which have often been followed to localize a suspected hormone within a given tissue include:

1) Surgical removal of the tissue which is presumed to be the site of hormone formation, concommittent with observations on the physiological effects of this procedure;
2) Transplantation of the excised tissue to an ectopic site while noting the reversal of the physiological change associated with the previous tissue removal;
3) Exogenous administration of extracts prepared from the potential hormone containing tissue to a suitable test animal (noting same parameter);
4) Purification and characterization procedures carried out on the active extracts which may lead to the isolation and positive identification of the new hormone.

Evidence for a "uterine luteolytic hormone"

A. Surgical removal of the uterus

Hysterectomy has been performed in at least 15 mammalian species and the effects of this procedure on the functional life-span of the corpus luteum have been determined. There is no change in the life-span of the corpora lutea following hysterectomy in many animals (ferret, cat, dog, mouse, oppossum, 13-lined ground squirell) [4]. However, in the hamster, rat and rabbit, although there is no effect on whatever estrous cycles may be present, removing the uterus results in a marked elongation of luteal function in pseudopregnant animals, the length of which then approximates the normal period of gestation [4]. Hysterectomy is unquestionably effective in the cow, sheep, pig, and guinea pig, as a means for increasing luteal life-span, again to a period which closely approximates the normal gestation length [2]. The situation in the primates, man and monkey, is as yet not clearly resolved. *Andreoli* [6] has reported that hysterectomy during the menstrual cycle has some effect on the duration of pregnandiol secretion in the urine of woman, but these studies have not been confirmed. In the monkey, one recent report would seem to place this species in the group which shows no effect of hysterectomy on the corpus luteum life-span, since the peripheral levels of progesterone throughout the menstrual cycle were the same as in intact animals [7]. This discrepancy between the various species studied leads some investigators to doubt the results in the several animals in which the evidence for a uterine involvement seems clear. However, there would be little reason to assume that the same mechanism for controlling luteal regression must be operating in all species, especially since there appears to be considerable species variation also in the mechanism involved in the formation of a fully secretory corpus luteum. We will, therefore, concentrate on the species that are markedly effected by hysterectomy and review the experimental procedures aimed at defining the role of the uterus in the regulation of normal ovarian periodicity in the sheep, pig, cow, hamster, guinea pig, and rat.

Even though total hysterectomy produces a uniform result in the above species, experiments involving the removal of various portions of the uterus indicate that in many animals there is a distinct "local" effect of the uterus. In the hamster, guinea pig, sheep, pig and cow, removing one uterine horn results in the prolonged maintenance of the corpora lutea present in the ovary without any uterine connection, while the corpora lutea next to the remaining uterine horn regress at the normal time [8]. The rat and rabbit show no such local utero-ovarian relationship [9, 10]. *Moor* and *Rowson* have demonstrated that this local control of the uterus over the sheep ovary is quantitative in nature since the life-span of the corpora lutea increased in proportion to the amount of uterine tissue that was surgically removed [11].

In superovulated ewes, with corpora lutea on both ovaries, the local effect is clearly evident as shown in Fig. 1 (4). There are those however who view the results of surgical procedures such as hysterectomy as meaningless from a physiological point of view [12]. Recently an abnormal sheep was described by this laboratory in which one uterine horn was found to be congenitally absent [13]. This animal had not exhibited behavioral estrus for over fifty days, whereupon a histological observation of the reproductive tract was made, which revealed the probable cause for the prolonged diestrous interval (Fig. 1). On the side of the ewe which had no direct connection between the ovary and the single uterine horn there was a large

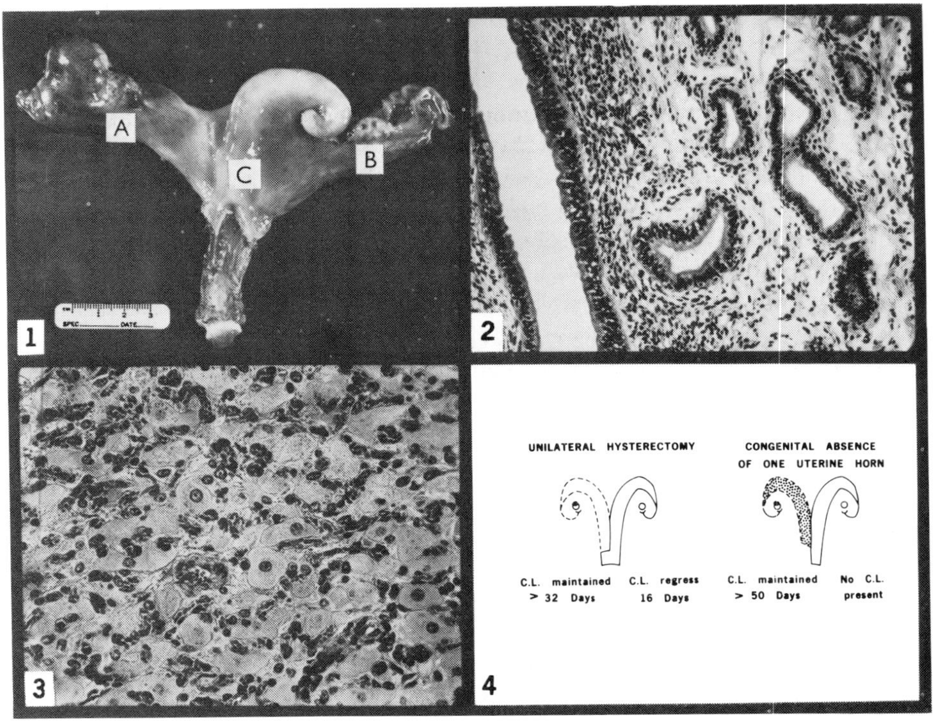

Fig. 1. Ewe with congenital absence of left uterine horn. (1) The excised reproductive tract showing; A, left ovary with large corpus luteum; B, right ovary with many small folicles and C, The single right uterine horn. (2) A section of endometrium from the right uterine horn 160 X; (3) A photomicrograph of a section of the large corpus luteum in left ovary 160 X; (4) Diagrammatic comparison of this animal with experimental results using hemihysterectomized animals.

corpus luteum which would have prevented this animal from returning to estrus. On the ovary which was connected to the remaining uterine horn, there was evidence of corpora albicantia tissue, but no luteal tissue. A comparison of this animal to the hemi-hysterectomized sheep gives further credence to the view that there is a local effect of the uterus on the ovary, and surgical intervention cannot be invoked as an explanation for the increased life-span of this corpus luteum. In the guinea pig [14], cow [15], and pig [16] there are similar morphological abnormalities of the uterus which result in prolonged luteal function.

Several investigators have reported that procedures which destroy the endometrium, such as the use of 10 % tannic acid, bouins fixitire, 12.5 % phenol, and 10 % silver nitrate also result in the maintenance of corpus luteum function in pigs [17]and guinea pigs [18]. However, these methods are particularly severe and should be viewed only in conjunction with the studies involving the surgical removal or spontaneous lack of the uterus.

Of major importance to the clarification of the local effect exerted by the uterus is the elucidation of the pathway(s) by which this luteolytic influence may be mediated to the adjacent ovary. In a recent review [3] this matter has been discussed fully, however, the local pathway remains unknown. From the work in several species it would appear that both a tissue and a vascular continuity is necessary between the ovary and uterus if normal ovarian periodicity is to take place. The definitive description of a portal blood or lymph system going from the uterus to ovary is still lacking, the discovery of which could add greatly to the feasibility of the "uterine luteolytic hormone" concept.

The day on which the uterus can be removed and still cause the corpora lutea to be maintained varies between the species. In sheep, hysterectomy can be performed up to day 15 of the 16 day estrus cycle, and still result in the maintenance of the corpus luteum as long as it was functional at the time of uterine removal [19]. A continuation of this work has established that not only are sheep corpora lutea maintained by late stage hysterectomy, but they are generally maintained at the same level of secretory function at which they existed just prior to the removal of the uterus [20]. This finding suggests that the luteolytic influence of the uterus is continuous in its action and that the process can be arrested as long as the degeneration of the luteal tissue has not progressed too far.

In the pig the effect of the uterus appears to operate over a much longer period of time. Hysterectomy as early as day 16 of the 21 day estrous cycle led to maintenance of the corpora lutea in only 5 of 11 pigs [21]. Another approach to determining the time during which the uterine luteolytic effect may be operating is by the induction of new corpora lutea by the exogenous administration of gonadotropins at various times during the estrous cycle. This procedure in sheep mirrored the hysterectomy work in that all corpora lutea present on the ovaries (spontaneous and induced) regressed at the normal time (regardless of the day on which the new C. L. had been induced [22]). In pigs, however, corpora lutea that were induced after day 8 would not regress at the same time as the original luteal tissue, but were maintained for a period in relation to the day on which they had been induced [23, 24]. Corpora lutea formed by the gonadotropins on day 6 did regress with the normal luteal tissue about day 19 [23], again suggesting that the action of the uterus in the pig must be over a prolonged period of time.

In the hamster the uterus seems to be "luteolytic" mainly between the afternoon of day 7, and the morning of day 8, in a 9 day pseudopregnant cycle [25]. The guinea pig has a 16 day estrous cycle and the uterus has its luteolytic activity between days 13 and 15 as determined again by late stage hysterectomy [26].

The overwhelming suggestion, therefore, from all of the experiments involving the loss of normal uterine tissue, is that the uterus exerts a direct control over the functional life-span of the corpus luteum in at least the above mentioned species.

B. Transplantation of uterine tissue to hysterectomized animals

Having noted a consistent physiological change associated with uterine removal, the second step is the attempted reversal of this parameter by the transplantation of the excized tissue. Fig. 2 depicts the effects of surgically placing the uterus, or endometrium in ectopic sites in the species which have thus far been established as

SPECIES	EFFECT OF HYST.	EFFECT OF UTERINE TRANSPLANTATION	TRANSPLANTATION PROCEDURE
SHEEP	C.L. MAINTAINED (> 145 days)	C.L. REGRESS (35-50 DAYS)	ENDOMETRIUM TO FLANK (A)
		C.L. REGRESS (17 DAYS)	UTERUS TO OMENTIUM (B)
		C.L. REGRESS (IRREGULAR)	UTERUS TO NECK (C)
		C.L. REGRESS (17 DAYS)	UTERUS AND OVARY TO NECK (D)
PIG	C.L. MAINTAINED (> 140 DAYS)	C.L. REGRESS (20-22 DAYS)	UTERUS TO ABDOMINAL OBLIQUE MUSCLES (E)
		C.L. REGRESS (21-23 DAYS)	UTERUS TO PERITONEUM (F)
		NO CYCLES CYCLES	UTERUS TO SMALL INTESTINE UTERUS TO BODY WALL (G)
HAMSTER	NORMAL ESTRUS CYCLES	C.L. REGRESS (13-15 DAYS)	UTERUS OR ENDOMETRIUM TO CHEEK POUCH (H)
	PSEUDOPREGNANCY EXTENDED - 18 DAYS	C.L. REGRESS (14-16 DAYS)	RAT UTERUS TO CHEEK POUCH (I)
		C.L. REGRESS (11-15 DAYS)	SHEEP ENDOMETRIUM TO CHEEK POUCH (J)
RAT	NORMAL ESTRUS CYCLES	C.L. REGRESS (15.5 DAYS)	UTERUS TO PERITONEAL CAVITY (K) UTERUS TO ABDOMINAL WALL (L)
	PSEUDOPREGNANCY EXTENDED - 22 DAYS	APPROX.	UTERUS TO KIDNEY (M)
GUINEA PIG	C.L. MAINTAINED (> 80 DAYS)	C.L. REGRESS (80 DAYS)	UTERINE AUTOTRANSPLANTS (N)
		C.L. REGRESS (20-30 DAYS)	UTERUS TO ABDOMINAL WALL (O) OR OBLIQUE MUSCLES (P)

Fig. 2. The effects of hysterectomy and uterine transplantation on corpora lutea life-span.

References:

A: [2,35]; B: [30]; C: [29,50]; D: [27,28]; E: [21]; F: [31]; G: [55]; H: [32,56]; I: [32,56]; J: [49]; K: [3]; L: [57]; M: [58]; N: [59]; O: [3]; P: [34].

being dependent on the uterus for normal ovarian periodicity. Although transplantation of the uterus to hysterectomized animals can effectively reduce the duration of corpus luteum function towards normal, usually the life span of the luteal tissue is significantly longer than normal. This failure to re-establish normal ovarian periodicity may be due to the fact that the local effect of the uterus is not operating in these experimental animals and only a systemic effect can be postulated to explain the effects of transplantation on luteal life-span. In one important case, however, autologous transplantation procedures are able to re-establish the normal duration of estrus. Sheep, in which the left ovary and left uterine horn are transplanted by vascular anastomosis to the jugular vein and carotid artery contained in a skin loop in the neck, exhibited normal estrus cycles [27, 28]. This result is in contrast to the findings when either a uterine horn, or an ovary are transplanted alone to the neck [29]. In these latter cases the cycle lengths are very elongated and irregular (see Fig. 3). The uterus has also been placed in the omentum of sheep, a procedure

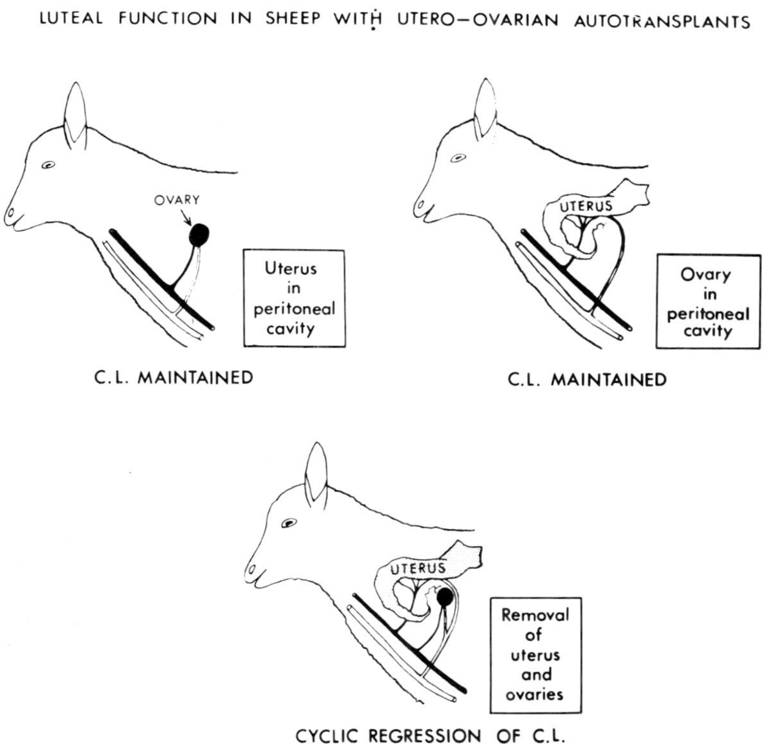

Fig. 3. Luteal function in sheep with utero-ovarian autotransplants.

which again results in normal cycle lengths [30] however, when endometrium is grafted into the flank of hysterectomized ewes the corpus luteum is still maintained significantly longer than normal. Luteal regression occured after about 30–40 days in these animals, as opposed to 145 days duration in hysterectomized animals without any uterine tissue present [2]. Similar uterine transplantation to ectopic sites in the pig [31], hamster [32], rat [33], and guinea pig [34] generally support the view that substances, like prostaglandin $F_2 \alpha$, may be the luteolysin acting to cause the regression of corpora lutea by specifically controlling the vascular bed

The corpus luteum itself has been the site to which endometrium has been transplanted in sheep [2, 35]. The hysterectomized sheep which received these grafts did not cycle, but a histological examination of the corpora lutea showed that endometrium taken from late in the estrus cycle had a local, graded, "lytic" effect on the luteal cells in closest proximity, however, there was no degenerative effect on the cells some distance away from the graft. Tissue taken from uteri early in the cycle had no effect, local or otherwise. Of some interest was the results of implanting to the corpus luteum small pieces of endometrium excized from the uterus while leaving the rest of the uterus intact. Even though the corpus luteum in these cases regressed at the normal time and were corpora alibicatia when observed up to thirty days later, the endometrial grafts were still vascularized and appeared normal under histological examination. This finding would not support the view that substances, like prostaglandin $F_2 \alpha$, may be the luteolysin acting to cause the regression of corpora lutea by specifically controlling the vascular bed of the luteal tissue [36]. It is difficult to reconcile the fact that the piece of endometrial graft continues to receive a good blood supply after the luteal tissue has totally regressed.

In two separate investigations results were obtained which did not, seemingly, support the view that the uterus and ovary must be directly connected for the normal cyclic regression of the corpora lutea. In sheep, following the autotransplantation of the uterus to the omentum the cycles were found to be very comparable with normal estrus cycle lengths [30]. Likewise in pigs, a two step procedure involving the revascularization of the uterus to the parietal wall of the peritoneum, resulted in the normal cyclic regression of the corpora lutea and cycle lengths of 21–23 days duration [31, 21]. In partial explanation for the lack of extended luteal function in these cases, the results of *Inskeep* may be appropriate. He found that suturing the uterus of sheep in such a manner as to prohibit any adhesion of the tissues resulted in prolonged luteal maintenance. However, if the uterus was severed and adhesions were allowed to form within the body cavity, then the estrous cycles were normal. Perhaps, then the formation of adhesions may be important in the re-establishment of normal ovarian periodicity in the two transplantation procedures described above.

The various surgical procedures involving total and subtotal hysterectomy indicated that the uterus most definitely has an effect on the ovary, and in many species the main action is local in nature. The transplantation experiments support the view that the anatomical relationship between the ovary and uterus is very important, but provide additional information that the uterus also has a systemic effect, since it is difficult to visualize any other pathway for the transplanted uterine tissue to exert control over the functional life span of the corpora lutea.

Uterine transplantation is, therefore, for the most part successful in reducing the prolonged luteal life-span following hysterectomy, which leads naturally into the use of extracts of uterine tissue to attempt a control of the corpus luteum function.

C. The effects of uterine extracts on corpus luteum function

In several species, extracts of uterine or endometrial tissue have been administered to hysterectomized animals (Fig. 4). To be sure the results are not conclusive, however, some of the procedures which have been followed and some of the test animals which have been used were not entirely suitable for these studies. In one investigation with sheep [37] crude extracts were prepared from the uterus during the estrus and mid-luteal phases of the cycle, and injected intramuscularly. The fact that the life span of the corpus luteum was not shortened is hardly a negative result since

1) the extracts were prepared from the uterus during a period when other work would suggest that there was unlikely any production of a "luteolytic" effect, and

2) the injections were I. M. and all previous work demonstrates that the initial effect of the uterus in sheep is local in nature, and the systemic effect is secondary, and operating at a much later period than studied by these investigators.

Another contrary report in pigs [38] was viewed with some understandable caution by the author who recognized the limitations of the test procedure. However, the infusion of endometrial filtrates into the uterine lumen of pigs was also ineffective in causing premature regression of the corpora lutea [39]. Once again, the procedures employed and the test animal used were not ideal.

Extracts of the uterus from guinea pigs were not able to alter the prolonged estrous cycle following hysterectomy in this species [40]. The hysterectomized rat was also used as an assay for uterine and endometrial extracts with mixed results. Extracts from the cow were unable to induce regression of the rat corpora lutea [41], nor did uterine extracts prepared from the rat uterus [42] have any effect. One report, however, suggested that the length of pseudopregnancy in hysterectomized animals was significantly reduced by the administration of rat endometrial suspensions [43].

TEST ANIMAL	EXTRACTS PREPARED FROM TISSUE	SPECIES	PARAMETER STUDIED	EFFECTS
HYSTERECTOMIZED HAMSTER	UTERUS (A)	HAMSTER	PSEUDOPREGNANCY	REDUCED
	ENDOMETRIUM (B)	SHEEP	" "	" "
	ENDOMETRIUM (C)	COW	" "	" "
	ENDOMETRIUM (D)	COW	" "	" "
	UTERINE (E) FLUSHINGS	PIG	" "	" "
HYSTERECTOMIZED RAT	ENDOMETRIUM (F)	RAT	PSEUDOPREGNANCY	REDUCED
	UTERUS (G)	RAT	" "	NONE
	ENDOMETRIUM (H)	COW	" "	NONE
HYSTERECTOMIZED RABBIT	UTERUS (I)	COW	PSEUDOPREGNANCY	REDUCED
HYSTERECTOMIZED SHEEP	UTERUS (J)	SHEEP	ESTRUS CYCLE	NONE
PIGS	ENDOMETRIUM (K)	PIG	ESTRUS CYCLE	NONE
	UTERINE (L) FLUSHINGS	PIG	C.L. WEIGHT & PROG. CONT.	NONE
HYSTERECTOMIZED GUINEA PIG	UTERUS (M)	G. PIG	ESTROUS CYCLE	NONE
	UTERUS (N)	G. PIG	C.L. SIZE	NONE

Fig. 4. The effects of various uterine extracts on corpus luteum life-span *in vivo*.

The most consistent findings have been reported in the rabbit and hamster. When acetone powder extracts were prepared from cow endometrium and injected into the pseudopregnant rabbit the corpora lutea regressed in 64 % of the animals [44]. Using the same extracts, these workers also injected hysterectomized hamsters and observed the length of pseudopregnancy. 70 % of the animals receiving injections on days 7, 8 and 9 of pseudopregnancy had cycle lengths that were significantly shorter than expected. However, extracts prepared from intestinal powder were even more effective in producing the shortened cycles [45].

References:

A: [32,56]; B: [49]; C: [47]; D: [56]; E: [38]; F: [43]; G: [42]; H: [41]; I: [44]; J: [37,42]; K: [39]; L: [38]; M: [40]; N: [3].

The hysterectomized hamster has been the most often used test animal and the results of various investigations are shown in figure 4. *Mazer* and *Wright* [46] injected extracts from hamster uteri and found that the duration of pseudopregnancy was significantly shortened only when the uteri were obtained from pseudopregnant animals on the 7th day following the onset of estrus and sterile mating. The controls showed no effect on pseudopregnancy length. *Mazer* has continued these investigations using extracts prepared from cow endometrium and has confirmed the findings of *Lukaszewka* and *Hansel* [47] that these extracts were able to cause a marked decrease in progesterone content and weight of the corpora lutea in hysterectomized hamsters [48]. *Caldwell, Moor* and *Lawson* recently reported that extracts from sheep endometrium taken late in the cycle were effective in shortening pseudopregnancy, as opposed to the lack of any effect due to extracts prepared from early in the estrous cycle [49]. These extracts had been previously tested on sheep granulosa cell monolyer cultures [2]; a technique that Dr. *Schomberg* will discuss in his presentation. The hysterectomized hamster was also the animal used to test pig uterine flushings that had been tested on granulosa cell monolayer cultures. The results were inconclusive, however, since extracts that were judged either "lytic" or "non lytic" by their assay on porcine granulosa cell monolayers both caused premature return to estrus in pseudopregnant animals [38].

The use of various extracts and homogenates of uterine tissue have certainly not been conclusive. One reason may be that there is no completely acceptable assay system that shows the specificity and sensitivity required. This laboratory is presently using the sheep which have been described by *McCracken,* in which an ovary is transplanted to the neck [50]. As yet the results are not completed for the first experiments, but we hope that this test animal may provide the necessary information to further characterize the possible effects of the sheep endometrial extracts. It is possible in these sheep to infuse the endometrial preparations directly into the ovary (via carotid artery) while taking periodic ovarian venous blood flow and steroid concentrating measurements in the unstressed animal (via jugular vein).

D. Purification and characterization of the uterine extracts

It would seem to many investigators that attempts to purify and characterize the active component in the uterine preparations are premature, and that the essential need is a sensitive, specific and reliable assay system, without which progress in this area will be slow. Nevertheless, there have been some interesting findings concerning the possible nature of the substance as to molecular weight and physical properties. The active component of the flushings from pig uteri may be a high molecular weight protein (from elution pattern on Sephadex G–200 column), which is nondializable, and heat stable at 56 °C but labile at a temperature of 90 °C for thirty minutes [5]. An endometrial factor from sheep which was "lytic" when applied to sheep granulosa monolayer cultures, was also found to be nondializable and heat stable at 50 °C, but its molecular weight was estimated to be quite low in a very inadequate Sephadex procedure [2]. The uterine extracts from the hamster which were active *in vivo,* were heat labile (70 ° for 15 min.), nondializable, and present in the submitochondrial fraction of the homogenates [46]. The highest degree of purification has been achieved with bovine endometrial extracts [48] which suggest that the active factor is a large molecular weight protein of the albumin class (Approx. 60,000). These workers eliminated the possible contamination of the extracts with lipids (and prostaglandins) by extraction with isooctane, finding the active component retained in the water soluble fraction. The latter workers used corpus luteum weights from hysterectomized hamsters injected with the fractions during pseudopregnancy, and compared the values against those recorded from animals which received control injections. However, once again the caution must be re-stated that lacking a thoroughly satisfactory assay system, these results are not conclusive.

The initial role of the conceptus in the maintenance of corpora lutea

If the uterus produces a luteolytic substance during the normal cycle then the embryo must either inhibit the formation of this substance, or secrete a luteotrophin to counteract the lytic influence. In sheep the lucid experiments of *Moor* and *Rowson* provide the most impressive evidence to date that the initial role of the developing conceptus is to prevent the uterus from acquiring its luteolytic properties [51]. The surgical transfer of embryos from a donor ewe to a recipient, at different times following mating in sheep, established the day on which the embryos must be present to prevent the regression of the corpus luteum. Fig. 5 is a summary of their work and clearly shows that the embryo first assumes its role of extending the life-span of the corpus luteum in sheep between the 12th and 13th day after

Fig. 5. The life-span of corpora lutea following the surgical removal or transfer of embryos in mated sheep at different intervals after estrus [51].

Day of Embryo Removal	Donor C. L. Regressed by Day 17 (%)	Day of Embryo Transfer	Recipient C. L. Maintained by Day 17 (%)
5	80	5	100
7	80	7	100
9	95	9	85
12	80	12	90
13	5	13	45
14	5	14	10

estrus, about 2–3 days before the lytic influence of the non-pregnant uterus becomes apparent. If an embryo was present in the uterus on or before day 12 the corpus luteum was maintained, however, the next 24 hours is crucial since embryos transferred later than day 12 had a much reduced effect on the corpus luteum. Embryos transferred on day 14 or later were for the most part unable to prevent the luteal tissue from regressing indicating that once the lytic process has begun the embryos are unable to alter the course of corpus luteum regression. The protective influence of this early stage embryo is more surprising since at 12–14 days there is no attachment to the maternal tissues, and the embryo occupies only a very small portion of the uterine lumen. *Moor* has further reported that homogenates of the day 14 sheep embryo are able to maintain the sheep corpus luteum when infused daily into the uterine lumen [52]. These results indicate that the active principal in the embryo can be as effective as the live embryo.

The protective "anti-luteolytic" effect of the embryo is local in nature in the same sense that the uterus was found to be luteolytic in a local manner. The transfer of embryos to an isolated uterine horn results in the maintenance of corpora lutea only on the ovary adjacent to the pregnant horn [53]. Fig. 6 shows the results of transferring embryos to a uterine horn that is surgically isolated from the other horn, and then inducing the formation of new corpora lutea in both ovaries by the exogenous administration of PMS and HCG [54]. It is evident from this work that the conceptusses retain their local protective influence for about fifty days during pregnancy. After this time the corpora lutea in both ovaries are maintained. The isolated, non-gravid horn remains luteolytic for about the first fifty days of pregnancy, therefore, after which it has been shown the ovary is no longer essential for maintenance of pregnancy in sheep [60].

Fig. 6. Local lytic effect of an isolated sterile horn during pregnancy in sheep [54].

No. of Sheep	New C. L. Induced (Day)	Day of Laparotomy	Corpora Lutea Present Adjacent to Sterile Horn-Pregnant Horn
8	0	20	0 – 8
9	30	50	0 – 9
6	50	70	5 – 6
8	70	90–110	8 – 8

Summary

The author recognizes that several other possible mechanisms have been suggested to account for the cyclic regression of the corpus luteum (see Fig. 7) [2]. The role of the pituitary in most species is unclear [12], and the participation of the nervous system is difficult to estimate but the transplantation studies would seem to diminish the importance of neural pathways in the control of lutea life-span. There is also the suggestion that the uterus may normally be responsible for the metabolism of a luteotropic substance (i.e. L. H., estrogens, or progesterone) the circulating level of which would then remain high following hysterectomy and result in the maintenance of the corpus luteum [4]. However, the apparent local effect of the uterus on the adjacent ovary would not support this concept. Of course the species variation must be kept in mind and one or more of the above mechanisms may be operating in certain animals.

From the results of the investigations described in this paper the following conclusion seem justified, with limited reservations.

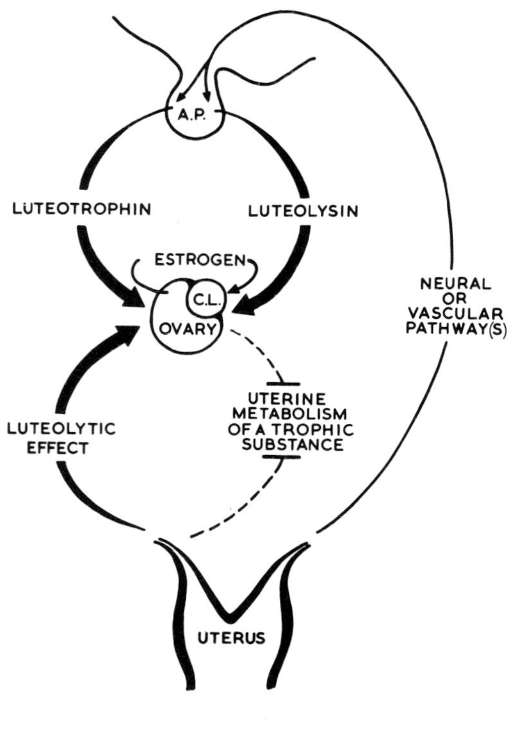

Fig. 7

Possible pathways for controlling the life-span of the corpus luteum, [27, 29].

1) Hysterectomy prolongs the life-span of the corpus luteum in many mammalian species;
2) Uterine transplantation can, at least partially, re-establish near normal ovarian periodicity;
3) Some extracts of uterine tissue are effective also in reversing to a degree the elongated luteal life-span following hysterectomy in several species.

To a large extent, the definitive establishment of a new hormone relies upon the description of an assay which is sensitive and specific enough to be a reliable indicator of the hormones action. The most pressing need for workers in this field is the development of such an assay. However, in sheep, where the evidence for a uterine involvement is most clear, the following working hypothesis is suggested to account for the control of luteal regression.

Under the proper sequence of circulating steroids from the ovary the endometrium develops the capacity to cause the regression of the corpus luteum between day 14–15. The regressing process is very rapid, acting over a period of about 24 hours and can be arrested in most stages by hysterectomy. The embryo must be present in the uterine lumen, although not attached, by day 12–13 to block the uterus from developing its lytic capabilities. The action of the uterus is primarily local in nature as is the protective influence of the embryo. This "anti-lytic" local effect of the embryo is apparent for about the first fifty days of pregnancy.

This scheme is presented as the most reasonable interpretation of the experimental results reported to date, recognizing, at the same time, that the unqualified existence of a "uterine luteolytic hormone" remains to be established.

References

[1] *Loeb, L.* (1923): Proc. Soc. Exptl. Biol. Med. **20,** 441.

[2] *Caldwell, B. V., Rowson, L. E. A., Moor, R. M. & Hay, M. F.* (1969): J. Reproduction and Fert. Suppl. 9.

[3] *Anderson, L. L., Bland, K. P. & Melampy, R. M.* (1969): Recent Progress in Hormone Research, Academic Press.

[4] *Melampy, R. M. & Anderson, L. L.* (1968): 8th Biennial Symposium on Animal Reproduction, Urbana,Illinois (in press). J. Ani. Sci. **27,** Suppl. 1, 77.

[5] *Schomberg, D. W.* (1969): In "The Gonads", *K. McKerns* (ed) Appleton-Century-Crofts, New York.

[6] *Andreoli, C.* (1965): Acta Endo. Copenh. **50,** 65–69.

[7] *Neill, J. D., Johansson, E. D. B. & Nobil, E.* (1969): Endocrinology **84,** 464.

[8] *Ginther, O. J.* (1967): J. Anim. Sci., **26,** 578.
[9] *Barley, D. A., Butcher, R. L. & Inskeep, E. L.* (1966): Endocrinology, **79,** 119.
[10] *Hunter, G. L. & Casida* (1967): J. Repro. Fert. **13,** 179.
[11] *Moor, R. M. & Rowson, L. E. A.* (1966): J. Repro. Fert. **11,** 307.
[12] *Nalbandov, A. V. & Cook, B.* (1968): Reproduction Ann. Rev. Physiol. **30,** 245–277.
[13] *McCracken, J. A. & Caldwell, B. V.* (1969): J. Repro. Fert. (in press).
[14] *Bland, V. P. & Donovan, B. T.* (1966): J. Endocrin. **34,** iii.
[15] *Ginther, O. J.* (1969): J. Am. Vet. Med. Assoc. (Dec.).
[16] *Wilson, R. F., Nalbandov, A. V. & Krider, S. L.* (1949): J. Anim. Sci., **8,** 558.
[17] *Anderson, L. L., Butcher, R. L. & Melampy, R. M.* (1961): Endocrinology, **69,** 571.
[18] *Butcher, R. L., Chu, K. I. & Melampy, R. M.* (1962): Endocrinology, **71,** 810.
[19] *Moor, R. M. & Rowson, L. E. A.* (1964): Nature, **201,** 522.
[20] *Moor, R. M., Hay, M. F., Short, R. V. & Rowson, L. E. A.* (1969): (In preparation).
[21] *Anderson, L. L., Butcher, R. L. & Melampy, R. M.* (1963): Nature, **198,** 311.
[22] *Inskeep, E. K., Olouja, M. M., Howland, B. E., Pope, A. L. & Casida, L. E.* (1963): J. Anim. Sci., **22,** 159.
[23] *Caldwell, B. V., Moor, R. M., Wilmot, I., Polge, C. & Rowson, L. E. A.* (1969): J. Repro. Fert. (in press).
[24] *Neill, J. D. & Day, B. N.* (1964): Endocrinology, **74,** 355.
[25] *Duby, R. T., McDaniel, J. W. & Black, D. L.* (1969): Acta Endocrinol. (in press).
[26] *Bland, K. P. & Donovan, B. T.* (1966): In Advances in Reproductive Physiology (*A. McLaren*, ed.) Vol. 1, 179 Logus Press, London.
[27] *Harrison, F. A., Heap, R. B. & Linzell, J. L.* (1968): J. Endocr. **40,** xiii (Abstract).
[28] *McCracken, J. A. & Caldwell, B. V.* (Unpublished observations).
[29] *Godding, J. R., Harrison, F. A., Heap, R. B. & Linzell, J. L.* (1967): J.Physiol. Lond. **191,** 129–130.
[30] *Niswender, C. D.* (1968): J. Animal Sci. **27,** Suppl. 1, 135.
[31] *DuMesnil du Buisson, F. & Rombarts, P.* (1963): Compt. Rend. **256,** 4984.
[32] *Caldwell, B. V., Mazer, R. S. & Wright, P. A.* (1966): Endocrin. **89,** 3, 477–482.
[33] *Hechter, O., Fraenkel, M., Lev, M. & Soskins, S.* (1940): Endocrin. **26,** 680.
[34] *Butcher, R. L., Chu, H. Y. & Melampy, R. M.* (1962): Endocrin. **70,** 442.
[35] *Rowson, L. E. A., Moor, R. M., Hay, M. F. & Caldwell, B. V.* (1969): J. Repro. Fert. (in press).
[36] *Pharriss, B. B., Wyngarden, L. J. & Guthnecht, G. D.* (1969): Proc. lst. Ann. Meeting Soc. Study Reprod., Vanderbilt.
[37] *Kiracofe, G. H. & Spies, H. G.* (1964): J. Anim. Sci., **23,** 908.
[38] *Schomberg, D. W.* (1968): Proc. 3rd Intern. Congr. Endocrinol. Mexico City, Exerpta Med. Intern. Cong. Ser. (in press).
[39] *Anderson, L. L. & Melampy, R. M.* (1962): J. Anim. Sci. **21,** 1018.
[40] *Howe, G. R.* (1965): Endocrin. **77,** 412.
[41] *Malven, P. V. & Hansel, W.* (1965): J. Repro. Fert. **9,** 207.
[42] *Kiracofe, G. H. & Spies, H. G.* (1966): J. Reprod. Fert. **12,** 217.

[43] *Bradbury, J. T., Brown, W. E. & Gray, L. A.* (1950): Recent Progr. Hormones Res. **5**, 151.

[44] *Williams, W. F., Johnston, J. O., Lauterbach, M. & Fagan, B.* (1967): J. Dairy Sci. **50**, 555.

[45] *Williams, W. F.* (Unpublished results).

[46] *Mazer, R. S. & Wright, P. A.* (1968): Endocrin. Nov. **83**, No. 5, 1065–1071.

[47] *Lukaszewska, Janina H. & Hansel, W.* (1968): Proc. lst Ann. Meeting and Symposium on Male Reproduction (in press). J. of Reprod.

[48] *Mazer, R. M. & Hansel, W.* (Publication in prep.).

[49] *Caldwell, B. V., Moor, R. M. & Lawson, R. A. S.* (1968): J. Repro. Fert. Dec. 1968.

[50] *McCracken, J. A. & Baird, D. T.* (1968): "The Gonads". *K. McKerns* (ed.) Appleton-Century-Crofts, N. Y. (in press).

[51] *Moor, R. M. & Rowson, L. E. A.* (1966): J. Endocrin. **34**, 497.

[52] *Moor, R. M.* (1968): J. Anim. Sci. **27**, Suppl. 1, 77.

[53] *Moor, R. M. & Rowson, L. E. A.* (1966): J. Repro. Fert. **12**, 539.

[54] *Moor, R. M., Rowson, L. E. A., Hay, M. F. & Caldwell, B. V.* (1969): J. Endocrin. (in press).

[55] *Du Mesnil du Buisson, F.* (1969): Arch. Anat. Microscop. Morphol. Exptl. 56. (in press).

[56] *Duby, R. T., McDaniel, J. W., Spillman, C. H. & Black, D. L.* (1969): Acta Endocrin. (in press).

[57] *Melampy, R. M., Anderson, L. L. & Kragt, C. L.* (1964): Endocrin. **74**, 501.

[58] *Anderson, L. L., Melampy, R. M. & Chen, C. L.* (1967): Arch. Anat. Microscop. Morphol. Exptl. **56**, 373.

[59] *Loeb, L.* (1927): Am. J. Physiol. **83**, 202.

[60] *Denamuor, R. & Martinet, J.* (1955): C. R. Seanc. Soc. Biol. **149**, 2105.

Discussion

D. R. S. Kirby: Luteolysin is, or you think it is a protein with a fairly respectable molecular weight. Have you or has anybody attempted to raise an antibody against it to see whether its activity can be influenced either *in vitro* or *in vivo?*

B. V. Caldwell: We are planning to do exactly this with Dr. *Bogden* of the Mason Research Institute in Worcester using our endometrial extracts. And it's a very good point, *David.* In fact, it is certainly the single most worrying point, from the practicle point of view for its potential use as a contraceptive in the human. If it is a high molecular weight substance, would the human, in which it is probably not present, reject it or form antibodies against the luteolysin which is present in the other species?

D. W. Schomberg: One of the last things I did, before having to leave Cambridge was to try to produce an antibody to the uterine substance. This work has not yet reached a final state and has not been confirmed or extended but it does look as though antiserum will diminish the luteolytic effect in the culture system.

F. G. Sulman: Do you see any possibility of studying the problem in tissue culture?

B. V. Caldwell: I think that you should put that question to Dr. *Schomberg.* He is the expert on the culture system, and he has an elegant presentation coming up.

L. Martini: I wanted to ask whether you have tried to treat sheep with high doses of estrogens and progestagens before preparing your uterine extracts. The reason why I am asking the question is you comment on the fact that only uterine transplants at a certain stage of the cycle appear to be active while other extracts are not.

B. V. Caldwell: No, we have not done that, and it is a good suggestion – to determine whether it's possible to induce the endometrium to produce a luteolytic substance.

R. G. Edwards: I listened very carefully to your comments on the human. You said specifically that there is no luteolysin. Is it possible that the lysin is produced from a different source?

B. V. Caldwell: Something must control the regression of the corpus luteum in the human. Anyone of, or several of the possible mechanisms which have been suggested may be operating in all species to one extent or another. I do not think there is any necessity for saying that a uterine luteolysin must be the dominant factor in all species, because certainly the luteotrophic processes also vary considerably among the species. There must be some control. It may not be in the uterus, but it has to be there.

R. J. Ericsson: I can enjoy the role of the Devil's Advocate when it comes to the corpus luteum . . . You build up a certain amount of hostility over the years listening to all these papers. It does not matter whether you want to call it a hormone or a luteolytic substance or what have you. Watching my colleagues' work over the last year and a half, one thing finally has convinced me. Data, $PGF_2\,\alpha$ works, regardless how it works, it works. It works in primates, it works in other laboratory animals, and it does work.

B. V. Caldwell: Pharmacologically, I agree entirely. It may work: however, the monkey data is especially unconvincing. I think Dr. *Johannsen* would be best able to answer this . . . but stress in the monkey can cause the corpus luteum to regress extremely rapidly, and 30 mg of Prostaglandin $F_2\alpha$ would, I feel, be very stressful. And I think this aspect should be completely studied before a particular claim of this nature for luteolysin is made.

R. J. Ericsson: I do not like the term luteolysin either. The work I refer to comes from a paper by Dr. *Kirton* (AAAS meeting, Dallas 1968). They have done additional work since then, and having been involved in work with this primate colony, I do not think stress is a big factor. I agree with you that working with monkeys may be a problem, but in this particular colony I do not think it is.

G. Hellinga: If you have a solution of gonadotropins and you put some endometrium or uterus tissue in it, the gonadotropins are consumed. Do you not think this is also a factor that plays a part here? When you have no uterus, then you have no consumption?

B. V. Caldwell: One group would avidly support the view that perhaps LH is metabolised or consumed or what have you by the uterus, and if you remove it you have a higher level of LH and therefore luteal maintainance. But how then do you explain the very clear local effects, because you obviously have a high level of LH circulating through the whole body, and yet only the one ovary in a hemi-hysterectomized animal will regress.

G. Hellinga: Would you say it 's not consumed on the spot?

B. V. Caldwell: Well, we must postulate a local pathway by which the luteolytic mechanism may operate, and so these people are very justified in saying that perhaps there is a very local protective influence against the LH. It s unanswered, I am sorry to say.

H. R. Lindner: I have just one comment concerning possible interspecific differences. The very clear-cut result that you obtained in unilateral pregnancy in the pig, – and I think *Anderson* and *Melampy* have similar results, – induced us to try the system in the rat and search for local effects of the putative luteolysin. We have used about half a dozen different methods of producing unilateral pregnancy in the rat and took out one ovary at the same time. In these animals, even where we left only the ovary on the contralateral side, that is the ovary in contact with the sterile uterine horn, all pregnancies went undisturbed to term (*J. F. Strauss,* III and *H. R. Lindner,* J. Reprod. Fertil. **18,** 129 (1969)). One would feel that if a local uterine action on the ovary exists in the rat, there must be an escape mechanism, possibly involving placental prolactin, that overrules it even at a very early stage of pregnancy.

B. Lunenfeld: When we worked with the labelled HCG, we found no evidence whatsoever that labelled HCG is taken up by the endometrium, whereas it was taken up by the ovaries.

B. V. Caldwell: There are people who now believe that LH is taken up by the uterus also.

B. Lunenfeld: Furthermore I would only like to make a small remark. In your paper you said that the uterus produces a hormonal substance involved in the regulation of luteal function. I would suggest to take out the word hormonal, since it has only a local effect, as you have shown before.

B. V. Caldwell: Well, it also must have a systemic effect since the corpus luteum will regress between 30 and 40 days, if you take out only one uterine horn. It's not maintained for 145 days, which it normally would be, so there must be some effect of the remaining uterine horn that gets into the bloodstream which ultimately has an effect on that corpus luteum. And the transplant studies where a piece of endometrium is placed in the flank also suggests the possible systemic effects. I would agree that there is no such thing as a hysterectomised sheep running around, and so probably there is no such thing as a systemic luteolysin in the sheep; the predominant activity must be local. Therefore, perhaps we should not call it a hormone.

Mechanism of Luteolysis: Prolactin and the Enzymatic Inactivation of Progesterone in the Corpus Luteum of the Rat

H. R. Lindner and S. A. Lamprecht

Department of Biodynamics, The Weizmann Institute of Science, Rehovot, Israel

Summary: Luteolysis may be defined in anatomical or in functional terms. The two processes thus defined do not necessarily coincide: some agents that terminate the progestational activity of the corpus luteum in the rat, paradoxically prolong its structural persistence.

A decline in the progestational function of the corpus luteum may be brought about in one of two ways: a decrease in the rate of progesterone synthesis, or an accelerated conversion of the nascent hormone to metabolites of lesser progestational potency. Several enzyme systems are now known to operate within the ovary of the rat that reduce either the side-chain or the A-ring of progesterone to yield less active compounds, such as 20 α-dihydroprogesterone (20 α-OH-P; *Wiest*), *allo*pregnanediol *(Mason)* or 3 β-hydroxy-5 α-pregnan-20-one *(Zmigrod* and *Lindner)*. Of these reductases, 20 α-hydroxysteroid dehydrogenase (20 α-SDH) has been most extensively studied.

Reduction of the 20-keto group of progesterone, as first shown by *Wiest,* is most prominent in the ovary during the estrous cycle, particularly at pro-estrus, and minimal during pregnancy or pseudopregnancy. 20 α-OH-SDH activity is chiefly located in regressing corpora lutea, while newly-formed corpora are almost devoid of histochemically demonstrable 20 α-OH-SDH activity and remain so, if pregnancy ensues, to near term; in unmated rats, the activity increases sharply towards the end of diestrus. It remains to be seen whether the ovarian ring-A reducing enzymes are also subject to changes related to the reproductive cycle or to the prepuberal maturation of the ovary.

Temporary suppression of prolactin secretion by the drug ergocornine during early gestation induces greatly enhanced synthesis of luteal 20 α-OH-SDH. As a result, ovarian secretion of the progestationally inactive compound 20 α-OH-P rises sharply, but progesterone release falls below the level required to maintain pregnancy. Exogenous prolactin prevents this effect and maintains the low enzyme level characteristic of pregnancy.

The luteotropic action of prolactin in the rat has hitherto only been demonstrated by use of biological responses interpreted as reflecting enhanced *de-novo* progesterone synthesis. We now suggest that the luteotropic action of prolactin may at least in part be due to a restraining influence on the synthesis of the luteal enzyme that converts progesterone to its less potent 20 α-dihydroderivative.

Hormones are generally thought of as stimulants of specific protein synthesis. The possibility that a pituitary hormone may exert one of its actions primarily by suppressing the synthesis of an enzyme protein is intriguing.

Manuscript received: 26 April 1969

Luteolysis can be defined either in anatomical terms, as the gradual atrophy and ultimate dissolution of luteal tissue, or in functional terms, as the cessation of the hormonal activity of the corpus luterum, in particular the loss of its ability to sustain a progestational endometrium. The two processes thus defined do not always coincide and may be under dissimilar control: under certain conditions hormones generally believed to stimulate steroidogenesis in the ovary, such as LH and prolactin, actually accelerate the regression of defunct corpora lutea [16, 20]; conversely, certain agents that terminate the progestational activity of the corpus luteum in the rat, such as the drug ergocornine (v.i.), paradoxically prolong its structural persistance [10]. I shall confine myself to the functional aspects of luteolysis.

A curious feature of the ovary in several species, including the human, the cow and the rat, is the presence of enzymes capable of diminishing the pregnancy-maintaining activity of the luteal hormone. A decline in the progestational function of the corpus luteum may therefore be brought about in one of two ways: a decrease in the rate of *de novo* progesterone synthesis, or an accelerated conversion of the nascent hormone to metabolites of lesser hormonal potency. Several enzyme systems are now known to operate within the ovary of the rat that reduce the side-chain or the A-ring of progesterone to yield compounds that lack significant progestational activity: 20 α-dihydroprogesterone (20 α-OH-P) is formed in the cytosol fraction of ovarian homogenates [28], 5 α-pregnane-3 α, 20 α-diol (*allo*pregnanediol) by ovarian slices [18] and 3 β-hydroxy-5 α-pregnan-20-one (3 β-OH-5 α P) was recently shown to be the major conversion product when rat ovarian microsomes are incubated with progesterone in the presence of NADPH [32]. Of these reductases, 20 α-hydroxysteroid dehydrogenase (20 α-OH-SDH) has been most extensively studied.

Enzymic reduction of the 20-keto group of progesterone is most prominent in the ovary during the estrous cycle, particularly at pro-estrus, and minimal during pregnancy or pseudopregnancy, as indicated by *in vitro* enzyme assay [28], ovarian content of progesterone and 20 α-OH-P [14] and the concentration of these steroids in ovarian venous blood [2, 4, 7, 24]. In pregnant animals, 20 α-OH-SDH activity increases strikingly near term [29].

Histochemical studies suggested that 20 α-OH-SDH is chiefly located in smaller, probably involuting corpora lutea [27, 29]. That this is indeed so could be clearly shown by assaying the enzyme, either by quantitative biochemical or by histochemical techniques, in corpora lutea of known age isolated by microdissection. The ovaries were marked by intravenous injection of trypan blue on the day of metestrus, so that corpora lutea newly formed at the next estrus could be recognized by their yolky colour on a bluish background. These newly-formed corpora were practically devoid of 20 α-OH-SDH activity (Fig. 1, A) and remained so, if pregnancy ensued, for at least eight days (Fig. 1, C); in unmated rats, considerable activity appeared in these corpora towards the end of dioestrus (Fig. 1, B). Corpora of the preceding cycle were always intensely positive. There was slight activity in thecal and interstitial elements.

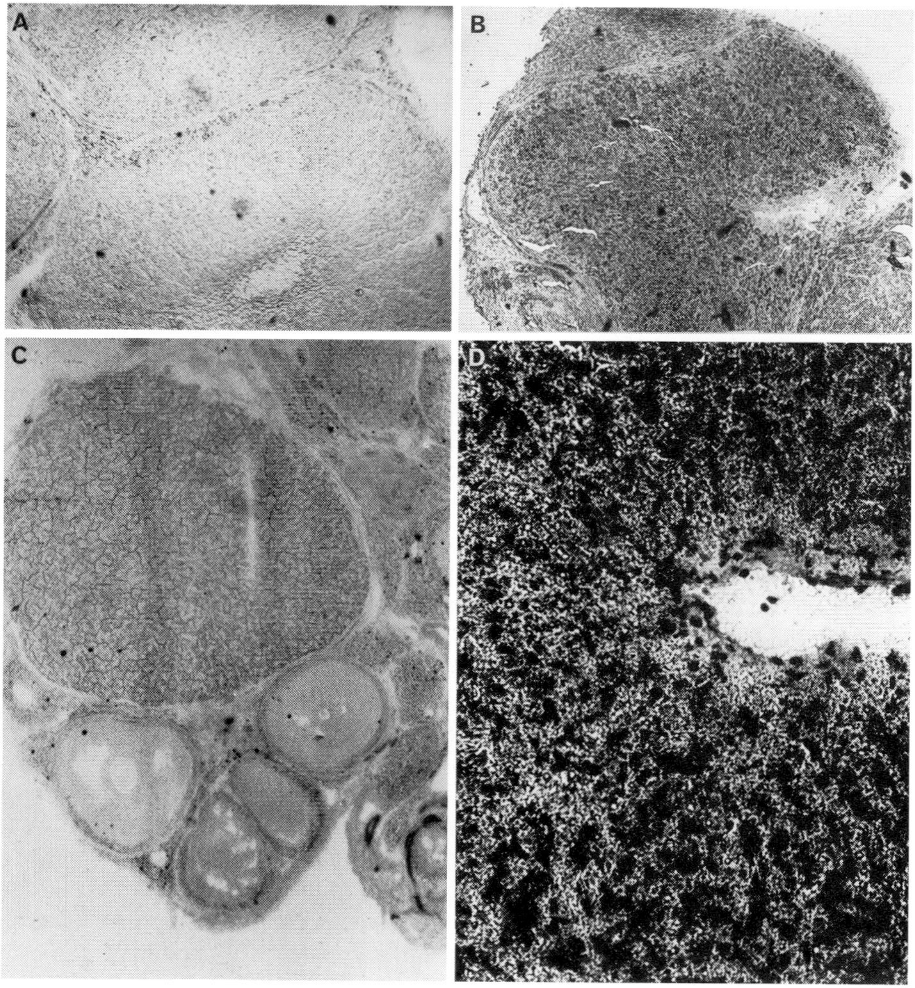

Fig. 1. *Histochemical demonstration of 20α-hydroxysteroid dehydrogenase.*

A-B, newly-formed cyclic corpora lutea (4-day-cycle rats): A; metestrus 3.30 p.m.; B: diestrus 3.30 p.m. (x 25.5). Only corpora unmarked by trypan-blue injected during the previous cycle were examined.

C: ovary on day 8 (p.c.) of pregnancy, untreated control (x 25.5).

D: corpus luteum on day 8 of pregnancy from rat treated with ergocornine on day 5 p.c. (x 255). Cryostat sections incubated in buffered medium containing 20α-OH-P, NADP and nitro-blue-tetrazolium.

We do not know whether the activity of the ovarian ring A-reducing enzymes bears a similar relationship to the reproductive cycle, yet early results *(Zmigrod & Lindner,* unpublished*)* indicate that the rate of formation of 3 β-OH-5 α P by ovarian microsomes is distinctly higher at prooestrus and during the first days of pregnancy than after the seventh day of gestation. Again we suspect that the activity may be localized in involuting corpora.

Against this background I would like to discuss some experiments in which ergocornine, a drug of the ergotoxine group of alkaloids, was used to suppress prolactin secretion, in order to explore the hormonal control of ovarian 20 α-OH-SDH synthesis. Briefly, a single injection of ergocornine methanesulphonate (Sandoz Ltd., Basle; ECO) during pseudopregnancy or the first seven days of pregnancy results in a return to estrus within 72–96 h; concurrent administration of progesterone or prolactin prevents this effect [21, 22]. ECO and certain other ergot alkaloids are also known to inhibit lactation in the rat [6, 19, 31] and mouse [17], and this action too is abolished by exogenous prolactin [31]. Since prolactin is believed to be the principal luteotropic hormone in the rat, these observations strongly suggest that ECO exerts its pregnancy-interrupting action by blocking the release of pituitary luteotropin, possibly by stimulating the secretion of an inhibitory hypothalamic neurohormone. Placental prolactin secretion is resistant to the drug [11, 30].

Gas chromatographic analysis of ovaries from ECO-treated pseudopregnant rats [13] indicated a sharp fall in progesterone content within 24–30 h, and a concomitant rise in the 20 α-OH-P content of the gland, so that the progesterone / 20 α-OH-P ratio was inverted. The total ovarian content of Δ^4-3-keto pregnenes remained unaffected, at least initially. In view of the known prolactin-release blocking action of ECO we interpreted these results as indicating that prolactin might act as an inhibitor of either the synthesis or the activity of ovarian 20 α-OH-SDH in the rat

More recent studies [12] provided direct evidence for this view. Ovarian 20 α-OH-SDH activity was assayed at varying time intervals following an injection of ECO to pseudopregnant or pregnant rats (1 mg on day 5 post-coitum). In control rats the enzyme activity was low (rate of progesterone formation from 20 α-OH-P 0.4–0.5 μg/mg tissue/min) and mainly attributable to involuting corpora persisting from the previous cycle. After ECO treatment, enzyme activity increased significantly within 24–48 h and by 72 h roughly equalled the level found at proestrus (about 4-fold; Fig. 1, D and Fig. 2). Concomitant prolactin treatment (2 mg ovine prolactin N.I.H.–P S-8, 28 I.U./mg, from day 4 to day 7 p.c.) completely prevented this effect, and 20 α-OH-SDH activity remained low (Fig. 2). The inductive effect of ECO appeared even more striking when 20 α-OH-SDH activity was assayed in isolated corpora lutea of pregnancy. True *corpora lutea graviditatis,* distinguished from involuting corpora of the previous cycle by lack of trypan-blue staining (v.s.),

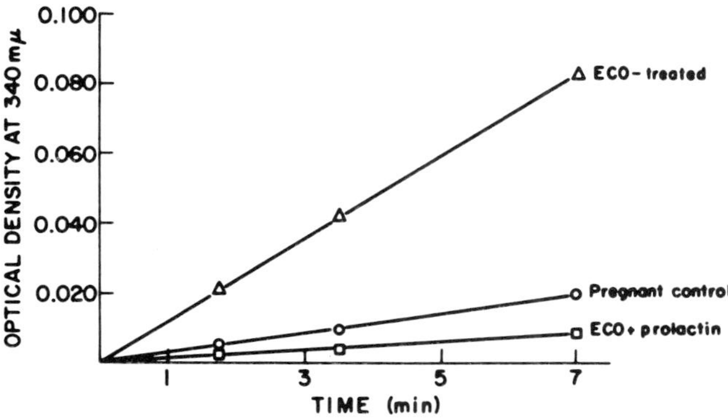

Fig. 2. *Effect of ergocornine on ovarian 20α-hydroxysteroid dehydrogenase and its reversal by exogenous prolactin.*

Supernates (20, 000 g) from homogenates of 1.25 mg tissue were incubated with NADP and 20α-OH-P at 37 °C, and absorption at 340 mμ was continuously recorded as a measure of NADPH generation. ○, ovaries from untreated pregnant rats on day 8 p.c.; △, ovaries collected 72 h after a single subcutaneous injection of ECO (1 mg) on day 5 of pregnancy; □, ovaries collected on day 8 of pregnancy from rats that received ECO (1 mg) 72 h earlier and daily injections of ovine prolactin (2 mg) from day 4 to day 7 p.c.

showed no 20α-OH-SDH activity in control animals on day 8 (p.c.) of pregnancy under the standard assay conditions (Fig. 3), though by using higher tissue concentrations a distinct basal enzyme level (progesterone generation from 20α-OH-P about 0.02 μg/mg/min) could be demonstrated. Following ECO treatment, a 100-fold rise, approximately, in enzyme activity was observed in such corpora (Fig. 3). Neither ECO nor prolactin affected 20α-OH-SDH activity when added *in vitro* to the assay system, excluding a direct modification of the enzyme protein as the basis for the *in vivo* action of these agents; rather does ECO appear to induce *de novo* enzyme synthesis, and prolactin would seem to suppress this.

Independent support for the notion that prolactin may suppress ovarian 20α-OH-SDH has been provided by the observation [29] that the rise in the activity of this enzyme produced by simultaneous hypophysectomy and partial dislocation of the placenta on day 13 of pregnancy was largely prevented by prolactin, and that 20α-OH-SDH activity was depressed during lactation. These authors, however, assign the major role in the regulation of 20α-OH-SDH activity to LH, which is believed to stimulate the synthesis of the enzyme.

The view that prolactin maintains luteal function in the rat has been based, in the past, exclusively on results of biological assays, such as the prevention of estrogen-induced vaginal cornification [1, 15] or the support of the deciduoma reaction in

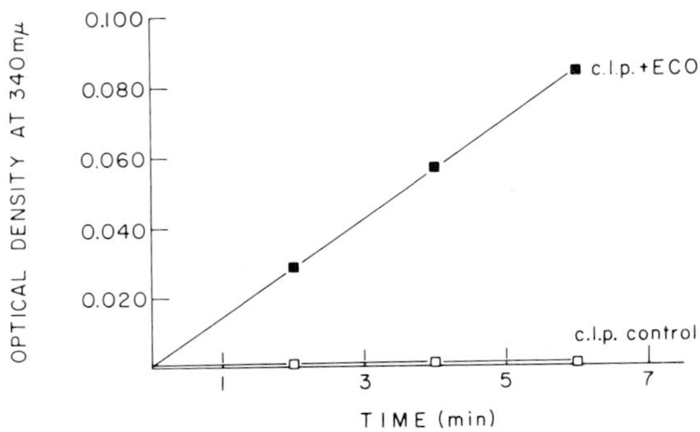

Fig. 3. *Effect of ergocornine treatment in-vivo on 20α-hydroxysteroid dehydrogenase activity in isolated corpora lutea of pregnancy.*

□, corpora lutea of pregnancy (c.l.p.) of untreated control rats on day 8 p.c.; ■, c.l.p. on day 8 p.c., after a single injection (1 mg) of ECO on day 5. Only corpora lutea unmarked by trypan-blue injection during previous cycle were dissected out for the assay. For conditions, see legend to Fig. 2.

hypophysectomized rats [3]. It was tacitly assumed that these effects reflected a stimulation of ovarian *de novo* progesterone synthesis. We now suggest that prolactin may in fact owe its luteotropic action mainly, if not exclusively, to an inhibitory effect on the synthesis of the ovarian enzyme that converts progesterone to its less potent 20 α-dihydro derivative, and perhaps of additional ovarian reductases. In particular, high prolactin levels, initially of pituitary and later of placental origin, may be the factor chiefly responsible for the low 20 α-OH-SDH levels obtaining during pregnancy.

Hormones, including prolactin [5, 25], are generally thought of as stimulants of specific protein synthesis, though a few exceptions to this generalization have recently been noted. The possibility that prolactin may exert one of its actions primarily by suppressing the synthesis of an enzyme protein is therefore of some basic interest.

The biological significance of the reducing enzymes present in the ovary remains a puzzle. Some of the reduced derivatives of progesterone may yet prove to have a specialized function. Thus 20 α-OH-P was shown to exert a positive feedback action on LH discharge in mated rabbits [8]. It is of interest that the testis of the immature rat, too, contains reductases that convert testosterone or its immediate precursors to ring-A reduced derivatives of lesser androgenic activity [9, 23]. The concentration of these enzymes falls abruptly before the onset of puberty, with a resultant rise in the androgenic potency of the testicular secretion. It is tempting to speculate that the rat ovary has adopted a similar regulatory mechanism by which the hormonal activity of the luteal secretion may be modified in accordance to physiological requirements through partial reduction of the progesterone formed [29, 32].

Coming here from the Middle East, an analogy comes to mind. During an uneasy armistice, a country may find it advantageous to let tank factories revert to tractor production, a minor and reversible adaptation, rather than to dismantle its entire armament industry. Similarly, a minor modification of the terminal steroid product of the corpus luteum in response to diminishing need for the hormone may offer advantages in speed, flexibility and economy over a regulatory mechanism that would bring about a reduction in hormone secretion by dismantling the entire steroidogenic enzymic apparatus. Moreover, should messenger molecules coding for steroidogenic enzymes in the luteal cell be long-lived (as is the haemoglobin messenger in erythrocytes), and synthesis of these enzymes be devoid of translational controls, there may not even exist a ready mechanism for abruptly shutting down progesterone production during the life-time of the luteal cell.

The rat is representative of those rodent species that are peculiar in returning to estrus with great speed when an opportunity for mating has been missed, – a feature of obvious survival value. This requires a prompt cessation of progesterone secretion by the still young luteal cells. It may be significant that only in species exhibiting this peculiar reproductive pattern (the rat, the mouse and perhaps the golden hamster; *van Tienhoven* [26]) has ovine prolactin thus far been shown to be luteotropic.

Wiest et al. [29] concluded that "progesterone catabolism to 20 α-OH-P constitutes a normal regulatory mechanism for reducing the progestational activity of rat ovarian secretions during pregnancy". For this to be an effective adaptive mechanism, induction of 20 α-OH-SDH should be a reversible process. Evidence to-date rather indicates that once induced by temporary prolactin deprivation, e.g. following a single injection of ECO or during the dioestrus of an uncovered cycle, elevated 20 αOH-SDH activity persists in the luteal cell for the remainder of its life-span and can then no longer be suppressed by prolactin; thus the cell's capacity to secrete progesterone as such is permanently impaired. In this sense the level of the enzyme may indeed be regarded as an index of the luteolytic process, as proposed by *Wiest* et al. [29].

Acknowledgement: This work was supported by the Ford Foundation and the Population Council. We are obliged to the N.I.H. Endocrine Study Section for a gift of purified ovine prolactin, and to Sandoz Ltd., Basle, for ergot alkaloids. Mr. *S. Gordon* provided valuable skilled assistance in the histochemical studies, and Mr. *Y. Shalom* in animal care and surgery.

References

[1] *Astwood, E. B.* (1941): Endocrinology **28**, 309.

[2] *Eto, T., Masuda, H., Shzuki, Y. & Hosi, T.* (1962): Jap. J. Animal Reprod., **8**, 34.

[3] *Evans, H. M., Simpson, M. E., Lyons, W. R. & Turpeinen, K.* (1941): Endocrinology, **28**, 933.

[4] *Fajer, A. B. & Barraclough, C. A.* (1967): Endocrinology, **81**, 617.

[5] *Falconer, I. R. & Fiddler, T. J.* (1968): Biochem. J., **110**, 56P.

[6] *Flückiger, E. & Wagner, H. R.* (1968): Experentia **24**, 1130.

[7] *Hashimoto, L., Henricks, D. M., Anderson, L. L. & Melampy, R. M.* (1968): Endocrinology, **82**, 333.

[8] *Hilliard, J., Penardi, R. & Sawyer, C. H.* (1967): Endocrinology, **79**, 268.

[9] *Inano, H., Hori, Y. & Tamaoki, B.* (1967): In: Endocrinology of the Testis, CIBA Foundation Colloquia on Endocrinology, Vo. 16, p. 105. Ed. *G. E. W. Wolstenholme & M. O'Connor,* London: J. & A. Churchill.

[10] *Kisch, E. S.* (1967): Doctoral Thesis, Weizmann Institute of Science, Rehovot, Israel.

[11] *Kisch, E. S. & Shelesnyak, M. C.* (1968): J. Reprod. Fertil. **15**, 401.

[12] *Lambrecht, S. A., Lindner, H. R. & Strauss III, J. F.* (1969): Biochim. Biophys. Acta. **187**, 133.

[13] *Lindner, H. R. & Shelesnyak, M. C.* (1967): Acta endocr., Copenh. **56**, 27.

[14] *Lindner, H. R. & Zmigrod, A.* (1967): Acta endocr., Copenh. **56**, 16.

[15] *MacDonald, G. J. & Greep, R. O.* (1968): Persp. Biol. Med. **11**, 490.

[16] *Malven, P. V. & Sawyer, C. H.* (1966): Endocrinology **79**, 268.

[17] *Mantle, P. G.* (1968): Proc. Roy. Soc., B. **170**, 423.

[18] *Mason, N. R.* (1968): Abstr. Brief Communications, IIIrd Intern. Congr. Endocrinology, Mexico D. F., p. 127, Amsterdam: Excerpta Medica Foundation.

[19] *Nordskog, A. W. & Clark, R. T.* (1945): Amer. g. Vet. Res. **6**, 107.

[20] *Rothchild, I.* (1965): Acta endocr., Copenh. **49**, 107.

[21] *Shelesnyak, M. C.* (1957): Recent Progr. Horm. Res. **13**, 269.

[22] *Shelesnyak, M. C.* (1958): Acta endocr., Copenh. **27**, 99.

[23] *Shikita, M. & Hall, P. F.* (1967): Biochim. biophys. Acta **136**, 484.

[24] *Telegdy, G. & Endröczi, E.* (1963): Steroids **2**, 119.

[25] *Turkington, R. W.* (1968): Abstr., Brief Communications, No. 180, IIIrd Intern. Congr. Endocrinology, Mexico D. F., Amsterdam: Excerpta Medica Foundation.

[26] *van Tienhoven, A.* (1968): Reproductive Physiology of Vertebrates, p. 131, Philadelphia: W. B. Saunders.

[27] *Turolla, E., Magrini, V. & Gaetani, M.* (1966): Experientia **22**, 675.

[28] *Wiest, W. G.* (1959): J. biol. Chem. **234**, 3115.

[29] *Wiest, W. G., Kidwell, W. R. & Balogh, K.* (1968): Endocrinology **82**, 844.

[30] *Zeilmaker, G. H.* (1968): Acta endocr., Copenh. **59**, 442.

[31] *Zeilmaker, G. H. & Carlsen, R. A.* (1962): Acta endocr., Copenh. **41**, 321.

[32] *Zmigrod, A. & Lindner, H. R.* (1969): Acta endocr., Copenh. **61**, 618.

Discussion

A. S. Goldman: I have a suggestion as to how the prolactin may be inhibiting in the enzyme. It comes from the work of *Margaret Wiener* and *S. H. G. Allen* (Steroids 9, 6, 1967) who showed with regard to these enzymes in the placenta, that there seemed to be a reciprocal relationship between 20 α-hydroxy-4-pregnene-3-one (20 α-HP) and estriol. In the one case 20 α-HP inhibited 3 β-hydroxysteroid dehydrogenase, and therefore blocked the production of progesterone. In the other case the estriol inhibited 20 α-reductase, blocking the production of 20 α-hydroxyprogesterone. In other words, she postulated that the levels of these enzymes are controlled by the peripheral concentrations of these steroids. It is possible that this kind of mechanism may be involved her. I am not sure whether estradiol and 20 α-progesterone could be shown to inhibit the ovarian enzymes, but it is a suggestion.

H. R. Lindner: It is an interesting suggestion. I cannot quite see yet how this would fit in. There is very high 3- (β)-ol-dehydrogenase activity of course in the corpus luteum, even in the face of 20 α-dihydroprogesterone concentrations.

A. S. Goldman: I am not sure that 20 α-progesterone is the only one capable of inhibiting the dehydrogenase. We found that estradiol inhibits the dehydrogenase too. The inhibitor may be one of the steroids, for which prolactin is responsible.

H. R. Lindner: Well, I will have to digest this further. So far we see little evidence for prolactin directly controlling the *de-novo* synthesis of any ovarian steroid, but we have not examined this ourselves.

R. I. Dorfman: I had just venture the guess that the amount of estradiol necessary for this inhibition may be very high. Estradiol is a competetive inhibitor; actually the isomerase reaction was demonstrated by *Talloway.* As I recall, the concentration was far beyond physiological limits.

A. S. Goldman: No, no, it is not, it is very low. Estrogen inhibits both the dehydrogenase and the siomerase.

R. I. Dorfman: At what concentrations?

A. S. Goldman: In the test tube?

R. I. Dorfman: Yes.

A. S. Goldman: Oh, very small. About 5 - 10 x 10^{-6} molar.

E. Steinberger: Does ergocornine have any effect on the luteinizing hormone?

H. R. Lindner: If it has any, it would seem to stimulate the secretion of luteinizing hormone, but this is a late effect, and we could not sort it out from an effect of progesterone withdrawal. These rats ovulate 72–96 hours after being given ergocornine, but since the progesterone level falls already 24–30 hours after giving the drug, we feel that the LH release responsible for this ovulation is probably due to release of the pituitary from negative feed-back following reduction of the progesterone level.

E. Steinberger: Could you test this substance in the more stable system, possibly in the male, where you donot have to interpret so much?

H. R. Lindner: In the male it does not affect fertility, this is all I know so far. I have not looked further.

L. Martini: Do you know, Dr. *Lindner,* what treatment with progesterone will do to your enzymes? If you give exogenous progesterone what will happen?

H. R. Lindner: We have not measured enzyme activity under these conditions. Concomitant progesterone administration will maintain pregnancy.

E. Steinberger: I am very much interested in 20 α-dehydrogenase. We find that this is a very "common" enzyme. I might say this enzyme is present in cells that look like fibroblasts. One may wonder if it is not rather a "non-specific" enzyme that is very easily induced in non-steroid producing type of cells.

H. R. Lindner: It may be so in other tissues, but in the ovary the fluctuations of this particular enzyme activity are very consistently related to the phases of the reproductive cycle. Therefore we felt that it may have some meaning and be part of a regulatory mechanism.

Effects of Uterine Flushings from Pigs on Porcine Granulosa Cells growing in Tissue Culture

D. W. Schomberg, J. A. Campbell, and G. D. Wilbanks

Departments of Obstetrics and Gynecology and of Physiology, Duke University Medical Center, Durham, North Carolina, USA

Summary: The results of earlier studies indicated that during days-12 to -18 of the estrous cycle, the uterus of the sow produced a high-molecular-weight substance in quantities sufficient to elicit marked destructive changes in tissue culture monolayers of porcine granulosa cells. This activity appeared in the uterine flushings during the period of the cycle characterized by declining progesterone levels and persisted during anatomical regression of the corpora lutea but had disappeared before the onset of the next estrus.

These studies have been extended using time-lapse cinematography. The results provide direct evidence that uterine flushings collected during this period of the estrous cycle induce a series of striking morphological changes which culminate in the death of the cells.

Introduction

Our recent studies originated with the observation by *Channing* [1] that equine granulosa cells growing as tissue culture monolayers continued to secrete progesterone and other steroids into the culture medium. It seemed that such cultures might serve as an *in vitro* test system for luteolytic substances, particularly since the experimental attempts to substantiate *Loeb's* [2, 3] concept of a uterine luteolytic hormone may have been hampered by the lack of an adequate assay system [4]. Furthermore, a sensitive and fairly rapid *in vitro* system serving as an index of luteolytic activity *in vivo* would be an invaluable aid in the isolation and purification of an active substance.

Previous studies: a summary

Previous studies demonstrated that monolayers of porcine granulosa cells also retained the ability to secrete progesterone in culture. The cultures were rapidly destroyed by a high-molecular-weight component(s) of uterine flushings from non-pregnant pigs which could be detected only during the period of the estrous cycle

---

Manuscript received: 28. April 1969

characterized by functional and anatomical regression of the corpora lutea. Flushings collected from pigs at other stages (days-0-10 and day-20) of the 21-day estrous cycle did not exhibit cytolytic activity. In many cases the 7-9-day old cultures were destroyed within 8 hours after replacement of the growth media by flushing media representing days-14-18 of the cycle. Histological examination of the cell clumps which remained attached to the coverslips gave the impression that the cell membranes had blistered and then ruptured. Studies of vital dye uptake, pH changes, and progesterone production all indicated that the cells had died but none of these indices provided absolutely unequivocal evidence of cell death by themselves. The active component(s) of the day-14-18 flushings could be consistently eluted from Sephadex G-200 columns and was heat-labile. It also demonstrated some degree of cellular specificity since it destroyed pig granulosa cell cultures much more rapidly than pig kidney cultures [4–6].

Current studies: applications of time-lapse cinematography

The above results have been confirmed and extended using time-lapse cinematography. The films provide direct evidence that uterine flushings from day-16 of the estrous cycle induce a series of striking morphological changes which culminate in the death of the cells.

Experimental procedures

The incubation conditions described earlier [4, 5] were modified to facilitate the time-lapse studies. The cultures appeared to grow equally well in either experimental situation. The harvested granulosa cells were deposited on a sterile glass coverslip in plastic petri dishes, 35 x 10 mm; 3 ml of media (fetal calf serum: minimal essential media plus phenol red and antibiotics, 10:90) were added; the cultures were incubated at 37 degrees C in a gas phase of 95 % air and 5 % CO_2 saturated with water. These same conditions were maintained in the plexiglass chamber housing the stage, condenser, and inverted objective lenses of a Nikon inverted microscope (Model M) with a Bolex camera attachment (Model H16). The petri dish was held in place by microscope stage clips. A particular group of cells could be photographed continuously throughout an experimental period even during a medium change by using a syringe and needle to withdraw and add the medium through a small hole produced previously in the plastic cover with a hot needle. This opening was covered by cellophane tape which could be easily sterilized with ethanol and perforated. In these experiments the growth media was changed at the beginning of an experiment. The cells were photographed for a 24-hour "control" period and then the growth media was replaced by media containing uterine flushings or aqueous myometrial extracts representing day-16 of the estrous cycle. The "experimental" observation period was 48 hours.

The exposure intervals were controlled by a time-lapse control panel; the exposure time which yielded the best results was 0.5 seconds with an interexposure interval of 59.5 seconds.

The developed films (Kodak Plus-x reversal movie film) were projected at a speed of 16 frames per second.

Results and discussion

In this particular experiment the 3 ml of flushing media representing day-16 of the cycle contained 0.75 mg of uterine protein and was added to a 13-day old culture 24 hours after beginning the time-lapse observations. During the 24-hour control period the degree of cellular migratory movement and pseudopodia activity were both moderate. The general cell morphology was epithelioid; no mitoses were noted. Almost immediately upon addition of the flushing media the movements of the pseudopodia over the cell surface and at the cell boundaries increased, giving the impression of heightened cellular activity although migratory activity did not change. Then, suddenly and without any prior indication, the cell cytoplasm snapped from its extended two-dimensional epithelioid form into a very much smaller spherical shape as the pseudopodia presumably became detached from the coverslip. Shortly after this, the cytoplasm ballooned to the point at which the cell membranes burst, with cellular organization disappearing as the dead cells contracted into a mass of amorphous debris. The time interval between detachment of the pseudopodia and rupture of the cell membranes varied from a few minutes to one hour.

An aqueous myometrial extract of the same uterus (obtained by hysterectomy) from which the flushing was obtained was used as the control preparation. Occasional mitoses were noted during the control period which in this case spanned days-7 and -8 of culture. The media containing the myometrial extract (0.68 mg total myometrial protein) produced no obvious detrimental effects. In fact, there was a noticeable increase in cell hypertrophy and hyperplasia with a marked increase in the number of mitoses after about 39 hours. The significance of these stimulatory effects is unclear, but it is clear that the myometrial extract did not exhibit cytolytic activity.

We are currently extending our observations on the cellular specificity of the cytolytic flushing component(s) since it was observed that monolayers of a well established cell line of embryonic human skin fibroblasts were destroyed much more rapidly by the day-16 flushing media than were the control pig granulosa cell monolayers. When this flushing media was tested upon primary cultures of embryonic pig skin fibroblasts and the effects compared to those observed in cultures of embryonic human skin fibroblasts and pig granulosa cells, it was found that the pig skin fibroblasts and granulosa cells were destroyed to approximately the same degree. However, the human skin fibroblasts were much more rapidly destroyed than either of the porcine cell types. The mechanism underlying these differences is at present unclear. One possibility is that the response of the established cell line is atypical due to chromosomal changes after several generations in culture. However, a more likely possibility, which does not necessarily exclude the former, is that these differences have an immunological basis. In view of this, it would probably be advisable to avoid species differences when one is comparing the effects of uterine components upon various cell types *in vitro.*

Specificity of the in vitro effect: considerations

Uterine flushings and endometrial extracts may well contain several components capable of inducing nonspecific cytolytic effects in a tissue culture system. Furthermore, a cytolytic agent of uterine origin may or may not preferentially destroy ovarian cell types *in vitro.* Looking at these two considerations another way, one can perhaps designate them as problems of substance specificity and cellular specificity.

Substance specificity

At the present time one cannot do much more than compile a list of identifiable compounds present in the endometrium which are known to affect cell morphology or function to a rather marked degree and then evaluate the effects of these agents in a tissue culture system. The alternative approach is to characterize an unidentified endometrial substance as completely as possible and to compare the properties of this component with those of other unidentified cytolytic substances extracted from other organs or tissues. The literature contains some very interesting references germane to both approaches. The relationships of these studies to our findings will be discussed.

Uterine luminal fluids or washings have been shown to contain several components ranging from whole cells [7] to cellular and leucocytic debris [8, 9] and soluble enzymes [10]. It is conceivable that enzymatic constituents of the leucocytic granules, e.g., acid phosphatase, β-glucuronidase, cathepsins, lysozyme, and ribonucleases [11], could be present in sufficient concentrations in uterine washings to produce significant detrimental effects *in vitro.* According to several workers, however, leucocytes are present in uterine washings from normally cycling, pregnant, or pseudopregnat animals in small or undetectable amounts [12–15]. Although there is no doubt that leucocytes are largely responsible for the bactericidal action of the uterus, there is some question as to their effectiveness in uterine fluids since *Broome* and *Lamming* [16] reported that contact between bacteria and living endometrium *in vivo* was necessary before bactericidal effects could be deomstrated. A portion of the bactericidal activity can be attributable to noncellular factors, presumably of leucocytic origin [17, 18]. Leucocytic infiltration of the endometrium is under steroidal control in the rabbit [19], sheep [20], cow [21], and pig [9], the uterus being most resistant to infection around the time of mating.

Assuming that the enzymes of the leucocytic granules could induce cytolysis of the granulosa cells in culture, it would appear from the available evidence that the maximum cytolytic effect should occur during the period of estrus. This was not the case [4, 5], but it would be worthwhile to test this assumption directly.

Ruptured *endometrial* cells present in luminal washings or intact endometrial cells themselves could also be sources of nonspecific enzymatic activity. Although one might expect that lysosomal enzymes originating from ruptured endometrial cells would predominate, it is also possible that other proteolytic enzymes could be secreted directly into the lumen since human endometrium in organ culture exhibits collagenase and other proteolytic activity [22, 23].

Many of the enzymes which have been indentified as lysosomal components or as secretory products of the endometrium have been used in tissue culture studies, primarily as agents for the dispersal of cell aggregates. *Grob* [24] used various proteolytic enzymes (pronase, trypsin, chymotrypsin, pancreatin, collagenase, elastase) to disperse the different cellular components of rat and mouse ovaries and achieved a stepwise dispersal of cells with 0.5 percent pronase in a Ca^{++}-Mg^{++}-free balanced salt solution. Only 10–15 % of the free cells were penetrated by a vital dye. *Essner, Suto,* and *Belkin* [25] found hyaluronidase, papain, lipase, beta-amylase, and lysozyme in concentrations of 1 mg/ml to be ineffective in separating cells in contrast to trypsin and chymotrypsin; the authors made no mention of toxic effects. Although *Rinaldini* [26] found collagenase to be detrimental to cells (in concentrations of 0.1 mg to 1.0 mg/ml), other workers [27, 28] have used it successfully. Ribonuclease inhibits the growth of cultured cells, but the effect is not rapid [29]. Cathepsins may be responsible for the degradation of cartilage matrix in bone rudiments *in vitro* [30], but their low pH optima (pH 3–4) could be cited as evidence against the possibility of their involvement in the cytolytic activity of the pig uterine flushings *in vitro. Rinaldini* cites other work indicating that the effects of cathepsins are much slower than that of trypsin.

A significant lysosomal enzyme, acid phosphatase, has apparently not been studied in the manner described above. The concentration of acid phosphatase in the endometrium of the pig changes very little throughout the estrous cycle [31].

In order to be sure that the enzymes referred to above are not responsible for the effects we have observed, their activity should be re-evaluated in the granulosa cell culture system; at the same time, enzymatic treatment of the flushing component(s) itself can provide important information about its chemical nature.

Unidentified large-molecular-weight substances which may be directly involved in the cellular response to injury [32] or function as anti-inflammatory factors, possibly inhibiting fibroblasts [33–35], could also theoretically induce cytolytic effects in a tissue culture system. Although this possibly may seem unlikely at first glance, it is of more than passing interest since the compound studied by *Menkin* [32, 36, 37], termed "necrosin", was believed to be identified with the "menstrual toxin" studied by *O. W.* and *G. S. Smith* [38–41] between 1940–1947. Injections of 0.05–0.2 ml of canine pleural exudate or menstrual discharge containing "necrosin" and the "menstrual toxin", respectively, killed 19–21 day old male or female rats within 48–96 hours. Both compounds were insoluble in saline and were believed to be identical immunologically [40, 41]. Also of interest in this regard are the studies of *Schneider* [42, 43] who obtained a nondialyzable toxic substance from the endometrium or placenta of several species (human, cow, sheep, rat, rabbit, and mouse). Extracts of progesterone dominated endometrium were more toxic when administered intravenously to mice than were extracts of estrogen dominated endometrium; small amounts of material with a similar effect were present in extracts of skeletal muscle.

All of these substances were characterized by extreme lability: Plasma or fluid containing the anti-inflammatory substance had to be used the same day [34]. Both the "menstrual toxin" and "necrosin" were extremely temperature-sensitive and were also rapidly destroyed if stored in aqueous solutions even after freezing [40, 41]. The compound(s) studied by *Schneider* [42] was also unstable, even at low temperatures, and was rapidly inactivated by serum. In contrast, the active component of the uterine flushings retains its activity after heating at 56 degrees C for 30 minutes and after several days of incubation at 37 degrees C in combination with serum and other constituents of the culture medium. The flushing solutions also retain their activity after repeated freezing and thawing and after at least 2 to 3 months of frozen storage [4, 5, 44].

Finally, there is the possibility that uterine flushings or endometrial extracts may contain nonspecific cytolytic components identical to those which can also be extracted from other organs or tissues. *Duncan, Bowerman, Anderson, Hearn* and *Melampy* [45] reported that progesterone synthesis by porcine luteal slices was inhibited by a dialyzable substance(s) present in endometrial filtrates representing days-16-18 of the cycle and in filtrates of striated muscle. The inhibitory activity of these substances was reduced by heating at 100 degrees C as well as lower temperatures for 15 minutes, but when the filtrates were ashed the residues exhibited inhibitory activity. Extracts of skeletal muscle are capable of inhibiting the growth of Ehrlich ascites tumour cells in suspension culture [46]. The inhibitory factor in the skeletal muscle extract was dialyzable and was removed by heating the original extract but not he dialysate suggesting that the active component was a heat stable, small-molecular-weight substance which was originally bound.

Cellular specificity

As pointed out above, uterine components may or may not affect ovarian cell types preferentially *in vitro.* The value of the *in vitro* system, however, will be determined entirely by the degree to which it exemplifies the *in vivo* situation. In this context, time-lapse cinematography may prove to be a unique tool for characterizing the effects of the uterine flushing component(s) upon different types of cell cultures.

Summary

The above considerations embody several questions, each concerning different aspects of the general question of *in vitro* specificity:

1) Is the action of the active component(s) of the flushings only that of a cytolysin ?
2) What are the effects of this component upon other types of cell cultures ?
3) What effects do extracts of other organs or tissues have upon the granulosa cell culture system ?

Perhaps with the techniques now available we can ask the proper questions, and in ways such that we will be able to obtain definite answers.

Acknowledgement: We are especially grateful to Dr. *L. C. Ulberg* and Mr. *Frank Fenton* of the Department of Animal Science, North Carolina State University, Raleigh, N.C., USA, for their interest and advice and for providing the uteri used in the time-lapse studies. This work was supported in part by research grants from the Duke Endowment and the Damon Runyan Memorial Fund for Cancer Research (Grant no. 888).

References

[1] *Channing, C.:* Progesterone biosynthesis by equine granulosa cells growing in tissue culture. Nature (London) **210**, 1266, (1966).

[2] *Loeb, L.:* The effect of extirpation of the uterus on the life and function of the corpus luteum in the guinea pig. Proc. Soc. Exp. Biol. Med. **20**, 441, (1923).

[3] *Loeb, L.:* The effects of hysterectomy on the system of sex organs and on the periodicity of the sexual cycle in the guinea pig. Am. J. Physiol. **83**, 202, (1927).

[4] *Schomberg, D. W.:* The concept of a uterine luteolytic hormone. In: The Gonads, *K. W. McKerns,* ed., Appelton-Century-Crofts, New York, (1969). In press.

[5] *Schomberg, D. W.:* A demonstration *in vitro* of luteolytic activity in pig uterine flushings. J. Endocr. **38,** 359 (1967).

[6] *Schomberg, D. W.:* Effects of pig uterine flushings on ovarian tissues ***in vitro*** and ***in vivo.*** Proceedings of the Third International Congress of Endocrinology, Mexico City, Mexico. Excerpta Medica Foundation, Amsterdam, (1969). In press.

[7] *Sedlis, A., Dacosta, A. & Connel, E.:* Cytochemistry of uterine washings. Acta Cytologica **11**, 353, (1967).

[8] *Olds, D. & Van Demark, N. L.:* Composition of luminal fluids in bovine genitalia. Fertil. Steril. **8**, 345 (1957).

[9] *Nellor, J. E. & Brown, J. E.:* Leucocyte-like cells of the vagina and uterus, and their modification during the normal estrous cycle and by progestin and estrogen treatment. Anat. Rec. **155**, 591, (1966).

[10] *Leone, E., Libonati, M. & Lutwak-Mann, C.:* Enzymes in the uterine and cervical fluid and in certain related tissues and body fluids of the rabbit. J. Endocr. **25**, 551, (1963).

[11] *Cohn, Z. A. & Hirsch, J. G.:* The isolation and properties of the specific cytoplasmic granules of rabbit polymorphonuclear leucocytes. J. Exp. Med. **112,** 983, (1960).

[12] *Parr, E. L., Schaedler, R. W. & Hirsch, J. G.:* The relationship of polymorphonuclear leukocytes to infertility in uteri containing foreign bodies. J. Exp. Med. **126**, 523, (1967).

[13] *Brown, J. E. & Nellor, J. E.:* A relationship between ovarian and genital tract eosinophilic granulated cells. Anat. Rec. **162,** 15, (1968).

[14] *Gerrits, R. J., Hawk, H. W. & Stormshak, F.:* Fertility and corpus luteum characteristics in pigs with plastic devices in the uterine lumen. J. Reprod. Fert. **17,** 501, (1968).

[15] *Marston, J. H. & Kelly, W. A.:* The effect of uterine anastomosis on the action of an intrauterine device in the rat. J. Endocr. **43**, 95, (1969).

[16] *Broome, A. W. J. & Lamming, G. E.:* Studies on the relationship between ovarian hormones and uterine infection. II. The effect of ovarian hormones on the anti-bacterial activity of the uterine environment. J. Endocr. **18,** 220, (1959).

[17] *Hawk, H. W.:* Investigations concerning bactericidal substances in rabbit uteri. Am. J. Vet. Res. **20,** 206, (1959).

[18] *Hawk, H. W., Turner, G. D. & Sykes, J. F.:* Noncellular and bactericidal factors in inflammatory exudates from the uterine and pleural cavities of estrous and pseudopregnant rabbits. Am. J. Vet. Res. **21**, 657, (1960).

[19] *Heap, R. B., Robinson, D. W. & Lamming, G. E.:* The relationship between ovarian hormones and uterine infection in the rabbit. A possible mode of action. J. Endocr. **23**, 351, (1962).

[20] *Brinsfield, T. H., Hawk, H. W. & Leffel, E. C.:* Control by ovarian hormones of the inflammatory response in the sheep uterus. J. Reprod. Fert. **6**, 79, (1963).

[21] *Black, W. G., Ulberg, L. C., Kidder, H. E., Simon, J., McNutt, S. H. & Casida, L. E.:* Inflammatory response of the bovine endometrium. Am. J. Vet. Res. **14,** 179, (1953).

[22] *Leroy, F.:* Analyse cytologique de cultures ***in vitro*** de l'endomètre humain et de ses constituants isolés. Bull. Soc. Roy. Belge Gyn. Obst. **32**, 483, (1962).

[23] *Kullander, S. & Källen, B.:* The fibrinolytic activity of human endometrium studied in tissue culture. Acta Obst. Gynec. Scand. **40,** 1, (1961).

[24] *Grob, H. S.:* Enzymatic dissection of the mammalian ovary. Science. **146,** 73, (1964).

[25] *Essner, E., Sato, H. & Belkin, M.:* Experiments on an Ascites Hepatoma. I. Enzymatic digestion and alkaline degradation of the cementing substance and separation of cells in tumor islands. Exp. Cell Res. **7**, 430, (1954).

[26] *Rinaldini, L. M. J.:* The isolation of living cells from animal tissues. Int. Rev. Cytology **7,** 587, (1958).

[27] *Lasfargues, E. Y.:* Cultivation and behavior ***in vitro*** of the normal mammary epithelium of the adult mouse. Exp. Cell Res. **13,** 553, (1957).

[28] *Brown, J. M. & Hilfer, S. R.:* Effects of the dissociating agents collagenase, pronase and tetraphenylboron on embryonic chick thyroid and heart cells. J. Cell Biol. **39,** 18a, (1968) (Abstract).

[29] *Firket, H., Chèvremont-Comhaire, S. & Chèvremont, M.:* Action of ribonuclease on living cells ***in vitro*** and synthesis of deoxyribonucleic acid. Nature (London) **176**, 1075, (1955).

[30] *Lucy, J. A., Dingle, J. T. & Fell, H. B.:* Studies on the mode of action of excess vitamin A. 2. A possible role of intracellular proteases in the degradation of cartilage matrix. Biochem. J. **79,** 500, (1961).

[31] *Akins, E. L., Morrissette, M. C. & Cardeilhac, P. T.:* Luteal and endometrial phosphatase activities during the porcine estrous cycle. J. Anim. Sci. **28,** 51, (1969).

[32] *Menkin, V.:* Studies on the isolation of the factor responsible for tissue injury in inflammation. Science **97,** 165, (1943a).

[33] *Di Pasquale, G. & Girerd, R. J.:* Anti-inflammatory properties of lyophilized inflammatory exudate. Am. J. Physiol. **201,** 1155, (1961).

[34] *Robinson, B. V. & Robson, J. M.:* Production of an anti-inflammatory substance at a site of inflammation. Brit. J. Pharmacol. **23,** 420, (1964).

[35] *Robinson, B. V. & Robson, J. M.:* Further studies on the anti-inflammatory factor found at a site of inflammation. Brit. J. Pharmacol. **26**, 372, (1966).

[36] *Menkin, V.:* Chemical factors in inflammation and cellular injury. New Eng. J. Med. **229,** 511, (1943b).

[37] *Menkin, V.:* The significance of biochemical units in inflammatory exudates. Science **101,** 421, (1945).

[38] *Smith, O. W. & Smith, G. S.:* Menstrual discharge of women. I. Its toxicity in rats. Proc. Soc. Exp. Biol. Med. **44**, 100, (1940).

[39] *Smith, O. W. & Smith, G. S.:* Further studies on the menstrual discharge of women. Proc. Soc. Exp. Biol. Med. **55**, 285, (1944).

[40] *Smith, O. W. & Smith, G. S.:* Evidence that menstrual "toxin" and canine "necrosin" are identical. Proc. Soc. Exp. Biol. Med. **59,** 116, (1945).

[41] *Smith, O. W.:* Menstrual toxin. I. Experimental studies. Am. J. Obstet. Gynec. **54,** 201, (1947).

[42] *Schneider, C. L.:* A toxic principle from progestational endometrium and placenta. Proc. Soc. Exp. Biol. Med. **62**, 322, (1946a).

[43] *Schneider, C. L.:* Liver necrosis in mice following injection of toxic extracts of progestational endometrium and placenta. Proc. Soc. Exp. Biol. Med. **62**, 325, (1946b).

[44] *Schomberg, D. W.:* Unpublished observations.

[45] *Duncan, G. W., Bowerman, A. M., Anderson, L. L., Hearn, W. R. & Melampy, R. M.:* Factors influencing ***in vitro*** synthesis of progesterone. Endocrinology **68,** 199, (1961).

[46] *Namba, T., Sato, T. & Grob, D.:* Inhibition of Ehrlich ascites tumour cells by skeletal muscle extracts. Brit. J. Exp. Path. **49**, 294, (1968).

Discussion

M. C. Chang: I would like to ask you, when you get this flush out from the uterus, this flush is contaminated with leukocytes and other components. Did you ever try to do an analysis of this uterine washing? According to my experience the uterine flushing contains various enzymes. What I mean is that there are many factors to influence your results rather than luteolysin alone.

D. W. Schomberg: This is quite possible. I did not specifically look for leukocytes. From other work that has been done, the highest concentration of leukocytes has been found at the time of estrus, and we did not observe a lytic effect at estrus. And it is also possible theoretically that by flushing the uterus with sterile-glass distilled water we could have ruptured the leukocytes, thereby liberating some types of proteolytic enzymes; however, when one flushes with sterile saline or sterile tissue culture medium this should not happen, and those flushings collected at the time of estrus did not induce cytolytic effects.

M. C. Chang: This could have an effect on your results, that is why it is of importance to be absolutely sure.

D. W. Schomberg: There are several questions, but to answer one of the three primary questions I mentioned we propose to test various of these possible enzymes in culture, and then use the same enzyme on the uterine component.

R. G. Edwards: I am worried about your cell cultures, because you are using here, essentially, a cell adhering to glass or plastic. Now it is well known that if you take primary cultures of cells you often find that they do not stick to glass very easily. The film you showed could have been explained by the fact that your cells are losing their adherence to the surface due to the compound you are adding. For example, you could get an effect with trypsin or collagenase, so that if you have different requirements for your different cell-control systems in their attachment to glass, you could get the kind of result that you have described.

D. W. Schomberg: This is what worried me so much about the haematoxylin and eosin stained preparations: was the substance simply cleaving the cells off the glass? However, I think with the time-lapse results where one can see the rupture of the cell membrane and then death, that even though the first action might be detachment from the glass, the cells do in fact rupture and die.

R. G. Edwards: If you try growing your cells in a suspension culture you might get a better effect, because in a suspension culture you do not demand this property of adherence to glass.

D. W. Schomberg: I did try this-not in a suspension culture but in roller tubes. I placed the granulosa cells immediately after harvest into "lytic" flushing media, and by histological criteria the effects were the same.

B. Lunenfeld: How do you obtain a pure population of granulosa cells?

D. W. Schomberg: I am partly assuming that we have not contaminated the cultures, because we do not damage the basement membrane separating the granulosa from the theca cells.

B. Lunenfeld: Could you give us some details concerning the technique.

D. W. Schomberg: We slit the follicle wall with a sterile scapel and then using a platinum loop of fairly small size, gently revolve this around the inside of the follicle.

Two such scrapings like this give us enough cells for a cover slip culture; a cover slip is 22 by 11 millimeters.

B. Lunenfeld: Did you ever try to digest the follicle scrapings with trypsin?

D. W. Schomberg: Yes, I have. In our routine work, one often harvests clumps of cells and it is impossible to count these in a cell-counting chamber. If one wants to do accurate steroid work for example, one would like a cell count. If the cells are treated with trypsin they can be dispersed, but for routine work, I did not use trypsin.

J. P. Bennett: My question really links up with that of Dr. *Edwards'*. I am very worried about your conclusions from the cell culture work because I cannot see that it specifically is a luteolytic model. There are so many other factors that might produce the same end result. Have you considered, for example, the sensitivity factor of effects of the differential cell layers you have studied; or hormonal regulation for example – maybe the cell explosions you show in your film occur because these cells are particularly sensitive to your dilutions and that the other non-exploding cells are not.

D. W. Schomberg: I did not mention this but we have tested dilution effects – not by measuring the osmolarity of the media – but by raising the concentration of the Hanks medium to tenfold. I have also studied the effect on these cells in distilled water plus serum, which should give us two extremes. In addition when one takes the sephadex purified material from which the salts have been removed and adds this to other consituents of the tissue culture medium, one still gets the same effect.

E. Steinberger: I am still somewhat worried about the problem of the cells com ing off the glass, because one may be able to release the cells from the glass by a non-specific mechanism and they may die. This non-specific reaction may cause a major problem in interpretation of your findings. Furthermore, I am concerned about the question whether you can tell *what* you are growing, what cell types. I would like to know what studies you have done to show that these are granulosa cells or at least steroid-producing cells. What is the ultrastructure of these cells? Most of the cells you have in culture are very rapidly growing mitotic cells, which you would not expect to show marked steroid activity or related activities. Also, I would like to ask what was the actual yield or percent conversion of pregnenolone to progesterone and how does this relate to rates of conversion of pregnenolone to progesterone by non-cultured granulosa cells?

D. W. Schomberg: We are planning ultra-structural studies on the tissue culture monolayers as well as on the cells after they have been exposed to the lytic flushings, but I have no data on this as yet. The progesterone production by individual cover slip cultures varies tremendously – we are using gas chromatography as our quantative step and the production per day (per 24 hours) may be as low as 20 nanograms, about the limit of sensitivity of our detection, to above 0.4 μg. This presumably depends upon the state of the animal at the time at which she is killed, whether the granulosa cells have been programmed, as it were, by FSH and LH. It would depend upon the quantity of circulating gonadotropins as well as the stage of development of the follicle. I think this is why we get such a tremendous variation.

The conversion of pregnenolone to progesterone was about 55 %. Using labelled acetate we demonstrated the synthesis of progesterone by re-crystallization. I did not actually calculate conversion figures, because, again, we have the problem of how much gonadotropin stimulation did these cells receive before they were harvested. I do not really think we could draw valid conclusions without larger numbers. The published estimates of the conversion of pregnenolone to progesterone by freshly isolated granulosa cells vary from 8.4 to 60 % (*Bjersing, L.* and *Carstensen, H.* J. Reprod. Fert. **14**, 101, 1967; *Ryan, K. J.* and *Petro, Z.* J. clin. Endocr. Metab. 26, 46, 1966).

E. Steinberger: Would that suggest, then, other metabolites?

D. W. Schomberg: No. The pig cells, as Drs. *Bjersing* and *Carstensen* have shown previously do not convert radioactive pregnenolone to radioactive estrogen (J. Reprod. Fert. **14**, 101, 1967).

P. Eckstein: Have you tried to grow theca interna cells by your technique?

D. W. Schomberg: No.

P. Eckstein: Have you attempted to either administer the flushings at the critical stage of the cycle directly to the corpus luteum, or, to infuse it through the ovarian artery?

D. W. Schomberg: I did not mention these experiments, because I wanted specifically to talk about the *in vitro* work. We have tried this in the hysterectomized pig by infusing lytic flushings into a branch of the middle uterine artery which anastomoses with the ovarian artery. We formed an infundibular pouch around the ovary – a "local application" – and we tried a systemic infusion. The quantity of flushings in terms of the number of animals used ranged from 15–20 uterine flushings. The duration of the infusions was from 3–7 days, and we found no effect whatsoever on peripheral blood progesterone concentrations or luteal concentrations of progesterone.

P. Eckstein: Could this have been due to a dilution factor, or did you try to concentrate the flushings?

D. W. Schomberg: I did concentrate them somewhat, but only to get a reasonable amount of fluid for the duration of these infusions. My explanation, which might be incorrect, is that we still need a lot more material.

P. Eckstein: But at the same time you got a positive effect in the tissue cultures.

D. W. Schomberg: Yes.

H. R. Lindner: Have you tried flushings from pregnant uterine horns on the tenth day or later?

D. W. Schomberg: We tried with one pig only on day 17, and this material was lytic. Many people have asked about this and I do not know the answer but, as Dr. *Caldwell* showed, there are lytic effects of a sterile uterine horn in pregnant animals. I do not know whether this allows one to infer that lytic material is present in the normal pregnant animal. Until we can identify something, it is difficult to know how much further we can proceed in this case.

Closing Address

The Past, Present, and Future of Fertility Control in Women

R. I. Dorfman

Syntex Research, Institute of Hormone Biology, Stanford Industrial Park Palo Alto, California, USA

I am honored to give this closing address which I hope does justice to the spirit of this fine symposium. My plan is to outline the successive advances in fertility control which have been employed so successfully up to the present, indicate some of their advantages and disadvantages, and look to possible future improved methods of contraception. The scientific bases for many of the future advances in the means of contraception have been discussed during the past days and have stimulated this essay.

Steroid hormones and related substances can be used to control fertility in women. This has been made possible by the development of highly active progestational agents and estrogens which are orally active and by significant advances in the field of physiology of reproduction in man. These two factors, together with certain clinical observations, have made possible the development of specific hormonal methods for control of fertility in women.

Fig. 1 summarizes the various methods and indicates their interrelation. Phase I of this highly important development resulted from the discovery that in amenorrheic women properly treated with estrogen and progesterone an artificial cycle can be produced and that estrogens, alone, administered from about the fifth to the twenty-fifth day of the cycle to normal women will inhibit ovulation. Further, that on cessation of estrogen treatment, a menstrual-like flow occurs within three to four days. This procedure was evolved perhaps more for its utility to treat dysmenorrhea than for contraception. The effectiveness for dysmenorrhea is believed to be due to the inhibition of ovulation.

The use of estrogen alone, as indicated in Phase I of the development of hormonal contraception, was unacceptable for the following reasons:

a) Excessive break-through bleeding and spotting.
b) Control of ovulation over and extended period of time was somewhat unreliable.
c) The need for relatively high doses of estrogens.

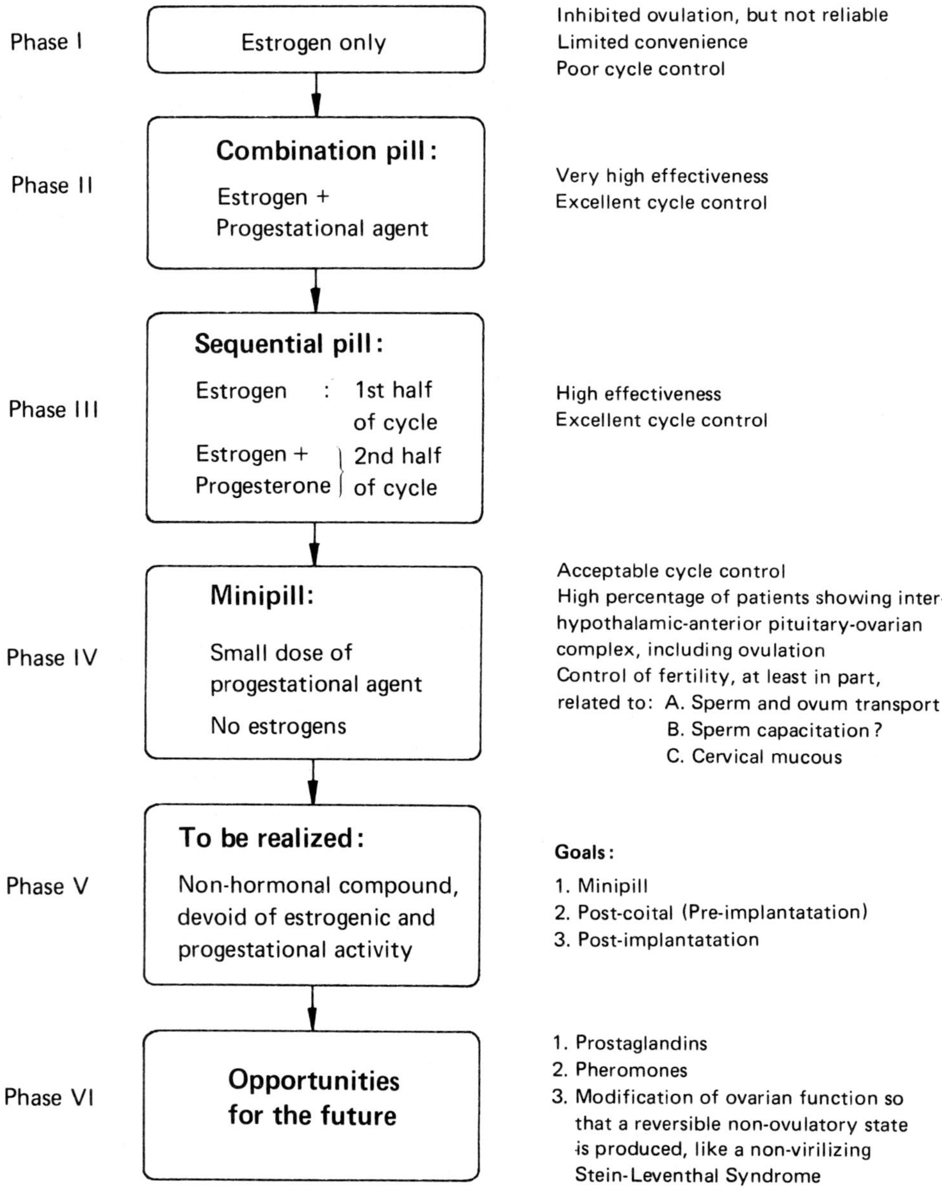

Fig. 1. Evaluation of hormonal control of contraception.

Many of the difficulties of the estrogen pill were overcome by the combination pill, which is indicated in Fig. 1 as Phase II. The basic development was the simultaneous administration of both an estrogen and a progestagen starting with day 5 of the cycle and continuing for a total of 20 consecutive days. Within 72 hours of cessation of treatment a menstrual flow results, not precisely like a normal flow but reasonably similar.

The combination pill afforded a contraceptive method with the following characteristics:

a) Minor break-through bleeding problems.
b) Excellent cycle control.
c) Very high effectiveness.
d) Duration of flow not unlike the natural menses.
e) Mechanism of action:
    1) Almost complete ovulation inhibition.
    2) Cervical mucous hostile to sperm.
    3) Uterine endometrium not ideal for implantation.
f) Excellent reversibility – withdrawal of pill leads to excellent fertility.

The combination pill was considered by some workers in the field as unphysiological on the basis that both estrogen and progesterone were administered during the first half of the menstrual cycle, a time during which only estrogen is present in the normal cycle. Accordingly, Phase III of hormonal contraception was introduced on the basis of estrogen administration during the first half of the cycle, reserving the combination of estrogen and progesterone for the second half or luteal phase of the cycle. This contraceptive regimen has many features in common with the combination pill but the efficacy is somewhat less.

The characteristics of the sequential pill follow:

a) Cycle control excellent.
b) Duration of flow not unlike the natural menses.
c) High effectiveness as a contraceptive method.
d) Mechanism of action.

Primarily inhibition of ovulation.

Although both the combination (Phase II) and sequential (Phase III) pills had extraordinarily fine properties for fertility control, further developments were indeed indicated. This was so since both procedures did have certain features that could be improved:

a) It would be of value to reduce the amount of estrogen.
b) Restraint of ovulation also called for rather intensive restraint of the hypothalamic-anterior pituitary-ovarian system.

To answer the objections to the use of the combination and sequential pill for fertility restraint in women, a new pill was developed consisting of continuous daily treatment with a progestagen and without estrogen. This form of contraception, control, labelled Phase IV in Fig. 1, has some particularly striking advantages: no exogenous estrogens, exceedingly low dose of the progestational agent, and minimum interference with the delicately balanced hypothalamic-anterior pituitary-ovarian organ system. The value of the low-dose progestational pill has been carefully established by studies in thousands of women over tens of thousands of cycles. The characteristics of this form of fertility restraint may be listed as follows:

a) High effectiveness as a method of contraception.
b) Cycle control acceptable but not excellent.
c) Absence of estrogens means less chance for certain metabolic effects, as influence on carbohydrate metabolism.
d) In the majority of cases the subject continues to ovulate and natural cyclicity is retained.
e) Mechanism of action:
   1) Hostile cervical mucous.
   2) Effect on ovum transport.
   3) Effect on sperm transport.
   4) Inhibition of sperm capacitation.

Phase IV leads logically to Phase V which may be defined as that stage of fertility control where mechanisms involving sperm and ovum transport, sperm capacitation and hostile cervical mucous are achieved by compounds devoid of hormonal activity. The compounds may be steroidal or non-steroidal but they should be non-estrogenic, non-progestational, and inactive on the hypothalamic-anterior pituitary-ovarian complex.

We would visualize the goals of Phase V compounds (Fig. 1) as being active as continuously administered minipills, post-coital agents administered for no more than two or three days after each contact, or agents active a few days to a few weeks after implantation.

Phase VI is really the future with some links to our present body of knowledge of reproductive physiology. Here we have the opportunities of modified prostaglandins having specific useful activities. Some discussions as to the utility of this class of compounds have been indicated during our discussions. The area of pheromones need only be mentioned in passing with the comment of a great opportunity for conception control.

The author is particularly impressed by the example of fertility control that nature has contributed. I am referring to the clinical state known as the Stein-Leventhal Syndrome and I am immediately aware of the fact that some of my clinical associates would take issue with the term but, on the other hand, they would not refute nature's experiment. There is a clinical condition in which cyclicity of ovarian function does not exist and the patient does not ovulate. It is true that some of these patients are hirsute and may even be virilized. But not all these patients show the effects of increased circulating androgens and in fact are well developed women.

Returning to the main point of this citation of the Stein-Leventhal Syndrome is the fact that by far the majority of these patients may by surgical or hormonal means be made fertile. In other words the sterility may be overcome and just as important is the fact that the new-found fertility may revert back to sterility. It is the speaker's feeling that a thorough study of the clinical state of the Stein-Leventhal syndrome may lead to a means of placing a young woman into a sterile state without influencing any other significant parameters of health or feminity for life. But that at various periods in a woman's life this restraint in fertility would be removed so that a child can be born and the woman again returned to the reproductive sterile state.

Having reviewed the facts of the past, reality of the present and indicated some dreams for the future it is time to close and in closing this highly significant, informative and wonderful symposium let me speak for the guests including myself by again thanking our host, Schering AG, and more specifically all the kind and thoughtful members of this firm who made this meeting possible. I could mention many deserving members of the Schering family and in my thoughts I do. At this time I want to single out particularly Prof. *Junkmann,* Prof. *Langecker,* Dr. *Raspé*, Dr. *Neumann,* and Dr. *Bernhard* to whom I give special thanks.

We all thank all the speakers and panel members for making this fine information exchange possible.

Lecturers

Adams, Cyril E., B. Sc., Ph. D.

1926 Born in Worthenbury, North Wales, United Kingdom, on February 1st
Education in Agriculture and Animal Physiology at the University of Wales and Cambridge University

Positions held:

Postdoctoral Research:

1952–1954 A.R.C. Unit of Animal Reproduction, Cambridge University, England

since 1954 A.R.C. Unit of Reproductive Physiology and Biochemistry, Cambridge, England

Fields of Interest:

Physiology and endocrinology of pregnancy, especially pre-implantation stages.

Austin, Colin R., D. Sc.

1914 Born in Sydney, Australia, on September 12th
Education in Biology at Sydney

Positions held:

Postdoctoral Research:

1954–1964 Member of the Research Staff, Medical Research Council, London

1964–1967 Professor
Department of Embryology, Tulane University, USA

since 1967 Professor
Department of Animal Embryology, Cambridge University, England

Fields of Interest:

Reproductive Physiology, Embryology, Electron Microscopy.

Bedford, John M., M. A., Vet. M. B., Ph. D.

1932	Born in England on May 21st Education in Veterinary Medicine at Cambridge and Physiology at London
	Positions held:
1959	Fellow Department of Surgery, University of Bristol
1959–1961	Research Associate Worcester Foundation, Shrewsbury
1961–1966	Lecurer Department of Physiology, Royal Veterinary College, London
1966–1967	Scientist Worcester Foundation, Shrewsbury
since 1967	Assistant Professor Columbia University

Fields of Interest:

Male reproduction – sperm, transport of gametes, fertilization – ovum physiology, membranes.

Bennett, John Ph., B. Sc. Ph. D.

1931	Born in Portsmouth, England, on June 12th Education in Reproductive Physiology at London and Cambridge
	Positions held:
1956–1963	Experimental Officer A.R.C. Unit of Reproductive Physiology and Biochemistry, Cambridge University, England
1963–1967	Head Department of Endocrinology, B.D.H. (Research) Ltd., Godalmina, England
since 1967	Head Department of Reproductive Physiology, Syntex Research, Palo Alto, California, USA

Fields of Interest:

Reproductive physiology and biochemistry, the control of fertility in man and animals.

Blye, Richard P., Ph. D.

1932	Born in W. Chester, Pennsylvania, on November 11th Education in Medicine at Pre-Med. Trinity College and Endocrinology at Rutgers

Positions held:

Postdoctoral Research:

1955–1956	Research Assistant Rutgers University
1956–1958	Teaching Assistant Rutgers University
1958–1959	Research Assistant Rutgers University
1959–1962	Associate Scientist Ortho Research Foundation, Raritan
1962–1963	Senior Scientist Ortho Research Foundation, Raritan
since 1964	Ortho Research Fellow Ortho Research Foundation, Raritan

Fields of Interest:

Physiology of reproduction, sex hormone assay, assessment of anti-fertility agents, mode of action of anti-fertility drugs.

Brackett, Benjamin G., D. V. M., Ph. D.

1938 Born in Athens, Georgia, on November 18th
Education in Veterinary Medicine and Biochemistry at the University of Georgia, Athens, Georgia

Positions held:

1962–1964 Fellow
A.V.M.A. and Morris Anim. Fed.
Department of Biochemistry,
University of Georgia

1964–1966 USPHS NICH and HD Postdoctoral Fellow
Department of Biochemistry
University of Georgia

1966–1968 Research Associate and Management Director
Primary College, University of Pennsylvania,
School of Medicine and School Veterinary Medicine

since 1968 Research Assistant Professor
and Management Director
Primary College, University of Pennsylvania,
School of Medicine and School Veterinary Medicine

Fields of Interest:

Mammalian reproduction, especially pre-implantation stages; veterinary medicine.

Brinster, Ralph L., V. M. D., Ph. D.

1932 Born in Montclair, New Jersey, on March 10th
Education in Veterinary Medicine and Physiology at the University of Pennsylvania

Positions held:

Postdoctoral Research:

1965–1966 Assistant Professor of Physiology
School of Veterinary Medicine,
Department of Animal Biology,
University of Pennsylvania

since 1966 Associate Professor of Physiology
Department of Animal Biology,
School of Veterinary Medicine,
University of Pennsylvania

Fields of Interest:

Biochemistry and physiology of the cleavage stages of mammalian embryos, differentiation of the early mammalian embryo, fertilization in mammals and invertebrates, mechanism of implantation, regulation of ovulation.

Caldwell, Burton V., Ph. D.

1938 Born in Boston, Massachusetts, on July 8th
Education in Endocrinology at New Hampshire, Boston and Tufts University

Positions held:

Postdoctoral Research:

1966–1967 N.I.H. Postdoctorate
Worcester Foundation, Shrewsbury

1967–1968 N.S.F. Postdoctorate
A.R.C. Unit of Reproductive Physiology and Biochemistry, Cambridge, England

since 1968 Staff Scientist
Worcester Foundation, Shrewsbury

Fields of Interest:

Steroidogenisis, utero-ovarian relationships, antibodies to steroid hormones.

Chang, M. C., B. Sc., Dipl. Agr. Sci., Ph. D.

1908 Born in Taiyuan, Shansi, China, on October 10th
Education in Biology and Agriculture at Tsing University, Peking, University of Edinburgh and Cambridge University

Positions held:

Postdoctoral Research:

1933–1938 Research Assistant and Instructor
Tsing Hua University, Peking, China

1941–1945 Research Worker
Cambridge University, England

1947–1950 Research Associate
Tufts Medical School, School of Agriculture

1945–1954 Research Associate
Worcester Foundation for experimental Biology

1954 Senior Scientist
Worcester Foundation for Experimental Biology

since 1961 Professor
Department of Biology, Boston University

Fields of Interest:

Mammalian Reproduction.

Dorfman, Ralph I., Ph. D.

1911	Born in Chicago, Illinois, on June 30th Education in Chemistry at the University of Illinois, Physiological Chemistry and Pharmacology at the University of Chicago
	Positions held:
	Postdoctoral Research:
1951–1956	Associate Director of Laboratories Worcester Foundation for Experimental Biology, Shrewsbury, Massachusetts
1951–1967	Research Professor Department of Biochemistry Boston University, Boston, Massachusetts
1956–1964	Professor of Chemistry (Affiliate) Clark University, Worcester, Massachusetts
1956–1964	Director of Laboratories Worcester Foundation for Experimental Biology
1964–1969	Director Institute of Hormone Biology, Syntex Research, Palo Alto, California
since 1964	Senior Vice President Syntex Research, Palo Alto, California
since 1967	Visiting Professor of Pharmacology Stanford University School of Medicine Palo Alto, California
	Fields of Interest:
	Biological activity of chlormadinone acetate.

Dukelow, W. Richard, Ph. D.

1936	Born in Princeton, Minnesota, on October 23rd Education in Reproductive Physiology and Biochemistry at St. Paul, University of Minnesota
	Positions held:
1960–1962	Instructor University of Minnesota, Grand Rapids
1962–1964	Assistant Professor University of Minnesota, Grand Rapids
1967–1968	Visiting Scientist Oregon Regional Primate Research Center, Beaverton
since 1965	Assistant Professor University of Georgia, Athens, Georgia
	Fields of Interest:
	Physiology and biochemistry of reproduction, capacitation, decapacitation factor, embryonic death, reproduction of primates.

Edwards, Richard G., Dr.

The ford Foundation Reader in Physiology

Being a bashful sort of chap I avoid curriculum vitae like the plague. I am a happy research worker contentedly sitting in traditional Cambridge, currently Reader in Physiology and having worked in various universities in Britain and America

Ericsson, Ronald J., Ph. D.

1935	Born in Belle Fourche, South Dakota, USA, on July 17th Education in Reproductive Physiology at University of Kentucky, Lexington, Kentucky
	Positions held:
	Postdoctoral Research:
1960–1963	Research Assistant University of Kentucky, Lexington, Kentucky
1963–1964	Postdoctoral Fellow University of Wisconsin, Madison, Wisconsin
since 1964	Senior Research Scientist The Upjohn Company, Kalamazoo, Michigan
since 1964	Adjunct Associate Professor Western Michigan University, Kalamazoo, Michigan
	Fields of Interest:
	Reproductive endocrinology, physiology of reproduction and specialty – male antifertility research.

Heller, Carl G., M. D. Ph. D.

1913 Born in Syracuse, New York, USA
Education in Zoology and Physiology at the University of Wisconsin

Positions held:

Postdoctoral Research:

1940–1941 Intern, Wisconsin General Hospital
1941–1944 Resident, Internal Medicine, City of Detroit, Rec. Hospital
1942–1944 Instructor and Assistant Professor Wayne State University
1944–1945 Assistant Professor Department of Physiology Wayne State University
1945–1948 Associate Professor Department of Physiology and Internal Medicine, University of Oregon, Medical School
1948–1957 Head Division of Endocrinology, University of Oregon, Medical School
since 1958 Director Division of Reproductive Physiology, Pacific Northwest Research Foundation

Fields of Interest:

Human reproductive physiology, endocrinology, radiation research.

Humphrey, Kevin W., B. V. Sc., Ph. D.

1939 Born in Melbourne, Australia, on January 27th
Education in Veterinary Science at Sydney, Ph. D. in Reproductive Physiology at Sydney

Positions held:

1962–1969 Teaching Fellow Sydney University

Fields of Interest:

Endocrinology of implantation and early pregnancy; ovum storage and transfer.

Kirby, David R. S., B. Sc. M. A. D. Phil.

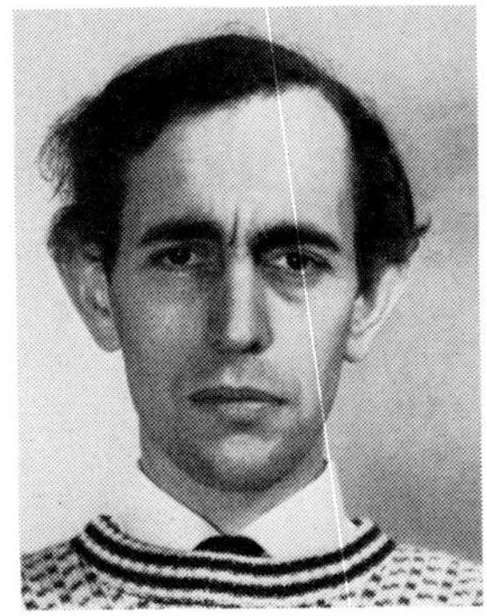

1932	Born in Eton, Bucks, England, on January 18th
	Education in Zoology at Oxford
	Positions held:
	Postdoctoral Research:
1965–1968	Senior Research Officer of Medical Research Council, Department of Zoology, Oxford University
since 1968	Lecturer Department of Zoology, Oxford University
since 1968	Fellow St. Catherine's College, Oxford
	Fields of Interest:
	Immunology of reproduction, particularly at ovo-implantation and placentation, hormonal control of early pregnancy.

Koester, Helmut, Dr., Priv.-Doz.

1928	Born in Münster/Westphalia, on August 25th
	Education in Medicine at the Universities Würzburg and Marburg
	Positions held:
1954	Dr. med., University Marburg Internship in Atlantic City, N. J., USA; assistant at the Dept. of Pathology, University Würzburg, and then Strahleninstitut, University Marburg.
1958	Assistant at the Universitäts-Frauenklinik, Gießen
1967	Habilitation (Physiology of Reproduction)
since 1965	Head-physician, Universitäts-Frauenklinik, Gießen
	Fields of Interest:
	Tubal secretions, early pregnancy, contraception, clinical themes of obstetrics and gynecology, especially treatment of carcinoma.

Lindner, Hans R., Ph. D.

1922 Born in Stettin, Germany, on April 21st
Education in Veterinary Science at Sydney, Biochemistry and Physiology at Cambridge

Positions held:

Postdoctoral Research:

to 1964 Principal Research Scientist
Division of Animal Physiology, Commonwealth Scientific Industrial Research Organisation, Sydney, N.S.W.

1964 Associate Professor
Department of Physiology, The Weizmann Institute of Science, Rehovot, Israel

since 1967 Head
Department of Biodynamics, The Weizmann Institute of Science, Rehovot, Israel
Part-Time Professor
School of Continuing Studies in Medicine, Tel-Aviv University

Fields of Interest:

Endocrinology and the biology of reproduction, particularly the biochemistry and physiology of the steroid hormones.

Lobel, Bertha, M. D.

1926 Born in Cologne, Germany, on December 31st
Education in Medicine at King's College, London and the University of Amsterdam, Holland, Medical School

Positions held:

Postdoctoral Research:

1958–1962 Department of Obstetrics and Gynecology, Albert Einstein Institute, Bronx, New York, USA

1963–1968 The Weizmann Institute of Science, Department of Biodynamics, Rehovot, Israel

Fields of Interest:

Reproductive physiology, cellular biology.

McCracken, John A., Ph. D.

1934	Born in Glasgow, Scotland, on August 16th Education in Veterinary Medicine at Glasgow, Cambridge and Birmingham
	Positions held:
1958–1959	Surgical Assistant University of Glasgow Baccalaureate in Vet. Medicine
1959–1960	Research Fellow University of Cambridge
1960–1963	Research Fellow University of Glasgow
1963–1964	Research Fellow University of Birmingham
1964–1969	Staff Scientist Worcester Foundation

Fields of Interest:

Physiology and endocrinology of reproduction: steroid biochemistry (determination of hormones in blood and tissues), organ transplantation (vascular transplantation of ovary or uterus with vascular anastomosis to the neck of sheep).

Psychoyos, Alexandre, M. D.

1927	Born in Piraeus Greece, on May 15th Education in Medicine at the University of Athens, Greece
	Positions held:
1955–1961	Attaché de R. College de France, Paris 5ème
1961–1964	Charge de R. College de France, Paris 5ème
1964–1967	Maître de R. College de France, Paris 5ème
1967–1969	Senior Scientist Worcester Foundation for Experimental Biology, Shrewsbury, Massachusetts, USA
since 1969	Maître de R. Institut National de la Santé et de la Recherche Medicale (INSERM)

Fields of Interest:

Hormonal requirements for implantation.

Schmidt-Mathiesen, Heinrich, Prof. Dr.

1923 Born in Witten/Ruhr, Germany, on March 28th
Education in Gynecology at Göttingen, Surgery at Heidelberg, Pathology and Histochemistry at Münster, Germany

Positions held:

Postdoctoral Research:
since 1963 Head
Department of Obstetrics and Gynecology, Göttingen University, Göttingen, Germany
since 1966 Associate Professor

Fields of Interest:

Histochemistry, physiology and patho-physiology of connective tissue, the cervical carcinoma, especially its system, reaction on current and the peculiarities of its invasion; physiology and patho-physiology of human endometrium, regarding functional hemorrhages as well as with respect to nidation (monography); hemorrhage syndrom, incl. tissue factors (monography); practical clinical themes of obstetrics and gynecology: birth mechanism, Roentgen diagnostic, additional therapy of carcinoma.

Schomberg, David W., Ph. D.

1939 Born in Lacrosse, Wisconsin, on November 6th
Education in Physiology at Duke University

Positions held:

1960–1965 Graduate Assistant
Purdue University, Lafayette, Indiana, USA
1965–1968 Postdoctoral Research:
University of Cambridge, England

Fields of Interest:

Biochemistry and physiology of utero-ovarian relationships, steroid biochemistry.

Schumacher, Gebhard F. B., Prof. Dr.

1924	Born in Osnabrück, Germany, on June 13th Education in Medicine at Göttingen and Tübingen
	Positions held:
	Postdoctoral Research:
1952–1954	Max Planck Institute for Biochemical and Virus Research, Tübingen
1954–1962	Assistant Professor Department of Obstetrics and Gynecology University of Tübingen
1962–1963	Research Associate Institute of Tuberculosis Research (Immunology), University of Illinois, Chicago
1963–1964	Assistant Professor Department of Obstetrics and Gynecology University of Chicago
1964–1965	Assistant Professor Department of Obstetrics and Gynecology University of Tübingen
1965–1967	Associate Professor Department of Obstetrics and Gynecology, Department of Biochemistry, Albany Medical College, Albany, New York
since 1967	Associate Professor Department of Obstetrics and Gynecology, University of Chicago, Illinois

Fields of Interest:

Protein chemistry, immunology, physiology of reproduction, fertility and sterility.

Steinberger, Emil, M. Sc. and M. D.

1928	Born in Berlin, Germany, on December 20th Education at State University of Iowa
	Positions held:
	Postdoctoral Research:
1958–1961	Research Associate Wayne State University
1961–1966	Associate Member Division of Endocrinology and Reproduction Albert Einstein Medical Center, Philadelphia
since 1967	Chairman Division of Endocrinology and Reproduction Albert Einstein Medical Center, Philadelphia

Fields of Interest:

Endocrinology and reproductive physiology.

Williams, William L., Ph. D.

1919 Born in St. Cloud, Minnesota, on June 14th
Education in Biochemistry and Bacteriology at St. Paul, Minnesota and Madison, Wisconsin

Positions held:

1942–1944 Biochemist
Fleischmann Laboratories, Stamford, Connecticut

1944–1945 Military (Lt. (jg))
Navy

1945–1946 Biochemist
Naval Medical Research Institute, Bethesda, Md.
Postdoctoral Research:

1949–1950 Assistant Professor
North Carolina State College, Raleigh, North Carolina

1950–1959 Group Leader
Lederle Laboratories, Pearl River, New York

since 1959 Research Professor
University of Georgia, Athens, Georgia

Fields of Interest:

Biochemistry of reproduction, effect of anticancer drugs on bacteria, nutrition and biochemistry of lactic acid bacteria.

Participants

Adams, C. E., B. Sc., Ph. D.
Animal Research Station, 307 Huntingdon Road, Cambridge, England

Albrecht, R., Dr.
Abteilung für Arzneimittelchemie der Schering AG

Apostolakis, M., Doz. Dr.
Hormonlabor der II. Medizinischen Universitätsklinik, 2 Hamburg 20, Martinistr. 52, Germany

Austin, C. R., D. Sc.
Physiological Laboratory, University of Cambridge, Cambridge, England

Baumgarten, Dr.
II. Universitäts-Frauenklinik
Spitalgasse 23, Wien, Austria

Bedfor, J. M., M. A., Vet. M. B., Ph. D.
College of Physicians and Surgeons of Columbia University, Department of Anatomy
630 West 168th Street, New York, N.Y. 10032, USA

Bennett, J. P., B. Sc., Ph. D.
Department of Reproductive Physiology
Syntex Research, Stanford Industrial Park
Palo Alto, California, USA

Bernhard, S., Dr.
Scientific Relations der Schering AG

Berswordt-Wallrabe, I. von, Dr.
3401 Mengershausen über Göttingen
Gut Heissenthal, Germany

Berswordt-Wallrabe, R. von, Dr.
Abteilung für Endokrinologie der Schering AG

Bettendorf, G., Prof. Dr.
Universitäts-Frauenklinik
2 Hamburg 20, Martinistr. 52, Germany

Bjφro, K., Dr.
Kvinnoklinikken, Rikshospitalet
Oslo, Norway

Blobel, R., Doz. Dr.
Universitäts-Frauenklinik
74 Tübingen, Germany

Blye, R. P., Ph. D.
Ortho Research Foundation, Division of Pharmacology
Raritan, New Jersey 08869, USA

Bottiglioni, F., Prof.
Clinica Ostetrica Ginecologica dell'Università
Bologna, Italy

Brackett, B. G., D. V. M., Ph. D.
University of Pennsylvania
Division of Reproductive Biology
36th and Hamilton Walk, Philadelphia, Pa. 19104, USA

Breckwoldt, M., Dr.
Universitäts-Frauenklinik
2 Hamburg 20, Martinistr. 52, Germany

Brinster, R. L., V. M. D., Ph. D.
Laboratory of Reproductive Physiology, School of Veterinary Medicine, University of Pennsylvania, Philadelphia, Pennsylvania, USA

Brosens, I., Dr.
22, Fonteinstraat, Heverlee, Belgium

Brosig, W., Prof. Dr.
Urologische Universitätsklinik und Poliklinik, Städt. Krankenhaus Westend
1 Berlin 19, Spandauer Damm 130, Germany

Buchholz, R., Prof. Dr.
Universitäts-Frauenklinik
355 Marburg, Pilgrimstein 3, Germany

Burger, H., Prof. Dr.
Städt. Frauenklinik
73 Esslingen/Neckar, Germany

Buschbeck, K., Dr.
Medizinisch-Wissenschaftliche Abteilung
der Schering AG

Bye, P. G. T., Dr.
Schering Chemicals Ltd., Pharmaceutical Division, Victoria Way, Burgess Hill, Sussex, England

Caldwell, B. V., Ph. D.
The Worcester Foundation for Experimental Biology, 222 Maple Ave., Shrewsbury, Massachusetts 01545, USA

Carlborg, L., Dr.
Hormonlaboratoriet, Akademiska sjukhuset
Uppsala 14, Sweden

Cecco de, L., Prof.
Clinica Ostetrica Ginecologica
dell'Università, Parma, Italy

Chang, M. C., B. Sc., Dipl. Agr. Sci., Ph. D.
Worcester Foundation for Experimental Biology, 222 Maple Avenue, Shrewsbury, Massachusetts 01545, USA

Cittadini, E., Prof.
Clinica Ostetrica Ginecologica
dell'Università, Palermo, Italy

Cohen, J., Dr.
11, rue Villebois-Mareuil
75 Paris XVIIe, France

Cremers, Dr.
Kastanjelaan 2, Hoogeveen, Netherlands

Dämmrich, K., Priv.-Doz. Dr.
Institut für Veterinärpathologie der Freien Universität, 1 Berlin 33, Drosselweg, Germany

Detering, K., Dr.
Medizinisch-Wissenschaftliche Abteilung
der Schering AG

Dietrich, W., Dr.
Medizinisch-Wissenschaftliche Abteilung
der Schering AG

Domenico, A.
Abteilung für Pharmakologie der Schering AG

Dorfman, R. I., Ph. D.
Syntex Research, Institute of Hormone Biology, Stanford Industrial Park
Palo Alto, California, USA

Eckstein, P., M. A., M. D., D. Sc.
University of Birmingham, Dept. of Anatomy, The Medical School, Birmingham 15, England

Edwards, R. G., Dr.
University of Cambridge, Physiological Laboratory, Cambridge, England

Eisenträger, H., Dr.
Klinische Forschung der Schering AG

Elger, W., Dr.
Abteilung für Endokrinologie der Schering AG

Eliasson, R., Prof. Dr.
Dept. of Physilogy, Karolinska Institutet, Faculty of Medicine, 10401 Stockholm, Sweden

Enigk, J., Dr.
Abteilung für Pharmakodynamik
der Schering AG

Ericsson, R. J., Ph. D.
Department of Fertility Research
The Upjohn Company, Kalamazoo, Michigan 49001, USA

Fezeu, P., Dr.
1058 Berlin, Gleimstr. 36, Germany/DDR

Fotherby, K., Prof. Dr.
Postgraduate Medical School of London
Ducane Road, Shepherd's Bush
London W. 12, England

Frenay, J., Dr.
R. Coles, Diegem/Bruxelles, Belgium

Friebel, J., Dr.
Medizinisch-Wissenschaftliche Abteilung
der Schering AG

Friedrichs, M., Dr.
Berlin Laboratories Inc., 375 Park Avenue
New York, N.V. 10022, USA

Fuchs, P., Dr.
Abteilung für Galenische Pharmazie
der Schering AG

Gemzell, C., Prof.
Akademiska sjukhuset, Uppsala, Sweden

Gerhards, E., Priv.-Doz. Dr. Dr.
Abteilung für Biologische Chemie
der Schering AG

Gibian, H., Dr.
Chemisch-Pharmazeutische Forschung
der Schering AG

Goldman, A. S., M. D.
Division of Experimental Pathology
The Children's Hospital of Philadelphia
1811, Fitzwater Street, Philadelphia,
Pa. 19146, USA

Goslar, H. G., Prof. Dr.
Anatomisches Institut der Universität
53 Bonn, Nußallee 10, Germany

Günzel, P., Dr.
Toxikologische Abt. der Schering AG
4619 Bergkamen, Germany

Hachmeister, Dr.
Pathologisches Institut der Universität
63 Gießen, Klinikstr. 32g, Germany

Hamann, H.-J.
Mitglied des Vorstandes der Schering AG

Hahn, J., Dr.
Klinik für Geburtshilfe und Gynäkologie
des Rindes der Tierärztlichen Hochschule
3 Hannover, Germany

Hahn, J.-D., Dr.
Abteilung für Endokrinologie der Schering AG

Hammerstein, J., Prof. Dr.
Klinikum der Freien Universität
Abteilung für Gynäkologische Endokrinologie
der Frauenklinik
1 Berlin 45, Klingsorstr. 95a, Germany

Haspels, A. A., Prof. Dr.
Royal Tropical Institute
63 Mauritskade, Amsterdam, Netherlands

Heinrichs, H. D., Priv.-Doz. Dr.
Universätits-Frauenklinik
5 Köln-Lindenthal, Robert-Koch-Str. 10,
Germany

Helge, H., Priv.-Doz. Dr.
Universitäts-Kinderklinik und -Poliklinik
1 Berlin 19, Heubnerweg 6, Germany

Heller, C. G., M. D., Ph. D.
Pacific Northwest Research Foundation
Laboratories for Medical Research
1102 Columbia Street
Seattle, Washington 98104, USA

Hellinga, G., Dr.
Stadionweg 68, Amsterdam-Z, Netherlands

Herken, H., Prof. Dr.
Pharmakologisches Institut der Freien
Universität, 1 Berlin 33, Thielallee 69,
Germany

Herlyn, U., Priv.-Doz. Dr.
Universätits-Frauenklinik
34 Göttingen, Humboldtallee 3, Germany

Heyns, W., Dr.
Akademisch Ziekenhuis St. Rafael
Kapucijnenvoer, Leuven, Belgium

Hlavka, E., Dr.
Schering Wien GmbH
Scheringgasse 2, Wien, Austria

Ho, S., Dr.
Medizinisch-Wissenschaftliche Abteilung
der Schering AG

Hocks, P., Dr.
Sonderlaboratorium der Schering AG

Hörmann, G., Prof. Dr.
Universitäts-Frauenklinik im Städt.
Krankenhaus Moabit, 1 Berlin 21, Turmstr. 21,
Germany

Hofmeister, H., Dr.
Abteilung für Steroidchemie der Schering AG

Hohlweg, W., Prof. Dr.
Universitäts-Frauenklinik
Prevenhueberweg 25, 8042 Graz, Austria

Humphrey, K. W., B. V. Sc., Ph. D.
The University of Sydney, Department of Veterinary Physiology
Sydney, N.S.W. 2006, Australia

Husslein, H., Prof. Dr.
II. Universitäts-Frauenklinik
Spitalgasse 23, Wien, Austria

Ijaiya, K., Dr.
104 Berlin, Novalisstraße, Germany/DDR

Johansson, E., Dr.
Kvinnoklin., Akademiska sjukhuset
75014 Uppsala, Sweden

Junkmann, K., Prof. Dr. Dr.
1 Berlin 33, In der Halde 14, Germany

Käser, O., Prof. Dr.
Universitäts-Frauenklinik
6 Frankfurt/Main, Ludwig-Rehn-Str. 14, Germany

Kaiser, R., Prof. Dr.
I. Universitäts-Frauenklinik
8 München 15, Maistr. 11, Germany

Karg, H., Prof. Dr.
Süddeutsche Versuchs- und Forschungsanstalt für Milchwirtschaft Weihenstephan
Institut für Physiologie, 8050 Freising, Germany

Kaspar, E., Dr.
Chemische Entwicklungs-Abteilung
der Schering AG, 4619 Bergkamen, Germany

Kewitz, H., Prof. Dr.
Institut für Veterinär-Pharmakologie und Toxikologie der Freien Universität
1 Berlin 33, Bitterstr. 14, Germany

Kieslich, K., Dr.
Abteilung für Chemische Mikrobiologie der Schering AG

Kirby, D. R. S., B. Sc., M. A., D. Phil.
Department of Zoology, University of Oxford, Parks Road, Oxford, England

Klink, G., Dr.
Abteilung für Pharmakodynamik der Schering AG

Koester, H., Priv.-Doz. Dr.
Universitäts-Frauenklinik
63 Gießen, Klinikstr. 32, Germany

Kolb, K. H., Dr.
Abteilung für Pharmakodynamik der Schering AG

Korte, R., Dr.
Abteilung für Toxikologie der Schering AG
4619 Bergkamen, Germany

Kracht, J., Prof. Dr.
Pathologisches Institut der Universität
63 Gießen, Klinikstr. 32g, Germany

Kramer, M., Prof. Dr.
Medizinisch-Experimentelle Forschung der Schering AG

Kremer, J., Dr.
The University of Groningen
Department of Obstetrics and Gynecology
Groningen, Netherlands

Lachnit, U., Dr.
Klinische Forschung der Schering AG

Langecker, H., Prof. Dr. Dr. Dr.
1 Berlin 65, Schöningstr. 1, Germany

Laron, Z., Prof. Dr.
Beilinson Hospital, P.O.B. 85
Petah Tiqva, Israel

Laschet, L., Dr.
Psychoendokrinologische Abteilung, Pfälzische Nervenklinik
5749 Landeck/Bad Bergzabern, Germany

Laschet, U., Dr.
Psychoendokrinologische Abteilung,
Pfälzische Nervenklinik
6749 Landeck/Bad Bergzabern, Germany

Laudahn, G., Priv.-Doz. Dr.
Klinische Forschung der Schering AG

Laurent, H., Dr.
Abteilung für Steroidchemie der Schering AG

Lax, H., Prof. Dr.
Universitäts-Frauenklinik
1 Berlin 19, Pulsstr. 4/14, Germany

Lembeck, F., Prof. Dr.
Pharmakologisches Institut der Universität
74 Tübingen, Wilhelmstr. 56, Germany

Lendle, L., Prof. Dr.
Pharmakologisches Institut der Universität
34 Göttingen, Geiststr. 9, Germany

Lengerke, J., Dr.
Klinische Forschung der Schering AG

Lindner, H. R., Prof. Dr.
The Weizmann Institute of Science
Department of Biodynamics, P.O.B. 26
Rehovot, Israel

Lobel, B. L., Dr.
The Weizmann Institute of Science
Department of Biodynamics, P.O.B. 26
Rehovot, Israel

Loeser, A., Prof. Dr. Dr.
Pharmakologisches Institut der Universität
44 Münster/Westfalen, Westring 12, Germany

Loewis of Menar, P. von, Dr.
II. Medizinische Klinik für Innere Medizin
der Freien Universität
1 Berlin 41, Klingsorstr. 95a, Germany

Loge, O., Dr.
Abteilung für Pharmakologie der Schering AG

Losert, W., Dr.
Abteilung für Pharmakologie der Schering AG

Lübke, F., Priv.-Doz. Dr.
Universitäts-Frauenklinik
1 Berlin 19, Pulsstr. 4/14, Germany

Lunenfeld, B., Prof.
Institute of Endocrinology, Tel-Hashomer
Government Hospital, Tel Aviv, Israel

Magnus, U., Dr.
Universitäts-Frauenklinik
1 Berlin 19, Pulsstr. 4/14, Germany

Martini, L., Prof. Dr.
Università degli Studi, Istituto di Farmacologia
e di Terapia, Via Vanvitelle 32, Milano, Italy

Martius, G., Prof. Dr.
1 Berlin 33, Wildpfad 1, Germany

McCracken, J. A., Ph. D.
The Worcester Foundation for Experimental
Biology, Shrewsbury, Massachusetts 01545,
USA

Mehring, M., Dr.
Abteilung für Endokrinologie der Schering AG

Merker, H.-J., Prof. Dr.
Forschungsabteilung für Elektronen-
mikroskopie der Freien Universität
1 Berlin 33, Königin-Luise-Str. 15, Germany

Meyhöfer, W., Prof. Dr.
Andrologische Abteilung der Universitäts-
Hautklinik, 63 Gießen, Gaffkystr. 14, Germany

Mikulaj, L., CSc., MUDr.
Institute of Endocrinology, Slovak Academy
of Sciences, Bratislava, CSR

Mittelstenscheid, K.-O., Dipl.-Kfm.
Mitglied des Vorstandes der Schering AG

Motta, M., Prof. Dr.
Università degli Studi, Istituto di Farmacologia
e di Terapia, Via Vanvitelli 32, Milano, Italy

Neumann, F., Dr.
Abteilung für Endokrinologie der Schering AG

Nevinny-Stickel, J., Prof. Dr.
Universitäts-Frauenklinik
1 Berlin 19, Pulsstr. 4/14, Germany

Niemann, H., Prof. Dr.
Universitäts-Hautklinik und Poliklinik
44 Münster, von Esmarch-Str. 56, Germany

Nieuweboer, B.
Abteilung für Biologische Chemie
der Schering AG

Nishino, Y., Dr.
Abteilung für Pharmakologie der Schering AG

Nocke, W., Dr.
Universitäts-Frauenklinik
355 Marburg, Pilgrimstein 3, Germany

Odeblad, E., Prof. Dr.
Department of Medical Physics,
University of Umeå, 90187 Umeå, Sweden

Overbeek, G. A., Dr.
N. V. Organon, Oss, Netherlands

Overzier, C., Prof. Dr.
II. Medizinische Universitäts-Klinik und Poliklinik, Abteilung für Klinische Endokrinologie, 65 Mainz, Langenbeckstr. 1, Germany

Pecile, A., Prof. Dr.
Università degli Studi, Istituto di
Farmacologia e di Terapia
Via Vanvitelli 32, Milano, Italy

Peeters, F., Dr.
Gynäkologische Abteilung
St. Elisabeth-Krankenhaus
Turnhout, Belgium

Pitchford, A. G., Dr.
Schering Chemicals Ltd., Pharmaceutical Division, Victoria Way, Burgess Hill, Sussex, England

Plotz, E. J., Prof. Dr.
Universitäts-Frauenklinik
53 Bonn-Venusberg, Germany

Prezewowsky, K., Dr.
Abteilung für Steroidchemie der Schering AG

Provè, A., Dr.
Sepps, B.P. 9253, 59 Roubaix, France

Psychoyos, A., Dr.
Hôpital Saint-Antoine
184, Rue du Faubourg Saint-Antoine
Paris XIIe, France

Pujol-Amat, P., Dr.
Ausias March, 125, Barcelona 13, Spain

Raspé, G., Dr.
Mitglied des Vorstandes der Schering AG

Rausch-Stroomann, J., Prof. Dr.
Medizinische Klinik, Klinikum Essen
der Ruhruniversität, 43 Essen, Hufelandstr. 55
Germany

Rauscher, H., Prof. Dr.
I. Universitäts-Frauenklinik, Hormonambulanz
Spitalgasse 23, Wien, Austria

Reichel, Prof. Dr.
Physiologisches Institut der Universität
2 Hamburg 20, Martinistr. 52, Germany

Richter, H., Dr.
Medizinisch-Wissenschaftliche Abteilung
der Schering AG

Richter, K. D., Dr.
Abteilung für Toxikologie der Schering AG
4619 Bergkamen, Germany

Richter, R. H. H., Dr.
Laboratorien der Universitäts-Frauenklinik
3012 Bern, Switzerland

Röpke, H., Dr.
Abteilung für Verfahrensentwicklung
der Schering AG

Ruiz, F., Dr.
Medizinisch-Wissenschaftliche Abteilung
der Schering AG

Runnebaum, B., Dr.
Universitäts-Frauenklinik
69 Heidelberg, Voßstr. 9, Germany

Salloch, R., Dr.
Abteilung für Endokrinologie
der Schering AG

Scharff, H. J., Dr.
Klinische Forschung der Schering AG

Schillinger, E., Dr.
Abteilung für Biologische Chemie
der Schering AG

Schirren, C., Prof. Dr.
Universitäts-Hautklinik und -Poliklinik
2 Hamburg 20, Martinistr. 52, Germany

Schmidt-Elmendorff, H., Dr.
Universitäts-Frauenklinik
4 Düsseldorf, Moorenstr. 5, Germany

Schmidt-Mathiesen, H., Prof. Dr.,
Oberarzt der Klinik, Universitäts-Frauenklinik, 34 Göttingen, Humboldtallee 3, Germany

Schomberg, D. W., Ph. D.
Duke University Medical Center
Department of Obstetrics and Gynecology
Durham, North Carolina, USA

Schreiner, W., Prof. Dr.
Universitäts-Frauenklinik, Kantonsspital
Rämistr. 100, 8006 Zürich, Switzerland

Schröder, E. K., Dr.
Abteilung für Arzneimittelchemie
der Schering AG

Schulze, P.-E., Dr.
Abteilung für Pharmakodynamik
der Schering AG

Schumacher, G. F. B., Prof. Dr.
431 N. Ashland Avenue, Homewood,
Ill. 60430, USA

Smidt, D., Doz. Dr.
Institut für Tierzucht und Haustiergenetik
der Universität, 34 Göttingen, Germany

Sommerville, J. F., M. D., Ph. D.,
Endocrine Laboratory Unit, Queen
Charlotte's and Chelsea Hospitals
Dovehouse Street, London S. W. 3, England

Sørensen, B., Dr.
Fødeafdeling A, Rigshospitalet
Blegdamsvej, 2100 Kopenhagen, Danmark

Soupart, P., M. D., Ph. D.,
Associate Professor, Vanderbilt University, Department of Obstetrics and Gynecology, Nashville, Tennessee 37203, USA

Staffeldt, Dr.
Universitäts-Frauenklinik
1 Berlin 19, Pulsstr. 4/14, Germany

Steeno, O., Dr.
Rega Institut, Laboratorium voor
Experimentele Geneeskunde, Leuven, Belgium

Stegner, R., Priv.-Doz. Dr.
Universitäts-Frauenklinik
2 Hamburg 20, Martinistr. 52, Germany

Steinbeck, H., Dr.
Abteilung für Endokrinologie der Schering AG

Steinberger, E., M. D., M. Sc.
Albert Einstein Medical Center, Research
Laboratories, Division of Endocrinology
and Reproduction, York and Tabor Roads
Philadelphia, Pa. 19141, USA

Stelling, U., Dr.
Medizinisch-Wissenschaftliche Abteilung
der Schering AG

Sulman, F. G., Prof. Dr.
Hebrew University of Jerusalem,
Department of Applied Pharmacology
P.O.B. 517, Jerusalem, Israel

Tamm, J., Prof. Dr.
II. Medizinische Universitäts-Klinik und
Poliklinik, 2 Hamburg 20, Martinistr. 52,
Germany

Taubert, H.-D., Prof. Dr.
Abteilung für Gynäkologische Endokrinologie
der Universitäts-Frauenklinik
6 Frankfurt/M., Ludwig-Rehn-Str. 14, Germany

Thomsen, K., Prof. Dr.
Universitäts-Frauenklinik
2 Hamburg 20, Martinistr. 52, Germany

Tonutti, E., Prof. Dr.
Medizinisch-Naturwissenschaftliche Hochschule, Abteilung für Klinische Morphologie, 79 Ulm/Donau, Parkstr. 10, Germany

Ufer, J., Dr.
Klinische Forschung der Schering AG

Vita, G., Dr.
Via L. Mancinelli 7, 20131 Milano, Italy

Voigt, K. D., Prof. Dr.
II. Medizinische Universitäts-Klinik und Poliklinik, 2 Hamburg 20, Martinistr. 52, Germany

Vokaer, R., Prof. Dr.
Université Libre de Bruxelles
Clinique de Gynecologie et d'Obstetrique
4, Place A. van Gehuchten, Bruxelles, Belgium

Vorbrüggen, H., Dr.
Abteilung für Arzneimittelchemie
der Schering AG

Wiechert, R., Dr.
Abteilung für Steroidchemie der Schering AG

Wiederholt, M., Dr.
Physiologisches Institut der Freien Universität
1 Berlin 33, Arnimallee 22, Germany

Williams, W. L., Ph. D.
The University of Georgia, Department of Biochemistry, Athens, Georgia 30601, USA

Witzel, H., Dr.
Mitglied des Vorstandes der Schering AG

Wyss, H., Priv.-Doz. Dr.
Kantonsspital Zürich, Universätits-Frauenklinik, Rämistr. 100, 8006 Zürich, Switzerland

Zöllner, G., Dr.
Abteilung für Pharmakodynamik
der Schering AG

Name Index

» Advances in the Biosciences 1

Schering Symposium on Endocrinology
Berlin, May 26 to 27, 1967

Edited by G. Raspé, Associate Editor: S. Bernhard. Braunschweig: Vieweg, 1969. 15,5 x 21 cm. VIII, 322 pages with 235 illustrations. Flexi-Cover
DM 32,00 – Bound £ 3.15/-, US $ 9.00

Contents:

» Advances in the Biosciences 3

Schering Workshop on Steroid Metabolism "In vitro versus in vivo" Berlin, December 13 and 14, 1968

Edited by G. Raspé, Associate Editors: S. Bernhard, M. Friedrichs, Braunschweig: Vieweg, 1969. 15,5 x 21 cm. VI, 217 pages with 139 illustrations. Flexi-Cover DM 24,00 – Bound £3.6/-, $ 8.00, DM 32,00

Excerpt from the Contents: